ARSENIC IN GROUND WATER

ARSENIC IN GROUND WATER

edited by

Alan H. Welch
U.S. Geological Survey

Kenneth G. Stollenwerk
U.S. Geological Survey

KLUWER ACADEMIC PUBLISHERS
Boston / Dordrecht / London

Distributors for North, Central and South America:
Kluwer Academic Publishers
101 Philip Drive
Assinippi Park
Norwell, Massachusetts 02061 USA
Telephone (781) 871-6600
Fax (781) 681-9045
E-Mail: kluwer@wkap.com

Distributors for all other countries:
Kluwer Academic Publishers Group
Post Office Box 322
3300 AH Dordrecht, THE NETHERLANDS
Telephone 31 786 576 000
Fax 31 786 576 474
E-Mail: services@wkap.nl

Electronic Services <http://www.wkap.nl>

Library of Congress Cataloging-in-Publication Data

Arsenic in ground water / edited by Alan H. Welch, Kenneth G. Stollenwerk.
p. cm.
Includes bibliographical references and index.
ISBN 1-40207-317-8 (alk. Paper)
1. Arsenic—Environmental aspects. 2. Groundwater—Pollution.
I. Welch, Alan H. II. Stollenwerk, Kenneth G.

TD427 .A77 .A75 2003
363.738'4—dc21

2002035708

Printed on acid-free paper.

Printed in the United States of America.

Contents

Contributors

C.A.J. Appelo
Hydrochemical Consultant
Amsterdam
The Netherlands

Nasiruddin Ahmad
Department of Public Health Engineering
Dhaka, Bangladesh

Kazi Matin Ahmed
Department of Geology
University of Dhaka
Dhaka, Bangladesh

Donald G. Archer
National Institute of Standards and Technology
Gaithersberg, MD

Laura M. Bexfield
U.S. Geological Survey
Albuquerque, NM

Simon Burden
British Geological Survey
Nottingham
UK

W.F. Cannon
U.S. Geological Survey
Reston, VA

John A. Colman
U.S. Geological Survey
Northborough, MA

Jeff Davies
British Geological Survey
Wallingford, Oxfordshire
UK

W.W.J.M. de Vet
Hydron-ZH
Gouda
The Netherlands

Lawrence S. Feinson
U.S. Geological Survey
Carson City, NV

Michael J. Focazio
U.S. Geological Survey
Reston, VA

Andrea L. Foster
U.S. Geological Survey
Menlo Park, CA

Philip Freiberg
Redwood National Park
Orick, CA

Irina Gaus
British Geological Survey
Wallingford, Oxfordshire
UK

M.B. Goldhaber
U.S. Geological Survey
Denver, CO

Madeline B. Gotkowitz
Wisconsin Geological And Natural History Survey
Madison, WI

S.K. Haack
U.S. Geological Survey
Lansing, MI

J.R. Hatch
U.S. Geological Survey
Denver, CO

S.M. Ihtishamul Huq
Department of Public Health Engineering
Dhaka, Bangladesh

M. –J Kim
Korea Institute of Science and Technology
Seoul, South Korea

David Kinniburgh
British Geological Survey
Wallingford, Oxfordshire
UK

Allan Kolker
U.S. Geological Survey
Reston, VA

R.C. Lee
U.S. Geological Survey
Denver, CO

Douglas K. Maurer
U.S. Geological Survey
Carson City, NV

Chris J. Milne
British Geological Survey
Wallingford, Oxfordshire
UK

Johnnie N. Moore
Department of Geology
University of Montana
Missoula, MT

D. Kirk Nordstrom
U.S. Geological Survey
Boulder, CO

Jerome Nriagu
University of Michigan
School of Public Health
Ann Arbor, MI

J.C. Pashin
Geological Survey of Alabama
Tuscaloosa, AL

L. Neil Plummer
U.S. Geological Survey
Reston, VA

Sarah J. Ryker
U.S. Geological Survey
(Present Address)
Carnegie Mellon University
Pittsburgh, PA

M.E. Schreiber
Department of Geological Sciences
Virginia Tech
Blacksburg, VA

J. Antonio Simo
Department of Geology and Geophysics
University of Wisconsin-Madison
Madison, WI

Kenneth G. Stollenwerk
U.S. Geological Survey
Denver, CO

Janice M. Trafford
Wallingford, Oxfordshire
UK

J. Treworgy
Earth Science Department
Principia College
Elsahy, IL

Jenny G. Webster
Environmental Chemistry/Water Quality
School of Environmental & Marine Science
University of Auckland
Auckland NZ

Alan H. Welch
U.S. Geological Survey
Carson City, NV

D.B. Westjohn
U.S. Geological Survey
Lansing, MI

William W. Woessner
Department of Geology
University of Montana
Missoula, MT

L.G. Woodruff
U.S. Geological Survey
Mounds View, MN

Preface

Interest in arsenic in ground water has greatly increased in the past decade because of the increased awareness of human health effects and the costs of avoidance or treatment of ground water supplies used for consumption. The goal of this book is to provide a description of the basic process*es* that affect arsenic occurrence and transport by providing sufficient background information on arsenic geochemistry and descriptions of high-arsenic ground water, both affected and unaffected by human activity.

An understanding of thermodynamics, adsorption, and the speciation of arsenic in solid phases, which are described in first three chapters, is needed to predict the fate of arsenic in ground water systems. Large-scale and deep movement of ground water can and has redistributed arsenic in the near surface environment, as described in the next two chapters. These large-scale systems can affect large volumes of both ground water and surface water, such as in the Yellowstone system, and can produce mineralised zones that subsequently release arsenic to ground water supplies. Regional identification of high-arsenic ground water and its consumption as described in the next three chapters clearly demonstrates a need for increased water-quality monitoring, particularly in south and southeast Asia. Chapters 9-11 provide examples of high arsenic ground water associated with sulfide mineral oxidation and alkaline conditions. Finally, smaller scale studies of the effects of human activities that have produced high-arsenic ground water and methods for attenuation of ground water are presented.

This volume would not have been possible without the financial support of the National Research and National Water-Quality Assessment Programs of the U.S. Geological Survey. The support by these programs is gratefully acknowledged. The able assistance of Nancy Damar, Teresa Foglesong, Chris Stone, and Angie Thacker in the preparation of this volume is greatly appreciated.

Finally, the editors dedicate this book to the victims of arsenic poisoning in the hope that it will help in some small way to lesson the impact of arsenic on humans.

Chapter 1

Arsenic thermodynamic data and environmental geochemistry

An evaluation of thermodynamic data for modeling the aqueous environmental geochemistry of arsenic

D. Kirk Nordstrom[1] and Donald G. Archer[2]

[1]*U.S. Geological Survey, Boulder, CO, USA*
[2]*National Institute of Standards and Technology, Gaithersburg, MD, USA*

Thermodynamic data are critical as input to models that attempt to interpret the geochemistry of environmentally important elements such as arsenic. Unfortunately, the thermodynamic data for mineral phases of arsenic and their solubilities have been highly discrepant and inadequately evaluated. This paper presents the results of a simultaneous weighted least-squares multiple regression on more than 75 thermochemical measurements of elemental arsenic, arsenic oxides, arsenic sulfides, their aqueous hydrolysis, and a few related reactions. The best-fitted thermodynamic database is related to mineral stability relationships for native arsenic, claudetite, arsenolite, orpiment, and realgar with pε-pH diagrams and with known occurrences and mineral transformations in the environment to test the compatibility of thermodynamic measurements and calculations with observations in nature. The results provide a much more consistent framework for geochemical modeling and the interpretation of geochemical processes involving arsenic in the environment.

1. INTRODUCTION

Aqueous geochemical models have become routine tools in the investigation of water-rock interactions (Alpers and Nordstrom, 1999; Drever, 1997; Langmuir, 1997; Nordstrom and Munoz, 1994; Parkhurst and Plummer, 1993), in the study of bioavailability and toxicity of contaminants to organisms (Morrison, 1989; Parker et al., 1995), in the prediction of arsenic behaviour in mining pit lakes in Nevada (Tempel et al., 1999), in the prediction of arsenic mobility from mine wastes (Doyle et al., 1994), in the prediction of ore deposit formation (Heinrich and Eadington, 1986), and in any quantitative interpretation of reactions in aqueous solution and natural water (Morel and Hering, 1993; Stumm and Morgan, 1996). As with all computerized models, the quality of the output depends on the quality of the input and thermodynamic data is one of the primary sets of data input to most geochemical codes. Unfortunately, the consistency and quality of thermodynamic data is not adequate for the wide variety of aqueous geochemical calculations needed for water quality investigations. Data for environmentally relevant arsenic species is a good example. Grenthe et al. (1992 p. 390), in their major critique of uranium thermodynamic data, stated that a complete re-analysis of thermodynamic data for arsenic species is necessary and data on uranium-arsenic complexes and compounds would necessarily be compromised. Nordstrom (2000) agreed with this conclusion and considered the consequences of estimating stability constants for some divalent and trivalent arsenate complexes on speciation of arsenate for some ground waters from Bangladesh. Those results demonstrated that speciation can change considerably but saturation indices are not significantly affected for these dilute waters. Presumably, waters of higher concentration would be affected more strongly.

Numerous compilations of thermodynamic data are available and many of these contain thermodynamic data for arsenic species. However, mere numbers of compilations do not provide any idea of the quality of the data nor the number or quality of the original measurements upon which the data are based. The presence of nearly identical property values in apparently different sources may give the erroneous impression that the properties for a particular substance are well determined when in fact they may be based on the same original source that, like a cousin, may be twice removed from the reference given. Most compilations of data for arsenic species cannot be considered reliable. The database of Sadiq and Lindsay (1981) has been used to speciate arsenic for waters and soils by Sadiq et al. (1983), and Sadiq (1990, 1997). Unfortunately, this database was not developed using critical evaluation procedures (e.g. see Ball and Nordstrom, 1998; Nordstrom, 2000) and several errors can be found there, including lack of consistency with thermodynamic relationships, inappropriate use of values from the literature, no evaluation of original sources, no evaluation of networks, and no consideration of temperature dependence. This database is not unique; there are many other similarly unevaluated compilations in the published literature. Many dangerous assumptions such as those outlined above can befall the unaware investigator.

Another factor that exacerbates this sort of problem is that some databases were republished at later dates without incorporation of changes in the literature that had occurred since the original publication date. The more recent publication dates might lead one to believe that these databases are based on more up-to-date information than might actually be the case. Examples involve publication series such as Wagman et al. (1968, 1982) or the series of Robie and Waldbaum (1968), Robie et al, (1978), and Robie and Hemingway (1995).

We illustrate these problems with a specific case. Young and Robins (2000) listed 13 values of the Gibbs energy of formation, $\Delta_f G$, of orpiment, As_2S_3(cr), each from a different literature source. These values ranged from -168.8 kJ·mol^{-1} to -82.4 kJ·mol^{-1} and, if examined closely, fell into groups of values. The first group contained two values, -168 and -169 kJ·mol^{-1}, and were cited as Wagman et al. (1982) and Robie et al. (1978). However, the Wagman et al. (1982) citation was a republication of values from an earlier publication (Wagman et al., 1968) whose properties for arsenic compounds were generated about 1964 and were not documented. Robie et al. (1978) took the 1964 value from Wagman et al. (1968), changed it slightly and republished it. Another group of values falls around -95 to -99 kJ·mol^{-1}. One of those values, -96.23 kJ·mol^{-1}, was obtained by Barton (1969) from examination of multi-phase equilibrium temperatures for liquid (arsenic + sulfur), realgar, and orpiment, assumptions about the fugacity of sulfur in mixed arsenic + sulfur melts, and extrapolations from the melting temperature to 298 K. Naumov et al. (1974) made a small change in this value and included it in their data compilation. Also in this group are two solubility studies in which the Gibbs energies of formation were calculated from the measured solubilities and, among other things, the Gibbs energy of formation of $H_3AsO_3{}^{o}{}_{(aq)}$, where the latter value was taken from Naumov et al. (1974). Naumov et al. based their value of $\Delta_f G$ for $H_3AsO_3{}^{o}{}_{(aq)}$, in part, on a value of $\Delta_f G$ for arsenolite calculated from an erroneous value reported by Beezer et al. (1965). Because the values calculated from the two solubility studies were based on erroneous auxiliary data, their agreement with Barton's (1969) equilibrium study is happenstance. Another group of values ranged from -90.7 to -91.5 kJ·mol^{-1} and was based on a fluorine combustion study from Johnson et al. (1980) combined with various auxiliary data. Finally, two additional, but smaller values (-86 and -82.4 kJ·mol^{-1}), were listed. One of these values was from Pokrovski (1996) who reanalyzed earlier solubility data using yet another set of auxiliary thermodynamic data, and the other was from Bryndzia and Kleppa (1988) who used direct synthesis calorimetry to determine the enthalpy of formation. Bryndzia and Kleppa (1988) reported $-$ 83.0 kJ·mol^{-1} for the enthalpy of formation of orpiment. To make matters worse, the more recent compilation of Robie and Hemingway (1995) lists the enthalpy as -91.6 kJ·mol^{-1} with Bryndzia and Kleppa (1988) as the source. Thus, we see examples of most of the problems mentioned above.

In this paper we present the results of an evaluation of selected thermodynamic data of arsenic species. The results are of two types with one that consists of data that have been fitted with a weighted least-squares regression, and a second that is derived from the first least-squares determined group by standard thermodynamic relationships. The results (in Tables 1a, 1b, and 2) and their implications for geochemistry and geochemical modelling are then discussed with known occurrences, observed mineral transformations in the environment and calculated pε-pH relationships. The objective is to provide a consistent and coherent framework of thermodynamic calculations and field relationships for mineral stabilities among arsenic species.

2. GENERAL APPROACH AND THE EVALUATION OF ELEMENTAL ARSENIC

From hundreds of papers that were found from the literature containing measurements on arsenic reactions from which thermodynamic data might have been extracted, measurements on 77 substances or reactions contained in 40 studies were selected for simultaneous least-squares regression. These substances or reactions were limited to elemental arsenic and simple compounds, the oxides and sulfides, their solubilities and hydrolysis products. Elemental arsenic was regressed separately because a different fitting procedure was used. This procedure is explained below. Details of the original papers that formed the basis for the least-squares regression of all the data and corrections that were made to the original measurements will be found in Archer and Nordstrom (2000). These details are beyond the scope of this paper.

Treatment of the measured reactions for arsenic species required adoption of thermodynamic properties of some other substances, e.g. $H_{2(g)}$, $Cl^{-}_{(aq)}$, etc. Most previous thermodynamic property compilations suffer from some type of networking problem. Consequently, such a choice cannot be made lightly. The most obvious choices could be either Wagman et al. (1968) or the CODATA values from Cox et al. (1989). The Wagman et al. (1968) values were generated quite some time ago and there are apparent problems with sulfate and carbonate species. These species are, of course, important in geochemical modeling of ground waters. On the other hand, there are problems with thermodynamic properties of at least some of the alkali halides given in the CODATA recommendations (Archer, 1992; Rard and Archer, 1995), species that are also important in geochemical modeling. We also have reactions that involve species not considered by Cox et al.

(1989) as being "key", e.g. $I_3^-{}_{(aq)}$. There are undisclosed optimized properties inherent in the CODATA recommendations, a fact that makes thermodynamic consistency based on the CODATA properties not possible (Archer, 2000). Hence, we chose to use the Wagman et al. (1968) properties, recognizing that its problems potentially affect the present work less than would the problems associated with the CODATA values. Properties given below are based on the common conventional standard pressure of 1 atm (101.325 kPa).

Thermodynamic properties for $As_{(\alpha,cr)}$ were calculated from a least-squares representation using a cubic-spline method described previously (Archer, 1992; Archer et al., 1996). Briefly, a function $f(T)$ was used, where:

$$f(T) = [T \cdot \{(C^{\circ}_{p,\mathrm{m}} - \gamma_{\mathrm{el}}T)/C^{\circ}_{p}\}^{-1/3} - bT]/T^{\circ} \quad (1)$$

and where T is temperature, T° was 1 K, $C^{\circ}_{p,\mathrm{m}}$ was the molar heat capacity, C°_{p} was 1.0 $\mathrm{J \cdot K^{-1} \cdot mol^{-1}}$, γ_{el} was the coefficient for the contribution to the heat capacity of the conduction electrons, and b was arbitrarily chosen to be 0.32 for the present case. The function (T) of equation (1) was fitted with a cubic spline using polynomials of the form:

$$f(T) = \mathrm{a_i}(T - T_{\mathrm{i}})^3 + \mathrm{b_i}(T - T_{\mathrm{i}})^2 + \mathrm{c_i}(T - T_{\mathrm{i}}) + \mathrm{d_i} \quad (2)$$

where the subscript i refers to the polynomial that contains the specified value of T and spans the temperature range T_{i} to $T_{\mathrm{i+1}}$. A particular (T_{i}, $\mathrm{d_i}$) pair is referred to as a "knot." A "natural spline" end condition (*i.e.* second derivative equal to 0) was imposed at the highest temperature end knot. The end condition imposed at the lowest temperature knot was a value of $-b$ (-0.2) for the first derivative. This approach was equivalent to assuming that the Debye temperature was independent of temperature near 0 K. (For the purpose of calculation: $T_{\mathrm{i+1}} > T > T_{\mathrm{i}}$). The calculated heat capacity was:

$$C^{\circ}_{p,\mathrm{m}}/C^{\circ}_{p} = \left(\frac{T}{T^{\circ} f(T) + \mathrm{b}T}\right)^3 + \gamma_{\mathrm{el}}T/C^{\circ}_{p} \quad (3)$$

Equation (3) was integrated numerically to obtain the enthalpy. The model was determined by fitting to the selected values with a nonlinear least-squares program. The vector of residuals was calculated using the numerical integration of equation (3) to obtain the enthalpy increments. Included in the representation were the enthalpy increment measurements from Klemm et al. (1963) and the heat capacity values given by Culvert

(1967), Paukov et al. (1969), and Anderson (1930). The heat capacity measurements from Culvert ranged from 0.7 K to 4 K. We have included only the measurements from 1 K to 4 K, as the lower temperature measurements are affected significantly by nuclear spin contributions. The heat capacity measurements from Paukov et al. and Anderson spanned the temperature ranges of 13.8 K to 289 K and of 57 K to 291 K, respectively. The enthalpy increment measurements reported by Klemm et al. (1967) were performed with a Bunsen ice calorimeter. Klemm et al. found eccentricities in several different properties of arsenic near 500 K, including the enthalpy, unit cell measurements, and electrical and magnetic properties. Taylor et al. (1965) measured the unit cell dimensions, and electrical and magnetic properties, as functions of temperature for arsenic crystals. They did not observe the effects reported by Klemm et al. Because the second-order transition reported by Klemm et al. was not reproducible, we did not include it in our representation of the thermal properties of arsenic. The results gave the standard state entropy at 298 K of 35.78 ± 0.18 $J \cdot K^{-1} \cdot mol^{-1}$ which compares very favourably to the value obtained by Ball et al. (1988) of 35.735 $J \cdot K^{-1} \cdot mol^{-1}$ by refitting the heat capacity data of Hultgren et al. (1973). The value given by Hultgren et al. (1973) is 35.69 ± 0.84 $J \cdot K^{-1} \cdot mol^{-1}$ and appears in Robie et al. (1978) and Robie and Hemingway (1995). The entropy value of 35.63 $J \cdot K^{-1} \cdot mol^{-1}$ that appears in Table 1a was obtained by taking the fitted elemental arsenic data and fitting it with all the other arsenic data. Even though this entropy value has decreased slightly in the overall fitting, the residual (0.145 $J \cdot K^{-1} \cdot mol^{-1}$) indicates that the entropy has a lower uncertainty than indicated by Hultgren et al. (1973) and the difference in the entropy from refitting is not significant.

3. RESULTS AND DISCUSSION

The results of simultaneous weighted least-squares regression of the data and some of the unfitted but derived quantities are shown in Tables 1a, 1b, and 2. Table 1a displays elemental arsenic, its simple oxides, and the reactions for arsenic oxidizing to arsenic trioxides. Table 1b introduces the hydrolysis species for As(III) and As(V) in solution, the hydrolysis reactions, and the solubility reactions for the simple oxides. Single species are shown at the top of each table with the reactions underneath. The following discussion describes some of the mineral occurrences for these substances, describes their relative stabilities from field observations, and considers the implications of the evaluated thermodynamic data in terms of these occurrences.

Table 1a. Arsenic Oxides

Species or reaction data for standard state conditions, 298.15 K and 1 atm	($\Delta_f G^o$; $\Delta_r G^o$) / $kJ \cdot mol^{-1}$	($\Delta_f H^o$; $\Delta_r H^o$) / $kJ \cdot mol^{-1}$	($\Delta_r S^o$; S^o) / $J \cdot K^{-1} \cdot mol^{-1}$	(C_P^o ; $\Delta_r C_P^o$) / $J \cdot K^{1} \cdot mol^{-1}$
$As(\alpha,cr)$	0.0	0.0	35.63	24.43
$O_2(g)$	0.0	0.0	205.029	29.355
$H_2(g)$	0.0	0.0	130.574	28.824
$H_2O(l)$	-237.178	-285.830	69.91	75.29
As_2O_3 (cubic, Arsenolite)	-576.34	-657.27	107.38	96.88
As_2O_3 (monoclinic, Claudetite)	-576.53	-655.67	113.37	96.98
$As_2O_5(cr)$	-774.96	-917.59	105.44	115.9
$As_2O_3(cubic) + O_2(g) = As_2O_5(cr)$	-198.62	-260.32	-206.97	-10.3
$2As(\alpha,cr) + 3H_2O(l) = As_2O_3(cubic) + 3H_2(g)$	135.19	200.22	218.11	-91.38
$2As(\alpha,cr) + 3H_2O(l) = As_2O_3(monoclinic) + 3H_2(g)$	135.00	201.82	224.10	-91.28

3.1 Minerals in the As-S-O system

There is no known mineral with the formula of As_2O_5 but polymorphic minerals have been identified for As_2S_3 and As_2O_3. These minerals are compiled in Table 3 along with their formulae, crystal class and space group, and references for their crystallography and occurrence. Additional arsenic sulfide minerals of different stoichiometries than As_2S_3 have been included in this list because they commonly occur with the other phases, although no thermodynamic data is known for them.

Arsenolite and claudetite have nearly identical free energies of formation, making it difficult to determine the most stable phase under standard state conditions. Wagman et al. (1982) give claudetite as the most stable

Table 1b. Solubility and Hydrolysis of Arsenic Oxides

Species or reaction data for standard state conditions, 298.15 K and 1 atm	$(\Delta_f G^o; \Delta_r G^o)$ / kJ·mol^{-1}	$(\Delta_f H^o; \Delta_r H^o)$ / kJ·mol^{-1}	$(\Delta_r S^o; S^o)$ / J·K^{-1}·mol^{-1}	Log K
$H_3AsO_3(aq)$	-640.03	-742.36	195.83	
$H_2AsO_3^-(aq)$	-587.66	-714.74	112.79	
$HAsO_3^{2-}(aq)$	(-507.4)			
$AsO_3^{3-}(aq)$	(-421.8)			
$H_3AsO_4(aq)$	-766.75	-903.45	183.07	
$H_2AsO_4^-(aq)$	-753.65	-911.42	112.38	
$HAsO_4^{2-}(aq)$	-713.73	-908.41	-11.42	
$AsO_4^{3-}(aq)$	-646.36	-890.21	-176.31	
$H_3AsO_3(aq) = H_2AsO_3^-(aq) + H^+(aq)$	52.37	27.62	-83.04	-9.17
$H_2AsO_3^-(aq) = HAsO_3^{2-}(aq) + H^+(aq)$	(80.3±1.0)			(-14.1)
$HAsO_3^{2-}(aq) = AsO_3^{3-}(aq) + H^+(aq)$	(85.6±1.5)			(-15.0)
$As_2O_3(cubic) + 3H_2O(l) = 2H_3AsO_3(aq)$	7.814	30.041	74.55	-1.38
$As_2O_3(monoclinic) + 3H_2O(l) = 2H_3AsO_3(aq)$	8.003	28.443	68.56	-1.34
$H_3AsO_4(aq) = H_2AsO_4^-(aq) + H^+(aq)$	13.10	-7.97	-70.69	-2.30
$H_2AsO_4^-(aq) = HAsO_4^{-2}(aq) + H^+(aq)$	39.93	3.02	-123.76	-6.99
$HAsO_4^{-2}(aq) = AsO_4^{-3}(aq) + H^+(aq)$	67.36	18.20	-164.88	-11.80
$H_3AsO_4(aq) + H_2(g) = H_3AsO_3(aq) + H_2O(l)$	-110.46	-124.74	-47.90	19.35

Table 2. Arsenic Sulfides

Species or reaction, data for standard state conditions, 298.15 K and 1 atm	$(\Delta_f G^o; \Delta_r G^o)$ / $kJ\cdot mol^{-1}$	$(\Delta_f H^o; \Delta_r H^o)$ / $kJ\cdot mol^{-1}$	$(\Delta_r S^o; S^o)$ / $J\cdot K^{-1}\cdot mol^{-1}$	$(C_P^o; \Delta_r C_P^o)$ / $J\cdot K^{-}\cdot mol^{-1}$	Log K
S(cr, rhombic)	0	0	31.8	22.6	
S_2(g)	79.33	128.37	228.07	32.47	
H_2S(aq)	-27.87	-40.21	121	167	
HS^-(aq)	12.05	-17.64	62.8	-92	
AsS(α, Realgar)	-31.3	-31.8	62.9	47	
AsS(β, Realgar)	-30.9	-31.0	63.5	47	
As_2S_3(α, Orpiment)	-84.9	-85.8	163.8	163	
As_2S_3(amorphous orpiment)	-76.8	-66.9	200		
$As_3S_4(SH)_2^-$(aq)	(-125.6)				
$AsS(OH)(SH)^-$(aq)	(-244.4)				
4AsS(α, cr) + S_2(g) = $2As_2S_3$(α, cr)	-123.9	-168.4	-148.1	106	21.7
As_2S_3(α, cr) + $6H_3O$(l) = $2H_3AsO_3$(aq) + $3HS^-$(aq) + $3H^+$(aq)	264.1	263.1	–3.2		-46.3
As_2S_3(α, cr) + $6H_2O$(l) = $2H_3AsO_3$(aq) + $3H_2S$(aq)	144.3	195.4	171		-25.3
As_2S_3(am) + $6H_2O$(l) = $2H_3AsO_3$(aq) +	136.2	176.5	136		-23.9
$1.5As_2S_3$(am) + $1.5H_2S$(aq) = $As_3S_4(SH)_2^-$(aq) + H^+(aq)	31.4				-5.5
$0.5As_2S_3$(am) + $0.5H_2S$(aq) + H_2O = $AsS(OH)(SH)^-$(aq) + H^+(aq)	45.1				-7.9

Table 3. Arsenic oxide and sulfide minerals

Mineral Name	Chemical Formula	Space Group	References
(Native) Arsenic	As, trigonal	$R\bar{3}m$	Anthony et al. (1990)
Arsenolamprite	As, orthorhombic	Bmab	Anthony et al. (1997)
Arsenolite	As_2O_3, cubic	Fd3m	Anthony et al. (1997)
Claudetite	As_2O_3, monoclinic	$P2_1/n$	Anthony et al. (1990)
Orpiment	α-As_2S_3, monoclinic	$P2_1/n$	Anthony et al. (1990), Stojanovic (1982)
Realgar	α-AsS, monoclinic	$P2_1/n$	Anthony et al. (1990), Stojanovic (1982)
Realgar	β-AsS, monoclinic	C2/c	Anthony et al. (1990), Yu and Zoltai (1972)
Pararealgar	AsS, monoclinic	$P2_1/c$	Roberts et al. (1980), Bonazzi et al. (1995)
Alacrinite	AsS, monoclinic	C2/c	Popova et al. (1986), Burns and Percival (2001)
Uzonite	As_4S_5, monoclinic	$P2_1/m$	Popova and Polyakov (1985)
Dimorphite	As_4S_3, orthorhombic	Pnma	Anthony et al. (1990)
Duranusite	As_4S, monoclinic	$P2_1/m$	Blackburn and Dennen (1997)

polymorph. Robie and Hemingway (1995), however, list arsenolite as the most stable phase, giving Wagman et al. (1982) as the only source for the free energy data yet reporting rather different numbers. Careful consideration of solubility, electrochemical, vapor pressure, heat capacity, entropy, and reaction enthalpy data along with the least squares weighted regression of the data leads to claudetite having the greater stability by –0.19 $kJ \cdot mol^{-1}$(Archer and Nordstrom, in press). One of the more definitive studies was by Kirschning and Plieth (1955) who determined the temperature of transition to be –33°C, above which claudetite was the more stable phase. Palache et al. (1944) report an observation that suggests claudetite forms at higher temperatures than arsenolite (>100°C), as well as paramorphism of arsenolite after claudetite but with such a small difference in free energy, small changes in temperature, pressure, grain size, humidity, and the salinity of solutions in contact could easily change the relative stability. It *is* clear that arsenolite and claudetite have formed as a secondary product of weathering. These dimorphs are typically white powdery coatings that have formed from the oxidation of arsenopyrite (Palache et al., 1944; Roberts and Rapp, 1965), from the oxidation of realgar (Beyer, 1989; Kelley, 1936;

Palache et al., 1944), from the oxidation of native arsenic (Clark, 1970; Palache et al., 1944), and from the weathering of scorodite (Eckel, 1997). They are often intimately associated with each other (Dana and Ford, 1949).

Orpiment appears to have a wide range of stability, being found as a hypogene mineral in epithermal ore deposits (i.e. formed from ascending hydrothermal solutions at temperatures of roughly 50-200oC, Lindgren, 1928; Park and MacDiarmid, 1975), as a precipitate in hot springs of warm to boiling temperatures (see Chapter 4, this volume), in fumarolic encrustations (Vergasova, 1983; White and Waring, 1963), as a sublimate from mine fires and burning coal seams (Lapham et al., 1980; Palache et al., 1944; Zacek and Ondrus, 1997), and, commonly, as a supergene mineral (formed under surface or near-surface weathering conditions) from the weathering of realgar and sometimes arsenopyrite (Dana and Ford, 1949; Palache et al., 1944). Realgar, however, seems to be more typical of hypogene mineralization. Beran et al. (1994), for example, found fluid inclusions in realgar from the Allchar deposit, Macedonia, that gave homogenization temperatures of 144-170°C. At the Mercur gold deposit in Utah, USA, Jewell and Parry (1988) found a calcite-realgar vein assemblage in which the fluid inclusion in the calcite gave formation temperatures of 150-190°C. Orpiment also occurs at Mercur in a separate vein assemblage with pyrite. Although realgar is found in fumarolic encrustations, mine fire sublimates, and hot spring environments (Clarke, 1924; Palache et al., 1944; Zacek and Ondrus, 1997), it is not commonly found as a direct precipitate from solution at temperatures less than 100°C like orpiment. As long ago as 1851, de Senarmount (1851) found that when either pulverized realgar or orpiment were heated in a sealed tube with Na_2CO_3 at 150°C they would recrystallize to realgar. Mixtures of realgar and orpiment were observed by the senior author to coat rocks from a hydrothermal steam vent of about 180°C at Solfatara, Italy (also see Sinno, 1951). Migdasov and Bychkov (1998) note that a zonation between orpiment and realgar seems to occur in the Uzon Caldera, Kamchatka, with realgar occurring between 70 and 95°C. This temperature range seems to be the lowest observed for the formation of realgar that we are aware of. It is also noteworthy that the occurrence of an amorphous (to X-ray diffractometry) orpiment phase is well known (Eary, 1992) but no one has reported an amorphous realgar phase. Amorphous phases commonly result when insoluble precipitates are formed at low temperatures (0-50°C). These observations suggest that realgar may be more stable at higher temperatures and orpiment is more stable at low temperatures. To test the consistency of this observation with thermodynamic data, we calculated the equilibrium boundaries between orpiment and α-realgar and between α-realgar and arsenic as a function of the partial pressure of $S_{2(g)}$, P_{S_2}, and temperature. It was assumed that the fugacity of $S_{2(g)}$ was equal to the partial pressure. The results shown in

Figure 1 indicate that although the stability field of orpiment decreases with increasing temperature, so does the stability field of realgar because of encroachment from the expanding field of native arsenic. Both orpiment and realgar require higher fugacities of sulfur to maintain equilibrium stability at higher temperatures. Although there are uncertainties in the thermodynamic properties for realgar and orpiment that need further refinement, the general features of this stability diagram are probably correct. Hence, both orpiment and realgar are stable over a wide range of temperature but realgar is stable over a narrow range of sulfur fugacity. Although this stability diagram does not show any increasing stability of realgar with temperature, it does point out an important aspect. Orpiment and realgar are often found together and sometimes one or the other is found alone. These observations can help to constrain the sulfur fugacity of the hydrothermal system. The coexisting orpiment/realgar assemblage provides a convenient buffer for the sulfur fugacity and may provide a fugacity estimate if the temperature is known or a geothermometer if the fugacity is known.

The formation of sulfide minerals such as realgar and orpiment is usually thought of as a simple inorganic precipitation phenomenon. Recent investigations by Huber et al. (2000), however, have found an anaerobic, hyperthermophilic, facultatively organotrophic archaeon that respires arsenate and precipitates realgar. *Pyrobaculum arsenaticum*, is the first microorganism to be reported that can precipitate realgar biologically over its temperature range of growth, 68 to 100°C. It grows chemolithoautotrophically with carbon dioxide as a carbon source, hydrogen as electron donor, and arsenate, thiosulfate, or sulfur as electron acceptors. It also respires selenate when grown organotrophically and forms elemental selenium. It was isolated from the Pisciarelli Solfatara, Italy. Newman et al. (1997) also found a bacterium, *Desulfotomaculum auripigmentum*, that precipitates orpiment both intracellularly and extracellularly.

Realgar, when exposed to solar radiation, transforms to a powdery material identified as pararealgar, a polymorph of realgar (Roberts et al., 1980). Pararealgar differs from realgar only in the manner in which the As_4S_4 molecules stack in the unit cell (Bonazzi et al., 1995). When different wavelengths of light were tested, Douglass et al. (1992) found that less than 670 nm was needed to detect the transformation and the lower the wavelength, the faster the reaction. Raman spectroscopy has been used to characterize pararealgar and it has become an important tool in the understanding of art history and art preservation (Trentelman et al., 1996). Earlier literature (Palache et al., 1944; Wells, 1962) describes realgar as breaking down to orpiment and arsenolite when exposed to light for long periods of time. This observation is interesting in that it suggests that the assemblage orpiment plus arsenolite may be more stable than realgar under standard state conditions.

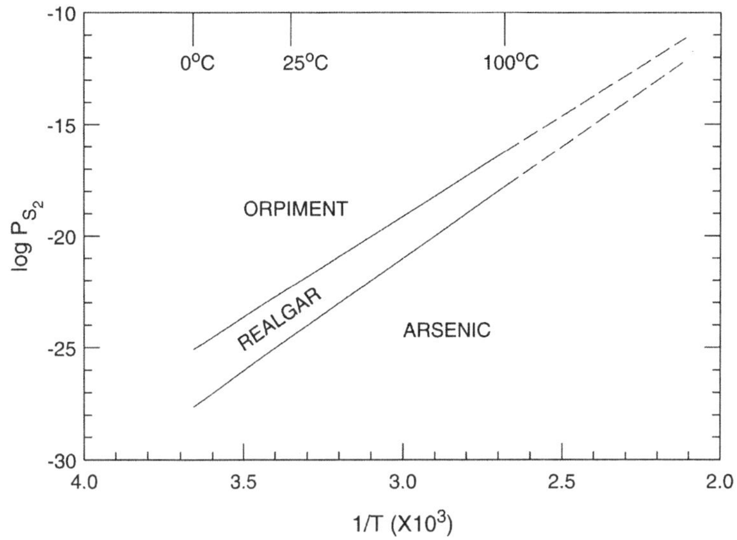

Figure 1. Univariant stable equilibria between orpiment, realgar, and arsenic as a function of P_{S_2} and 1/T where T is in degrees Kelvin.

Apparently alacrinite is also a low-temperature polymorph that simply stacks differently as well (Burns and Percival, 2001). We are not aware of any thermodynamic data on either pararealgar or alacrinite but since the crystal structures are identical except for molecular stacking, the free energies should be very similar to within the accuracy of calorimetric or solubility measurements. Duranusite and dimorphite (and possibly uzonite) are secondary alteration products of realgar (Clark, 1972; Marquez-Zavalia et al., 1999).

Native arsenic has been reported from numerous arsenic-rich mineral deposits (Clarke, 1924; Dana and Ford, 1949; Eckel, 1997; God and Zemann, 2000) but the descriptions of small aggregates, crystalline, massive, and botryoidal arsenic are consistent with a hypogene origin (e.g. Clark, 1970; Paronikyan and Matevosyan, 1965). It is commonly found when there is an abundance of arsenic sulfide minerals in a hydrothermal deposit. Hence, native arsenic is a product of low-temperature epithermal mineralization (50-200°C) under sulfur-deficient and strongly reducing conditions rather than a weathering product. Where native arsenic is found exposed to weathering, under dry conditions, it has usually developed a coating of arsenolite or claudetite or both (Clark, 1970; Palache et al., 1944). In low-temperature sedimentary conditions (0-50°C) there is frequently much more sulfur and iron available than arsenic so that the arsenic is incorporated into arsenian pyrite or orpiment rather than occurring in the native form. Arsenolamprite is a dimorph of arsenic that is very rare and is known to occur in carbonate-hosted mineralized zones. It may be stabilized by high concentrations of trace elements because the original type material contained up to 3% Bi.

Many other arsenic-containing minerals are known and have some thermochemical measurements but they are beyond the scope of this paper.

3.2 pε-pH diagrams for arsenic

The tabulated thermodynamic data can be used to develop the pε-pH diagrams for arsenic, shown in Figures 2 and 3, that summarize the predominance fields for aqueous species and the mineral stability fields, respectively. These diagrams also help to focus the discussion on environmentally relevant geochemical processes.

The predominance area diagram of Figure 2 shows that, under oxidizing conditions, arsenate hydrolyzes to four possible species for the range of pH encountered in surface and ground waters, although the fully dissociated arsenate ion would be rare because very few waters reach a pH greater than 11.5. Under reducing conditions, the fully protonated arsenite species is predominant over a wide range of pH and because it is not ionized and adsorbs less strongly than arsenate species, dissolved arsenite tends to be much more soluble than arsenate. Hence, reducing conditions usually lead to increased concentrations of arsenic in ground waters provided that arsenic is available in the aquifer or the sediments. Arsine, AsH_3, was estimated to be at the same pε and pH conditions as the formation of hydrogen, i.e. the lower limits for water. Hence, it does not show on this diagram.

Figure 3 shows the sequence of stable minerals from fully oxidized arsenic pentoxide to fully reduced native arsenic in the presence of 10^{-4}m total dissolved sulfide. Native arsenic has a narrow stability field only under the strongest reducing conditions, consistent with field observations except that in the field it seems to form at higher temperatures than 25°C. No mineral corresponds with As(V) oxide because it is extremely soluble (about 40 grams per 100 grams of solution, Menzies and Potter, 1912) and the addition of the type of divalent cations commonly found in surface and ground waters would promote the precipitation of metal arsenates that are less soluble than the pentoxide (e.g. calcium arsenate precipitation, Nishimura and Robins, Dunning, 1988; Nishimura and Robins, 1998).

The stability field for orpiment occurs between realgar and claudetite and the stability field for realgar is barely visible at all in contrast to other diagrams such as that found in Brookins (1986). The relative stability field for realgar can change significantly with small changes in free energy for either (or both) orpiment and realgar and further refinement of the data

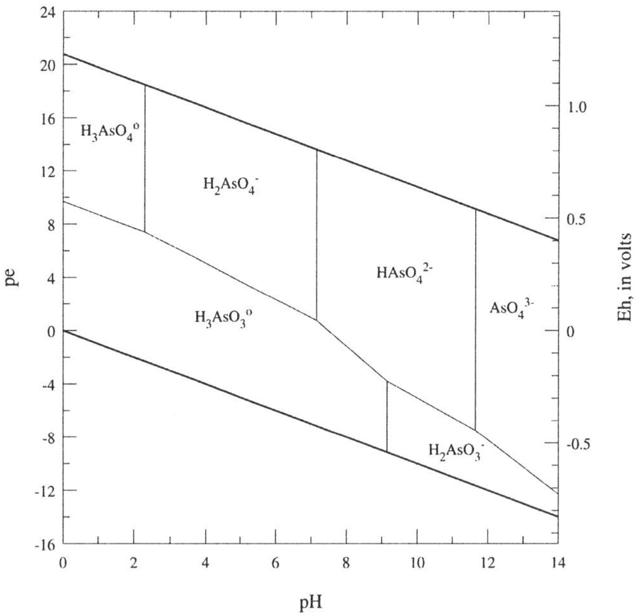

Figure 2. pε-pH diagram for predominant aqueous species of arsenic at equilibrium and 298.15 K and 1 atmosphere pressure.

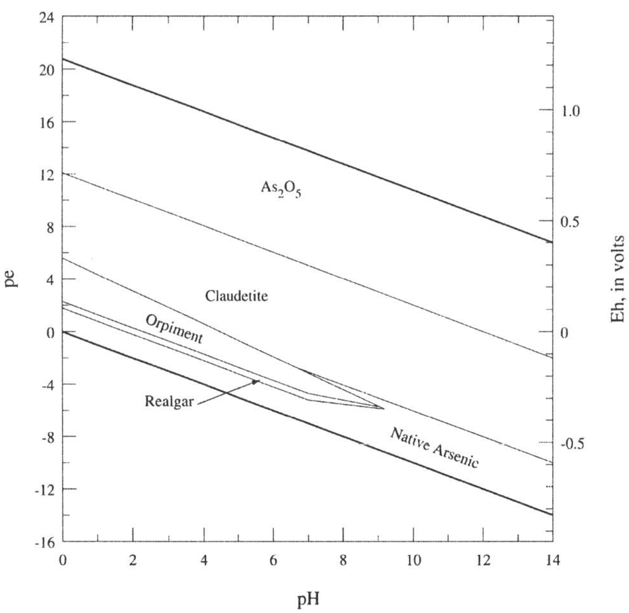

Figure 3. pε-pH diagram for equilibrium mineral stabilities in the As-O-S-H_2O system at 298.15 K and 1 atmosphere pressure. Total dissolved sulfur = 10^{-4} m.

would help eliminate these ambiguities. An earlier attempt to fit the thermodynamic data resulted in an impossible chemography in which realgar was never stable over any temperature range relative to native arsenic and orpiment. Errors in enthalpy, entropy, and free energy are too large at the moment to get a better estimate for these stability fields than what we have tabulated.

3.3 Matters of stoichiometry

The first problem has to do with the appropriate stoichiometry of a compound. For example, arsenic (III) oxide appears in the literature as As_2O_3 or As_4O_6. Similarly, the formula for realgar appears variously as As_4S_4, As_2S_2, or AsS. The higher order stoichiometries reflect the unit structure of bonding in dominantly covalent compounds. They are useful from a bonding and crystal structure viewpoint but serve no purpose in chemical thermodynamics and only the simplest unit stoichiometry is used in this paper: As_2O_3 for arsenolite and claudetite, AsS for realgar, and As_2S_3 for orpiment.

Several different formulae were given in the earlier literature to describe the species formed when As_2O_3 was dissolved in water under neutral or mildly acidic conditions. In fact, one finds several of these listed in Wagman et al. (1968; 1982) which gave properties for both $H_3AsO_{3(aq)}$ and $HAsO_{2(aq)}$. Those species were redundant to each other in Wagman et al. (1982), as were the species AsO_2^-(aq) and $H_2AsO_3^-{}_{(aq)}$. Redundant species cannot be included in model calculations, if one expects to obtain correct results from the simulation. These two redundancies were not identified as such in the original tables but were noted by Wagman et al. (1982) in an introduction to the 1982 publication. Raman spectra (Loehr and Plane, 1968) were consistent with H_3AsO_3 being the predominant form of dissolved As(III) in dilute neutral solution. We adopted this formulation here and did not consider the other earlier formulae.

3.4 Polymerization in aqueous solution

At higher concentrations of dissolved As(III), polymeric species may form in solution and the implications from their formation must be understood. A particular equilibrium constant for an aqueous reaction will depend on what species, or what other reactions, are presumed to exist in the solution. It sometimes happens that tables of thermodynamic properties may have been generated by averaging equilibrium constants for a particular reaction or by averaging other thermodynamic properties derived therefrom. In some instances, appropriate attention had not been given to the fact that the equilibrium constants that were being averaged had been obtained originally from different assumptions regarding the species present in the solution. Such averaged equilibrium constants will introduce errors in the overall model assumed for the system. An example of indiscriminate averaging of equilibrium constants for the generation of tabulated thermodynamic properties was given previously by one of the authors (Archer, 1998).

Such a problem arises with respect to the first dissociation constant of arsenious acid, $H_3AsO_{3(aq)}$. Table 4 gives some of the equilibrium constant values found in the literature for the first deprotonation step. The values in the table are for 298.15 K and all are standard-state values; this means that adjustments were made to some of the values given in the literature. The values in the table fall, more or less, into two groups of values. The first four values were determined from potentiometric titration followed with a glass electrode. The value from Garrett et al. (1940) was obtained from a study of the pH and the solubility of As_2O_3 in solutions containing different amounts of sodium hydroxide or hydrogen chloride. The two values from Ivakin et al. (1976) were determined from 1) a study of the absorptivity of arsenic-containing solutions and 2) potentiometric titrations of arsenious acid with an alkaline solution of arsenious acid.

Two distinctly different model assumptions characterize the two groups of values. The first group was obtained with an assumption that the only reaction that occurred in the solution was the deprotonation reaction as in equation 4.

$$H_3AsO_{3(aq)} = H_2AsO_3^-{}_{(aq)} + H^+ \qquad (4)$$

The second group of values came from studies where it was assumed that polymerization reactions occurred, such as the formation of $H_5As_2O_6^-{}_{(aq)}$, in addition to the deprotonation reaction. For chemical and mathematical reasons, the dissociation constant calculated from a set of measurements becomes smaller as one introduces polymeric anions into the model. The differences of the models chosen, at first appearance, could serve to explain the differences of the equilibrium constants given in the previous table. Unfortunately, the situation, from the perspective of data evaluation, is more complex. In principle, there should be a sufficient dilution of arsenious acid for which one would not expect the formation of a significant proportion of species like $H_5As_2O_6^-{}_{(aq)}$ upon addition of base. For such a condition, the equilibrium constant determined assuming that only the monomer exists, should approach that determined for the multi-species model. Britton and Jackson (1934) performed potentiometric titration at two concentrations of arsenious acid (0.0170 and 0.0914 molar) and obtained essentially the same value. Therefore, all of the variation of the equilibrium constants is not yet explained by the assumptions of different models.

Table 4. Equilibrium constants for the first deprotonation step of arsenious acid

K	Source
7.4×10^{-10}	Britton and Jackson (1934)
7.04×10^{-10}	Antikainen and Tevanen (1961)
7.08×10^{-10}	Vesala and Saloma (1977)
5.25×10^{-10}	Salomaa et al. (1969)
2.4×10^{-10}	Garrett et al. (1940)
6.8×10^{-10}	Ivakin et al. (1976)
5.4×10^{-10}	Ivakin et al. (1976)

Raman spectroscopic studies of aqueous solutions of arsenious acids have been conducted for water solutions saturated with arsenic oxide and for ratios of $[OH^-]/[As(III)]$ ranging from 3.5 to 15 (Loehr and Plane, 1968). The spectra obtained in those studies were interpreted as being consistent with the formation of mono-arsenic species only. No peak was observed that was attributed to either an As-As bond or to an As-O-H-O-As species, where a delocalization occurs about the hydrogen atom. Ivakin et al. (1976) assumed that one of the species that existed in aqueous solution was $H_6As_2O_{6(aq)}$. However, colligative property determinations, e.g. the freezing-point depression measurements from Roth and Schwartz (1926), were not consistent with significant dimerization of the arsenious acid. Thus, one is left with a quandary. One can accept the conclusion from the Raman spectroscopy and the larger value of *K* for eq 4. But then one has no way of explaining the greatly increased solubility of As(III) in basic solutions over that predicted by assumption of the monomeric-only model. Conversely, if one accepts the formation of poly-As(III) species in solution, then how does one interpret the Raman spectroscopy results? (There is a recent theoretical study that suggests the existence of circular trinuclear arsenic species in neutral solution (Tossell, 1997). Additionally, a recent study by Pokrovski et al. (1996) may also support the existence of polynuclear arsenic species in solution.). The problem of which value or which model to accept for the first deprotonation step, is not merely academic. This is because the Gibbs energy of formation of $H_2AsO_3^-{}_{(aq)}$ is a primary contribution to the Gibbs energies of formation of the aqueous arsenite ion and crystalline arsenite compounds.

Our interest here is the generation of a set of properties that are useful for modeling the geochemistry of clean and contaminated waters, industrial process and waste fluids, and related problems. These fluids are generally dilute in arsenic and do not contain the high levels of hydroxide considered by Garrett et al. (1940). Therefore, we have not included in our tables, or method, the polymeric forms suggested by Garrett et al. (1940), Ivakin et al. (1976), and Pokrovski et al. (1996).

3.5 Complications in chloride media

Wagman et al. (1968, 1982) gave a value for the Gibbs energy of formation of undissociated arsenic acid. That value was obtained from values reported by Foerster and Pressprich (1927) and auxiliary values. Foerster and Pressprich (1927) measured the electromotive force (emf) of cells in which the electrochemical reaction was presumed to be:

$$H_3AsO_{4(aq)} + H_{2(g)} = H_3AsO_{3(aq)} + H_2O_{(l)} \quad (5)$$

To suppress dissociation of the arsenic acid, the measurements were conducted in solutions acidified with hydrochloric acid of (1 to 6) M. The presence of large amounts of chloride in the acidified arsenic solutions is problematic. The formation of $As(OH)_xCl_{(3-x)(aq)}$ for x from 0 to 2, has been proposed to explain the observed increases of solubility of As(III) in hydrochloric acid (Garrett et al., 1940).

Arcand (1957) determined distribution coefficients of As(III) between a hydrochloric acid phase and a dichlorodiethyl ether phase. He used his measurements, determined at different concentrations of HCl(aq), to argue that complexes of the type $As(OH)_xCl_{(3-x)}$ existed in those solutions for x from 0 to 3. He used solubility data of arsenic trioxide in $HCl_{(aq)}$ and $HClO_{4(aq)}$ to argue the existence of $As(OH)_2{}^+{}_{(aq)}$ in acid solutions also. He calculated approximate equilibrium constants for formation of $As(OH)_2{}^+{}_{(aq)}$, $As(OH)_2Cl_{(aq)}$, $AsOHCl_{2(aq)}$, and $AsCl_{3(aq)}$. Finally, he showed that the assumptions of species and values of the equilibrium constants nearly quantitatively represented the dependence of Foerster and Pressprich's (1927) emf measurements on the concentration of hydrochloric acid in the solution. Consequently, there were unaccounted species with different activities in Foerster and Pressprich's (1927) measurement cells and the lack of accounting of these species may have compromised the determination of $\Delta_r G$ determined for eq 5 in some thermodynamic compilations. The significance of a lack of confidence in the thermodynamic properties of arsenic acid is that they contribute to the formation properties of the arsenate ions, $H_2AsO_4{}^-{}_{(aq)}$, $HAsO_4{}^{2-}{}_{(aq)}$ and $AsO_4{}^{3-}{}_{(aq)}$ and also to the thermodynamic properties of crystalline arsenate compounds.

3.6 Discrepancies in the arsenic sulfide system

Some tabulations of thermodynamic properties of arsenic compounds list the enthalpies of formation for $AsS_{(cr)}$ and $As_2S_{3(cr)}$ to be of the order of –71 and –169 $kJ{\cdot}mol^{-1}$, respectively. These values, when accompanied with reasonable values of the entropies of the crystalline phases, and the elements, lead to Gibbs energies of formation of the same magnitude. Those enthalpies were obtained from oxygen combustion measurements performed by Britzke et al. (1933). The products of the combustion were a mixture of arsenic oxides and SO_x compounds. (It was recognized some time ago that there are two crystalline forms of each of $AsS_{(cr)}$ and $As_2S_{3(cr)}$ and that each pair has an equilibrium interconversion temperature substantially above room temperature.) Barton (1969) examined the phase behavior of the As+S system as a subset of his work on the Fe+As+S system and concluded that the enthalpies of formation were much too negative and that better values were on the order of –36 and –95 $kJ{\cdot}mol^{-1}$ for $AsS_{(cr)}$ and $As_2S_{3(cr)}$, respectively. Johnson et al. (1980) measured the enthalpy of fluorine combustion of samples of the high-temperature form of AsS and of glassy As_2S_3. They combined those values with other measured properties and some estimates to obtain enthalpies of formation of –33±1.6 and –91.6±4.8 $kJ{\cdot}mol^{-1}$ for $AsS_{(\alpha,cr)}$ and $As_2S_{3(\beta,cr)}$, respectively. Subsequently Bryndzia and Kleppa (1988) used direct-synthesis calorimetry and obtained enthalpies of formation of –28.1 ± 0.38 and –83.0 ± 3.8 $kJ{\cdot}mol^{-1}$ for $AsS_{(\alpha,cr)}$ and $As_2S_{3(\beta,cr)}$, respectively. The differences between these two more recent enthalpy determinations for these two arsenic sulfide minerals, (–5 and –9) $kJ{\cdot}mol^{-1}$, remain large for the purpose of calculating equilibria in a system containing arsenic, sulfur, water and other potential components. Bryndzia and Kleppa (1988) observed previously that enthalpies of formation of other materials determined by means of fluorine combustion disagreed with both observed phase behavior and high-temperature solution calorimetry, e.g., the case of chalcopyrite. Nonetheless, the fluorine reactions for these materials apparently went to completion with well-defined end products and a cavalier dismissal of them would not be wise.

Fortunately, for at least orpiment, there is additional information available that helps to clarify the picture. In a separate study, Webster (1990) measured the solubility of synthetic orpiment in water and her value gives $\Delta_r G = -131.91\ kJ{\cdot}mol^{-1}$ for the reaction:

$$0.5As_2S_{3(\forall,cr)} + 3H_2O = H_3AsO_{3(aq)} + 1.5HS^-_{(aq)} + 1.5H^+_{(aq)} \qquad (6)$$

at 298.15 K. This value is in very good agreement with the enthalpy of formation determined by Bryndzia and Kleppa (1988). Based on the lack of agreement of the Gibbs energy of orpiment determined from fluorine combustion measurements with that obtained from 298.15 K equilibrium data, we gave insignificant weight to the fluorine combustion results for both orpiment and realgar.

Helz et al. (1995) interpreted differences between the solubility determinations made independently by Webster (1990) and by Mironova et al. (1984) as being due to differences in the Gibbs energies of formation of the crystalline orpiment samples used in the respective studies. Webster (1990) used a synthetic orpiment and Mironova et al. (1984) used a naturally occurring orpiment crystal. Helz et al. (1995) concluded that the sample used by Webster (1990) was less stable than the naturally occurring orpiment by approximately 3 $kJ \cdot mol^{-1}$. The Gibbs energy of formation obtained from Webster's solubility determinations for the synthetic crystal is, however, in very good agreement with the enthalpy of formation determined by Bryndzia and Kleppa (1988) and the entropy determined from low-temperature adiabatic calorimetry. The latter two values were obtained from natural orpiment crystals. We also note the following. Bryndzia and Kleppa (1988) determined the enthalpy of fusion of natural orpiment to be 23.33 $kJ \cdot mol^{-1}$ at 298.15 K. Myers and Felty (1970) determined the enthalpy of fusion of natural orpiment to be 28.7 ± 1.3 $kJ \cdot mol^{-1}$ at 588 K. Adjustment of that value to 298.15 K gave 21.6 ± 2.6 $kJ \cdot mol^{-1}$, using the enthalpy increment for vitreous As_2S_3 from Johnson et al. (1980) and the enthalpy increment for orpiment from Blachnik et al. (1980). Blachnik et al. (1980) measured the enthalpy of fusion of a synthetic orpiment. Their enthalpy of fusion, when adjusted to 298.15 K, was 23.34 ± 2.5 $kJ \cdot mol^{-1}$. These results suggest thermodynamic equivalence of some synthetic orpiments with some natural orpiments within about ± 2 $kJ \cdot mol^{-1}$, or less. Finally, we note that the uncertainty in the Gibbs energy of reaction for 298.15 K from Mironova et al. (1984) was ± 3 $kJ \cdot mol^{-1}$, and thus represents the entire difference between the Gibbs energies of reaction from Mironova et al. (1984) and that from Webster (1990). Thus, it is not so clear that the differences in the solubility determinations must have arisen from differences in the stabilities of the different crystal samples rather than having arisen from differences in the solubility determinations per se (i.e. systematic error).

For orpiment, the solubility data and the entropy determined from calorimetric measurements support Bryndzia and Kleppa's (1988) determination of its enthalpy of formation. This might lead one to accept their value for the enthalpy of formation of realgar also. However, if one also takes Bryndzia and Kleppa's value for the enthalpy of formation for realgar and an entropy value determined from calorimetric values, the resulting Gibbs energy of formation of realgar has a fugacity of $S_{2(g)}$ that is larger than that of orpiment, which is the opposite of what is expected on the basis of chemography or on the basis of Barton's (1969) phase diagram for the As + S system. Therefore, calorimetric measurements failed to provide information of sufficient accuracy to predict a reasonable phase diagram for the arsenic + sulfur phase diagram and other considerations must be applied. We obtained thermodynamic properties of the realgar phases by accepting Barton's value for the chemical potential of the reaction:

$$4AsS_{(\alpha,\ realgar)} + S_{2(g)} = 2As_2S_{3(\alpha,\ orpiment)} \qquad (7)$$

the entropy of $AsS_{(\beta,\ realgar)}$ from Weller and Kelley (1964), and the enthalpy of transition for the transition of $AsS_{(\alpha,\ realgar)}$ to $AsS_{(\beta,\ realgar)}$ at 540 K. The enthalpy of formation calculated by this method is 3.2 $kJ{\cdot}mol^{-1}$ more exothermic than that obtained by Bryndzia and Kleppa (1988). These calculations presume that the entropy obtained from Weller and Kelley's (1964) calorimetric measurements from 52 K to 300 K, the assumption of a non-degenerate ground state for the β-form of realgar at 0 K, and the assumption that no phase transition occurs between 0 K and 52 K are correct.

Eary (1992) prepared an amorphous sample of As_2S_3 by precipitation of the sulfide from an aqueous solution of $NaAsO_2$, buffered at pH 4.0, to which was added an excess of $Na_2S{\cdot}9H_2O$. The precipitate was aged 1 to 3 days, and washed with water to remove the buffer. X-ray analysis showed no realgar, orpiment, crystalline sulfur or crystalline arsenic. Eary determined the solubility of this material at 298.15 K, 313.15 K, 333.15 K, and 363.15 K in solutions with the presence and absence of added sulfide. He gave equilibrium constants for the reaction:

$$0.5As_2S_{3(am)} + 3H_2O = H_3AsO_{3(aq)} + 1.5H_2S_{(aq)} \qquad (8)$$

at the four temperatures. The heat capacity of the reaction was estimated, $\Delta_r C_p = (180 \pm 100)\ J{\cdot}K^{-1}{\cdot}mol^{-1}$, from values of the heat capacities given in Archer and Nordstrom (in press), and where we have assumed that the heat capacity of the amorphous material is approximately equal to that of the crystalline material. These values were used to calculate:

$$\Delta_r G(T) - \Delta_r C_p\,\{T - T_r - T\ln(T/T_r)\} = \Delta_r G(T_r) - \Delta_r S(T - T_r) \qquad (9)$$

This relation was used to determine the entropy of reaction, $\Delta_r S = 66 \pm 30\ J{\cdot}K^{-1}{\cdot}mol^{-1}$. The entropy of reaction corresponds to an entropy of $As_2S_{3(am)}$ of $(200 \pm 60)\ J{\cdot}K^{-1}{\cdot}mol^{-1}$. We note that there is a lower bound to the uncertainty of the amorphous material, that being the entropy of crystalline orpiment, approximately 165 $J{\cdot}K^{-1}{\cdot}mol^{-1}$. Therefore, a symmetrical uncertainty cannot be assigned in this case and we give $S(As_2S_{3(am)},$ 298.15

K) = (200 +60/-35) $J \cdot K^{-1} \cdot mol^{-1}$. The Gibbs energy of reaction was $\Delta_r G(298.15\ K) = 68.08 \pm 0.85\ kJ \cdot mol^{-1}$, which when combined with other Gibbs energies of formation for the species in equation (9) leads to $\Delta_f G = -76.8\ kJ \cdot mol^{-1}$.

3.7 Thioarsenites

There has been some debate as to the identity of the aqueous arsenic species that are present in an aqueous solution in equilibrium with As_2S_3 where that solution contains an excess of sulfide over that which would be present from dissolution of As_2S_3 alone (see e.g. Krupp, 1990; Spycher and Reed, 1989). We had believed this controversy to have been settled recently by Helz et al. (1995) who utilized spectroscopy, molecular orbital calculations, and the solubility studies to arrive at the conclusion that the principal aqueous species in excess-sulfidic solutions are $H_2As_3S_6^-(aq)$ and $AsO(SH)_2^-{}_{(aq)}$.

Helz et al. (1995) fitted the solubility measurements determined by Eary (1992) to obtain pK = 5.5 for the reaction:

$$1.5As_2S_{3(am)} + 1.5H_2S_{(aq)} = H_2As_3S_6^-{}_{(aq)} + H^+{}_{(aq)} \qquad (10)$$

This value leads to $\Delta_f G(As_3S_4(SH)_2^-{}_{(aq)}) = -125.6\ kJ \cdot mol^{-1}$. They also obtained p$K$ = 7.9 for the reaction:

$$0.5As_2S_{3(am)} + 0.5H_2S_{(aq)} + H_2O_{(l)} = H_2As_3S_4(SH)_2^-{}_{(aq)} + H^+{}_{(aq)} \qquad (11)$$

This value leads to $\Delta_f G(H_2AsS_2O^-{}_{(aq)}) = -244.4\ kJ \cdot mol^{-1}$. Clarke and Helz (2000) analyzed phase behavior in the copper + arsenic + sulfur+ water system and obtained a somewhat different value for the equilibrium constant for eq 11. They obtained pK = -8.23 ± 0.32. This value leads to $\Delta_f G(H_2AsS_2O^-{}_{(aq)} = -242.5\ kJ \cdot mol^{-1}$. The minimum uncertainty in this value, and presumably the immediately previous value is at least 1.8 $kJ \cdot mol^{-1}$. We took the former value that arose solely from the treatment of Eary's solubility determinations over those obtained in the mixed arsenic + copper aqueous sulfidic system, because of difficulties in establishing the activity of arsenic sulfide in the latter system.

3.8 Hydrolysis constants for arsenious acid

Ivakin et al. (1979) determined the second and third ionization constants of arsenious acid in NaCl and KCl solutions. The pK values were determined by two methods, potentiometric and spectrophotometric. The solutions had ionic strength 1 M and the temperature of the measurements was 293.15 K. The individual pK values were:

$PK_2 = 13.54 \pm 0.03$ $\quad pK_3 = 13.99 \pm 0.04$ NaCl(aq) potentiometric (12)

$PK_2 = 13.65 \pm 0.05$ $\quad pK_3 = 13.75 \pm 0.04$ KCl(aq) spectrophotometric (13)

$$H_2AsO_3^-{}_{(aq)} = HAsO_3^{2-}{}_{(aq)} + H^+{}_{(aq)} \qquad (14)$$

$$HAsO_3^{2-}{}_{(aq)} = AsO_3^{3-}{}_{(aq)} + H^+{}_{(aq)} \qquad (15)$$

where the subscript to the equilibrium constant identifies the deprotonation step. We adjusted these values for the ionic strength effect, obtaining $pK_2 = 14.2 \pm 0.1$ and $pK_3 = 15.1 \pm 0.22$. We used estimates of the entropies of reactions (14) and (15) (–125 and –167 $J \cdot K^{-1} \cdot mol^{-1}$, respectively), to calculate the Gibbs energies for 298.15 K. These values led to Gibbs energies of reaction of 80.3 ± 0.1 $kJ \cdot mol^{-1}$ and 85.6 ± 1.5 $kJ \cdot mol^{-1}$, for reactions (14) and (15), respectively. These values are entered in Table 1b parenthetically because there is only one investigation and estimates had to be used to derive the values.

4. CONCLUSIONS

Sufficient physicochemical measurements exist to evaluate the thermodynamic properties for the simple arsenic minerals: native arsenic, arsenolite, claudetite, α-realgar, β-realgar, crystalline orpiment, and amorphous orpiment. The evaluation has been done by examining the available literature, screening and selecting the data, and performing a simultaneous weighted multiple least-squares regression on substances and reactions of arsenic, its oxides, its sulfides, and their aqueous hydrolysis products. These results have been combined with known occurrences and mineral transformations observed in nature to provide a coherent framework for mineral reactions in the As-S-O-H_2O system. The results show the following:

1. Claudetite is the most stable phase of arsenic trioxide under standard state conditions.
2. Claudetite and arsenolite are products of weathering of several arsenic sulfide minerals, of native arsenic, and of scorodite.

3. Claudetite and arsenolite are stable in equilibrium with waters of high pH.
4. Orpiment is more stable than realgar at standard state conditions and is stable in equilibrium with waters of low pH.
5. Orpiment and realgar can form under a wide variety of conditions that include hydrothermal mineralization, hot spring environments, mine fire sublimates, and fumarolic encrustations.
6. Realgar appears to be more stable than orpiment in high temperature environments (100-200°C).
7. Native arsenic is stable only under strongly reducing conditions and although it appears as a stable phase under standard state conditions, field observations indicate it only forms under hydrothermal conditions.
8. New polymorphs of realgar, pararealgar and alacrinite, have been characterized and reported in the literature and appear to form under earth's surface conditions. Pararealgar forms by visible light radiation of realgar. No thermodynamic data exist for these phases.

ACKNOWLEDGEMENTS

The senior author wished to acknowledge the support of the National Research Program of the USGS and the support of the EPA to actually sample for thioarsenite species at Yellowstone National Park (unpublished data).

Chapter 2

Spectroscopic Investigations of Arsenic Species in Solid Phases

Andrea L. Foster

Mineral Resource Program, U.S. Geological Survey, Menlo Park, CA

Many of the important chemical reactions controlling arsenic partitioning between solid and liquid phases in aquifers occur at particle-water interfaces. Several spectroscopic methods exist to monitor the electronic, vibrational, and other properties of atoms or molecules localized in the interfacial region. These methods provide information on valence, local coordination, protonation, and other properties that is difficult to obtain by other means. This chapter synthesizes recent infrared, x-ray photoelectron, and x-ray absorption spectroscopic studies of arsenic speciation in natural and synthetic solid phases. The local coordination of arsenic in sulfide minerals, in arsenate and arsenite precipitates, in secondary sulfates and carbonates, adsorbed on iron, manganese, and aluminium hydrous oxides, and adsorbed on aluminosilicate clay minerals is summarized. The chapter concludes with a discussion of the implications of these studies (conducted primarily in model systems) for arsenic speciation in aquifer sediments.

1. INTRODUCTION

Potable ground water supplies in many countries (including Bangladesh, India, Taiwan, Mongolia, Vietnam, Argentina, Chile, Mexico, Ghana, and the United States) contain dissolved arsenic (As) in excess of 10 μg/l, the maximum level recommended for potable waters by the World Health Organization (WHO, 1993). The primary source of As is natural (derived from interactions between ground water and aquifer sediments) in these locations, and not anthropogenic (Azcue and Nriagu, 1994; Cebrian et al., 1994; Chen et al., 1994; Nickson et al., 2000; Welch et al., 1988; Welch et al., 2000). Aquifer sediments are composed of inorganic and organic particles of various size and chemical reactivity that undergo a multitude of simultaneous chemical reactions, each occurring at a unique rate. These reactions occur almost exclusively at particle-water interfaces (Stumm, 1992) and because of this, spectroscopic techniques that are sensitive to physical or chemical properties of specific atomic and molecular species in the bulk phase or at particle surfaces are needed. Such investigations provide unique information that is difficult or impossible to obtain by other methods.

The designation of an aquifer as either a net "source" or "sink" with respect to As depends on the overall mass of As transferred from solid phases to the dissolved phase (and vice versa) in all interfacial reactions. The most important interfacial reactions involving As species are adsorption and precipitation reactions (resulting in arsenic retention by the solid phase), desorption and dissolution reactions (resulting in arsenic release from the solid phase), and oxidation/reduction reactions (resulting in retention or release, depending on the particular chemistry of reactants and products; (Welch et al., 1988; Welch et al., 2000). Other reactions, such as ion exchange, hydrolysis, complexation, and polymerization that are important for trace metal cations (Davis and Kent, 1990) have not been shown to be influential in the partitioning of trace metalloid oxoanions such as arsenic in the environment, but may be critical in specific systems.

The particles comprising aquifer sediments range in size from sand (2-0.5 mm diameter; (Allaby and Allaby, 1990) to colloid (between 1 nm and 1 μm in diameter; (Stumm, 1992). Smaller-sized particles and/or those with complex shapes have a higher surface area-to-volume ratio and more reactive surface area (i.e., the fraction of total surface area that participates in interfacial reactions) per unit mass than larger, simply shaped particles (Parks, 1990). As a result, the smallest particles may dominate the overall mass transfer in aquifer sediments. These particles are commonly x-ray amorphous, meaning that their structure cannot be resolved using standard x-ray diffraction techniques. However, they are partially ordered, with crystalline domains typically < 15 Å in diameter (Waychunas et al., 1996). Commonly, the smallest particles are also physically or chemically attached to the surface of larger particles.

Mineral or particle surfaces are enriched with As due to several processes that are collectively referred to as sorption (Parks, 1990), but the chemical properties of surface-associated As have been difficult to study directly. Outer-sphere, or physisorption, describes weak, long range, attractive forces between the surface and sorbing As; inner-sphere, or chemisorption, refers to the formation of chemical bonds between the surface and adsorbing As. Stronger adsorption is expected by the formation of a bidentate (two bond) adsorbed complex rather than a monodentate (1 bond) complex. Selective chemical extraction methods have been useful for empirical determination of the dominant chemical/mineralogical compartments retaining As in aquifer sediments and soils, but it is generally agreed that these techniques have limited applicability (Davis and Kent, 1990; Gruebel et al., 1988) and do not provide the kind of information about interfacial reactions (particularly the composition and geometry of adsorbed As complexes) that geochemical modelers need for predictive purposes (Parks, 1990).

Spectroscopic techniques (particularly infrared, x-ray photoelectron, and x-ray absorption spectroscopy) have been applied to fill the information gap about chemical speciation and interfacial reactions of As in model and natural materials. They have been used to determine the structure of x-ray amorphous particles involved in interfacial reactions, to identify the types of sorption reactions occurring in simplified model systems containing As and one or more phases, and to identify the valence and speciation of predominant As species present in natural, heterogeneous materials. This chapter summarizes much of the recent spectroscopic information on arsenic speciation in minerals and other solid phases that are analogous to phases present in aquifer sediments. These data are primarily derived from analysis of synthetic samples or natural model compounds.

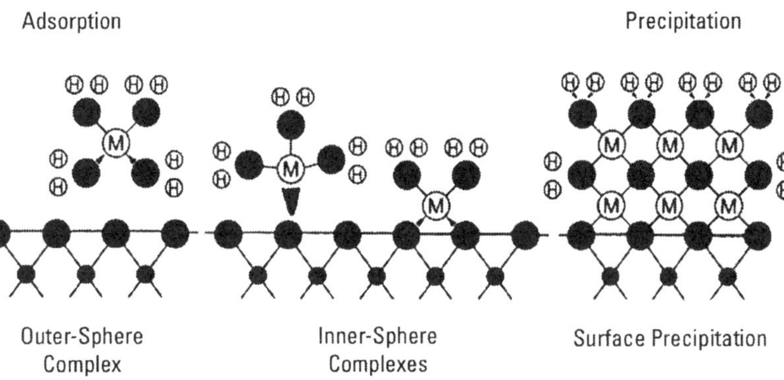

Figure 1. Schematic representation of possible arsenic sorption complexes on mineral surfaces (Modified from Brown, 1990). Outer-sphere (physisorbed) complexes, in which As is fully coordinated by water molecules, are bound to the mineral surface by weak electrostatic forces. Inner-sphere (chemisorbed) complexes are characterized by the formation of one or more chemical bonds between the sorbing As oxoanion and the mineral surface. Surface precipitation refers to the formation of a new phase on the mineral surface. Reprinted with permission.

2. MACROSCOPIC AND MOLECULAR-SCALE TECHNIQUES FOR THE CHARACTERIZATION OF ARSENIC IN SOLID PHASES

Solid-phase As can be found in many different forms in aquifer sediments: examples include: (1) stoichiometric arsenic minerals; (2) solid solution of arsenic in minerals or x-ray amorphous phases from trace (< 1000 ppm) to atom percent levels; (3) coprecipitation of As with minerals during their formation; and (4) adsorption of As on particle surfaces. This section gives a brief background on the macroscopic and spectroscopic techniques that are commonly used for ascertaining As species in these phases.

2.1 Extraction Methods of Arsenic Speciation in Solid Phases

Extraction methods were developed to obtain empirical information about the chemistry of a particular element in heterogeneous material that may contain multiple species of the element of interest. Species of an element partitioned into solution by a particular extraction solution are grouped into compartments named according to the dominant minerals or phases present in that compartment. When extraction solutions are applied in order from the least to most chemically aggressive, the classic compartmentalization, (in order from most to least available) is obtained: a water-exchangeable fraction, a carbonate fraction, a fraction associated with organic matter, a fraction retained on iron, aluminum, and/or manganese oxides or oxyhydroxides, and a residual fraction (usually assumed to represent As hosted by silicate or sulfide minerals; (Tessier et al., 1979). Refinement of the original sequential extraction techniques now allows for the discrimination of As associated with crystalline Fe hydroxides from As associated with x-ray amorphous Fe hydroxides and the specific identification of sulfide-associated As (Keon et al., 2000). However, the standard interpretation of sequential extraction data presupposes that there is no mass transfer between compartments during extraction, an assumption that has not proved valid in some empirical tests (Gruebel et al., 1988; Ostergren et al., 1999). As an example, during the extraction of As associated with Fe oxyhydroxides, the As released was partially readsorbed by aluminosilicate clay minerals (initially free of arsenic) that were not attacked by the extraction solution. This lead to an underestimation of the fraction of As associated with the Fe oxyhydroxides, and an overestimation of the fraction associated with the aluminosilicate residual (Gruebel et al., 1988).

Another difficulty in the interpretation of sequential extraction data is that the availability of As from most stoichiometric As minerals does not fit neatly into one of the predefined chemical/mineralogical compartments, because it will be a function of the mineral solubility and dissolution kinetics under the chemical conditions of the extraction solution(s). For example, Voigt and Brantley (1997) determined that As in the mineral hoernesite ($Mg_3(AsO_4)_2 \bullet 8\ H_2O$) was released during the exchangeable, carbonate, and Fe oxyhydroxide extraction steps before it was completely dissolved, leading to an overestimation of the fraction of As in ion-exchangeable sites and As associated with carbonate minerals.

2.2 Spectroscopic Methods for Arsenic Speciation in Solid Phases

Spectroscopic analysis involves the measurement of energy absorbed or emitted by a system in response to an external perturbation. The absorption and/or emission of energy results from changes in quantized magnetic, rotational, vibrational, or electronic energy levels of atoms or molecules in the system. The external perturbation usually (but not always) takes the form of electromagnetic radiation. The widely used spectroscopic techniques of x-ray fluorescence and atomic absorption spectrometry provide quantitative information about the elemental composition of a sample; other techniques, such as infrared, x-ray photoelectron, and x-ray absorption spectroscopy, provide quantitative or semi-quantitative elemental speciation of a sample. The latter three techniques are briefly reviewed below. For a detailed explanation of general spectroscopic principles, the reader is referred to Levine (1975).

2.2.1 Infrared (IR) and Raman Spectroscopy

IR and Raman are sensitive to the rotation and vibration of molecules in solid phases (crystalline or x-ray amorphous). Molecular units of similar structure and composition absorb IR radiation in the same energy range, usually independent of the larger structure of the material; this property makes IR spectroscopy useful for studying molecules in the interfacial region such as surface hydroxyl groups and As oxoanions on mineral surfaces, and for fingerprinting the local environment of As in crystalline and x-ray amorphous solids. Some IR techniques can be conducted *in situ* (in the presence of adsorbed water), and others are solely *ex-situ* (the latter experimental condition is less representative of the hydrated conditions present at mineral surfaces in the aquifer environment, but has analytical benefits). The interaction of arsenic oxoanions with surface hydroxyl groups on mineral surfaces can be studied by examination of changes in the IR signal of arsenic group vibrations and/or surface hydroxyl group vibrations as a function of arsenic sorption. IR spectra of natural materials are typically complex and difficult to interpret quantitatively due to overlap of molecular vibrations arising from all the phases in the sample (Smith, 1996). For this reason most IR studies of arsenic species have been conducted in synthetic, simplified systems.

Ex situ IR data are collected on dried, diluted powder films in a low vacuum environment or one purged with a dry gas such as N_2. Attenuated total reflectance (ATR)-IR spectroscopy provides surface-sensitive IR measurements and can be used for *in situ* studies of sorption phenomena. Raman spectroscopy is a related vibrational spectroscopy that provides complimentary information to IR. It can also be used to collect vibrational spectra of aqueous samples. Typical data reduction for vibrational spectra involves subtraction of a background spectrum collected under identical conditions from the raw, averaged sample spectrum. Data analysis usually consists of an examination of changes in peak position and shape and peak fitting (Smith, 1996). These and other spectral parameters are tracked as a function of macroscopic variables such as pH, adsorption density, and ionic strength.

2.2.2 X-ray Photoelectron Spectroscopy (XPS)

During an XPS analysis the sample is irradiated by photons in the soft x-ray energy range of the electromagnetic spectrum (usually from a Mg or Al x-ray source), and the signal intensity is measured as a function of binding energy of the escaping photoelectrons. This technique is surface-sensitive (usually < 100 Å; (Hochella, 1995). With an x-ray source, the photoelectron spectrum is most sensitive to changes in electronic states, and can be used to determine both valence and speciation information. Laboratory XPS experiments are conducted under *ex-situ*, mid-to-high vacuum conditions (10^{-5}-10^{-9} torr; Demuth and Avouris, 1983) due to the need to characterize surfaces before and after reaction with specific agents (and so that the photoelectron signal is not significantly attenuated by gases and water vapour in air).

XPS spectra can be collected from powders or single crystals. The broadscan (low resolution) XPS spectrum, usually spanning ~1,000 eV, shows several peaks corresponding to the elements present in the sample superimposed upon a background. When high resolution (narrow scan) spectra are collected over smaller intervals spanning a peak of interest, a single peak is usually observed to be composed of several peaks. These peaks (called chemical shifts) are due multiple species of a single element present in a sample. Chemical shifts arise from coordinative differences, as in surface vs. bulk atoms, and/or the presence of multiple oxidation states. Peak positions are calibrated by reference to an appropriate standard (usually gold) collected under identical conditions as the sample. A background function is subtracted from the photopeaks of interest, and peak fitting using appropriate lineshape functions is applied. Details of XPS data collection and analysis procedures as applied to arsenic can be found in Nesbitt et al. (1998; 1995; 1999). Similar to IR spectroscopy, XPS analysis is not an ideal technique for the study of arsenic in complex, heterogeneous natural materials due to peak overlap from all the elements present. However, it has been very useful for the analysis of arsenic valence in model systems and single phases, as will be shown by examples in the sections to follow.

2.2.3 X-ray Absorption Fine Structure (XAFS) Spectroscopy

In an As K-edge XAFS experiment, both the source radiation and the absorbed/emitted radiation are in the hard x-ray region of the electromagnetic spectrum. The technique is sensitive to the scattering of electrons ejected from the innermost atomic core levels, and in contrast to IR or XPS, is exclusively element-specific. Quantitative oxidation state information is routinely determined from x-ray absorption near-edge spectroscopy (XANES) spectra of materials with As concentrations as low as 50 ppm, and under optimal conditions, and given plenty of time for data collection, detection limits can be lowered. Quantitative information about the number, distance, identity, and disorder (static and thermal) of atoms 5-7 Å away from the average As atom is derived from analysis of extended x-ray absorption fine structure (EXAFS) spectra in materials with As concentrations as low as 100 ppm, but commonly greater than 500 ppm. Spectra can be collected under *in situ* or *ex situ* conditions. The element specificity of XAFS spectroscopy makes it the technique of choice for examining trace element species (particularly valence states) in complex natural materials such as aquifer sediments. However, as with the other spectroscopic methods, the identification of As species (as opposed to simply determining oxidation states) present in natural materials is complicated by the presence of multiple species, because the bulk XAFS spectrum is a weighted sum of all species of a particular element present in a sample. Nevertheless, it is often possible to identify the predominant As species or at least eliminate potential species as major components (> 10%) of the spectrum by fitting the unknown XANES or EXAFS spectrum with a linear combination of model component spectra.

XAFS spectra are collected from concentrated or dilute specimens under ambient conditions with, in many cases, no sample pretreatment (a great benefit for the analysis of natural materials). Spatially-resolved XAFS techniques exist, but they are either in development or have limited availability; as a result, the majority of XAFS investigations to date have been carried out in bulk mode. Details of XAFS data analysis methods can be found in (Hayes and Boyce, 1982; Sayers and Bunker, 1988). The XANES portion is energy calibrated to a model standard (for arsenic, this is typically the elemental form, assigned an energy of 11,867 eV) for oxidation state determination. Quantitative speciation information can be obtained from the EXAFS spectrum by fitting using a non-linear, least-squares algorithm employing theoretical backscattering phase and amplitude functions for model absorber (As)-backscatterer pairs (e.g., As-O, As-Ca, or As-Fe). These functions are derived from theoretical EXAFS spectra generated by *ab initio* calculations of photoelectron scattering in known structures.

3. ARSENIC SPECIATION IN NATURAL AND SYNTHETIC PHASES

Aquifer sediments are ultimately derived from rock weathering. Arsenic concentrations in unmineralized rocks of the United States are highest in sedimentary units (particularly marine shales), but are also elevated in volcanic rocks and in metamorphic rocks derived from sedimentary units (Welch et al., 1988). Although mineral sources of As have been directly identified in only a few aquifer sediments to date, pyrite, arsenopyrite, and/or unspecified sulfide minerals are often proposed to be the primary source of As in ground waters (Gotkowitz et al., 2000; Nickson et al., 2000; Peters et al., 1999; Robinson et al., 2000; Serfes et al., 2000; Smedley and Kinniburgh, 2001).

3.1 Arsenic Speciation in Sulfide Minerals

Arsenic-bearing sulfides typically form by crystallization from hydrothermal solutions interacting with sediments, and/or by direct crystallization during late-stage pegmatite formation (Peters et al., 1999). Secondary (authigenic) sulfides may also be important in As sequestration (Morse, 1994). Although As-bearing sulfides and arsenides containing Pb, Cu, Ni, and Fe are known, arsenopyrite, (FeAsS) and arsenian pyrite (FeS_2, containing As in trace to atom percent levels) are quantitatively the most important As-bearing sulfides reported from mineralized areas and may the most important As-bearing phases in the lithologic units that were eroded to form aquifer sediments in regions where ground water As concentrations are detrimental to human health (Smedley and Kinniburgh, 2001). In particular, arsenian pyrite is much more commonly found than the stoichiometric As sulfides, and may be the single most important reservoir for reduced arsenic in sediments.

In soils and sediments, importance has been attached to valence assignment because valence is commonly used to infer reactivity and toxicity of As. The assignment of valence in sulfides is complicated by strong covalent bonding, and is perhaps not very reliable except in XPS studies where the valence of S and other elements can be determined simultaneously. XPS and XAFS analysis indicate that the average arsenic valence in arsenopyrite is -1, as would be calculated for Fe with a valence of –2 and S with a valence of -1 (Nesbitt et al., 1995; Savage et al., 2000). However, up to 15% As(0) was observed on freshly cleaved arsenopyrite surfaces in vacuum by XPS (Nesbitt et al., 1995), an indication that the bulk valence assignment may reflect a composite of multiple reduced As species (either formed by oxidation or present as small amounts of other minerals, particularly native As, within arsenopyrite). Nickel (Ni) is typically assigned a valence of +2 in oxides, hydroxides, and silicates; however, the XPS-determined valence of Ni in niccolite (NiAs) is +1, and the valence of arsenic is –1 (Nesbitt and Reinke, 1999). The nominal valence of As in arsenian pyrite has been determined to be between 0 and –1 by XAFS spectroscopy (Savage et al., 2000).

Light or electron microscopic identification of As-hosting sulfides in aquifer sediments is usually precluded by the low modal abundance of sulfide phases and/or As concentrations below detection limits of conventional energy dispersive detectors. However, identification of As-bearing minerals is relatively simple using XAFS spectroscopy (when one type of mineral is the predominant host). XAFS spectra can be used both quantitatively and as chemical fingerprints to discriminate between arsenian pyrite in mine tailings and bulk arsenopyrite (Savage et al., 2000; Fig. 2a and 2b). As-bearing sulfide phases can also be identified based on bond distances: The shortest As-S bonds (between 2.23 and 2.28 Å) are found in arsenian pyrite and in the arsenic sulfosalts orpiment and realgar (Table 1). At 2.34 Å, the shortest As-S bond distance in arsenopyrite is considerably

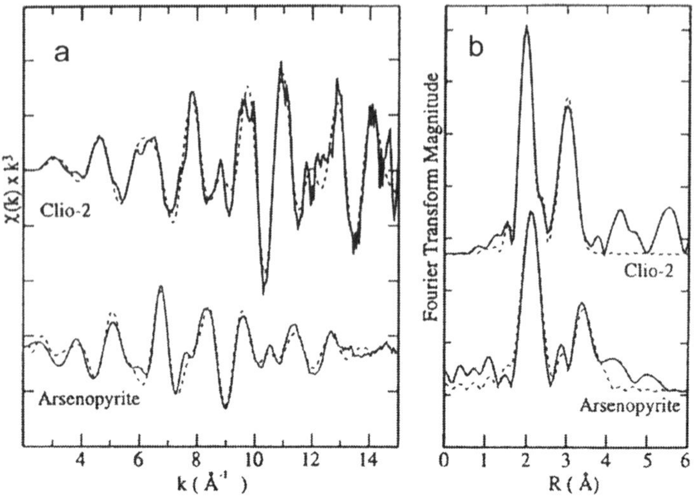

Figure 2. Normalized, k^3-weighted As-EXAFS spectra (a) and Fourier transforms (FT), b) of natural arsenian pyrite sample Clio-2 and model arsenopyrite (modified from Savage et al., 2000). Solid line is experimental data, dashed line is least squares fit. The coordination environment of As in the two phases is quite different, as indicated by the magnitude and position of oscillations in the As-EXAFS spectra and by the magnitude and position of peaks in the FTs. Peak positions in FTs are not corrected for phase-shift effects, and are therefore approximately 0.5 Å shorter than the true distance. Reprinted with permission.

longer than in these phases. There are also quantitative differences in the As-Fe and As-As bond lengths in these minerals that can also be used for identification (Table 1).

Another common concern in the examination of natural materials is the crystallinity of the As-hosting phase: because metastable x-ray amorphous materials are often finer-grained, they are probably more reactive than their crystalline counterparts (if all properties such as the type and proportion of surface functional groups are equivalent in the materials). Helz et al. (1995) compared arsenic speciation in natural orpiment, natural realgar, and synthetic x-ray amorphous arsenic sulfide. Based on similarities in the primary As-S shell at 2.38 Å and the As-As shell at ~ 3.5 Å, amorphous arsenic sulfide was found to have a structure more similar to orpiment than realgar. However, S atoms between 3-4 Å that can be detected in the XAFS spectrum of orpiment could not be detected in the XAFS spectrum of the amorphous analog. This discrepancy might indicate a high degree of positional disorder in the amorphous phase, which might mask scattering from S atoms more effectively than scattering from As atoms, since the latter have about twice the number of electrons. The formation of amorphous As sulfide has been documented in pure cultures of anerobic, dissimilatory arsenic-reducing bacteria (Newman et al., 1997) and XANES spectroscopy

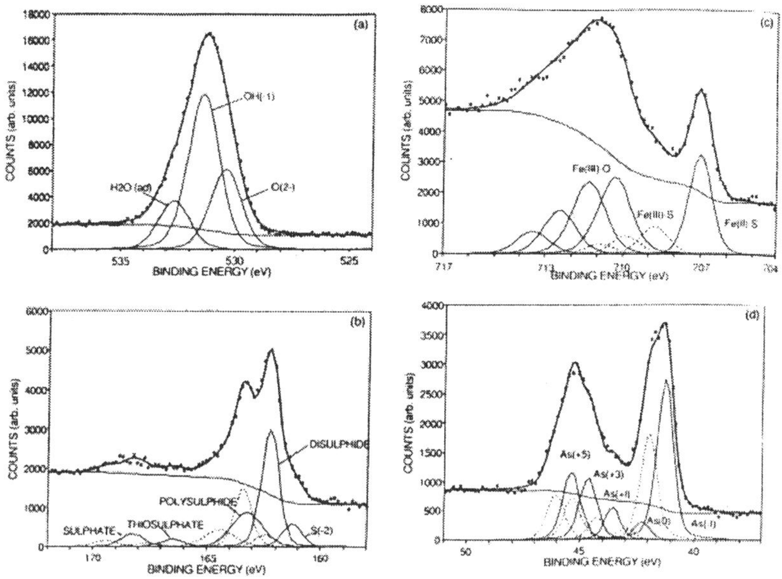

Figure 3. X-ray photoelectron spectra of a fractured surface of arsenopyrite after 8 hours reaction with distilled water. Narrow scan spectra are presented for (a) O(1s), (b) S($2p_{3/2}$), (c), Fe($2p_{3/2}$), and (d), As($3d_{5/2}$). XPS data are the black squares, the fit is the thick black line, the background and XPS peaks by thin solid lines, and peak doublets by dashed lines. Reprinted from Nesbitt et al. (1995) with permission.

has been used to identify orpiment-like phases in a wetland sediment (La Force et al., 2000), suggesting a re-evaluation of the importance of authigenic sulfide minerals that sequester As under anoxic conditions.

The reactivity of As-bearing sulfides in the subsurface is not well understood. For example, under equivalent geochemical and hydraulic conditions, will an aquifer containing arsenopyrite have greater dissolved As concentrations than one containing arsenian pyrite? Field observations suggest that arsenian pyrite weathers faster than arsenic-free pyrite [Savage et al., (2000) and references therein], and an XPS study demonstrates that arsenopyrite oxidizes relatively slowly upon exposure to air or water (Richardson and Vaughan, 1989). More recent XPS work contradicts the earlier study, indicating that 24-hour air exposure and 8-hour water exposure produces extensive oxidation of arsenopyrite surfaces and formation of As(V), As(III), and As(I) ions (Nesbitt et al., 1995) Fig. 3). Nevertheless, the newer study still indicates that As oxidation is more rapid than S oxidation on fracture surfaces of arsenopyrite (Nesbitt et al., 1995; Nesbitt and Reinke, 1999). This observation may hold true for As-bearing sulfides generally, and may be related to differences in electronic structure (the

Table 1. Summary of XAFS Studies of Arsenic Speciation in Metal Sulfides.

Mineral	Absorption Edge (keV)*	EXAFS-determined Average Local Coordination to 4 Å**		Mode of Arsenic Incorporation	Reference
Arsenian Pyrite (natural)	11.866	1 S 3 Fe 4 S 3 As 4 Fe	2.25 Å 2.32 Å 3.10 Å 3.17 Å 3.50 Å	Coprecipitation; ~ 30% of 1st shell S replaced by As; ~ 43% of 2nd shell S replaced by As	Savage et al. (2000)
Elemental As (natural)	11.867	3 As 3 As 6 As	2.50 Å 3.11 Å 3.74 Å	Stoichiometric precipitation	Foster et al. (1998)
Arsenopyrite (natural)	11.867	3 Fe** 1 As 2 As 4 Fe	2.36 Å 3.08 Å 3.36 Å 3.76 Å	Stoichiometric precipitation	Foster et al. (1998)
Orpiment§ (natural)	11.869	3 S 1 As 3 S 3 As	2.28 Å 3.21 Å 3.41 Å 3.52 Å	Stoichiometric precipitation	Helz et al (1995) ; Foster et al. (1998)
Realgar §§ (natural)	--	2 S 1.7 As 2.5 As 1 S 1.3 As	2.23 Å 2.56 Å 3.46 Å 3.59 Å 3.55 Å	Stoichiometric precipitation	Helz et al (1995)
Amorphous arsenic sulfide (synthetic)	--	3.5 S 3.0 As	2.28 Å 3.55 Å	Stoichiometric precipitation	Helz et al (1995)

**calibrated on As(0) with value of 11.8670 keV; error = ± 0.0005 keV*

***average error in coordination number =±20%; average error in distance = ±0.02 Å.*

§EXAFS results averaged over 2 different As sites in orpimen

§§ EXAFS results averaged over 4 different As sites in realgar

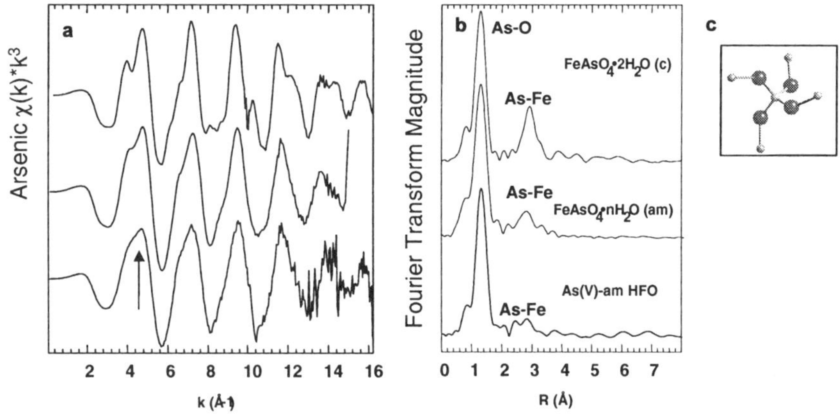

Figure 4. Normalized, k^3-weighted As-EXAFS spectra (a) and uncorrected Fourier transforms (FTs) (b) of scorodite (a crystalline ferric arsenate), an x-ray amorphous analog, and As(V) sorbed to amorphous hydrous ferric oxide (HFO). The EXAFS spectra can clearly be used to distinguish the coordination environment of arsenic in each of the materials. The highly symmetric local environment of arsenic in scorodite is shown in (c); each arsenic is surrounded by 4 Fe neighbors at a distance of 3.34 Å. (see Table 2 for details of precipitate fits, and Table 3 for details of sorption sample fit). Reprinted from Foster (1999). The arrow highlights the region of particular difference among the three spectra. Peak positions in FTs are not corrected for phase-shift effects, and are therefore approximately 0.5 Å shorter than the true distance.

increased p-character of the As bonds relative to S bonds) and/or to substitutional characteristics (As appears to exhibit clustered substitution in hydrothermal arsenian pyrite; Savage et al., 2000).

A recent report provides evidence that under anoxic laboratory conditions, MnO_2 solids can enhance the oxidative attack of FeS_2 and FeS by dissolved Fe(III), whereas in the absence of MnO_2 (and under sterile conditions), FeS_2 is not affected by dissolved Fe(III) or even NO_3 (Shippers and Jorgensen, 2001). This reaction does not need molecular oxygen to proceed, and could potentially take place even in anoxic aquifers. Further study is needed on the stability of sulfides generally, and As-bearing sulfides in particular, under anoxic and suboxic conditions.

3.1.1 Arsenic Speciation in Arsenates and Arsenites.

Of the greater than 320 known As minerals, fewer than 10 are commonly identified in the environment, and with the exception of scorodite, the most commonly reported arsenic minerals are arsenic oxides or arsenic sulfides (Welch et al., 1988). Very little structural or thermodynamic data exists for arsenate and arsenite minerals, and even less spectroscopic information. Hess and Blanchar (1976) determined solubility products for synthetic Al,

Ca, Fe, Mn, and Pb arsenate phases; of these, the Pb and Mn arsenates were predicted to be stable under the geochemical conditions prevailing in two Missouri orchard soil slurries. Sadiq (1997) calculated solubility isotherms as a function of pH for several metal arsenates, and concluded that Cd, Ni, Pb, Cu, and Zn arsenates were quite soluble and not expected to form in oxidized soil systems. In contrast, Sadiq (1997) determined that "$Fe_3(AsO_4)_2$", "$Ca_3(AsO_4)_2$", and "$Mn_3(AsO_4)_2$" (structural information not given) have similarly low solubilities, and therefore could be stable in alkaline oxidizing systems. However, under acidic oxidizing conditions, "$FeAsO_4$" and "$AlAsO_4$" were predicted to be the more stable phases. Rochette et al. (1999) found that under anoxic conditions (redox potential < 0 mV), the experimental solubility of arsenate phases followed the following order: $FeAsO_4{\bullet}2H_2O$ (scorodite) $\geq$ $CaHASO_4$ (weilite) = $Na_2HAsO_4{\bullet}7H_2O$ > $AlAsO_4{\bullet}2H_2O$ > $MnHAsO_4$. In addition, Cummings et al. (1999) showed that dissimilatory metal reducing bacteria could liberate As(V) from scorodite. The As(V) was not reduced, but once liberated formed a new ferrous arsenate phase with Fe(II) produced from the reductive dissolution of scorodite. The precipitate was x-ray amorphous, but the ratio of Fe to As was estimated at 1:1 by SEM-EDS. Further complicating the problem, little thermodynamic data for metal arsenite salts exists, to our knowledge.

Despite thermodynamic predictions and direct observations that metal arsenates control dissolved As concentrations under certain geochemical conditions, minerals containing As oxoanions have not been directly documented microscopically or spectroscopically in aquifer sediments, to our knowledge. Most locations where metal arsenates have been identified have the common characteristic that As porewater concentrations are extremely high. Scorodite ($FeAsO_4{\bullet}2H_2O$) and various Ca-Fe(III) arsenate phases have been identified in seasonally dry and continuously inundated gold mine tailings (Paktunc et al., 2000), and an unusual Mg-arsenate phase has been described from a tidally-influenced site contaminated by industrial processes (Voigt and Brantley, 1997). Dissolved As concentrations in As-impacted aquifers rarely reach the levels needed to induce the precipitation of metal arsenates. However, scorodite is typically found in close association with sulfide phases such as arsenical pyrite or arsenopyrite (Dove and Rimstidt, 1985), suggesting precipitation due to local supersaturation rather than precipitation from the bulk aqueous medium was the important process in its formation.

Although formation of arsenic oxoanion minerals may be uncommon in the subsurface, many other minerals in which oxoanions are the fundamental structural unit (most commonly sulfates, phosphates, and carbonates) can assimilate trace amounts of arsenic via adsorption and coprecipitation processes. By virtue of their ubiquitous nature in sediments, these phases may serve as significant repositories of arsenic even when bulk arsenic concentrations are at trace levels. Adsorbed As is subject to desorption due to changing geochemical conditions, including phase dissolution resulting from changes in pH and competitive desorption (especially by high concentrations of inorganic ligands such as OH^-, PO_4^{3-}, $CO_3^{2-}$and SO_4^{2-} and by simple organic ligands), but the availability of coprecipitated arsenic depends more on the bulk mineral solubility, because a greater proportion of the arsenic is retained in the interior of the phase rather than at the surface.

The results of several EXAFS studies of arsenic species in minerals and synthetic phases are collected in Table 2. The studies indicate that arsenic sorption on these minerals primarily occurs by two mechanisms: (1) Substitution for other oxoanions (such as carbonate or sulfate) during coprecipitation, and (2) Bonding through exchange with surface hydroxyl groups to structural metal ions during chemisorption. At least one study by Myneni (1995) suggests that mechanism (1) predominates when As is present during mineral formation, whereas mechanism (2) predominates when minerals are exposed to As-bearing solutions after their formation.

Substitution of arsenate for sulfate in jarosite (K, Na, NH4, H3O)Fe3[SO4]2(OH)6; Savage et al., 2000) and in ettringite(Ca6Al2(SO4)3(OH)12 ·26 H2O; Myneni et al., 1997), has been documented by EXAFS spectroscopy. In ettringite, as in other salt phases, significant distortion of the bonds around the substituted As are observed because of the size mismatch between the anions (Table 2; Myneni,1995), and if substitution is significant, may decrease the stability of the substituted mineral relative to the pure phase. Although equivalent spectroscopic evidence of arsenite substitution for other oxoanions in minerals is lacking, substitution of the structurally analogous selenite oxoanion has been described in calcite (along with the selenate oxoanion, which is structurally analogous to arsenate; Cheng et al., 1997; Reeder et al., 1994).

Spectroscopic evidence for mechanism (2), arsenic adsorption on oxoanion precipitates, has been provided by Myneni (1995) and Myneni et al. (1997) in the arsenate-ettringite system. Arsenate was found to adsorb to different sites as a function of sorption density. At low sorption density, As oxoanions exchange for surface hydroxyl groups bonded to Al atoms and at higher sorption densities As oxoanions exchange for surface hydroxyl groups bonded to Ca atoms, with minor bonding to Al atoms (Table 2).

In a separate study, Myneni et al. (1998) used IR data to examine the coordination of As(V) in Mg, Co, Na, and Fe arsenate salts. The authors determined that hydrogen bonding, which is substantial in these materials, had little effect on the position of bands corresponding to vibrations within the arsenate tetrahedron. However, the authors determined that the position of the peak assigned to the symmetric stretch of the As-OH bond increases with successive protonation of the arsenate tetrahedron. Myneni et al (1998) also determined that the arsenate tetrahedron is very distorted in Mg, Co, Na, and Fe salts, as well as in aqueous solution.

Foster (1999) compared the local structure of arsenate in scorodite with that of an x-ray amorphous synthetic ferric arsenate. The synthetic compound had the characteristic pale-green/yellow color characteristic of scorodite, and not the rust-brown or orange color characteristic of Fe oxyhydroxides (which could have formed during the synthesis), but its measured stoichiometry was not reported. XAFS analysis indicates that the x-ray amorphous compound has fewer Fe atomic neighbors than scorodite, and the number and position of Fe shells is more similar to As(V) sorbed on Fe oxyhydroxides than to scorodite. However, the EXAFS spectrum of scorodite is clearly distinguished from that of the x-ray amorphous ferric arsenate, which is itself different from As(V) adsorbed to Fe oxyhydroxide (Fig. 4).

The valence of arsenic associated with natural precipitates has been determined in surface sediments and mine tailings using XAFS spectroscopy. Jarosites from mining environments predominantly contain As(V) oxoanions substituting for sulfate and adsorbed to the jarosite surface (Foster et al., 1998; Savage et al., 2000). Hoernesite, a magnesium-arsenate phase abundant in a contaminated surface soil also contains only As(V) (detection limit: 5-10 %; Foster, 1999; see Table 2). As(III) species have not been observed adsorbed to or coprecipitated in salt phases in natural samples to date, but they presumably undergo similar substitution or sorption reactions on oxoanion precipitates. Their spectroscopic observation may increase as since sample handling, preparation, and analysis techniques have improved enormously in recent years.

3.2 Arsenic Speciation in Crystalline and X-ray Amorphous Metal Hydroxides.

Hydrous metal oxides are quantitatively the most important source/sink terms for arsenic in aquifer sediments (Smedley and Kinniburgh, 2001). These very fine-grained solid phases commonly occur as coatings on other minerals and are often x-ray amorphous. Hydrous ferric oxides (HFOs) are common products of sulfide mineral oxidation and incongruent dissolution of minerals such as scorodite. Hydrous manganese oxides (HMOs) and hydrous aluminum oxides (HAOs) may be important for sequestering As, particularly where these oxides are very abundant relative to HFOs and under chemical conditions where HFOs are unstable (Rochette et al., 1998). Arsenic can be associated with hydrous metal oxides as both adsorbed and coprecipitated species. As was discussed in the previous section, the availability of coprecipitated As is probably lower than adsorbed As because a higher fraction of the total As is not available for reaction with aqueous solution.

3.2.1 The molecular-scale structure and chemistry of crystalline and X-ray amorphous metal oxide precipitates

HFOs, HMOs, and HAOs are composed of octahedrally coordinated metal atoms that share edges to form chains; these in turn are linked together to form 2-D and 3-D structures. For example, HFO phases consist of chains of edge-sharing Fe(III) octahedra that are linked to opposing chains in three predominant ways: by face-sharing (hematite and feroxyhyte), by edge-sharing (lepidocrocite), and by corner-sharing (goethite, akaganeite; Manceau and Combes, 1988; Waychunas et al., 1993). The common HAO phases gibbsite, boehmite, and diaspore contain similar linkages and have structures similar to the HFOs because the Al(III) atom has the same charge and a nearly identical ionic radius as the Fe(III) atom (Waychunas, 1991). The numerous HMO phases found in nature commonly form more complex tunnel and layer structures than HFOs or HAOs (Manceau and Combes, 1988). HMO phases also have the added complexity of multiple oxidation states; several investigators have determined that synthetic HMO phases such as vernadite and birnessite have up to 30% Mn(II+III) (Nesbitt et al., 1998; Scott and Morgan, 1995). XAFS, IR, and XPS spectroscopic techniques are sensitive to different types of linkages and oxidation states, and can be used to identify these phases in natural and synthetic systems.

X-ray amorphous hydrous metal oxides are also extremely important as transitory reservoirs of As, but have been difficult to characterize using traditional methods. Waychunas et al. (1993) used XAFS spectroscopy to examine structural differences between crystalline HFO phases and synthetic 2-line ferrihydrite. Fe-O and Fe-Fe bond distances in the ferrihydrite indicated that this phase had a structure similar to goethite, being composed of double chains of corner-sharing octahedra linked primarily by other corner-sharing octahedra. However, in ferrihydrite the average double chain length was shorter and there were fewer cross-chain linkages than in the crystalline HFOs (Waychunas et al., 1993). Therefore, for HFO phases, as well as HMOs and HAOs, x-ray amorphous phases are not truly amorphous, but simply have smaller crystalline domains than their well-crystalline counterparts.

The structure of x-ray amorphous synthetic hydrous manganese oxide (vernadite) and poorly crystalline birnessite was examined using XAFS spectroscopy by Manceau and Combes (1988) and Manceau et al. (1992a; 1992b). XAFS results were in agreement with the structure of synthetic

Table 2. Summary of XAFS Studies of Arsenate co-precipitation in and adsorption on natural and synthetic materials.

Mineral	EXAFS-determined Average Local Coordination to 4 Å*		Mode of Arsenic Incorporation	Reference
Scorodite (natural)	4 O 4 Fe 1 As	1.68 Å 3.35 Å 4.21 Å	Stoichiometric	Waychunas et al. (1993); Foster et (1999)
Ferric Arsenate, x-ray amorphous (synthetic)	4 O 1.6 Fe 0.4 Fe	1.69 Å 3.33 Å 3.63 Å	Stoichiometric (short range order)	Foster (1999)
Hoernesite (natural) $[Mg_3(AsO_4)_2 \bullet 8H_2O]$	5.4 O 5 Mg 1 As	1.69 Å 3.31 Å 3.70 Å	Stoichiometric	Foster (1999)
Jarosite (K, Na, NH_4, $H_3O)Fe_3[SO_4]_2(OH)_6$	4.2 O 1.8 Fe 1.3 K	1.70 Å 3.32 Å 3.04 Å	Substitution for sulfate	Savage et al. (2000)
Rauenthalite $(Ca_3(AsO_4)_2 \bullet 10\ H_2O)$ Pharmacolite, $(CaHAsO_4 \bullet 2H_2O)$ Hadingerite $(CaHAsO_4 \bullet H_2O)$	4 O 1 Ca 1-3 Ca	1.69 Å 3.22-3.32 Å 3.38-3.64 Å	Stoichiometric	Myneni (1995)
Synthetic Ettringite coprecipitate $[Ca_6Al_2(SO_4)_3(OH)_{12} \bullet 26\ H_2O]$	4 O $1Ca_1$ 1-2 Ca_2 1 Ca_3	1.69 Å 3.18 Å 3.21-3.74 Å 4.09-5.09 Å	Substitution for sulfate	Myneni et al. (1997)
Synthetic Ettringite sorption low loading < 0.65 mol As/kg high loading (> 0.65 mol As/kg)	4 O 0.5-1.3 Al 0.5-2 Ca 4 O 0.5-3Ca	1.69 Å 3.17-3.21 Å 3.18-3.70 Å 1.69 Å 3.20-4.51 Å	Sorption to surface Al sites (low sorption density) and/or Ca sites (high sorption density)	Myneni et al. (1995)
Poorly crystalline Mn(II)-As(V) phase (XRD-pattern similar to geigerite)	4.0 O 6.6 Mn	1.68 Å 3.45 Å	Stoichiometric	Foster et al. (submitted)
Fe(II)-Arsenate $(FeHAsO_4 \cdot xH_2O)$	4.0 O 1.3 Fe (II) 3.4 Fe (III) 0.6 As	1.68 Å 3.10 Å 3.31 Å 4.19 Å	Stoichiometric	Cummings et al. (1999)

**average error in coordination number = 20%; average error in distance =0.02 Å.*

birnessite as determined from Rietveld refinement (Post and Veblen, 1990) which has a layered structure built from linkages between edge-sharing Mn octahedra, but suggested that synthetic vernadite contains a greater proportion of corner-sharing Mn octahedra than birnessite. This conclusion was reached based on the XAFS spectral similarites to todorokite, which has a tunnel structure formed by corner-sharing Mn octadedra (Post and Bish, 1988).

X-ray photoelectron studies of single mineral crystals indicate that a hydrous metal oxide surface cleaved in perfect vacuum contains structural oxygen atoms, structural hydroxyl molecules (if any), and metal atoms. Upon exposure to air and water, oxygen atoms react to form OH^- molecules, and OH^- molecules can react to form adsorbed H_2O. The structural arrangement of surface metal atom coordination polyhedra and the pH, ionic strength, and identity of counterions in the surrounding aqueous milieu determines the reactivity of these surface hydroxyl and water groups. Macroscopic sorption studies, surface complexation modeling, and bond valence calculations predict that the affinity of As for surface functional groups on metal oxide surfaces should vary with the distribution and type of surface functional groups available.

3.2.2 Arsenic sorption to and coprecipitation with hydrous metal oxides

Surface and bulk functional groups on hydrous metal oxide phases can be easily distinguished using infrared spectroscopy. Considering the phases of concern to this report (Fe, Al, and Mn hydrous metal oxides), surface OH groups fall into three categories: singly coordinated ("A-type") OH groups, doubly coordinated ("B-type) OH groups, and triply coordinated ("C-type") OH groups. Knowledge of the specific surface functional groups that participate in As chemisorption is important because it can provide better constraints on the equations describing chemisorption of As that are used in surface complexation models.

Chemisorption or coprecipitation is inferred from IR spectroscopy either by the shift, splitting, or disappearance of peaks assigned to A, B, or C-type hydroxyl groups involved in bonding to As oxoanions, or from the shift, splitting, or disappearance of peaks assigned to the As oxoanion itself (Myneni et al., 1998; Sun and Doner, 1996). Chemisorption of As is usually inferred from an XAFS sample spectrum by the presence of one or more neighboring surface metal atoms at a radius of 3-4 Å from the central As atom. If some or all the neighboring atoms are As atoms, then formation of multimeric complexes or precipitates is usually inferred. If a precipitate is

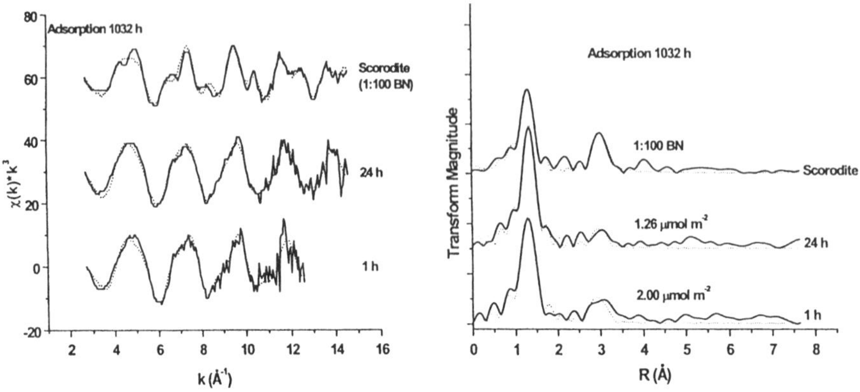

Figure 5. Normalized k^3-weighted As-EXAFS spectra (a) and Fourier transforms (FTs) (b) of As(V) sorption on goethite (total reaction time = 1032 hours) after 1 hour and 24 hours desorption of As(V) by 6 mM phosphate. The spectrum of scorodite is plotted for comparison. Solid lines are experimental data; dotted lines are least-squares fits. Peak positions in FTs are not corrected for phase-shift effects, and are therefore approximately 0.5 Å shorter than the true distance. Reprinted from O'Reilly et al.(2001) with permission.

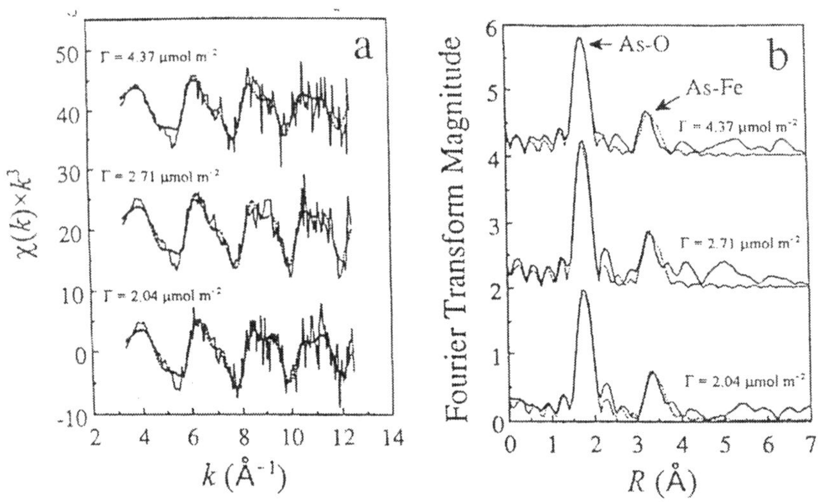

Figure 6. Normalized k^3-weighted As-EXAFS spectra (a) and uncorrected Fourier transforms (FTs) (b) of As(III) sorption on goethite at three surface coverages, pH 7.2-7.4. Solid lines are experimental data, and dotted lines are least-squares fits. Peak positions in FTs are not corrected for phase-shift effects, and are therefore approximately 0.5 Å shorter than the true distance. Reprinted from Manning et al. (1998) with permission.

suspected, then direct identification by electron diffraction or other microbeam analysis is the best means of verification. If there are no neighboring atoms detected, then the adsorbed complex is inferred to be outer-sphere, or physisorbed, with no direct bonds between the As oxoanion and the mineral surface (Brown, 1990).

3.2.3 Hydrous Ferric Oxides

As(III) and As(V) sorption on to synthetic goethite (α-FeOOH) and ferrihydrite has been investigated by XAFS spectroscopy; the quantitative fit results of several studies are summarized in Tables 3 [As(V)] and 4 [As(III)]. The presence of peaks in the Fourier transformed EXAFS spectrum beyond the first shell of oxygen/hydroxyl atoms indicates chemisorption. Such peaks are evident in the FTs of both As(V) sorbed to HFOs (Fig. 5) and As(III) sorbed to HFOs (Fig. 6). Due to differences in the oxoanion structure [As-(O,OH) bond distances in the As(V) tetrahedron are approximately 0.5 Å shorter than As-(O,OH) distances in the As(III) trigonal pyramid], the predominant sorption complexes formed by the two species differ in their characteristic interatomic As-Fe distance: As(V)-Fe(III) = 3.25-3.30 Å (Fendorf et al., 1997; Manceau, 1995; Waychunas et al., 1993), and As(III)-Fe = 3.38 Å (Manning et al., 1998). Additional (but minor) peaks at 3.5-3.6 Å (Waychunas et al., 1993) and ~ 2.8 Å (Manceau, 1995) were also reported for As(V) sorption on HFOs. Similar minor peaks were not reported in the XAFS study of As(III) sorption on HFOs (Manning et al., 1998). Minor differences in the sorption behavior of As(V) on crystalline HFOs and x-ray amorphous HFOs (i.e., ferrihydrite) have been observed under conditions of low to medium sorption density, but overall the complexes formed by As(V) on HFOs are very similar (Table 3; Waychunas et al., 1993). The interatomic As(V)-Fe distance is also invariant as a function of time and desorption by phosphate (O'Reilly et al., 2001; Waychunas et al., 1993 ; Table 5).

Several potential geometries of chemisorbed As(V) and As(III) complexes can be eliminated because the resultant As-Fe interatomic distances do not agree with EXAFS-derived distances (Waychunas et al., 1993). On this basis, the most likely coordination geometry for the predominant sorption complexes is one in which each As(V) tetrahedron and As(III) trigonal pyramid forms two bonds with the apices of adjacent, edge-sharing Fe octahedra (i.e., a bidentate geometry; Fig. 7). Another possible complex is the formation of a face-sharing arrangement between one As(V) tetrahedron and one Fe(III) octahedron (i.e., a tridentate complex; Fig. 7). This geometry yields the identical interatomic distance as the bidentate complex, but there is only one Fe nearest neighbor, which does not agree with the

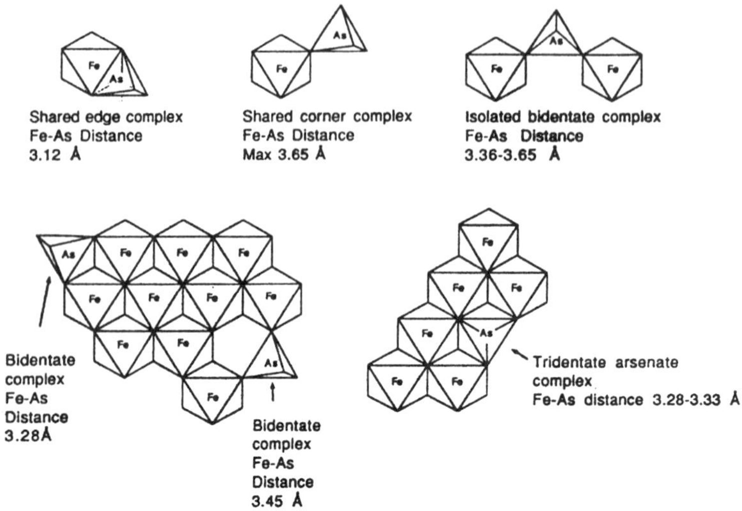

Figure 7. Possible bonding sites of As(V) on HFO phases such as goethite and resultant As(V)-Fe bond distances for comparison with XAFS-derived bond distances and coordination numbers. Both bidentate complexes and tridentate complexes give similar XAFS-derived As(V)-Fe bond distances of ~ 3.28 Å. Modified from Waychunas et al. (1993). Reprinted with permission.

EXAFS data. However, XAFS spectroscopy is not very sensitive to coordination numbers, so independent verification is needed. The As(V) species with the characteristic As-Fe distance of 3.5-3.6 Å can be matched by a geometry in which only one bond is formed between an As(V) tetrahedron and an Fe(III) octahedron (i.e., a monodentate coordination geometry, Fendorf et al., 1997; Manceau, 1995; Waychunas et al., 1995; Waychunas et al., 1996; Waychunas et al., 1993, table 4).

Several investigators have used IR spectroscopy to study As(III) and As(V) sorption on goethite and ferrihydrite (Tables 3 and 4). The studies (in both wet and dry modes) suggest that As(V) and As(III) form chemisorbed surface complexes on the HFOs, in agreement with XAFS analyses. However, Goldberg and Johnston (2001) suggest that at pH 5, a component of surface-associated As(III) is physisorbed on am-HFO, rather than chemisorbed. The IR studies also suggest that $HAs(V)O_4^-$ and $H_2As(III)O_3^-$ are the predominant surface-bound As complexes on HFOs (Goldberg and Johnston, 2001; Myneni et al., 1998; Suarez et al., 1999). As explained in Suarez et al. (1999), use of the $H_2As(III)O_3^-$ species in the chemisorption equation of As(III) on am-HFO results in no *net* hydroxyl release and no shift in the point of zero charge, which is concordant with macroscopic titration and electrophoretic mobility measurements of As(III) on HFOs.

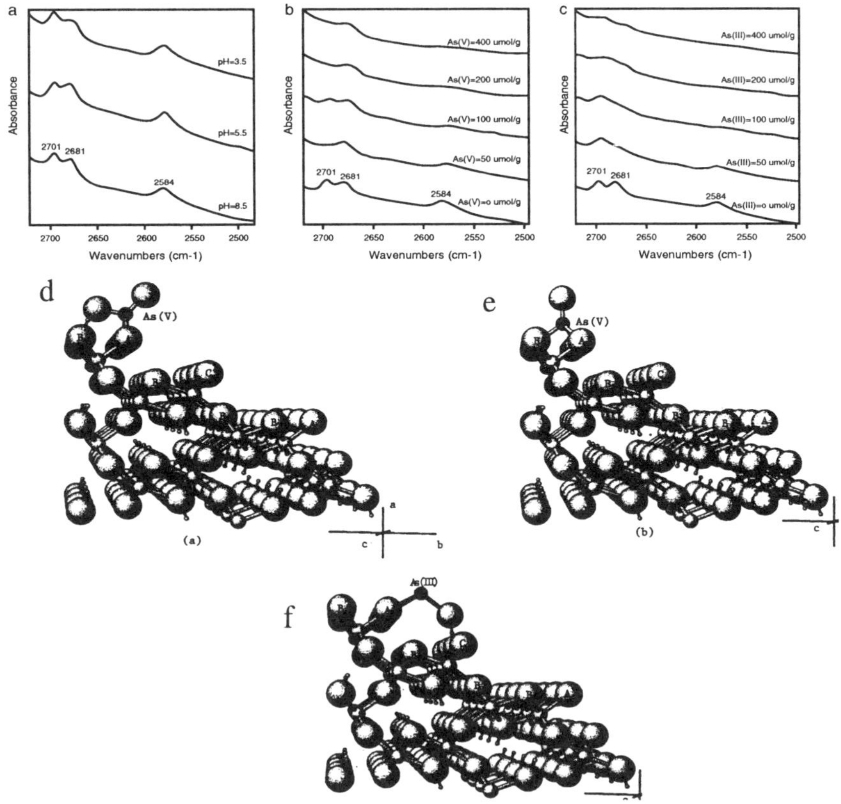

Figure 8. The ATR-FTIR absorbance spectra of surface duterated goethite without arsenic (a), and goethite reacted with As(V) at different loadings (b), and reacted with As(III) at different loadings (c). The band at 2584 wavenumbers is assigned to A-type (singly-coordinated) OD groups, the band at 2681 wavenumbers is assigned to C-type (triply-coordinated) OD groups, and the band at 2071 wavenumbers is assigned to B-type (doubly-coordinated) OD groups. Also shown are schematic representations of possible bonding sites of As(V) and As(III) on (110) faces of goethite as derived from IR and XAFS studies. (d) binuclear complex of As(V) with two A-type hydroxyls, and hydrogen bonding with b-type hydroxyl. (e) trinuclear complex of As(V) with two A-type hydroxyls and one B-type hydroxyl. (f) binuclear complex of As(III) with two A-type hydroxyls and hydrogen bonding to one C-type hydroxyl. Modified from Sun and Doner (1996); reprinted with permission.

Without spectroscopic information, these macroscopic observations might be taken as indicating physisorption rather than chemisorption.

Sun and Doner (1996) used deuterated goethite to examine the affect of As(III/V) adsorption on surface hydroxyl (OH) groups. Deuterium exchanges for H^+ in surface OH groups, forming OD groups on the surface. The peaks corresponding to OD vibrations are shifted relative to OH bands, which in goethite represent both surface and bulk OH groups. In this way the technique provides surface sensitive information. The authors concluded that both As(III) and As(V) preferentially bind to A-type hydroxyl groups on goethite at low sorption densities (50-100 μmol/g), as noted by the decreased intensity or disappearance of the peak corresponding to A-type OD groups (Fig. 8 a, b, c). By similar interpretation, they concluded that As(V) modifies B-type surface OD groups at mid-level sorption densities (100-200 μmol/g), but As(III) oxoanions react preferentially with C-type OD groups. At the highest sorption densities (400 μmol/g), As(V) sorption affects C-type OH groups, but As(III) sorption affects B-type groups (Sun and Doner, 1996). These observations suggest differences in bonding sites of As(III) and As(V) oxoanions at mid to high surface coverages, but overall indicate the predominance of A-type hydroxyl groups in As oxoanion sorption. The A-type hydroxyl groups could participate in both the monodentate and bidentate bonding that is observed by XAFS spectroscopy. Monodentate complexes were observed on goethite only for As(V) and then only at low sorption densities, where they could constitute up to 30% of the total sorbed As(V) (Waychunas et al., 1995; Waychunas et al., 1993).

Surface complexes between As(V) or As(III) and B- and C-type hydroxyls are observed in IR studies but have not been verified by XAFS spectroscopy, probably because hydrogen bonding dominates the interaction between As and B- or C-type OH moieties (Sun and Doner, 1996) Fig. 8 d, e, and f). Hydrogen bonding is a long-range attractive force between atoms that does not involve direct bonding, and therefore would be classified as physisorption as opposed to chemisorption, which involves the formation of a chemical bond.

3.2.4 Hydrous Aluminum Oxides (HAOs)

The As-HAO system presents special difficulties for IR and XAFS spectroscopic analysis. In an XAFS spectrum, the magnitude of peaks in the Fourier transformed EXAFS spectrum is a function of several variables, two of which are atomic number (z) and distance from the central As atom. With only half as many electrons as Fe, the scattering power of Al is weak, therefore peaks representing As-Al scattering in the Fourier-transformed EXAFS are smaller and more difficult to interpret. IR and Raman spectra of As(V) sorbed on gibbsite are difficult to interpret for an entirely different reason: substantial overlap of peaks representing Al(V)-O/Al-OH vibrations and As(V)-O/As(V)-OH vibrations (Myneni et al., 1998).

Notwithstanding the added ambiguity in data interpretation due to the system being studied, both XAFS and IR analyses indicate fairly clearly that As(V) forms chemisorbed species on HAO phases. XAFS spectra of As(V)

Table 3. Spectroscopic Studies of Arsenic(V) Sorption on Hydrous Ferric Oxides. Data from Harrison and Berkheiser (1982) and Lumsdon et al. (1984) are presented as interpreted by Myneni et al. (1998).

XAFS	Bond Type	N (±20%)	R (±0.02 Å)	Conclusions	References
Goethite	As(V)-Fe	2.0-3.0	3.25	Bidentate, corner	Waychunas et al. (1993)
Ferrihydrite	As(V)-Fe	2.0-3.0	3.25	Bidentate, corner	Waychunas et al. (1993)
Lepidocrocite	As(V)-Fe	2.4	3.29	Bidentate, corner	Waychunas et al. (1993)
Akaganeite	As(V)-Fe	2.4	3.26	Bidentate, corner	Waychunas et al. (1993)
Goethite	As(V)-Fe	0.5-1.3	2.83	Bidentate, edge	Manceau, (1995); Fendorf et al. (1997)
Goethite	As(V)-Fe	0.4-1.7	3.57-3.65	Monodentate, corner	Waychunas et al. (1993)
IR	**Peak Pos'ns ($\pm \sim 5\ cm^{-1}$)**	**Assignment***		**Conclusions**	**References**
am-HFO	700 802 875	ν(As-OH) of $HAsO_4^{2-}$ν(As-OFe), ν (As-O) of AsO_4^{3-}ν (As-OFe) of $HAsO_4^{2-}$		Primary adsorbed species: $HAsO_4^{2-}$ Minor component of AsO_4^{3-}	Harrison and Berkheiser (1982)
am-HFO pH 5 and 9	480 (pH 5) 605 (pH 5) 700 (pH 5) 820 (pH 9), 834 (pH 5) 872 (pH 9), 880 (pH 5)	ν(As-O,OH) unspecified ν(As-O,OH) unspecified unassigned ν(As-O) (tentative) ν(As-O) (tentative)		$HAsO_4^{2-}$ is primary species, can only form monodentate adsorption complexes AsO_4^{3-} species: can only form bidentate sorption complexes (conclusions based also on data from electrophoretic mobility and sorption expts.)	Suarez et al. (1999)**
am HFO	817, pH 9 824, pH 5 854, pH 9 861, pH 5	ν(As-OFe) ν(As-OFe) ν(As-O) ν(As-O)		Splitting of bands (e.g., 817 and 854, pH 9) indicates direct bonding to surface, potentially two species of As(V) sorbing?	Goldberg and Johnston (2001)

Table 3 (cont'd).

R	Peak Pos'ns ($\pm$~5 cm^{-1})	Assignment*	Conclusions	References
goethite	719 730 ~810 834 938 1016	ν(As-OH) of $HAsO_4^{2-}$ ν(As-OH) of $HAsO_4^{2-}$ ν (As-OFe) of AsO_4^{3-} ν(As-OX) of $HAsO_4^{2-}$ ν(As-OX) of $HAsO_4^{2-}$ unassigned	Primary adsorbed species: $HAsO_4^{2-}$ Minor component of AsO_4 X = Fe or H	Lumsdom (1984)
goethite (duterated) pH 5.5	2713, 2686 2584	B and C-type OD groups A-type OD groups	As(V) interacts primarily with A-type OH groups, next with B-type (hydrogen bonding?) Results consistent with bidentate adsorption of As(V) as suggested by EXAFS	Sun and Doner (1996)

ν(As-X) denotes asymmetric or symmetric stretching or bending modes of bond As-X.
***peak positions reported for the As(V)-Fe(OH)[3] system described in Suarez et al. (1999) are not identical to those reported in the text of Suarez et al. (1999); we have listed the peak positions and assignments as reported in the text.*

sorption complexes on gibbsite [γ-Al(OH)$_3$] differ substantially from those of As(V) in aqueous solution, and are instead similar to a spectrum of As(V) coprecipitated with am-HAO; both observations are indirect indications of chemisorption (Fig. 9; Foster, 1999; Ladeira et al., 2001). The predominant As(V) sorption coordination geometry as derived from least-squares fits to XAFS spectra of As(V) sorbed to gibbsite is ~ 2 Al atoms at 3.21 Å. Both the number and distance of Al neighbors is consistent with either a chemisorbed complex or a coprecipitate with am-HAO, since the predominant sorbed As(V) species in both cases appear to be indistinguishable by XAFS (Table 5; Foster, 1999). Note that a similar conclusion was reached by Waychunas et al. (1993) in a comparison of As(V) adsorbed on ferrihydrite versus As(V) coprecipitated with ferrihydrite. Formation of a coprecipitate on dissolving gibbsite surfaces during the sorption reaction cannot be ruled out.

IR studies indicate that As(V) is chemisorbed to HAO, as indicated by changes in the vibrations assigned to the As(V)(O,OH) tetrahedron (Table 5). Myneni et al. (1998) found similarities in the position of ν(As-O) peaks and ν(As-OAl) peaks in a poorly crystalline Al arsenate and in samples of As(V) adsorbed on gibbsite (but many peaks were obscured, as noted above). Goldberg and Johnston (2001) concluded that As(V) chemisorbs on

am-HAO after observing that: (1) the positions of the ν(As-O) modes of sorbed As(V) were indifferent to pH, (2) the Raman spectra of As(V) in

Table 4. Spectroscopic Studies of Arsenic(III) Sorption on Hydrous Ferric Oxides.

XAFS	**Bond Type**	**N (±20%)**	**R (±0.02 Å)**	**Conclusions**	**References**
Goethite	As(III)-Fe	2.3	3.38	Bidentate, corner	(Manning et al., 1998)
IR/Raman**	**Peak Pos'n (± ~5 cm^{-1})**	**Assignment***		**Conclusions**	**References**
am-HFO pH 5	453 630 794 794	minor ν(As-OH) ν(As-OH) ν(As-O)		Primary adsorbed species: $H_2AsO_3^-$	Suarez et al. (1999)
am-HFO pH 5 and 10.5	783 (pH 5)	ν (As-O)		Evidence for inner-sphere surface complex, but Significant overlap of goethite and As(III) bands in 600 cm^{-1} IR region precludes further analysis	Goldberg and Johnston (2001)
goethite (duterated) pH 5.5	2584 2681 2701	A-type OD groups C-type OD groups B-type OD groups		As(III) binds primarily with A-type OH groups, next with C-type (hydrogen bonding?) Results consistent with bidentate adsorption of As(III) as suggested by EXAFS	Sun and Doner (1996)

**ν(As-X) denotes asymmetric or symmetric stretching or bending modes of bond As-X.*

aqueous solution and As(V) adsorbed to am-HAO are significantly different at pH 5 and 9; and (3) the splitting between IR and Raman-active ν(As-O) bands decreases in the sorbed species indicating a decrease in symmetry (increase in tetrahedral distortion) that is consistent with chemisorption.

Most spectroscopic analyses of the As(III)-HAO system conclude that As(III) also chemisorbs to HAOs, but at least one report comes to a different conclusion. XAFS analysis of As(III) sorbed to gibbsite indicates that the majority of As(III) atoms are within 3.33 Å of neighboring Al atoms. A sample prepared to be a model for As(III) coprecipitated with am-HAO is thought to have actually produced a poorly crystalline, uncharacterized Al-As(III) phase, because the number of Al neighbors is higher, and the As-Al bond distance longer than in sorption samples of As(III) on gibbsite (Fig. 9, Table 6). Suarez et al. (1999) concluded that $H_2As(III)O_3^-$ was the chemisorbed species on am-HAO, based on equivalent peak positions in the sorbed and the aqueous forms of this species. In contrast, to the spectroscopic studies indicating As(III) chemisorption on HAO phases, Goldberg and Johnston (2001) found no evidence for ν(As-O) bands in samples of As(III) sorbed to am-HAO; they concluded that the lack of signal indicated physisorption of As(III).

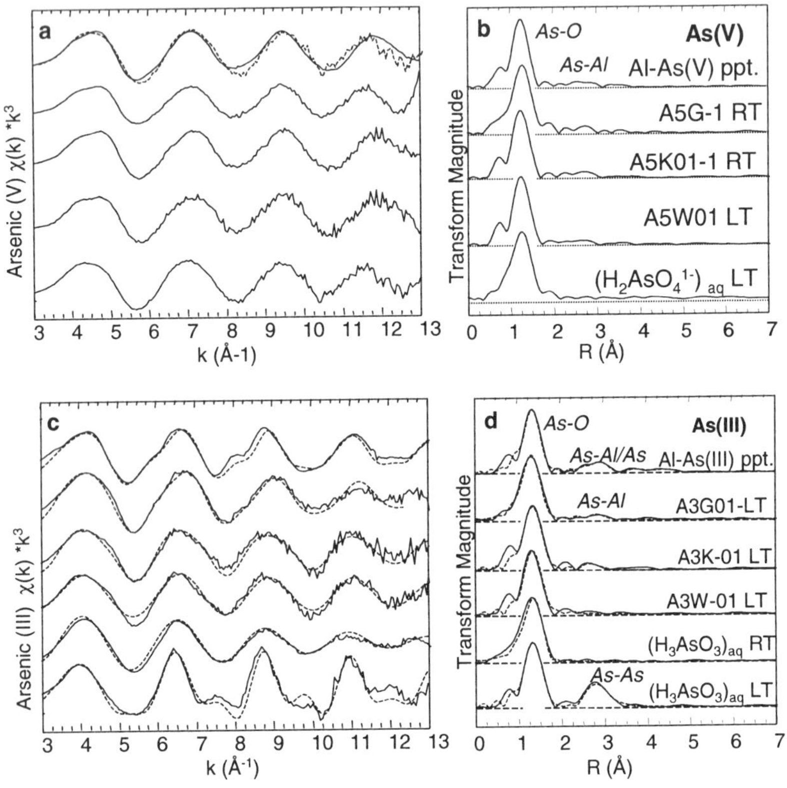

Figure 9. XAFS spectra (a,c) and Fourier transforms (FTs; b,d) of As(V) and As(III) sorbed to various HAO and aluminosilicate clay phases. Sample nomenclature is as follows: "A5" and "A3" = As(V) amd As(III), respectively; "ppt." = coprecpitate with amorphous HAO; "G" = gibbsite; "K" = kaolinite; "W" = smectite; "1" or "01" = concentration of 0.1 or 0.01 M $NaNO_3$ or NaCl, respectively; "LT" = cryogenic data collection conditions (10-50 K); "RT" = ambient data collection conditions. Peak positions in FTs are not corrected for phase-shift effects, and are therefore approximately 0.5 Å shorter than the true distance. From Foster (1999).

The IR/Raman studies to date do not draw conclusions as to the geometric configuration of the As(V) sorbed on HAO phases. However, by construction of geometric models and elimination of those inconsistent with the XAFS-derived As-Al bond distance, Foster (1999) concluded that adsorbed As(III) probably adopts the same type of bridging bidentate complexes analogous to that found on HFO phases for adsorbed As(III) and As(V) (Figs. 7 and 10). It appears that more study is needed on this system to determine the bonding and protonation of As(III) species sorbed to HAOs.

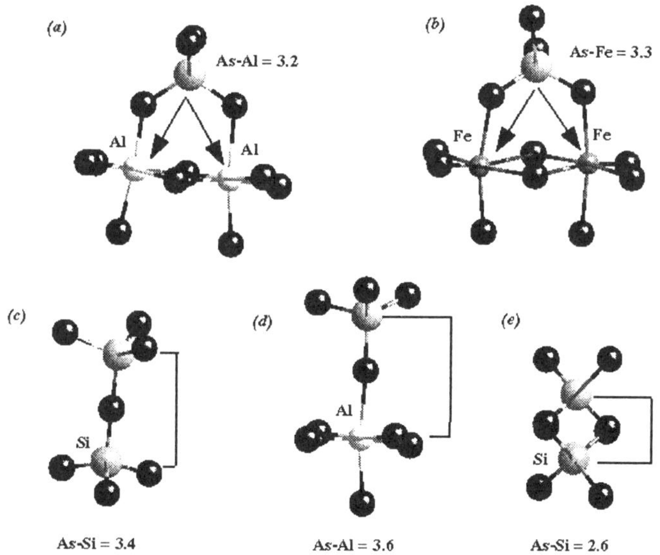

Figure 10. Comparison of As(V)-metal distances obtained for different binding modes of As(V) with (a) corner-sharing Al octahedra as found in gibbsite or in the gibbsite sheet of kaolinite; (b) corner-sharing Fe octahedra as found in goethite; (c) a Si tetrahedron; (d) an aluminum octahedron; and (e) an aluminum tetrahedron.

3.2.5 Layered Aluminosilicates

Layered aluminosilicates include the micas and clay minerals; they share the characteristic of having alternating layers of a gibbsite-like "aluminol" sheet and a silicon oxide (silanol) sheet, but differ in the stacking of these sheets, the presence of additional cations in various structural locations, and the amount of structural water and hydroxyl groups. By virtue of analogy with amorphous silica and quartz, the silanol sheets are not

Table 5. Spectroscopic Studies of Arsenic(V) Sorption on Hydrous Aluminum Oxides and Aluminosilicate Clay Minerals.

XAFS	Bond Type	N (±20%)	R (±0.02 Å)	Conclusions	References
Gibbsite	As(V)-Al	2.0	3.19-3.21	Bidentate, corner-sharing with Al octahedra	Foster (1998); Ladeira et al. (2001)
Am-HAO coprecipitate	As(V)-Al	2.0	3.22	Coprecipitate has very similar As (V) bonding environment to adsorption samples	Foster (1999)
Kaolinite	As(V)-Al	2.0	3.19	Bidentate, corner-sharing with Al octahedra	Foster (1999)
Smectite	As(V)-Al	--	--	Physisorption indicated by lack of second shell	Foster (1999)

IR/Raman**	Peak Pos'ns* (± ~5 cm^{-1})	Assignment	Conclusions	References
Gibbsite	880	ν(As-O)	Significant overlap between gibbsite vibrations and As(V) tetradedral vibrations limited analysis	Myneni et al. (1998)
Poorly-crystalline Al-arsenate	740 887	ν(As-OAl) ν(As-O)	Useful as model for chemisorbed As on gibbsite, but since structure is unknown, can't rule out presence of protonated As(V).	Myneni et al. (1998)
As(V)-am HAO	853_R* pH=5 845_R* pH = 9 $856\text{-}862_{IR}$ (pH 5 and 9)	ν(As-O) ν(As-O) ν(As-O)	No splitting or pH variation in ν(As-O) modes indicates chemisorption of As(V).	Goldberg and Johnston (2001)

** peak positions derived from Raman data are indicated by a subscript "R" Raman peaks are shifted relative to IR peaks.*

predicted to react with arsenic because they possess a net negative charge. Negative charge is also accumulated on basal (001) planes of the aluminol sheet by cation substitution, therefore these faces are also assumed to have a low density of surface sites with an affinity for the negatively charged As oxoanions. In contrast to the basal planes, the edges of layered aluminosilicates contain amphoteric aluminum octahedra that are expected to be the primary sites for arsenate and arsenite oxoanion bonding on the aluminosilicate minerals.

Table 6. Spectroscopic Studies of Arsenic(III) Sorption on Hydrous Aluminum Oxides and Aluminosilicate Clay Minerals.

XAFS	**Bond Type**	**N (±20%)**	**R (±0.02 Å)**	**Conclusions**	**References**
Gibbsite	As(III)-Al	1.2	3.33	Bidentate, corner-sharing with Al octahedra	Foster (1999)
Coprecipitate with am-HAO	As(III)-Al	3.4	3.50	Previously undescribed Al-As(III) phase mixed with pure adsorbed As(III) ??	Foster (1999)
Kaolinite	As(III)-Al/Si	1.0	3.22	Small second-shell peak could represent Al scatterers, but could also arise from multiple scattering within As(OH)3 unit or from small amount of As(V) Chemisorbed As(III) and physisorbed As(III)	Foster (1999)
Smectite	As(III)-Al/Si	--	--	Spectrum similar to dissolved As(III); indicates physisorption	Foster (1999)
IR	**Peak Pos'ns ($\pm \sim 5$ cm^{-1})**	**Assignment**		**Conclusions**	**References**
As(III)-am HAO	397 pH 8 610 (pH 8), 600 (pH 5) 669 pH 8 823 (pH 8), 828 (pH 5)	unspecified unspecified unspecified ν(As-X)		Chemisorption of $H_2AsO_3^-$ species X = O or OH of $H_2AsO_3^-$	Suarez et al. (1999)
As(III)-am HAO	None reported			No indication of ν(As-O) modes for $As(OH,O)_3$ indicates physisorption	Goldberg and Johnston (2001)

Spectroscopic studies of As sorbed on aluminosilicates suffer from all the complications described for the As-HAO system in the previous section. There is, however, an additional difficulty: elements differing by < 2 in atomic number cannot be distinguished by XAFS spectroscopy (Sayers and Bunker, 1988), and therefore neighboring Al atoms cannot be distinguished from Si. However, Al and Si bond lengths are considerably different (geometric models indicate that As-Al and As-Si distances should differ by more than 0.2 Å), and therefore should be discriminated by EXAFS spectroscopy (Fig. 10).

Several studies show that As sorbed to aluminosilicate clay minerals is more readily extracted than As sorbed to HAOs or HFOs (Christensen et al., 1989; Hamdy and Gissel-Nielsen, 1977), so As associated with these phases is assumed to be fairly labile. XAFS spectroscopy of As(V) and As(III) adsorbed to model aluminosilicate clays (kaolinite and smectite) in the laboratory confirms these macroscopic studies, indicating that nearly all As(III) and As(V) sorbed to smectite clay is physisorbed, but that at least a portion of As(III) and As(V) sorbed on kaolinite is chemisorbed. Evidence for physisorption on smectite comes from the lack of second-shell neighbors in smectite sorption spectra and from similarity between smectite sorption spectra and the corresponding spectra of aqueous As(III) and As(V) (Fig. 9). Evidence for chemisorption of As(V) and As(III) on kaolinite is indicated by a shell of 1-2 Al neighbors at a distance of 3.16-3.20 Å for As(V) and 3.22 Å for As(III) (Foster, 1999). This distance is too short for monodentate linkages to aluminum polyhedra (Fig. 10), and suggests that bidentate complexes similar to those reported for As(V) adsorption on HFOs and HAOs from on kaolinite, presumably at Al edge sites. Of course, the number of such complexes formed would be far lower on clay minerals, which typically have a bulk surface area < 30 m^2/g, than the number formed on gibbsite or amorphous Al oxyhydroxide, which typically have bulk surface areas > 100 m^2/g.

3.2.6 Hydrous Mn oxides (HMOs)

HMO phases undergo both sorption and electron transfer reactions with As species (Chiu and Hering, 2000; Oscarson et al., 1981b; Scott and Morgan, 1995; Sun et al., 1999; Thanabalasingam and Pickering, 1986). Heterogeneous oxidation of As(III) by HMO phases is coupled to the reductive dissolution of the HMO: the released Mn(II) can be readsorbed and re-oxidized heterogeneously via abiotic and biotically-mediated processes (Emerson et al., 1982; Junta and Hochella, 1994; Murray, 1975; Stone and Ulrich, 1989). On the basis of macroscopic data, Scott and Morgan (Scott and Morgan, 1995) hypothesized that As(III) forms a chemisorbed surface complex with two Mn(IV) atoms. Unfortunately, As(III) oxidation by HMOs is too rapid under ambient conditions to allow XAFS spectroscopic determination of sorbed As(III) species on this material (Manning et al., 2002).

Nesbitt et al. (1998) used XPS data collected on the As(III)/birnessite system to formulate a two-step mechanism for the process whereby As(III) donates one electron to each of two surface Mn(IV) octahedra. These Mn(IV) octahedra are reduced, forming Mn(III), which is subsequently attacked by more As(III), and reduced to Mn(II). Below pH 6, Mn(II) is not appreciably sorbed by Mn(IV) hydroxides, and therefore is preferentially partitioned into solution. The rate of Mn(IV) reduction to Mn(III) is faster than the transformation of Mn(III) to Mn(II); as a result, the authors suggested that a short-lived Mn(III)-rich hydroxide phase forms on the HMO surface during the reaction (Fig. 11).

Despite observations that As(V) does not adsorb to HMOs in the presence of As(III) (Scott and Morgan, 1995), XAFS spectroscopy indicates that in As(III)-free systems As(V) forms chemisorbed complexes on synthetic vernadite (δ-MnO_2) and birnessite (type formula: $Na_4Mn_{14}O_{27}{\bullet}9H_2O$; McKenzie, 1989).The complexes are bidentate and binuclear with a characteristic As-Mn distance of ~ 3.16 Å (Foster et al., submitted; Fig. 12, Table 7). Geometric arguments suggest that the structure of this complex is the same as was previously described for As(V) on HFOs and HAOs. As previously mentioned, synthetic birnessite and vernadite samples typically contain up to 30% Mn(III) and/or Mn(II) due to incomplete oxidation of Mn(II) during synthesis, and this lower-valent Mn could provide binding sites for As(V). A survey of As-Mn(II) interatomic distances in crystalline precipitates indicates that typical distances are approximately 0.3 Å longer than the As(V)-Mn interatomic distance deduced from EXAFS analysis (Foster et al., submitted; Table 7). However As(V)-Mn(III) distances vary widely, and linkages to Mn(III) octahedra cannot be ruled out. However, precipitate formation with Mn(II) arising from dissolution of HMO phases is unlikely, based on comparison to the XAFS spectrum of a poorly crystalline Mn-arsenate (Fig. 12, Table 7).

4. CONCLUSION: IMPLICATIONS FOR AQUIFER SEDIMENTS

There have been very few spectroscopic studies of As species in aquifer sediments, because they are complex heterogeneous materials that probably contain many different As species, and because the typical bulk concentrations of As are low. XAFS spectroscopy is one of the best methods for analysis of As species in natural samples even though the bulk spectrum is the weighted sum of all the species present in the sample.

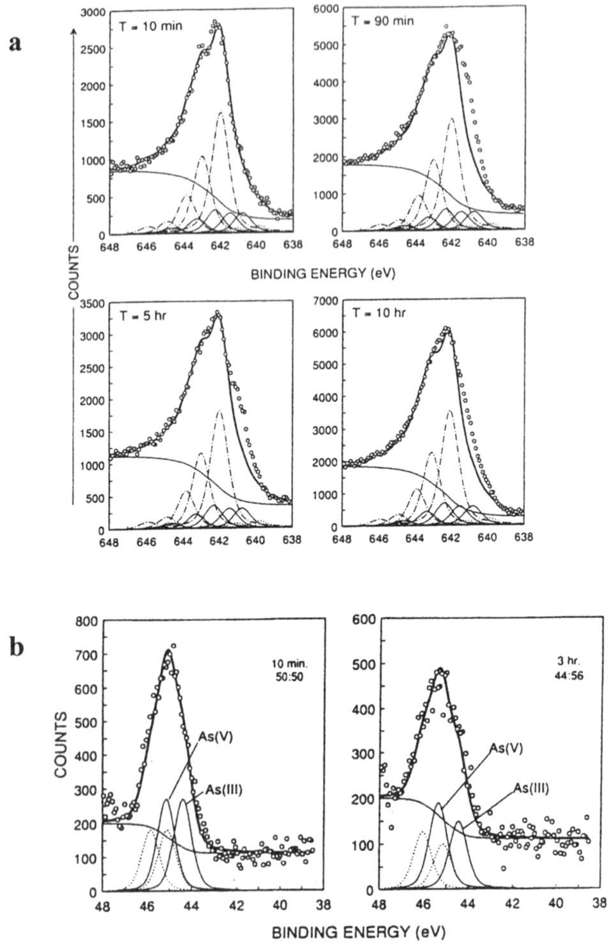

Figure 11. *(a) X-ray photoelectron narrowscan spectra of the Mn(2p$_{3/2}$) peaks of 7 Å birnessite reacted with 10 µM As(III) for various times. Open circles are data, and the thick, solid lines in panels T = 10 min - 10 hr were calculated for the fixed proportions of Mn(IV):Mn(III):Mn(II) at 70:25:5 (the same as determined in an unreacted birnessite sample). After 10 minutes reaction, no change in the proportion of Mn species is observed, as evidenced by the excellent agreement between data and calculation, but after 90 minutes of reaction, a significant misfit is observed on the low energy side of the spectrum. The misfit diminishes with increasing reaction time of 5 and 10 hours. The authors interpret the misfit to represent the formation and subsequent destruction of an Mn(III) hydroxide intermediate during the oxidation of As(III) to As(V). (b) Evidence for mixed As species on the surface of birnessite after 10 minutes and 3 hours of reaction. Symbols are same as (a), except now As(3d$_{5/2}$) peaks and multiplets are shown. Reprinted from Nesbitt et al. (1998) with permission.*

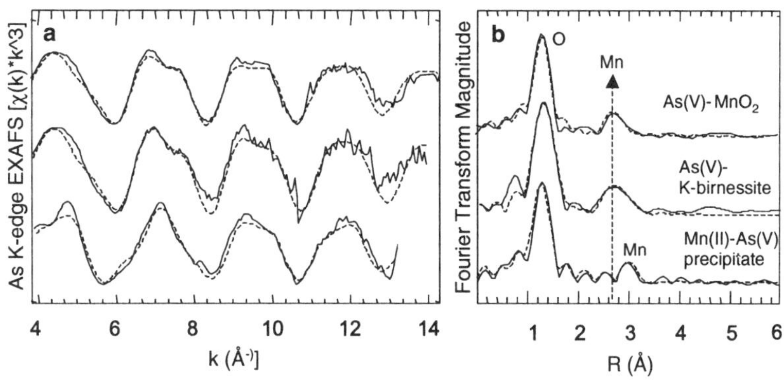

Figure 12. Normalized, k^3-weighted arsenic EXAFS spectra (solid lines in a) and Fourier transforms (solid lines in b) of As(V) sorbed to δ-MnO_2 and birnessite and in an Mn(II)-As(V) precipitate. Non-linear, least-squares fits to raw EXAFS data are plotted as dotted lines. The spectra of all sorption samples appear very similar, but sorption sample spectra are distinct from the model precipitate. Peak positions in FTs are not corrected for phase-shift effects, and are therefore approximately 0.5 Å shorter than the true distance. From Foster et al. (submitted); reprinted with permission.

However, collection of good-quality EXAFS spectra under standard conditions at synchrotron facilities is currently limited to samples containing more than ~100 ppm As. XANES spectra can be collected even for samples containing very low arsenic, as was recently shown for aquifer sediment samples from east-central Bangladesh (Foster et al., 2000). In that study, As-XANES spectra were collected from samples containing as little as 4 ppm As. Oxidation states were easily quantified, and correlated very well with the sediment redox conditions [as indicated by color; (Foster et al., 2000). Qualitative chemical speciation was determined by a "fingerprinting" method using a linear least-squares routine to fit the sample spectrum to model spectra such as those shown in Fig. 13. Preliminary XANES evaluation of the As species present suggests that As(V) sorbed to HFOs, As(III) sorbed to HFOs, and As(III) sorbed to HAOs (probably precipitated on the edges of mica minerals) are the predominant species present.

Although Foster et al. (2000) conservatively fit only 1 or 2 different model As spectra to each sample spectrum, least-squares fitting does not impose limitations on the number of models used in the fit, nor is their any parameter besides the fit residual to guide selection of the models used in the fits. Well-constrained fits can be obtained using principal component analysis (PCA) to guide selection of the type and number of components used in linear least-squares fits (Ressler et al., 2000). However, use of PCA

Table 7. Spectroscopic Studies of Arsenic(V) Sorption on Hydrous Manganese Oxides.

XAFS	Bond Type	N (±20%)	R (±0.02 Å)	Conclusions	References
Birnessite	As(V)-Mn*	1-2	3.15-3.20	Bidentate, corner-sharing with Mn octahedra	Manning et al. (2002); Foster (submitted)
vernadite	As(V)-Mn	2-3	3.16	Bidentate, corner-sharing with Mn octahedra	Foster et al. (submitted)
Poorly crystalline Mn-Arsenate	As-Mn(II)	6.6	3.45	Structure similar to geigerite [$Mn_5(H_2O)_8(AsO_3OH)_2(AsO_4)_2 \bullet 2H_2O$] Model for As(V) bound to Mn(II)	Foster et al. (submitted)

**bond distances represent As(V) bonded to Mn(IV), Mn(III), and Mn(II), but the exact proportions are unknown. XAFS and XPS evidence suggests that the relative proportion of Mn valence states in synthetic birnessite is approximately 70% Mn(IV), 25% Mn(III), and 5% Mn(II).*

does not remove all ambiguity from fits to As-XANES spectra; although many As-XANES spectra of As(III) and As(V) models are clearly distinguished on the basis of lineshape (Fig. 13), other model spectra in which the near (~ 4Å) -coordination environment of As is very similar are practically indistinguishable, as illustrated in Fig. 14. Note that in contrast to the XANES spectra, EXAFS spectra of As(V) in scorodite, in x-ray amorphous ferric arsenate, and sorbed to HFO are easily distinguished (Fig. 4). *Ab initio* calculations of As-XANES spectra in model systems are needed to explain these observations.

Based on spectroscopic and macroscopic studies to date, the chemistry of As in aquifer sediments is probably dominated by species partitioned in primary or authigenic sulfides and species partitioned in oxyhydroxides and/or other products of sulfide oxidation. In sulfides, reduced As species can be found from trace to atom percent levels; however, due to a lack of thermodynamic equilibrium, dissolved As oxoanions may be in contact with and partitioned onto sulfide minerals (Bostick and Fendorf, 1999; Smedley and Kinniburgh, 2001). Oxidative attack on arsenic-bearing sulfide minerals, usually through localized aquifer drawdown or by gases traveling along fractures, is currently accepted as the primary means of release of As from sulfides, but sulfide oxidation may occur under anoxic conditions to a greater extent than was previously supposed, through the action of electron

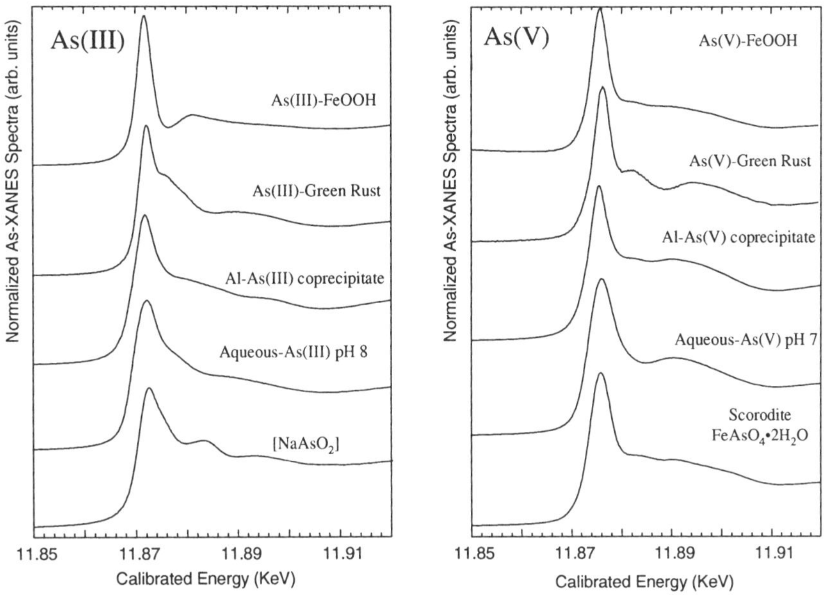

Figure 13. Normalized, As-K edge XANES spectra of (a) As(III) and (b) As(V) in different model coordination environments, illustrating the power of XANES spectroscopy for chemical fingerprinting. Often the low bulk concentrations of As in aquifer sediments preclude a quantitative EXAFS analysis; fortunately, quantitative oxidation state and semi-quantitative speciation information can be obtained from high-quality XANES spectra.

such as dissolved nitrate and even MnO_2 solids (Shippers and Jorgensen, 2001).

As-bearing precipitates commonly form in close association with sulfide minerals, probably because the free energy of formation of these phases is exceeded locally near the site of sulfide dissolution. For example, scorodite and As-bearing jarosite were produced during the oxidation of arsenical pyrite by mixed cultures of moderately thermophilic bacteria (Carlson et al., 1992). If As is present during the formation of these phases, it usually substitutes for a structural oxoanion such as sulfate, carbonate, or phosphate. However, based on results previously described for the As(V)-ettringite system, if As interacts with carbonate, sulfate, or phosphate minerals after their formation, adsorption complexes are probably formed with surface metal atoms instead of oxoanion substitution. The availability of arsenic from secondary phases is usually dependent on solubility; solubility, in turn, usually increases with increasing anoxic and extreme pH conditions, or with decreasing crystallinity (Rochette et al., 1998).

Sorption of arsenic to hydrous metal oxide phases (usually found as coatings on other mineral phases, or as clay-to-colloidal-sized particles) is an

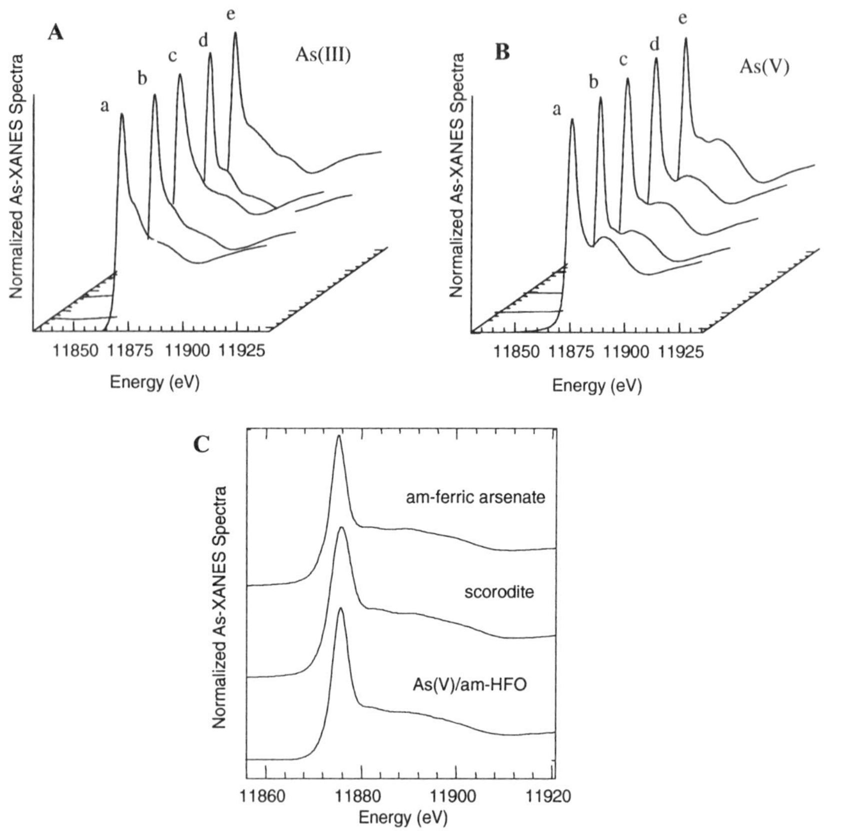

Figure 14. Normalized, As-K edge XANES spectra of ***(A)*** *As(III) and* ***(B)*** *As(V) in coordination environments that appear similar by XANES spectroscopy: (a) in aqueous solution, (b) sorbed to smectite clay, (c) sorbed to kaolinite clay, (d) sorbed to gibbsite, and (e) coprecipitated with amorphous HAO. Panel* ***(C)*** *shows that similar ambiguity exists among the XANES spectra of As(V) in coordination environments containing Fe(III), such as precipitated in an amorphous ferric arsenate, precipitated in scorodite, and sorbed to amorphous HFO. In contrast to these XANES results, the coordination environments As(V) species shown in panel (c) can be easily discriminated using EXAFS spectroscopy (see Fig. 4).*

extremely important process in nature both for scavenging As and for releasing As to the environment. Hydrous ferric oxides probably dominate As oxoanion sorption in most aquifer sediments (Welch, 1999), but hydrous manganese oxides (HMOs) and hydrous aluminum oxides (HAOs) can also accumulate significant amounts of As species. One example is in the aquifer sediments of La Pampa Province, Argentina. There it is estimated that half of the 10-50% of As associated with metal oxyhydroxides is partitioned onto HMO (and possibly HAO) phases (Smedley and Kinniburgh, 2001).

Heterogeneous electron transfer reactions can also influence the partitioning of As in aquifer sediments. As(III) is rapidly oxidized to As(V) in the presence of HMO phases. HMOs and HFOs are also susceptible to reductive dissolution by microorganisms, therefore the fate of As associatedwith HFOs and HMOs depends on the prevailing geochemical conditions. HAO phases are not susceptible to redox attack, but microbes specializing in respiration of As(V) to As(III) can produce the selective release of As from these phases (Zobrist et al., 1998). Aluminosilicate clay and mica minerals rarely contain arsenic in greater than trace amounts, and do not sorb arsenic in great quantities. However, they may be significant agents of As sorption due to secondary coatings of metal hydroxides on their surfaces. These coatings can be important sites of arsenic accumulation.

Although the chemistry of hydrous metal oxides varies dramatically, geometric arguments based on XAFS spectroscopic evidence indicate that chemisorbed As(V) and As(III) oxoanions primarily form bidentate complexes regardless of the identity of the hydrous oxide. Singly-coordinated surface hydroxyl groups are probably the primary sites for As(V) and As(III) chemisorption. IR and Raman spectroscopies corroborate this evidence, and further provide detailed information about the protonation of sorbed As complexes that can be used to provide constraints on the chemical reactions used in surface complexation models. Additional spectroscopic studies on natural aquifer sediments are needed to further test prevailing hypotheses about As chemistry in solid phases from the subsurface, because these ideas were primarily generated from observations of As chemistry in solid phases from near-surface environments.

ACKNOWLEDGEMENTS

The majority of the XAFS experiments cited herein (and all of the XAFS work by Foster and colleagues) were conducted at the Stanford Synchrotron Radiation Laboratory, a national user facility operated by Stanford University on behalf of the U.S. Department of Energy, Office of Basic Energy Sciences. Work by AF cited in this chapter was supported by the Department of Energy through grant DE-FG03-93ER14347-A008, by the National Science Foundation through NSF Grant CHE-0089215, by the Corning Foundation Graduate Science Fellowship program, by the U.S. Geological Survey Mineral Resources Program, and by the U. S. State Department.

Chapter 3

Geochemical Processes Controlling Transport of Arsenic in Groundwater: A Review of Adsorption

Kenneth G. Stollenwerk
US Geological Survey, Denver, CO USA

Adsorption is the predominate mechanism controlling transport of arsenic in many groundwater systems. Hydrous oxides of iron, aluminum, and manganese, and clay minerals are commonly associated with aquifer solids and have been shown to be significant adsorbents of arsenic. The extent of arsenic adsorption is influenced by the chemistry of the aqueous phase including pH, arsenic speciation, and the presence and concentration of competing ions. Under moderately reducing conditions, trivalent arsenite is stable and adsorption increases with increasing pH. In an oxidizing environment, arsenate is stable and adsorption decreases with increasing pH. The presence of phosphate, sulfate, carbonate, silica, and other anions have been shown to decrease adsorption of arsenic to varying degrees. The effects of complex aqueous and solid phase chemistry on arsenic adsorption are best simulated using surface complexation models. Coupling of such models with hydrologic solute transport codes provide a powerful method for predicting the spatial and temporal distribution of arsenic in groundwater.

1. INTRODUCTION

Long-term exposure to arsenic (As) in drinking water has been implicated in a variety of health concerns including several types of cancer, cardiovascular disease, diabetes, and neurological effects (National Research Council, 1999). To protect people against the effects of long-term chronic exposure to As, the United States Environmental Protection Agency has proposed lowering the maximum contaminant level (MCL) for As in drinking water from 50 micrograms per liter (μg/L) to 10 μg/L (U.S. Environmental Protection Agency, 2001). The World Health Organization has set a provisional guideline concentration for drinking water of 10 μg/L (WHO, 1993).

Potable water supplies are compromised by As contamination of groundwater in many countries of the world, including Bangladesh, western India, Taiwan, Mongolia, Vietnam, Argentina, Chile, Mexico, Ghana, and the United States. In all of these locations, the source of As is primarily natural and not anthropogenic or geothermal (Ayotte et al., 1998; Azcue and Nriagu, 1994; Chen et al., 1994; Das et al., 1996; Focazio et al., 2000; Kolker et al., 1998; Korte, 1991; Matisoff et al., 1982; Nickson et al., 2000; Nimick, 1998; Panno et al., 1994; Robertson, 1989; Schlottmann and Breit, 1992; Schreiber et al., 2000; Smedley et al., 2002; Sonderegger and Ohguchi, 1988; Welch et al., 1988).

Higher concentrations of As tend to be found more in groundwater than in surface water sources of drinking water; therefore, the emphasis of this paper is on the geochemical processes that affect As transport in groundwater. However, many of the processes discussed are also applicable to surface waters and the variably saturated zone above the groundwater table.

The geochemistry of As can be complex. Important processes that control the partitioning of As between aqueous and solid phases include mineral precipitation/dissolution, adsorption/desorption, oxidation/reduction, and biological transformation. Adsorption/desorption is the most significant process controlling As concentrations in many groundwater environments, and is the focus of this review. Adsorption processes are controlled by pH, redox potential, As concentration, the concentration of competing and complexing ions, aquifer mineralogy, and reaction kinetics. Because speciation of As is important in adsorption, a discussion of the aqueous chemistry of As precedes the review of adsorption. This review is also limited to geochemical environments that are oxidizing to slightly reducing.

2. AQUEOUS CHEMISTRY OF ARSENIC

2.1 Speciation

Dissolved As speciation is important in determining the extent of reaction with the solid phase and therefore the mobility of As in groundwater. Arsenic is generally present as arsenate [As(V)] or arsenite [As(III)] for Eh conditions prevalent in most groundwaters (Fig. 1). Arsenic metal rarely occurs, and the -3 oxidation state is found only in very reducing environments. Arsenite has been considered to be the more toxic oxidation state (U.S. Environmental Protection Agency, 1976) however, more recent studies have shown that most ingested As(V) can be reduced to As(III).

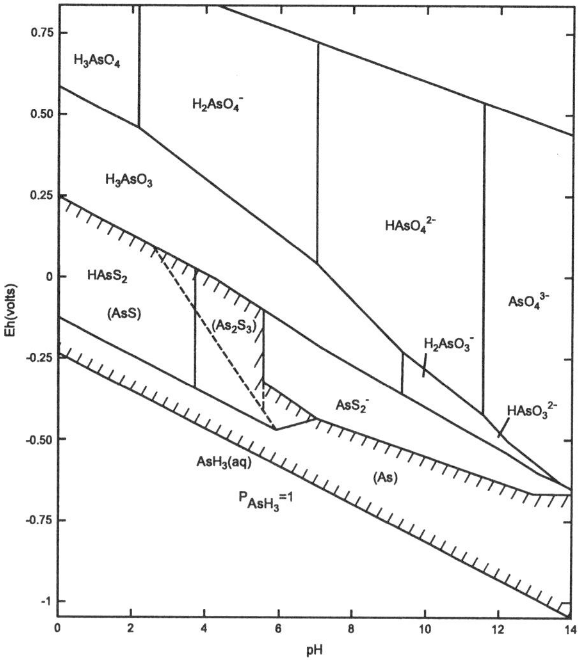

Figure 1. The Eh-pH diagram for As at 25°C and one atmosphere with total arsenic 10^{-5} mol l^{-1} and total sulfur 10^{-3} mol l^{-1}. Solid species are enclosed in parentheses in cross-hatched area, which indicates solubility less than $10^{-5.3}$ mol l^{-1}. Reprinted from Ferguson and Gavis (1972) with permission from Elsevier Science.

Thus, exposure to both forms of As may result in similar toxicological effects (National Research Council, 1999).

Both As(III) and As(V) form protonated oxyanions in aqueous solutions, the degree of protonation dependent on pH. Arsenate is stable in oxidizing environments. For pH values common in groundwater, the predominate As(V) species in solution are $H_2AsO_4^-$ between pH 2.2 and 6.9, and $HAsO_4^{-2}$ between pH 6.9 and 11.5. Arsenite is stable in moderately reducing environments; $H_3AsO_3^o$ predominates up to pH 9.2, and $H_2AsO_3^-$ from pH 9.2-12 (Ferguson and Gavis, 1972). If conditions are favorable, biomethylation can form a variety of methylated species of As(III) and As(V) that are stable under oxidizing or reducing conditions. The more common methylated species include methylarsonic acid $[CH_3AsO(OH)_2^o]$, dimethylarsinic acid $[(CH_3)_2AsO(OH)^o]$, methylarsonous acid $[CH_3As(OH)_2^o]$, and dimethylarsinous acid $[(CH_3)_2AsOH^o]$ (Braman and Foreback, 1973; Cullen and Reimer, 1989; National Research Council, 1999).

2.2 Redox transformation

Natural geochemical and biological processes play critical roles in controlling the fate and transformation of As in the subsurface. Arsenite is thermodynamically unstable in aerobic environments and should oxidize to As(V); however, studies have shown that this reaction proceeds slowly when oxygen is the only oxidant. Cherry et al. (1979) measured a slow rate of As(III) oxidation in water saturated with pure O_2 at pH 7. After 6 months about 25% of the As(III) had been oxidized compared with no oxidation in deoxygenated water. Eary and Schramke (1990) measured a half-life of 1 to 3 years for oxidation of As(III) in water equilibrated with atmospheric oxygen. The rate of oxidation in these experiments was a minimum at pH 5.5 and increased for both lower and higher pH values. Manning and Goldberg (1997a) reported a significant increase in the rate of oxidation of As(III) to As(V) in alkaline solutions (pH > 9) under atmospheric conditions.

The presence of other redox sensitive species increases the rate of As(III) oxidation. Addition of ferric iron [Fe(III)] to oxygenated water increased the rate of As(III) oxidation below pH 7 (Cherry et al., 1979) in agreement with thermodynamic predictions (Scott and Morgan, 1995). Oxidation of As(III) by Fe oxyhydroxide and goethite has been observed by some investigators (Haury et al., 2000; Sun and Doner, 1998) but not by others (Oscarson et al., 1981a). The reaction appears to be catalyzed by light. DeVitre et al, (1991) measured only 17% recovery of As(III) that had been added to Fe hydroxide for 10 days in the presence of light, but 91% recovery in the absence of light.

Manganese (Mn) oxides are commonly associated with aquifer solids and have been shown to be important oxidants of As(III). Oxidation of As(III) by Mn oxides is thermodynamically favorable over a wide range of pH (Scott and Morgan, 1995), and several studies have provided direct and indirect evidence for this reaction. Oxidation of As(III) by synthetic birnessite (δ-MnO_2) and manganite (γ-MnOOH) has been measured in laboratory experiments (Chiu and Hering, 2000; Moore et al., 1990; Oscarson et al., 1981a). The reaction is rapid with 80% of the As(III) oxidized to As(V) within an hour at pH values between 4-8 (Scott and Morgan, 1995; Sun and Doner, 1998). Chiu and Hering (2000) measured three times more As(III) oxidation by manganite at pH 4.0 than at pH 6.3, possibly a result of greater As(III) adsorption at the lower pH. The hypothesized reaction sequence is adsorption of As(III), followed by reduction of Mn(IV) or Mn(III) to Mn(II) and release of As(V) to solution (Scott and Morgan, 1995). Incorporation of As(V) into the Mn oxide structure has also been reported (Moore et al., 1990). The degree of Mn oxide crystallinity has a significant effect on the rate of As(III) oxidation. Highly ordered, low surface area pyrolusite (β-

MnO_2) oxidized As(III) at a much slower rate than poorly crystalline, high surface area birnessite and cryptomelane (α-MnO_2) (Oscarson et al., 1983).

Oxidation of As(III) to As(V) by lake sediments has been attributed to the presence of Mn oxides (Oscarson et al., 1980). In these experiments, elimination of dissolved oxygen and suppression of microbial activity had no effect on the rate of oxidation. However, treating the sediments with hydroxylamine hydrochloride or sodium acetate to remove Mn oxides significantly decreased oxidation of As(III) (Oscarson et al., 1981b).

Oxidation of As(III) has also been observed in groundwater. Arsenite was injected into oxic [200-300 μM (micromoles per liter) O_2] and suboxic (1-6 μM O_2) zones of groundwater at a site on Cape Cod, Massachusetts, USA. Significant oxidation of As(III) to As(V) along a 2.2 meter flow path was observed in both the oxic and suboxic zones (Kent et al., 2001). Manganese oxides are present in the mixutre of oxides that coat these aquifer solids (Fuller et al., 1996) and are the most likely electron acceptor in both the low O_2 suboxic zone and possibly the oxic zone.

Clay minerals have also been reported to be capable of oxidizing As(III). Lin and Puls (2000) observed that, under a nitrogen atmosphere, oxidation of adsorbed As(III) by halloysite, kaolinite, illite, illite/montmorillonite, and chlorite increased with reaction time; oxidation was 100% after 75 days. The cause of oxidation was postulated to be the presence of Fe oxides or trace amounts of impurities such as iodide in the clay minerals. Manning and Goldberg (1997a) postulated that minor amounts of Mn oxides caused the observed oxidation of As(III) by kaolinite and illite in their experiments. Work by Foster et al. (1998) has shown that titanium can mediate the transfer of electrons from As(III) to O_2. They found that oxidation of As(III) by a Georgia kaolinite (KGa-1b), containing 1.76 weight percent TiO_2, was dependent on O_2 concentration. These results were identical to separate experiments using pure TiO_2. Manganese oxide impurities were ruled out in these experiments because of the low Mn concentration in this kaolinite and the observation that oxidation of As(III) by MnO_2 was independent of O_2 concentration.

Oxidation of As(III) to As(V) by microorganisms using oxygen as the terminal electron acceptor has been reported in laboratory cultures (Osborne and Ehrlich, 1976; Phillips and Taylor, 1976). This process has also been observed in natural settings. Wilkie and Hering (1996) found that As(III) from geothermal sources entering a stream was rapidly oxidized to As(V) by bacterial colonies associated with macrophytes.

While As(III) can be metastable to varying degrees in aerobic environments, As(V) tends to be more rapidly reduced under anaerobic conditions. There have been numerous studies in recent years that have identified several different strains of bacteria capable of As(V) reduction. Reaction products include dimethylarsine (McBride and Wolfe, 1971) and inorganic As(III) (Freeman, 1985; Laverman et al., 1995; Newman et al., 1997; Zobrist et al., 2000). Some microbes apparently reduce As(V) that has entered the cell to As(III), which is then excreted as a detoxification mechanism (Ji and Silver, 1995; Rosen et al., 1995). Other microbes can utilize As(V) as an energy source if a more preferable electron acceptor such as oxygen is not available (Macy et al., 1996; Newman et al., 1997) (Newman et al., 1998; Stolz and Oremland, 1999).

Arsenate reduction under abiotic conditions also has been measured. Both H_2S and H_2 are capable of reducing As(V) to As(III) (Cherry et al., 1979). Arsenate reduction by H_2S is rapid, and the rate increases with decreasing pH (Rochette et al., 2000). Intermediate by-products include arsenic sulfide species ($H_xAs_3S_{3x}^{-3}$) which dissociate to $H_3AsO_3^{o}$ within several days. Reduction of As(V) by aqueous Fe(II) was not observed in laboratory experiments (Zobrist et al., 2000); however, reduction of As(V) to As(III) was observed in the natural gradient tracer experiment of (Kent et al., 2001). In this experiment, As(V) was injected into an anoxic groundwater zone containing high concentrations of dissolved Fe(II). Arsenate may have been reduced directly by Fe(II), with the aid of microorganisms, or As(V) may have been reduced by low concentrations of organic carbon in this zone. Abiotic As(V) reduction by humic substances also has been reported (Haury et al., 2000).

3. ADSORPTION/DESORPTION

Arsenic concentrations observed in most groundwaters are orders of magnitude less than the solubilities of most As-bearing minerals. Thermodynamic predictions show that some As minerals could potentially control As concentrations in groundwater under common geochemical conditions; however, these minerals are rarely if ever documented outside of mineralized area (Foster, this volume). Therefore, adsorption reactions between As and mineral surfaces are generally the most important control on the dissolved concentration of As in groundwater environments where water supply is an issue. Adsorption of As is a complex function of the interrelationship between the properties of the solid surface, pH, the concentration of As and competing ions, and As speciation.

3.1 Mechanism

Oxides of iron, aluminum (Al), and manganese are potentially the most important source/sink for As in aquifer sediments because of their chemistry, widespread occurrence, and tendency to coat other particles. Upon exposure to water, metal ions on the oxide surface complete their coordination shells with OH groups (Hingston et al., 1972; Fig.2). Depending on pH, these OH groups can bind or release H^+, resulting in the development of a surface charge. The adsorption properties of oxides are due to the presence of these OH_2^+, OH, and O^- surface functional groups (Sposito, 1984).

There are two widely accepted mechanisms for adsorption of solutes by a solid surface. Outer-sphere surface complexation, or non-specific adsorption, involves the electrostatic attraction between a charged surface and an oppositely charged ion in solution (Fig. 3). The adsorbed ion resides at a certain distance from the mineral surface. Inner-sphere complexation, also termed specific adsorption, involves the formation of a coordinative complex with the mineral surface (Hingston et al., 1972; Fig. 3). Inner-sphere complex bonds are more difficult to break than outer-sphere complex bonds and result in stronger adsorption of ions.

Arsenic adsorbs by ligand exchange with OH and OH_2^+ surface functional groups, forming an inner- sphere complex. This type of adsorption requires an incompletely dissociated acid (i.e., $H_2AsO_4^-$) to provide a proton for complexation with the surface OH group, forming H_2O and providing a space for the anion (Hingston et al., 1972). These reactions can be written several ways. For example:

$$=\text{SurfOH} + H_3AsO_4^o = =\text{SurfH}_2AsO_4^o + H_2O \tag{1}$$

$$=\text{SurfOH} + H_2AsO_4^- = =\text{SurfHAsO}_4^- + H_2O \tag{2}$$

$$=\text{SurfOH} + HAsO_4^{-2} = =\text{SurfAsO}_4^{-2} + H_2O \tag{3}$$

$$=\text{SurfOH} + H_3AsO_3^o = =\text{SurfH}_2AsO_3^o + H_2O \tag{4}$$

$$=\text{SurfOH} + H_2AsO_3^- = =\text{SurfHAsO}_3^- + H_2O \tag{5}$$

$$=\text{SurfOH} + \text{HAsO}_3^{-2} = \ =\text{SurfAsO}_3^{-2} + \text{H}_2\text{O} \quad (6)$$

where, =SurfOH represents the structural metal atom and associated OH surface functional group, and, $=\text{SurfHAsO}_4^-$ (for example) is the surface-arsenic complex, (Dzombak and Morel, 1990). The energy required to dissociate the weak acid at the oxide surface varies with pH. As a result, the amount adsorbed varies with pH. For example, the dissociation constant for

$$\text{H}_3\text{AsO}_3^{o} = \text{H}_2\text{AsO}_3^- + \text{H}^+ \quad (7)$$

is at pK_{a1} 9.2 (Cherry et al., 1979). Therefore, at a pH of 9.2 the free energy required to dissociate $H_3AsO_3^o$ is a minimum and adsorption should be a maximum.

Chemical evidence of inner-sphere complex formation between As and oxide surfaces comes from experiments designed to evaluate the effects of ionic strength (I) and pH on As adsorption. Changes in I affect the electrostatic forces near the mineral surface (Davis and Kent, 1990). Anions that form inner- sphere complexes coordinate directly with the oxide surface in a manner that is relatively independent of I. In contrast, the formation of outer-sphere complexes by weakly adsorbed anions such as sulfate and selenate is sensitive to I. Experimental data from Hsia et al. (1994) and Grossl and Sparks (1995) show essentially no change in the concentration of As(V) adsorbed by ferrihydrite and goethite for I that ranged from 0.005 to 0.1 M, consistent with an inner-sphere surface complex. Other investigators have demonstrated that adsorption of As(III) by amorphous $Al(OH)_3$, kaolinite, and illite was independent of I (Manning and Goldberg, 1997a). Adsorption of methylated As(V) species by ferrihydrite and Al oxide were also relatively insensitive to changes in I, indicating formation of inner-sphere surface complexes (Cox and Ghosh, 1994).

Inner-sphere surface complexation of As has also been inferred from changes in the isoelectric point (IEP) of the adsorbing solid. The IEP of a solid is the solution pH at which the net charge on a particle is zero (Sposito, 1984). At pH values below the IEP, net surface charge is positive due to adsorption of excess H^+. At pH values greater than the IEP, net surface charge is negative due to desorption of H^+. Formation of inner-sphere surface complexes of anions has been shown to increase the negative charge of solid surfaces, thereby decreasing the IEP to lower pH values. The amount of shift in the IEP depends on the ion and its concentration, as well as the particular solid surface. Adsorption of both As(III) and As(V) species have been found to lower the IEP of various oxides including goethite, ferrihydrite, gibbsite, and amorphous $As(OH)_3$ (Anderson and Malotky, 1979; Ghosh and Teoh, 1985; Hingston et al., 1972; Jain et al., 1999;

Manning and Goldberg, 1996b). The magnitude of the decrease depends on As concentration. For example, Hsia et al. (1994) measured a decrease in the IEP for ferrihydrite from pH 8 to pH 4 as As(V) concentration increased from 0 to 100 μM.

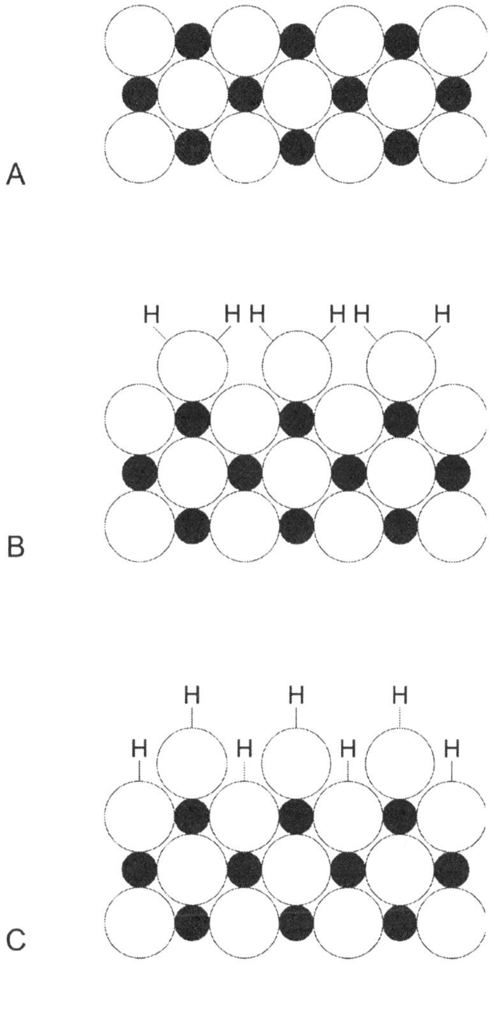

Figure 2. Cross section of the surface layer of a metal oxide: Metal ions are filled circles, oxide ions are open circles:

(a) Surface ions are coordinatively unsaturated.

(b) In the presence of water, the surface metal ions may coordinate H_2O molecules.

(c) Dissociative chemisorption leads to a hydroxylated surface.

Reprinted from Schindler (1981) *with permission from Butterworth-Heinemann.*

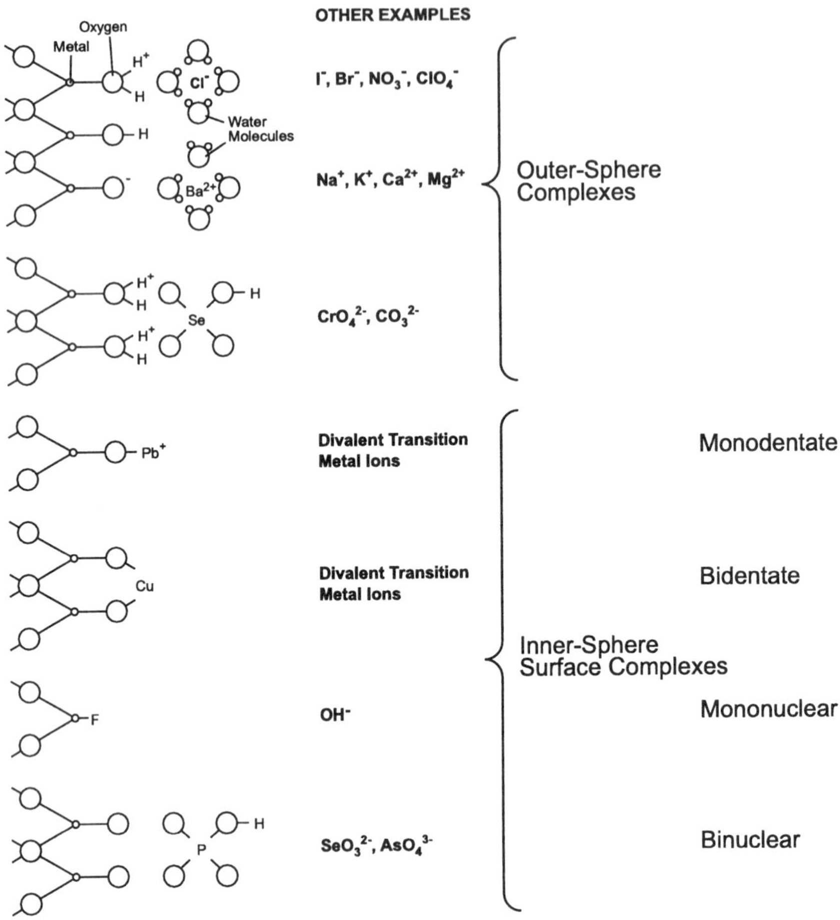

Figure 3. Schematic representation of coordinative surface complexes and ion pairs formed between inorganic ions and hydroxyl groups of an oxide surface in the triple layer model. Reprinted from Hayes (1987) with permission.

Using Extended X-ray Absorption Fine Structure (EXAFS) spectroscopy, Waychunas et al. (1993) determined the geometry of As(V) adsorbed on ferrihydrite and goethite and confirmed the formation of a binuclear, and to a lesser extent, monodentate inner-sphere surface complex. Manning et al. (1998) used EXAFS to show that As(III) also forms inner-sphere surface complexes on goethite. (Note, a thorough review of the molecular-scale structure and chemistry of As adsorbed by crystalline and x-ray amorphous metal oxides is presented by Foster (this volume).

3.2 Solid Phase Adsorption Properties

Aquifer solids are rarely composed of discrete mineral phases such as quartz, feldspar, or clay minerals. Over time, weathering reactions produce complex mixtures of different minerals of varying degrees of crystallinity. Since adsorption implies a surface or near surface phenomena, it is the mineral surfaces of aquifer solids that are of principal concern. These surfaces can consist of mixtures and coatings of various oxides, alumino-silicate minerals, carbonates, and organic matter (Coston et al., 1995; Farmer et al., 1991; Jenne, 1976; Taylor, 1987).

Many studies that have investigated the effects of different geochemical properties on adsorption of As species have used controlled laboratory experiments to evaluate adsorption of As by pure mineral phases such as Fe, Al, and Mn oxides, and clay minerals. Since these mineral phases are often components of aquifer solids, the data from such experiments can be used to help interpret the behavior of As in groundwater. Such an approach has commonly been used to assess As adsorption by soils and lacustrine sediments (Belzile and Tessier, 1990; Bowell, 1994; Fordham and Norrish, 1979; Jacobs et al., 1970).

3.3 Iron oxides

Iron oxides of varying crystallinity and composition are one of the most common phases associated with aquifer solids. They can be present as discrete particles or as coatings on other mineral surfaces. Many laboratory investigations of As adsorption have used unaged hydrous Fe(III) oxide formed by rapid precipitation of Fe(III) from solution, or oxidation of Fe(II) followed by precipitation at the oxic/anoxic boundary. This phase has a stoichiometry near $Fe_2O_3 \cdot 2H_2O$ and has been termed ferrihydrite (Eggleton and Fitzpatrick, 1988). Ferrihydrite is very poorly crystalline and has a large surface area which results in significant adsorption capacity (Dzombak and Morel, 1990). The adsorption behavior of ferrihydrite is considered to be analogous to that of recently formed, poorly crystalline Fe oxyhydroxides that are often associated with aquifer solids.

One of the most important controls on adsorption of As(III) and As(V) is pH; pH has a major influence on aqueous As speciation and on the composition of surface functional groups through protonation and deprotonation reactions. At pH values less than the IEP of a solid, the net charge is positive, facilitating adsorption of anions. At pH values greater than the IEP of a solid, the net charge is negative, facilitating adsorption of cations. However, adsorption of anions can occur when net surface charge is negative, and cations can adsorb when net surface charge is positive.

The effect of pH on As adsorption differs between As(III) and As(V). Data for the adsorption of As(III) and As(V) by ferrihydrite that are plotted in Fig. 4 demonstrate characteristics of As adsorption commonly observed in laboratory batch experiments (Bowell, 1994; Hsia et al., 1992; Jain and Loeppert, 2000; Manning et al., 1998; Meng et al., 2000; Pierce and Moore, 1982; Wilkie and Hering, 1996). The initial As concentrations used in these experiments were 1.33 and 13.3 μM and are within the range found in natural settings of As contaminated groundwater. Arsenate adsorption was greatest at low pH values and decreased with increasing pH. Greater adsorption of As(V) at low pH is attributable to more favorable adsorption energies between the more positively charged surface and negatively charged $H_2AsO_4^-$, the predominate As(V) species between pH 2.2 and 6.9. As pH increased above 6.9, $HAsO_4^{-2}$ became the predominate aqueous species, surface charge became less positive, and adsorption was less favorable. However, even at pH 10 there were some surface functional groups capable of exchanging for As(V), as shown in Fig. 4, and in studies by Hsia et al. (1994) and Jain and Loeppert (2000).

Quantitative adsorption of As(V) as a function of pH is also influenced by the concentration of adsorption sites relative to the aqueous concentration of As (Fig. 4). At the lower initial concentration of 1.33 μM As(V), there was an excess of adsorption sites and As(V) adsorption was independent of pH up to a pH of 6. When the concentration of As(V) was increased to 13.3 μM, available sites become filled at a lower pH and As(V) adsorption began to decrease immediately with any further increase in pH.

The pH dependent adsorption of As(III) differs from As(V). The data in Fig. 4 show that at lower pH values, more As(V) was adsorbed than As(III). However, as pH increased above about 6.5, adsorption of As(III) became greater than As(V). The predominate As(III) species in solution at pH values less than 9.2 is the neutral $H_3AsO_3^o$. At higher pH values, neutral $H_3AsO_3^o$ is more readily able to donate a proton to the surface OH group than the negatively charged As(V) species. The pH where adsorption of As(III) becomes greater than As(V) is affected by the surface charge of the solid and generally decreases as the IEP of the sorbent decreases.

The pH dependent adsorption of As(III) and As(V) have been compared under transport conditions. Joshi and Chaudhuri (1996) eluted either As(III) or As(V) through columns packed with Fe oxide-coated sand at pH 7.6. Breakthrough curves exhibited slightly greater retardation of As(III) than As(V). In a column experiment with Fe oxide-coated sand at pH 5.7, Gulens et al. (1979) observed that As(III) moved 5- 6 times faster than As(V).

The data in Fig. 4 represent separate adsorption experiments with either As(III) or As(V). When As(V) and As(III) were present in solution together (2.08 mM each), Jain and Loeppert (2000) found that As(III) had little effect on adsorption of As(V) below pH 6, but caused a decrease in As(V) adsorption of as much as 25% at pH greater than 6 due to strong adsorption of As(III). Arsenate resulted in a decrease in adsorption of As(III) of 15-25% between pH 4 and 6; however, by pH 9, As(V) decreased adsorption of As(III) by less than 5%.

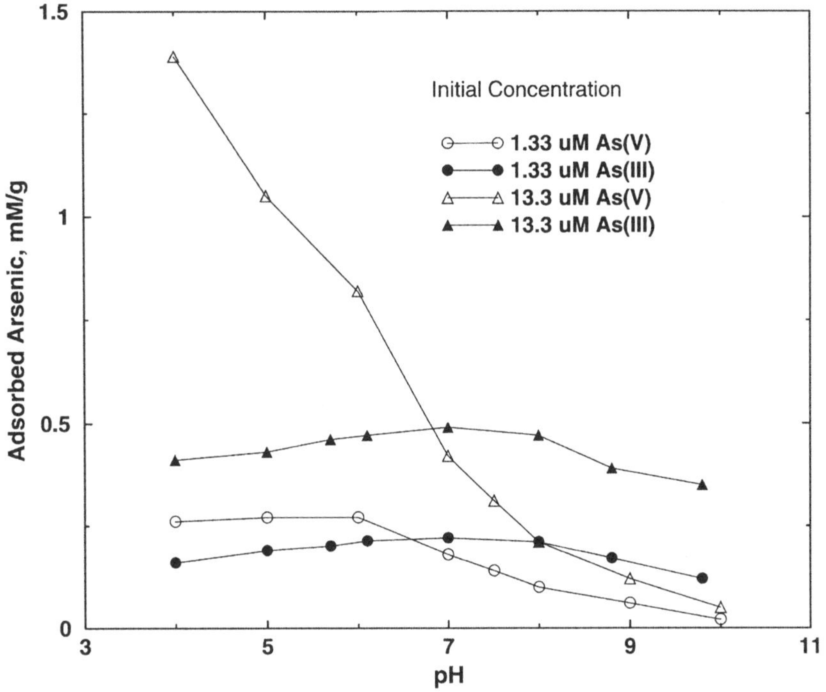

Figure 4. Adsorption of As(V) and As(III) by ferrihydrite as a function of pH. Ionic strength, 0.01 M; concentration of ferrihydrite, 0.00445 g/L. (Modified from Pierce and Moore, 1982).

Adsorption of the methylated As(V) species, $[CH_3AsO_2(OH)^-]$ and $[(CH_3)_2AsO(OH)^o]$, by ferrihydrite followed a pattern similar to As(V). Maximum adsorption of both species occurred at lower pH values, and the amount adsorbed sharply decreased for pH values greater than 7 (Cox and Ghosh, 1994). For any given pH, the negatively charged $[CH_3AsO_2(OH)^-]$ adsorbed to a greater extent than the neutral $[(CH_3)_2AsO(OH)^o]$.

As ferrihydrite ages, crystallite growth occurs resulting in a decrease in surface area and the number of surface complexation sites. Fuller et al. (1996) measured a 20% decrease in adsorption of As(V) by ferrihydrite that had been aged for six days compared to ferrihydrite that had aged for only a few hours. Given sufficient time and appropriate conditions, ferrihydrite usually recrystallizes into goethite.

Goethite is a common Fe oxide component of aquifer solids and has also been synthesized in the laboratory for use in As adsorption experiments. The pH dependent characteristics for adsorption of As(V) and As(III) by goethite are similar to ferrihydrite (Grossl and Sparks, 1995; Hingston et al., 1971) (Manning et al., 1998; Matis et al., 1999; Sun and Doner, 1998). However, there is generally less adsorption of As by goethite than ferrihydrite on a per gram basis because the greater degree of crystallinity of goethite, results in lower concentrations of surface-complexation sites. For example, at an initial As concentration of 250 μM/g and a pH of 6, Sun and Doner (1998) measured about 40% less adsorption of As(V) and As(III) by goethite (surface area = 80 m^2/g) than Raven et al. (1998) measured for ferrihydrite (surface area = 200 m^2/ g).

Hematite generally consists of more densely packed larger crystals than goethite and has a lower surface area (Schwertman and Taylor, 1989). Bowell (1994) reported that adsorption of As(V), As(III), MMA, and DMA by hematite was about half as much as goethite. Other investigators have also measured lower concentrations of As(V) and As(III) adsorbed by hematite (Singh et al., 1988; Van der Hoek et al., 1994; Xu et al., 1988) than were reported for goethite and ferrihydrite in the previously cited studies.

3.4 Aluminum oxides and oxyhydroxides

The Al(III) atom has the same charge and a nearly identical radius as the Fe(III) atom. As a result, the common hydrous Al oxide phases are structurally similar to hydrous Fe oxides (Waychunas, 1991). Oxides and hydroxides of Al also have significant adsorption capacity for As, and their pH dependent adsorption isotherms are similar to those for Fe oxides and hydroxides; both have similar IEP's (Anderson et al., 1976; Parks, 1965). Arsenate, $CH_3AsO(OH)_2^o$, and $(CH_3)_2AsOOH^o$ were strongly adsorbed up to a pH of about 7 by amorphous $Al(OH)_3$, crystalline $Al(OH)_3$ (gibbsite), α-Al_2O_3, and γ-Al_2O_3. Adsorption significantly decreased at higher pH values (Anderson et al., 1976; Cox and Ghosh, 1994; Ghosh and Teoh, 1985; Ghosh and Yuan, 1987; Halter and Pfeifer, 2001; Hingston et al., 1971; Xu et al., 1988). Arsenite adsorption increased to a maximum at pH 8, then decreased at higher pH values (Ghosh and Teoh, 1985; Gupta and Chen, 1978). In all of these studies, the amount of adsorption of As species increased as initial aqueous concentration increased until sites became saturated.

The degree of crystallinity of Al oxides and hydroxides also is important for As adsorption. For comparable experimental conditions, x-ray amorphous $Al(OH)_3$ adsorbed more As(V) per gram than crystalline γ-Al_2O_3 and gibbsite (Anderson et al., 1976; Anderson and Malotky, 1979; Ghosh and Teoh, 1985; Hingston et al., 1971).

Experiments that have compared As adsorption by Fe and Al minerals with similar structures are few. Gulledge and O'Conner (1973) reported slightly greater adsorption of As(V) per gram of hydrous Fe oxide than hydrous Al oxide at pH values of 5-8. Surface areas in these experiments were not reported. Manning and Goldberg (1996b) compared adsorption of As(V) by goethite (44 m^2/g) and gibbsite (45 m^2/g). In these experiments, the initial concentration of either solid and aqueous As(V) were identical, allowing a direct comparison of the results. Between pH 5 and 9, gibbsite adsorbed about 17% less As(V) than goethite. Hingston et al. (1971) measured four times more adsorption of As(V) by goethite (60 m^2/g) than by gibbsite (31 m^2/g) between pH 5 and 9. When the initial aqueous As(V) and solid concentration for these experiments are normalized to surface area, the initial suspensions contained about 4 μM As(V) per m^2 of either goethite or gibbsite per liter. These results suggest that surface area alone may not always be a good indicator of the number of reactive surface-complexation sites.

3.5 Manganese oxides

Manganese oxides in natural systems are often complex minerals characterized by poor crystallinity and mixed oxidation states (McKenzie, 1989). In addition to being electron acceptors in the oxidation of As(III) to As(V), Mn oxides can also adsorb As. Direct evidence for adsorption of As by Mn oxides comes from laboratory experiments in which pure mineral phases have been used.

The birnessite group are the most common Mn oxides formed from weathering processes and have a low IEP, which results in a net negative surface charge at pH values common in groundwater. Arsenate adsorption by synthetic birnessite (IEP ~ 2.5) was reported to be negligible (Oscarson et al., 1981a; Oscarson et al., 1983; Scott and Morgan, 1995). At the pH values used in these experiments (4-7), the negatively charged species, $H_2AsO_4^-$, apparently could not overcome the energy barrier required to displace a functional group from the negatively charged birnessite surface. However, As(III) was adsorbed by birnessite (164 μM/g MnO_2) at pH 7; the neutral charge on $H_3AsO_3^o$ results in a lower energy barrier to exchange with surface functional groups (Oscarson et al., 1983). Chiu and Hering (2000) reported about 30% greater adsorption of As(III) by manganite (IEP = 6.2) at pH 4.0 than at pH 6.3. These results are contrary to most observations of As(III) adsorption by Fe and Al oxides where adsorption of As(III) increases between pH 4.0 and 6.3.

Adsorption of both As(III) and As(V) were reported by the manganese oxides cryptomelane (IEP = 2.9) and pyrolusite (IEP = 6.4) (Oscarson et al., 1983). Differences in the degree of crystallinity between these two minerals had a significant effect on As adsorption. Cryptomelane, with a greater surface area (34.6 hm^2/kg) adsorbed 637 μM/g MnO_2 compared to only 32 μM/g adsorbed by pyrolusite (0.8 hm^2/kg).

Bajpai and Chaudhuri (1999) evaluated As adsorption by Mn dioxide-coated sand in column experiments. The eluent contained equal concentrations (6.7 μM) of As(III) and As(V) in groundwater at pH 7.6. There was complete removal of As for about 135 pore volumes, equivalent to about 190 μM As/g MnO_2.

3.6 Silica oxides

Oxides of Si have an IEP near pH 2. As a result, they have a net negative charge at pH values common in water and are not good adsorbents for anions (Stumm and Morgan, 1981). Adsorption of As(V) by quartz was negligible for pH values above 3 (Xu et al., 1988). Transport of As(V) through a column packed with acid purified silica sand exhibited minimal adsorption (Darland and Inskeep, 1997a). Complete breakthrough of As(V) occured at pore volume 1.5, versus the expected pore volume of 1.0 for a nonreactive solute.

3.7 Clay minerals

Aluminosilicate clay minerals are composed of alternating layers of silica oxide and, most commonly, aluminum oxide. There are several different types of sites on clay minerals that can potentially adsorb ions. By analogy with amorphous silica and quartz, OH groups of silica sheets have a negative charge above pH 2 and are not predicted to react with As. Only the OH groups associated with Al ions exposed at the edges of clay particles are considered to be proton acceptors and able to complex anionic species of As (Davis and Kent, 1990). Thus, the same geochemical properties that control adsorption of As by Al oxide minerals are also important for clay minerals, specifically pH, As concentration and speciation, and the concentration of adsorption sites.

Results from experiments evaluating As adsorption by different clay minerals have some general commonalities. In all cases, adsorption per gram of solid of both As(III) and As(V) increased with initial solution concentration of As. Maximum adsorption of As(V) by kaolinite, montmorillonite, illite, halloysite, and chlorite occurred up to pH values near 7, then decreased with further pH increases (Frost and Griffin, 1977; Goldberg and Glaubig, 1988b; Lin and Puls, 2000; Manning and Goldberg, 1997a). Adsorption of As(III) by these same clay minerals was a minimum at low pH and increased with increasing pH. Arsenate adsorbed to a greater extent than As(III) on all clay minerals at pH <7. At higher pH values, adsorption of As(V) and As(III) were more comparable, and in some cases As(III) adsorption exceeded that of As(V) (Frost and Griffin, 1977; Lin and Puls, 2000).

Surface area can be an important factor in adsorption of As by clay minerals. Frost and Griffin (1977) found that montmorillonite adsorbed about twice as much As(III) and As(V) as kaolinite. The surface area of montmorillonite was 2.5 times greater than kaolinite. In experiments using a poorly crystallized kaolinite with a surface area comparable to montmorillonite, the amount of As(V) adsorbed was about the same (Goldberg and Glaubig, 1988b). Halloysite and chlorite were found to adsorb As(V) to a much greater extent than kaolinite, illite, and montmorillonite (Lin and Puls, 2000). The authors attributed this behavior to the much larger surface area of the highly disordered halloysite. The chlorite used was an iron rich species, which could have provided additional surface functional groups at Fe-OH sites. However, results from Manning and Goldberg (1996a) show that surface area is not always a good indicator of As(V) adsorption by clay minerals. They found that kaolinite (9.1 m^2/g) adsorbed 30% more As(V) than illite (18.6 m^2/g) and 50% more than montmorillonite (24.2 m^2/g).

3.8 Carbonates

There is evidence that carbonate minerals could be important in controlling the aqueous concentrations of As, especially at higher pH values. Oscarson et al. (1981b) measured no adsorption of As(III) by $CaCO_3$ at pH 7. Goldberg and Glaubig (1988b) observed that adsorption of As(V) by calcite increased from a minimum of 0.7 mM/kg at pH 6 to a maximum of 2.0 mM/kg at pH 11. La Force et al. (2000) observed that As was associated with the carbonate extractable fraction of sediment in a wetland influenced by mine wastes. Cheng et al. (1999) allowed a solution of As(III) to react with calcite. Using data from x-ray standing wave diffraction they concluded that As(III) was removed from solution and occupied the carbonate sites on the calcite surface. These studies indicate that adsorption of As in carbonate groundwater systems could be important.

3.9 Humic acids

Organic compounds such as humic acid can adsorb on aquifer solids or be present in aquifers as a result of depositional history. These compounds contain surface functional groups which can adsorb ions from solution (Thurman, 1985).

Adsorption of As by two humic acids was a function of pH, As speciation, and the humic acid composition (Thanabalasingam and Pickering, 1986). For both humic acids, adsorption of As(V) was slightly greater than As(III). However, the pH effect differed with humic acid composition. Arsenate adsorption by the humic acid with a higher ash and Ca content was a maximum at pH 6, whereas As(III) adsorption was a maximum at pH 8.5. For the lower ash and Ca humic acid, both As(V) and As(III) exhibited broad adsorption maxima between pH 5.5 and 7.5.

3.10 Natural materials

Most studies of As adsorption by natural materials have been conducted on soils and lacustrine sediments (Aggett and O'Brien, 1985; Elkhatib et al., 1984a; Manning and Goldberg, 1997b). The surfaces of soil and lacustrine sediment particles tend to contain mixtures of various oxides and clay minerals that can be similar to those found on aquifer solids, although the organic carbon content of soils and lacustrine sediments is generally greater. Chemical extraction methods have been used to qualitatively identify the mineral phases in soils associated with As adsorption. Generally, adsorption of As(III), As(V), $CH_3AsO(OH)_2^{\circ}$, and $(CH_3)_2AsOOH^{\circ}$ have been positively correlated with the Fe and Al oxide and clay mineral content of soils (Elkhatib et al., 1984b; Fordham and Norrish, 1979; Jacobs et al., 1970; Livesey and Huang, 1981; Sakata, 1987; Wauchope, 1975; Wauchope and McDowell, 1984).

Carrillo and Drever (1998) used batch reactors to investigate adsorption of As by aquifer material collected from the bottom of wells. Adsorption decreased from about 2.8 mM/M Fe at pH 7 to 0.7 mM/M Fe at pH 11 and was correlated with Fe oxide content. Arsenic speciation was not determined in these experiments.

3.11 Ionic Competition

Many inorganic and organic aqueous species have an influence on As adsorption. The following discussion deals with those elements that are commonly found in groundwater. Solutes can compete directly with As for available surface binding sites and can indirectly influence adsorption by alteration of the electrostatic charge at the solid surface. Both processes are influenced by pH, solute concentration, and the intrinsic binding affinity of the solid (Davis and Kent, 1990).

3.12 Phosphate

Spectroscopic studies have shown that the surface species formed by phosphate [P(V)] are identical to those formed by As(V) (Hiemstra and Van Riemsdijk, 1999; Waychunas et al., 1993). Adsorption experiments have demonstrated that the affinity of P(V) for surface sites is similar to As(V) (Gao and Mucci, 2001; Hingston et al., 1971; Manning and Goldberg, 1996a; Manning and Goldberg, 1996b; Manning and Goldberg, 1996b; Ryden et al., 1987). Experimental data from Jain and Loeppert (2000), replotted in Fig. 5, show the influence of P(V) on adsorption of As(V) and As(III) by ferrihydrite as a function of pH. The P(V)/As ratios (1:1 and 10:1) used in these experiments are within the range reported for many groundwaters. Adsorption of both As(V) and As(III) decreased with increasing P(V) concentration. For As(V), the decrease was significant over the entire pH range. Competition for adsorption sites between As(V) and P(V) involves species with the same charge. At pH values less than about 7, $H_2AsO_4^-$ and $H_2PO_4^-$ predominate. At higher pH values, $HAsO_4^{-2}$ and HPO_4^{-2} predominate. Phosphate had the greatest effect on As(III) adsorption at lower pH values. At pH 9, adsorption of As(III) decreased by only a few percent, even at the highest P(V) concentration. Apparently, the neutral $H_3AsO_3^o$ was better able to compete for surface-complexation sites with HPO_4^{-2} at higher pH.

Phosphate competition with As(V) has been observed for other adsorbents. Hingston et al. (1971) measured an 85% decrease in As(V) adsorption by goethite when the P(V)/As(V) ratio was increased from zero to 12:1. Manning and Goldberg (1996b) reported that a P(V)/As(V) ratio of 1:1 caused a 30% decrease in As(V) adsorption by both goethite and gibbsite at pH values less than 8, compared to P(V)-free solutions. Similar effects of P(V) on As(V) adsorption were observed for kaolinite, montmorillonite, and illite (Manning and Goldberg, 1996a). Adding P(V) to natural systems such as soils has also been found to increase the mobility of As (Darland and Inskeep, 1997a; Melamed et al., 1995; Peryea, 1991; Roy et al., 1986a, b).

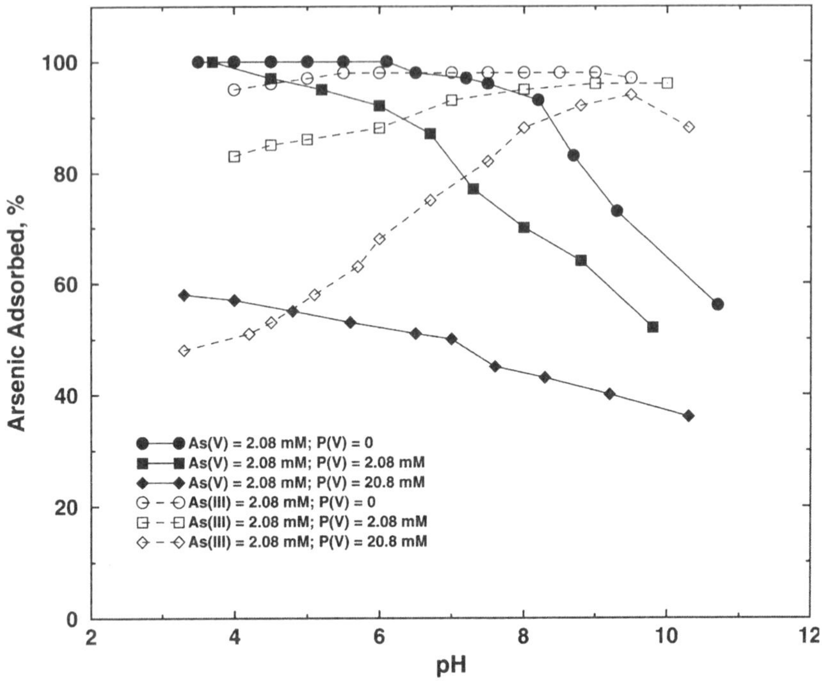

Figure 5. Effect of phosphate on adsorption of As(V) and As(III) by ferrihydrite. Ionic strength, 0.1 M; concentration of ferrihydrite, 2 g/L (Modified from Jain and Loeppert, 2000).

3.13 Sulfate

Macroscopic chemical evidence indicates that sulfate [S(VI)] adsorbs via electrostatic attraction as an outer-sphere complex and should therefore have much less effect on As adsorption than inner-sphere complex formers such as P(V). Sulfate would only be expected to adsorb when the net surface charge is positive, i.e., at pH values below the IEP of the solid (Hingston et al., 1972). This is consistent with experimental data for S(VI) adsorption by synthetic ferrihydrite (Dzombak and Morel, 1990) and natural aquifer solids (Stollenwerk, 1995). Experimental data from Jain and Loeppert (2000), replotted in Fig. 6, show the influence of S(VI) on adsorption of As(V) and As(III) by ferrihydrite as a function of pH. The S(VI)/As ratios for Fig. 6, 10:1 and 50:1 reflect the fact that S(VI) concentrations in groundwater are generally much greater than As. Sulfate had essentially no effect on As(V) adsorption over the entire pH range. At the highest S(VI) concentration, adsorption actually increased by a few percent, which the authors attributed to a possible ionic strength effect. There was a decrease in As(III) adsorption for both S(VI) concentrations at pH <7 but no effect at higher pH.

Similar results were observed by Wilkie and Hering (1996). For a S(VI)/As(III) ratio of 4,000:1, As(III) adsorption by ferrihydrite was suppressed by 20% at pH 7 and 70% at pH 4. However, As(V) adsorption decreased by only a few percent at pH 5.

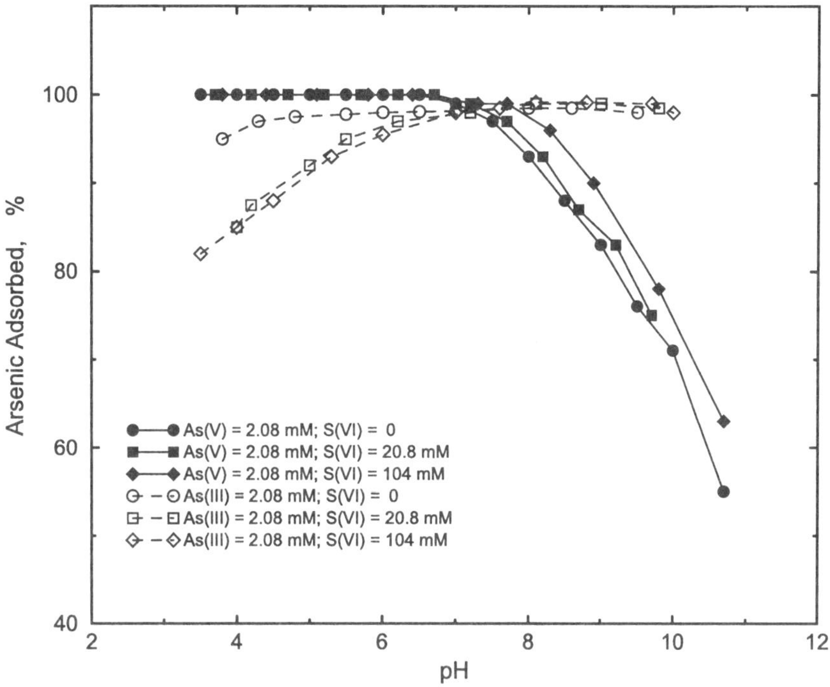

Figure 6. Effect of sulfate on adsorption of As(V) and As(III) by ferrihydrite. Ionic strength, 0.1 M; concentration of ferrihydrite, 2 g/L. (Modified from Jain and Loeppert, 2000).

3.14 Carbonate

Limited evidence indicates that carbonate species may have some effect on adsorption of As. For the pH range of interest in groundwater, all three carbonate species, $H_2CO_3^o$, HCO_3^-, and CO_3^{-2} can be present. No attempt is made here to distinguish between the adsorbing species, and CO_3^{-2} is used for discussion purposes. Carbonate is a weakly adsorbing anion, but its adsorption mechanism is not well understood. Outer-sphere complexation has been assumed by many investigators. However, Su and Suarez (1997) found that adsorption of CO_3^{-2} lowered the electrophoretic mobility and decreased the IEP of amorphous Al and Fe oxides, gibbsite, and goethite, implying inner-sphere complex formation. Spectroscopic evidence suggested formation of a monodente inner-sphere complex. Adsorption of CO_3^{-2} has been reported to decrease adsorption of outer-sphere complexing anions such as Cr(VI) by Fe oxides (Van Geen et al., 1994; Zachara et al., 1987). However, CO_3^{-2} has also been found to increase adsorption of weakly adsorbing sulfate and selenate by aluminum oxide in the pH range of 6 to 8 (Wijnja and Schulthess, 2000). This promotive effect was a maximum at 0.3 mM CO_3^{-2} and decreased at higher CO_3^{-2} concentrations. The authors suggested that coadsorption of H^+ increased the positive charge on the mineral surface, facilitating adsorption. There was no effect of CO_3^{-2} on strongly adsorbing P(V).

Meng et al. (2000) measured essentially no effect of CO_3^{-2} on adsorption of As(III) and As(V) by ferrihydrite. These experiments were conducted in equilibrium with air and total CO_3^{-2} concentration was 0.01 mM. Fuller et al. (1993) reported little effect of 3 mM CO_3^{-2} in stream water on adsorption of As(V) at pH 8. In a column experiment using sand that contained Fe oxide coatings, there was essentially no difference (<5%) in transport of As(V) at pH 8.5 between solutions buffered with CO_2 (1 mM CO_3^{-2}) and N_2 (<0.1 mM CO_3^{-2}) (Darland and Inskeep, 1997a). Other evidence indicates that CO_3^{-2} may have some effect on adsorption of As(III) but not As(V). Wilkie and Hering (1996) found that 1 mM CO_3^{-2} caused about a 10% decrease in As(III) adsorption by ferrihydrite at pH 9 but no effect at pH 6. Arsenate adsorption at pH 9 decreased by only a few percent. Appelo and de Vet (this volume) presented evidence which suggested that adsorption of CO_3^{-2} by ferrihydrite decreased adsorption of As(III), and to a lesser extent As(V), in a shallow aquifer; CO_3^{-2} concentration in groundwater was 5 mM and pH values were between 7.1 and 7.7. Results from these experiments suggest that higher concentrations of CO_3^{-2} often found in groundwater may have an effect on As adsorption.

3.15 Silica

Silicic acid (H_4SiO_4) has also been shown to effectively complete with As for adsorption sites (Davis et al., 2001). Meng et al. (2000) evaluated the effect of Si and As adsorption in experiments as ferrihydrite was precipitating from solution. Adsorption of As(III) (300 μg/L) and As(V) (500 μg/L) at pH 6.8 began to decrease in the presence of 1 mg/L Si; at 10 mg/L Si, adsorption decreased by 70% for As(V) and 80% for As(III). Results from experiments using ferrihydrite that had been aged for 18-24 hours prior to adding Si showed that 3.1 mg/L Si had little effect on adsorption of As(III) and As(V) at pH values <8. At pH >8, As(V) adsorption was inhibited by as much as 40% (Swedlund and Webster, 1999). In the presence of 56 mg/L Si, As(III) adsorption decreased between pH 4 and 10 by as much as 35%. Adsorption of As(V) was only inhibited at pH >6, but the decrease was as much as 60%. Swedlund and Webster (1999) attributed the decrease in As adsorption at high Si concentrations to a combination of competition for surface sites and polymerization of Si resulting in an increase in negative surface charge.

3.16 Organic compounds

Dissolved organic solutes can also compete for sites with As. Fulvic acid decreased adsorption of As(V) by Al oxides. The effect was most pronounced at pH values <7 but occurred to some extent even at higher pH (Xu et al., 1988). Fulvic acid decreased adsorption of As(V), As(III), $CH_3AsO(OH)_2$, $(CH_3)_2AsO(OH)$, by Fe oxides in the pH range 4-8 (Bowell, 1994). Hering et al. (1997) found little effect of natural organic matter (1-5 mg/L) on adsorption of As(V) (10 μg/L) by ferrihydrite at pH 6. However, 4 mg/L organic matter did decrease adsorption of As(III) (10 μg/L) between pH 4 and 6 by more than 50%

3.17 Calcium and magnesium

Some solutes have been reported to enhance adsorption of As. For example, adsorption of Ca by ferrihydrite at pH 9 resulted in an increase in As(V) adsorption (Wilkie and Hering, 1996), apparently by increasing the amount of positive charge on the oxide surface. Adding Ca and Mg to suspensions of ferrihydrite negated part of the competitive effect of Si on As adsorption (Meng et al., 2000). The presence of Ca also enhanced adsorption of As(V) by Al oxide at pH >8 (Ghosh and Teoh, 1985).

3.18 Kinetics

The chemical reaction step in adsorption of As(III) and As(V) by metal oxyhydroxides has been reported to be rapid, with greater than 90% adsorption occurring within a few hours (Anderson et al., 1976; Darland and Inskeep, 1997b; Elkhatib et al., 1984b; Fuller et al., 1993; Grossl and Sparks, 1995; Singh et al., 1988; Van der Hoek et al., 1994; Xu et al., 1988). This initial rapid adsorption step is followed by a slower approach to equilibrium which may take up to several days. The slow rate of approach to equilibrium suggests that diffusion-controlled mass transfer of As to adsorption sites located in the internal porosity of minerals and mineral aggregates may be the rate limiting step (Fuller et al., 1993; Willett et al., 1988). Experimental results from Fuller and Davis (1989) support this hypothesis. They measured As(V) adsorption as ferrihydrite precipitated from solution, ensuring that As(V) was in contact with surface sites as they formed. Equilibrium was attained within one hour. Waychunas et al. (1993) used EXAFS to study the time dependence of As(V) adsorption by ferrihydrite. They observed no change in the bonding mechanism with time and found no evidence for formation of an As(V) mineral phase.

Speciation and pH also affect As adsorption rates. Arsenate adsorption was generally faster than adsorption of As(III), especially at lower pH (Ghosh and Teoh, 1985; Pierce and Moore, 1982). At a higher pH of 9, the rate of As(V) adsorption was slower and similar to As(III) (Ghosh and Teoh, 1985; Ghosh and Yuan, 1987; Raven et al., 1998).

Breakthrough curves from column experiments have been used to provide evidence for diffusion of As to adsorption sites as a rate-controlling mechanism. Darland and Inskeep (1997b) found that adsorption rate constants for As(V) determined under batch conditions were smaller than those necessary to model breakthrough curves for As(V) from columns packed with iron oxide coated sand; the rate constants needed to model the breakthrough curves increased with pore water velocity. For example, at the slowest velocity of 1 cm/h, the batch condition rate constant was 4 times smaller than the rate constant needed to model As adsorption in the column experiment. For a velocity of 90 cm/h, the batch rate constant was 35 times smaller. These results are consistent with adsorption limited by diffusion of As(V) from the flowing phase to sites within mineral aggregates. Puls and Powell (1992) also measured more retardation and smaller rate constants for As(V) at slower flow velocities where there was sufficient time for diffusion to adsorption sites.

3.19 Reversibility

Desorption of strongly-bound ions often takes longer to attain equilibrium than adsorption. Several factors may contribute to adsorption/desorption hysteresis including differences in the aqueous composition of the adsorbing and desorbing solution, diffusion, time, and the nature of the adsorption bond.

Rates of As desorption are affected by pH. Darland and Inskeep (1997a) conducted column experiments using sand that contained Fe oxide coatings. The pH during both the adsorption and desorption parts of these experiments was maintained at a constant value of 4.5, 6.5, or 8.5. One pore volume of water containing 133 μM As(V) was eluted through the column followed by about 10 pore volumes of As(V)-free water. Arsenate adsorption was lowest at pH 8.5; the As(V) pulse was only retarded by 1 pore volume and the peak concentration was 110 μM. Elution of As(V)-free water at pH 8.5 desorbed more than 95% of the added As(V) within 10 pore volumes. Arsenate adsorption was greater at pH 4.5 and 6.5; As(V) broke through about 1 pore volume later than in the pH 8.5 experiment, and peak concentrations were only about 31 μM. Recovery of As(V) after 10 pore volumes was 35% at pH 4.5 and 40% at pH 6.5.

Fuller et al. (1993) measured desorption rates for As(V) from ferrihydrite, aged for 24 h, as a result of increasing pH. Arsenate was first equilibrated with ferrihydrite for 144 h at pH 8.0 The molar ratio of As(V) adsorbed to Fe in ferrihydrite was 0.10 (Fig. 7). Desorption was initiated by rapidly increasing the pH to 9.0. Within a few hours, the molar ratio of As(V)/Fe had decreased to about 0.08. The rate of desorption then slowed as the rate became limited by diffusion of As(V) from pores within mineral aggregates. Within 96 h, the concentration of As(V) still adsorbed was only about 5% greater than the adsorbed concentration of As(V) determined in a separate adsorption experiment at pH 9.0. Similar desorption behavior for As(V) from ferrihydrite were measured by Fuller and Davis (1989).

Addition of a strongly competing ion can enhance desorption of As (Darland and Inskeep, 1997b). In a column experiment at pH 4.5, a one pore volume pulse of 143 μM As(V) was added to a column packed with sand, followed by elution with 20 pore volumes of As(V)-free water. About 20% of the As(V) was recovered. The column was then continuously eluted with 1420 μM P(V), which resulted in an additional 40% recovery of As(V).

Lin and Puls (2000) investigated the effects of aging on desorption of As(III) and As(V) by clay minerals. Solutions containing either As(III) or As(V) were equilibrated at pH 5.5 for 24 h, centrifuged and filtered. For desorption experiments, the clays were either unaged or aged for 30 and 75 days. Desorption was accomplished by adding a solution containing 1 mM KH_2PO_4/K_2HPO_4 at pH 7 to the clay for 10 h. Arsenic desorption varied with the type of clay mineral, As species, and in most cases time of aging. Only a few examples of the results are given here. For halloysite and kaolinite, desorption of As(III) was almost 100% for the unaged experiments. After aging for 75 days, only about 15% As(III) was desorbed from halloysite, and about 30% from kaolinite. Desorption of As(V) from halloysite decreased from 85% in the unaged experiment to 65% after 75 days of aging. For kaolinite, As(V) desorption decreased from about 98% in the unaged experiment to 70% after 75 days of aging.

Initial As concentration was also a factor in the amount desorbed. Puls and Powell (1992) measured desorption of As(V) from Fe_2O_3 in batch experiments by repeated mixing with As-free solutions. A pH of 7 was used during both the adsorption and desorption experiments. They found that the percent desorbed was directly proportional to the initial As(V) concentration. For an initial As(V) concentration of 2.7 μM, about 16 μM/g was adsorbed and only 2% of this was desorbed. For an initial concentration of 40.8 μM, about 82 μM/g was adsorbed and 20% was desorbed. Results indicated a decline in the energy of adsorption as surfaces became increasingly saturated with As(V).

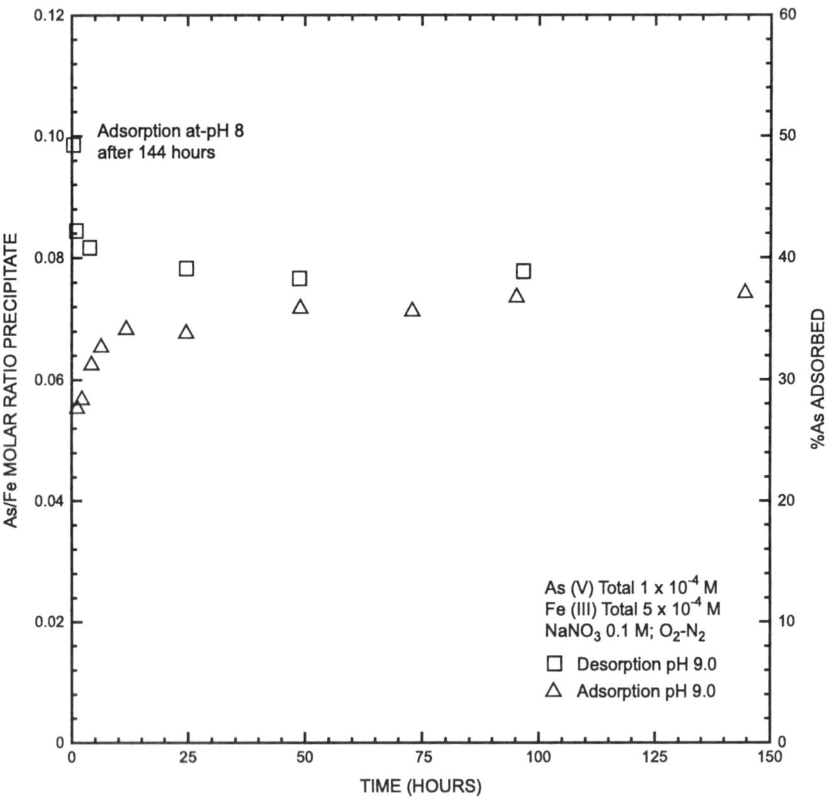

Figure 7. Rate of As(V) desorption from ferrihydrite upon pH increase to 9.0, after adsorption at pH 8.0 for 144 h. Ferrihydrite precipitated and aged at pH 8.0 for 24 h prior to As(V) addition. Rate of As(V) adsorption at pH 9.0 shown for comparison. Reprinted from Fuller et al. (1993). with permission from Elsevier Science.

The type of bond formed between As and the solid surface has also been shown to be important for desorption. Grossl and Sparks (1995) attributed the slow desorption of As(V) from the goethite surface to a two-step reaction process. Using pressure-jump relaxation techniques, they concluded that in the initial adsorption step, a monodente surface complex was rapidly formed by ligand exchange with OH groups. This was followed by a slower, second ligand exchange reaction resulting in formation of a bidente complex. Thus, the adsorption rate constant reflected the rapid monodente step, whereas the desorption rate constant was controlled by the slower breaking of the bidente bond. Formation of a binuclear inner sphere complex, and to a lesser extent, a monodente inner sphere complex between As(V) and the ferrihydrite surface has also been confirmed by EXAFS (Waychunas et al., 1993).

The hysteresis often observed in the adsorption and desorption of As has important implications for remediation strategies designed to decrease As concentrations to levels that meet drinking water supply criteria. The initial efficiency may be high as the readily desorbable fraction is removed from solution; however, efficiencies may decrease significantly if As adsorption is not readily reversible. Changing the chemistry of the desorbing solution can facilitate or inhibit As desorption.

4. ADSORPTION MODELS

Two general types of models have been developed to describe adsorption/desorption reactions at mineral surfaces: (1) empirical models that are based on the partioning relationships of a solute between the aqueous and solid phase, and (2) conceptual models for surface complexation that treat adsorption reactions similar to ion association reactions in solution. All of these models assume that adsorption reactions are at equilibrium.

4.1 Empirical Adsorption Models

The most widely used empirical adsorption models include the Distribution Coefficient:

$$\overline{C} = K_d C \tag{8}$$

the Langmuir Isotherm:

$$\overline{C} = K_L S C/(1 + K_L C) \tag{9}$$

and the Freundlich Isotherm:

$$\overline{C} = K_F C^a \tag{10}$$

where; $\overline{C}$ is the adsorbed concentration, C is the aqueous concentration at equilibrium with the solid phase, S is the adsorption site concentration, K_d is the distribution coefficient, K_L is the Langmuir equilibrium constant, K_F is the Freundlich equilibrium constant, and "a" is the Freundlich exponent (Zheng and Bennett, 1995).

Equilibrium constants for empirical models are determined from measurements of the activity of a solute in water and the equilibrium adsorbed concentration. Experiments usually consist of a series of batch reactors where a solid phase is suspended in solutions that have a range of known solute concentrations. Equilibrium constants are then derived from various types of plots of the adsorbed and aqueous concentrations. These equilibrium constants are highly dependent on solution and solid composition. For example, values of K_L and K_F for As adsorption have been shown to be a function of pH, the concentration of competing ions, and the mineralogy of the adsorbent (Darland and Inskeep, 1997a; Ghosh and Yuan, 1987; Hingston et al., 1971; Hsia et al., 1992; Pierce and Moore, 1982; Sakata, 1987). Therefore, empirical models are limited to the specific experimental conditions used to determine the Log K's.

Modifications to the Langmuir and Freundlich isotherm equations have been made to extend their flexibility for accomodating the effects on As adsorption of pH variability (Anderson et al., 1976) and ionic competition (Hingston et al., 1971; Roy et al., 1986a). However, even these modified equations are limited in their ability to simulate As adsorption in complex natural environments.

4.2 Surface Complexation Models

Surface complexation models (SCM's) provide a rational interpretation of the physical and chemical processes of adsorption and are able to simulate adsorption in complex geochemical systems. Chemical reactions at the solid-solution interface are treated as surface complexation reactions analogous to the formation of complexes in solution. Each reaction is defined in terms of a mass action equation and an equilibrium constant. The activities of adsorbing ions are modified by a coulombic term to account for the energy required to penetrate the electrostatic-potential field extending away from the surface. Detailed information on surface complexation theory and the models that have been developed, can be found in (Stumm et al., 1976;

1980), Westall and Hohl (1980), Davis and Kent (1990), Dzombak and Morel (1990) and Stumm (1992). A brief overview is presented here.

Several SCM's have been described in the literature. The more commonly used models include the Constant Capacitance Model (Schindler and Stumm, 1987), the Diffuse Double Layer Model (Stumm et al., 1970) and the Triple Layer Model (Davis et al., 1978; Yates et al., 1974). All are based on electric double layer theory but differ in their geometric description of the oxide-water interface and the treatment of the electrostatic interactions.

Application of SCM's to describe adsorption requires accurate solution speciation of all elements that might have an effect on the activity of the adsorbing solute of interest. Incorporation of SCM's in geochemical equilibrium models such as MINTEQA2 (Allison et al., 1991) and PHREEQC (Parkhurst and Appelo, 1999) have greatly facilitated such calculations.

4.3 Model parameters

Methods for measurement of parameters used in SCM's have been described in the literature. Only a brief summary is presented here. Surface complexation model parameters that can be measured directly include, (1) the solid concentration, (2) surface site density, (3) surface area, and (4) equilibrium constants for the mass action equations describing all relevant adsorption reactions. The relation between surface charge and potential is calculated in geochemical equilibrium models.

Measurement of solid concentration is straightforward given esimates of aquifer porosity and the density of solids. Surface area for these solids can be measured by several techniques (Dzombak and Morel, 1990). All have their advantages and disadvantages. The BET, N_2 gas method is one of the most widely used techniques (see for example Coston et al., 1995).

Reasonable estimates of surface site density are important for accurate modeling of solute adsorption (Davis and Kent, 1990). The most common methods for estimating surface site density include measurement of adsorption maxima of protons, cations, or anions at pH values where adsorption of the specific solute is greatest. Most measurements of surface site density for Fe, Al, and Mn oxides range from 1 to 7 μmoles/m^2 (Davis and Kent, 1990). Dzombak and Morel (1990) reviewed the adsorption literature for ferrihydrite and selected a surface site density of 3.84 μmoles/m^2 in constructing a coherent thermodynamic database for adsorption of cations and anions by ferrihydrite.

Natural systems contain complex mixtures of minerals and it is difficult to quantify the concentration of surface sites for individual minerals. For consistency in applying the surface complexation modeling approach to natural systems, Davis and Kent (1990) recommended that equilibrium constants for strongly binding solutes be derived using the site density from Dzombak and Morel (1990) of 3.84 µmoles/m^2. This appears to be a reasonable choice for systems where adsorption is dominated by metal oxides. Stollenwerk (1995) measured the surface site density of aquifer solids from a sand and gravel outwash aquifer by titration of the solids with H^+ to a pH of 3.5. The average surface site density for 51 core samples was 3.3 µmoles/m^2 (range 1.6-4.2 µmoles/m^2). The predominate sorbents were hydrous Fe and Al oxide coatings on quartz and feldspar (Coston et al., 1995; Fuller et al., 1996).

Protonation/deprotonation reactions are included in SCM's to describe the amphoteric behavior of oxide surfaces:

$$SOH + H^+ = SOH_2^+ \tag{11}$$

$$SOH = SO^- + H^+ \tag{12}$$

Accurate measurement of surface acidity constants for reactions 11 and 12 can be difficult (Davis and Kent, 1990; Dzombak and Morel, 1990), especially for natural systems (Stollenwerk, 1995). Acidity constants are often derived from acid-base titration data (Parks and de Bruyn, 1962), and constants for several metal oxides have been published. These published values have been used as initial estimates for modeling adsorption in natural systems with complex mineralogy. The acidity constants have then been optimized simultaneously with equilibrium constants for other solutes to give the best fit to experimental data (Goldberg and Glaubig, 1988b; Kent et al., 1995; Kent et al., 2001).

Quantifying adsorption of As requires mass action equations such as those listed in an earlier section of this chapter (reactions 1-6). Multiple surface species are needed to describe the pH dependent adsorption of As. For example, the surface species $SH_2AsO_4^o$, $SHAsO_4^-$, and $SAsO_4^{-2}$ (where S = surface) are applicable to As(V) adsorption over the pH range of groundwater (Hsia et al., 1992; Manning and Goldberg, 1996b). Comparable surface species that have been used for adsorption of As(III) are $SH_2AsO_3^o$, $SHAsO_3^-$, and $SAsO_3^{-2}$ (Manning and Goldberg, 1997a). Equilibrium constants for these reactions are generally fit to data from batch reactor experiments similar to those used to quantify empirical models. It is important that the Log K's for these reactions be calibrated for the range of As concentrations expected in a particular study area.

Because the various SCM's have different formulations for treating adsorption reactions and the electrostatic terms, parameters fit to one model may not be applicable to other models (Morel et al., 1981). For example, Gao and Mucci (2001) determined different Log K's for As(V) adsorption by goethite when the data were fit to the Constant Capacitance Model, the Basic Stern Model, and the Triple Layer Model.

The presence of solutes that compete for adsorption sites with As should also be included in SCM's. For example, the competitive effects of the pH dependent adsorption of P(V) can be modeled using:

$$SOH + H_3PO_4^{o} = SH_2PO_4^{o} + H_2O \qquad (13)$$

$$SOH + H_2PO_4^{-} = SHPO_4^{-} + H_2O \qquad (14)$$

$$SOH + HPO_4^{-2} = SPO_4^{-2} + H_2O \qquad (15)$$

(Manning and Goldberg, 1996b). Similar reactions can be written for as many solutes as needed; however, mass action equations for each additional solute requires a corresponding increase in the number of laboratory experiments needed to quantify adsorption.

Deriving equilibrium constants for surface complexation reactions based on experimental data can be facilitated with paramater optimization programs such as FITEQL (Westall, 1982). Ideally, surface site density is known; however, it can also be adjusted simultaneously with LogK's to obtain the best match with experimental data. Parameter estimation procedures have inherent limitations due to the covarience of estimated parameters (Hayes et al., 1990). For example, Log K's and surface site density can be adjusted simultaneously to achieve the best fit to experimental adsorption data. However, changing the site density necessitates a new set of Log K's.

Surface complexation models have been successfully used to describe the pH dependent adsorption of As(III) and As(V) by several minerals including ferrihydrite, goethite, amorphous As(OH)3, gibbsite, kaolinite, montmorillonite, and illite (Goldberg and Glaubig, 1988b; Hsia et al., 1992; Manning et al., 1998; Manning and Goldberg, 1997a). Because of differences in adsorption properties between these oxides, different Log K's were determined for each mineral.

Use of Log K's derived from experimental data for single anion systems have met with mixed success in describing the effects of anion competition on As adsorption. Swedlund and Webster (1999) were able to model competitive adsorption between H_4SiO_4 and As(III) or As(V) using Log K's derived separately for each solute. Equilibrium constants for As(V) and P(V) adsorption by goethite and gibbsite derived from single anion experimental data were able to match As(V) adsorption reasonable well for a range of P(V) concentrations in binary adsorption experiments (Hiemstra and Van Riemsdijk, 1999; Manning and Goldberg, 1996b). However, in other studies single anion Log K's had only limited success in modeling competitive adsorption of As(V) and P(V) by kaolinite, montmorillonite, and illite (Manning and Goldberg, 1996a), and goethite (Gao and Mucci, 2001).

Other studies have shown that SCM's were more successful in modeling competitive adsorption when the Log K's were determined for binary systems. For example, determining Log K's for As from experiments in the presence of P (Goldberg, 1986).

Accurate predictions of the transport of As in groundwater requires site specific data to model adsorption/desorption reactions. In complex mixtures of minerals, it may not be possible to quantify the adsorption properties of individual minerals. Therefore, it has been suggested that adsorption properties of composite materials should be characterized as a whole (Davis and Kent, 1990). Previously published data for adsorption by pure mineral phases such as the surface complexation database for adsorption by ferrihydrite (Dzombak and Morel, 1990) can be a useful starting point for modeling adsorption of solutes in groundwater; however, these equilibrium constants may not reflect the adsorption properties of composite oxide coatings on aquifer solids. For example, incorporation of Si, and to a lesser extent, Al into Fe oxyhydroxides has been shown to decrease adsorption reactivity towards anions (Ainsworth et al., 1989; Anderson and Benjamin, 1990; Anderson et al., 1985). Therefore, equilibrium constants will likely need to be modified for site-specific studies.

Results from two field investigations illustrate this point. Kent et al. (1995) modeled adsorption of Cr(VI) in the presence of S(VI) by natural aquifer solids using equilibrium constants that were 1.75 Log K units lower than those used to model adsorption by ferrihydrite (Dzombak and Morel, 1990). To model competitive adsorption of Mo(VI) and P(V) by these same aquifer solids, Stollenwerk (1995) decreased the Dzombak and Morel (1990) equilibrium constants for both anions by 1.2 Log K units. Results from both of these studies indicate that the dominant adsorbents had lower affinities for adsorption of anions than pure ferrihydrite.

5. MODELING TRANSPORT OF ARSENIC

Predictions of the spatial and temporal distribution of As in groundwater are achieved by coupling adsorption models with solute transport models.

Empirical adsorption models have been widely used to model transport of reactive solutes in groundwater. The main advantage of these models is their mathematical simplicity which readily allows incorporation in solute transport models (Grove and Stollenwerk, 1987). However, these models are limited in their ability to describe As adsorption in complex geochemical systems.

A more recent approach has been the coupling of geochemical speciation models with advective-dispersive solute-transport models. The chemical reactions for each cell in the model are solved separately by the geochemical model. Species are then transported one time step by the transport model. This process is repeated sequentially until the end of the simulation. Several such codes have been developed. These include, but are not limited to, TRANQL (Cederberg et al., 1985), HYDROGEOCHEM (Yeh and Tripathi, 1991), MST1D (Engesgaard and Kipp, 1992), PHAST (Kipp and Parkhurst, in press; Parkhurst et al., 1995), MOC PHREEQC (Engesgaard and Traberg, 1996), and PHREEQC (Parkhurst and Appelo, 1999). This coupled approach has been successfully applied to groundwater investigations (Keating and Bahr, 1998; Stollenwerk, 1996; Stollenwerk, 1998).

6. SUMMARY

Adsorption of As by metal oxyhydroxides and clay minerals is likely to be a major control on the transport of As in groundwater. Studies have shown that adsorption of As is a complex function of the interrelationships between the properties of the solid surface, pH, the concentration of As and competing ions, and As speciation.

Arsenic is generally present as As(V) in oxidizing environments and as As(III) in moderately reducing environments. However, the rate of As(III) oxidation can be relatively slow in the presence of oxygen and As(III) can coexist with As(V). The rate of As(III) oxidation is greatly increased by the presence of other oxidants such as manganese oxides.

The extent of As adsorption and desorption is largely controlled by pH; As(III) and As(V) respond differently to changes in pH. Arsenate adsorbs over the pH range found in most groundwaters; however, adsorption is much stronger at lower pH values. As pH increases above about 7, the increasing negative charge of the solid surface and the aqueous As(V) species results in significantly less adsorption. Arsenite also adsorbs over a large range in pH. In contrast to As(V), As(III) adsorption increases with pH, attaining a maximum at about pH 8 or 9. As a result of this contrasting behavior, As(III) is generally more mobile than As(V) in acidic environments and less mobile in alkaline environments.

Many of the solutes commonly present in groundwater can influence adsorption of As. Anions compete directly with As for surface complexation sites, the stronger the anion adsorption the greater the competition. Adsorption of cations can increase adsorption of As by increasing the surface charge of a solid.

The kinetics of As adsorption is rapid, with greater than 90% adsorption occurring within a few hours. However, several days may be required for complete equilibrium because of diffusion-controlled mass transfer of As through water in the internal porosity of minerals and mineral aggregates. Some studies have shown that As adsorption is reversible under comparable adsorption and desorption conditions; however, other studies have demonstrated hysteresis leading to incomplete desorption. Desorption can be enhanced by altering pH or the addition of other strongly adsorbing solutes. This observation has important implications for remediation strategies designed to decrease As concentrations to levels that meet drinking water supply criteria.

Surface complexation models are the most versatile for modeling As adsorption for a wide range of geochemical conditions. Coupling of these models with solute transport models provides a powerful tool for predicting the spatial and temporal distribution of As in groundwater.

ACKNOWLEDGEMENTS

The author would like to thank Chris Fuller and Don Thorstenson for reading and providing a constructive review of this manuscript.

Chapter 4

Geothermal Arsenic

The source, transport and fate of arsenic in geothermal systems

Jenny G. Webster[1] and D. Kirk Nordstrom[2]
[1]*University of Auckland, Auckland, New Zealand*
[2]*US Geological Survey, Boulder, CO, USA*

The release of arsenic from geothermal systems into surface and ground waters compromises the use of these waters as drinking water resources. In surface waters, As contamination can also adversely affect aquatic ecosystems, accumulating in sediments and plants. This review examines the source of arsenic in geothermal areas, its transport and speciation in geothermal fluids and receiving waters, as well as the deposition and removal mechanisms occurring in both natural environments and waste or water treatment systems. The effect of microorganisms on As mobility, and the opportunities that exist for further research in this field, are discussed. The review focuses on two geothermally active regions which have been intensively studied: Yellowstone National Park in the USA, and the Taupo Volcanic Zone in New Zealand, and their associated catchments.

1. INTRODUCTION

Arsenic (As) is a ubiquitous component of active and fossil geothermal systems. It occurs together with other environmental contaminants such as mercury (Hg), antimony (Sb), selenium (Se), thallium (Tl), boron (B), lithium (Li), fluoride (F) and hydrogen sulfide (H_2S). These elements are now recognised as being a typical "geothermal suite" of contaminants (Webster-Brown, 2000). High temperature geothermal systems occur throughout the world, generally in one of three tectonic settings. One such setting is on or close to tectonic plate boundaries. The edge of the Pacific Plate, for example, is defined by geothermal fields in New Zealand, Papua New Guinea, Philippines, Indonesia, Japan, Kamchatka, Alaska, western USA, Mexico, central America and Chile. Another setting is the Earth's "hot spots", where local magma chambers rise from near the mantle to shallow depths in the earth's crust, such as in Hawaii, Yellowstone and the Azores.

Although surface expression of recent volcanism is often associated with geothermal activity, it is not a prerequisite. The Larderello (Italy) and the Geysers (USA) geothermal steam fields, for example, are associated with instrusive activity in areas of fissured sedimentary and metamorphic rocks (Ellis and Mahon, 1967). The third setting is in rift zones where the tectonic plates diverge, such as the Gregory Rift Valley in Ethiopia, Kenya and Tanzania, the Rio Grande Rift valley in Colorado and New Mexico, and in the rift system in Iceland. Ellis and Mahon (1977) provide a more comprehensive description of geothermal settings.

Many geothermal fields have been developed, or are targeted for development, to generate energy from the steam and hot water reservoirs beneath the earth's surface.

1.1 Geothermal arsenic toxicity

The acute and chronic toxic effects of As are well documented. Recently As has also been declared a human carcinogen, contributing to a high incidence of skin and other cancers in populations exposed to high levels of As in drinking water (WHO, 1993). The WHO drinking water guideline for As was lowered from 0.05 mg/L to 0.01 mg/L in 1993, and the new value has since been adopted by many countries as a drinking water standard. In January 2001 the USEPA lowered the USA drinking water limit from 0.05 to 0.01 mg/L (Federal Register, 2001). However, this new standard was challenged by the new federal administration and further review was mandated. The results from three new scientific reports conclude that the new standard should be less than 0.01 mg/L.

Arsenic concentrations in natural surface drainage systems frequently exceed 0.01 mg/L in areas of geothermal activity. The Madison River, which drains part of Yellowstone National Park (Fig. 1), for example, contains ~ 2mg/L As near West Yellowstone (Thomson, 1979). Symptoms of chronic As poisoning such as skin lesions and high As concentrations in hair and nails have been reported from geothermal areas (Chen et al., 1999), but this may not always be as a direct consequence of drinking water contamination. At the Mt. Apo geothermal field in the Philippines, for example, the two rivers draining the field carry elevated As concentrations due to hot spring activity in the river beds. During development of the field, alternative clean drinking water supplies were provided and used by local residents, but the symptoms of high As exposure appeared to persist (Webster, 1999). The accumulation of As in edible aquatic plants is likely to have been to blame, as high levels of As have been reported in aquatic weeds in other river systems receiving geothermal fluids (e.g., Reay, 1972).

This chapter reviews what is currently known of the source and mobility of As in geothermal systems, as well as its fate and toxicity in the near-surface environment. The discussion focuses on two of the most thoroughly studied geothermally active regions: Yellowstone National Park in the US, which is mainly drained by the Madison and Yellowstone Rivers, and the Taupo Volcanic Zone in New Zealand, which is drained by the Waikato River (Fig. 1).

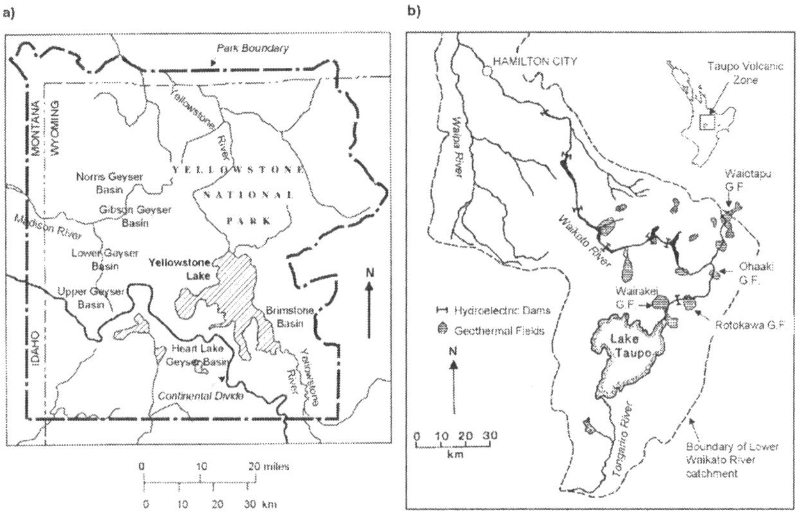

Figure 1. a) Yellowstone National Park (USA) and b) Taupo Volcanic Zone (NZ) and their surface drainage systems.

2. THE SOURCE OF ARSENIC IN GEOTHERMAL SYSTEMS

2.1 The source and nature of geothermal fluids

Geothermally active areas occur where an unusually high geothermal gradient allows hot water or steam to emerge at the earth's surface. The temperature of geothermal fluids may be elevated by only a few degrees, or by hundreds of degrees, above ambient (e.g., Ellis, 1967; White et al., 1971). Heat sources may be related to volcanic or magmatic activity, metamorphism, faulting and radioactivity.

Although geothermal fluids were initially thought to be of magmatic origin, isotopic analyses over the last 50 yrs (e.g., Craig et al., 1956; Giggenbach, 1971) have indicated that they are most commonly derived from meteoric waters. Deuterium-hydrogen and oxygen-isotope ratios are typically similar to those of local surface waters, although a minor contribution from magmatic fluids is a possibility at some fields (e.g., White, 1974). Irrespective of the heat source, deeply circulating ground waters are conductively heated in the crust. The hot fluids are of lower density than surrounding waters, and rise through the host rock, to complete the circulation system (Fig. 2).

The chemical composition of this fluid, which effectively constitutes the "geothermal reservoir", is a function of fluid temperature and host rock composition. As a solvent, hot water has a high dielectric constant, enabling it to dissolve ionic crystalline minerals in the host rock, forming new hydrated minerals, and transporting solutes away from the reaction site. A series of high temperature water-rock interaction experiments were undertaken in the 1960s to characterise the interactions between thermal waters and a variety of rock types including basalts, andesites, rhyolites and greywackes. The experiments were conducted at temperatures of 150 – 600°C and at various pressures (Ellis and Mahon, 1977; Ellis and Mahon, 1964) and confirmed that Cl, B and F could be rapidly leached into solution, even before significant alteration of the host rock had occurred. Other elements such as Li, Rb and Cs, which had previously been assumed to derive *only* from magmatic fluids, were also released into the hot water leachate during mineral alteration.

Such experiments under conditions of controlled gas fugacities and/or leachate chemistry, confirmed the importance of hot water-mineral equilibria in determining geothermal fluid compositions. Together with detailed analysis of mineral alteration assemblages in geothermal host rocks, they led to the development and refinement of a variety of geothermometers. The most applicable of these are the Na/K, Na/K/Ca and SiO_2 geothermometers, which are based on the temperature dependent solution equilibria with feldspars and micas, and with quartz respectively (e.g., Fournier and Potter, 1982; Fournier and Truesdell, 1973; Giggenbach, 1988). The concentrations of dissolved CO_2 and H_2S are also a function of mineral solubility, controlled mainly by equilibria with carbonate and sulfide minerals.

As geothermal fluids rise through the crust, the pressure decrease allows the single-phase fluid to separate into two phases: steam and water. This "boiling" process usually occurs at a shallow depth, but can occur wherever there is a sudden decrease in pressure due to rock fracture or fissure. The surface features in areas of geothermal activity can be grouped into two main

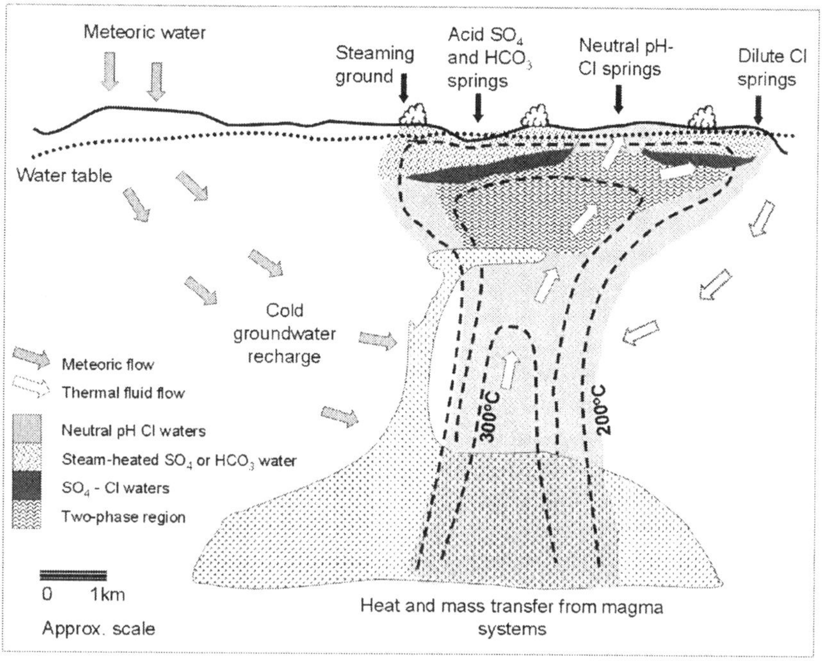

Figure 2. Generalized structure of a geothermal system (modified from Henley, 1985a, b).

types, on the basis of their relationship to the rising geothermal fluids (Fig. 2).

Hot water springs, rich in chloride and silica and with a near-neutral pH, represent direct discharge of the hot water phase. Steam vents and features such as acidic, sulfate-rich springs are formed by the interaction of the steam phase with shallow aquifer waters, which leads to precipitation and sublimation of elemental sulfur, and subsequent microbial oxidation to sulfuric acid. The intense argillic alteration of the host rock associated with these more corrosive fluids can lead to an unstable ground surface, and the formation of bubbling "mudpools" where alteration has rendered the host rock down to a slurry of clay minerals (e.g., Shoen et al., 1974). Surface features of intermediate chemistry can also occur. Geysers and carbonate-rich springs are also hot water discharges, for example, but include steam-heating and steam- phase mixing respectively.

Development of a geothermal field for power generations tends to increase the rate and volume of geothermal fluids reaching the surface. The reservoir fluid is intercepted by drilling and brought to the surface under pressure where it is "flashed" to a desirable temperature to generate the steam to run steam turbines. The extent of steam and water separation (boiling) can be artificially manipulated to maximise plant efficiency. The water phase formed at this separation becomes a waste product, although further heat can sometimes be extracted. This wastewater often has higher contaminant concentrations than analogous natural hot spring water because the processes that remove or immobilise contaminants in natural geothermal features, such as the precipitation of mineral-rich sinters, have been bypassed. Disposal of these wastewaters can be problematic. At most modern geothermal power stations, they are re-injected into the field. However, at some older fields such as the Wairakei Geothermal Field in New Zealand (Fig. 1b), these waters are still discharged into surface drainage systems.

2.2 The source of As in geothermal fluids

The presence of As in geothermal fluids has been known since the mid-19th century. Lefort reported traces of As in acid waters in the crater of Popocatepetl volcano in Mexico in 1863 (Clarke, 1924). The first measurements of As in hot springs and geysers of Yellowstone National Park were reported by Gooch and Whitfield (1888). Yellowstone National Park constitutes one of the largest geothermal systems in the world and has been intensely studied, particularly over the last 40 years. Arsenic concentrations in the thermal features of this field generally range from <0.1 mg/kg to 10 mg/kg, a range that is observed in most active geothermal fields (Table 1). Notable exceptions include the geothermal fields in Iceland and Hawaii where As concentrations of <0.10 mg/kg are consistently reported (Arnorsson and Linvall, 2001; Gudmundsson and Arnorsson, in press; D. Thomas, pers. comm.). These low As levels are attributed to the prevalence of fresh basaltic host rocks, low in arsenic, in the reservoirs of the respective geothermal fields. At the other extreme, As concentrations of >20 mg/kg are not uncommon in geothermal well fluids (Table 1). Geothermal fluid As concentrations are therefore usually three orders of magnitude greater than those in uncontaminated surface and ground waters.

2.2.1 Arsenic from host rock leaching

In the hot water-rock leaching experiments of Ellis and Mahon (1964; 1967), appreciable quantities of As were leached from unmineralised andesite yielding As concentrations of 1.3 mg/kg in the leachate. Subsequent experimental leaching of greywacke with hot water also indicated significant leaching of As, together with Sb, Se and S, at temperatures less than those in many geothermal fields (Ewers and Keays, 1977). On this basis, it can be assumed that neither magmatic fluid input, nor

Table 1. Arsenic (mg/kg) in hot springs and in production or exploration well brines at various geothermal fields.

Field	As, mg/kg	Reference
Hot springs:		
Yellowstone Nat. Park, US	0.16 – 10	Stauffer and Thomson (1984); Unpublished data, USGS
Wairakei, NZ	0.23 - 3.0	Ritchie (1961)
Waiotapu, NZ	0.71 - 6.5	Webster (1990)
Ohaaki/Broadlands, NZ	1.0	Ellis and Mahon(1977)
Valles Caldera, New Mexico	0.021 - 2.4	Criaud and Fouillac (1989)
Los Azufres, Mexico	0.001 – 4.0	Birkle andMerkel (2000)
Mt Apo, Philippines	3.1 - 6.2	Webster (1999)
Tamagawa, Japan	2.3 - 2.6	Noguchi and Nakagawa (1969
Japan (national survey)	0.001 - 9.5	Sakamoto et al. (1988)
Salton Sea, USA	0.03 - 12	White (1968)
Kamchatka, Russia	2.0 - 3.6	Karpov and Naboko (1990)
Phlegraean Fields, Italy	0.012 - 5.6	Celico et al. (1992)
Iceland	0.001 - 0.048	Arnorsson and Linvall (2001)
Well fluids:		
Wairakei, NZ	1.0 - 5.2	Ritche (1961)
Waiotapu, NZ	2.1 - 3.9	Ellis and Mahon (1977)
Ohaaki/Broadlands, NZ	5.7 -9.0	Ewers and Keays (1977)
Tongonan, Philippines	28 (mean)	Darby (1980)
Hawaii	<0.01 – 0.07	D. Thomas (pers. comm.)
Los Azufres, Mexico	5.1 - 24	Birkle and Merkel (2000)
Lassen Nat Park, US	2 -19	Thompson (1985)
El Tatio, Chile	45 - 50	Ellis and Mahon (1977)
Kamchatka/Kuriles	2 - 30	Goleva (1974)
Ebeko volcano, Siberia	0.19-28	Khramova (1974)

As mineralisation at depth, is a prerequisite for high As concentrations in geothermal fluids. Most reservoir fluids are undersaturated with respect to arsenopyrite and other arsenic minerals (Ballantyne and Moore, 1988) and hence As leaching, rather than As precipitation, is predicted to occur in the reservoir.

Early analyses of As in hot springs of the Taupo Volcanic Zone in New Zealand by Ritchie (1961) revealed a positive correlation between As and Cl which has since been confirmed for most geothermal fields. However, the interpretation of trends in the As/Cl ratio requires caution (Ballantyne and Moore, 1988), as the association is more a function of common behaviour, than of a common source or direct chemical association. The source of the Cl ion may be gaseous HCl associated with magma intrusion, host rock leaching or seawater, whereas As is derived mainly by host rock leaching. However, both elements remain in the fluid phase during sub-surface boiling and phase separation. Consequently, As concentrations are invariably higher in the neutral pH-chloride hot springs than in the acid-sulfate features. Likewise during the separation of steam and bore water in geothermal power generation, As is retained in the waste bore water. This high As content is one of the main problems in the disposal of this water.

The As/Cl ratio can be used as a tool to trace the effects of geothermal fluid dilution, concentration or mixing within a geothermal field. In Yellowstone National Park, for example, Stauffer and Thompson (1984) showed that the ratio could be used to identify simple mixing trends in different geyser basins. Many thermal waters mix with cool meteoric ground waters on their way to the surface and show linear trends of As vs Cl with a dilute end-member having As and Cl concentrations of ~0. A more recent study of As vs Cl concentrations in Yellowstone National Park thermal waters is shown in Figure 3a-c (Nordstrom et al., 2001). Linear mixing trends were identified in at least two pools; Cinder Pool and Cistern Spring in Norris Geyser Basin (Fig. 3a). Cistern Spring was recognised by Fournier et al. (1986; 1992) as a spring that was very sensitive to mixing. The mixing behaviour is caused by changes in subsurface flow patterns following climatic or seasonal changes in weather.

Other thermal waters from around Norris and Upper Geyser Basin tend to lie along the mixing trends defined by extensions of the Cinder Pool and Cistern Spring mixing lines (Fig. 3b). This suggests that the solutes had a similar source but that the Norris waters had undergone adiabatic cooling (boiling) which effectively increased the solute concentrations along sub-horizontal flow paths. Fournier et al. (1986) estimated that the Cl content of the deep thermal reservoir(s) at Yellowstone is about 400 mg/L, so that concentrations greater than this must reflect a concentration mechanism such as boiling. Other conservative solutes also follow this trend. These waters have subsequently mixed with dilute shallow ground waters as they made their way to the surface.

A more general survey of Yellowstone thermal waters (Fig. 3c) shows that many springs are enriched in As relative to Cl. Stauffer and Thompson (1984) showed a direct relationship between high CO_2 concentrations and a high As/Cl ratio. A fluid with high concentrations of oxyanions at depth, especially carbonate ions, will tend to react more rapidly with the host rocks, hydrolysing and solubilising other oxyanions such as arsenite. Nordstrom et al. (2001) also provide evidence for localized enrichment of As relative to Cl through accumulation of arsenic in precipitates near the surface or through dynamic geyser activity.

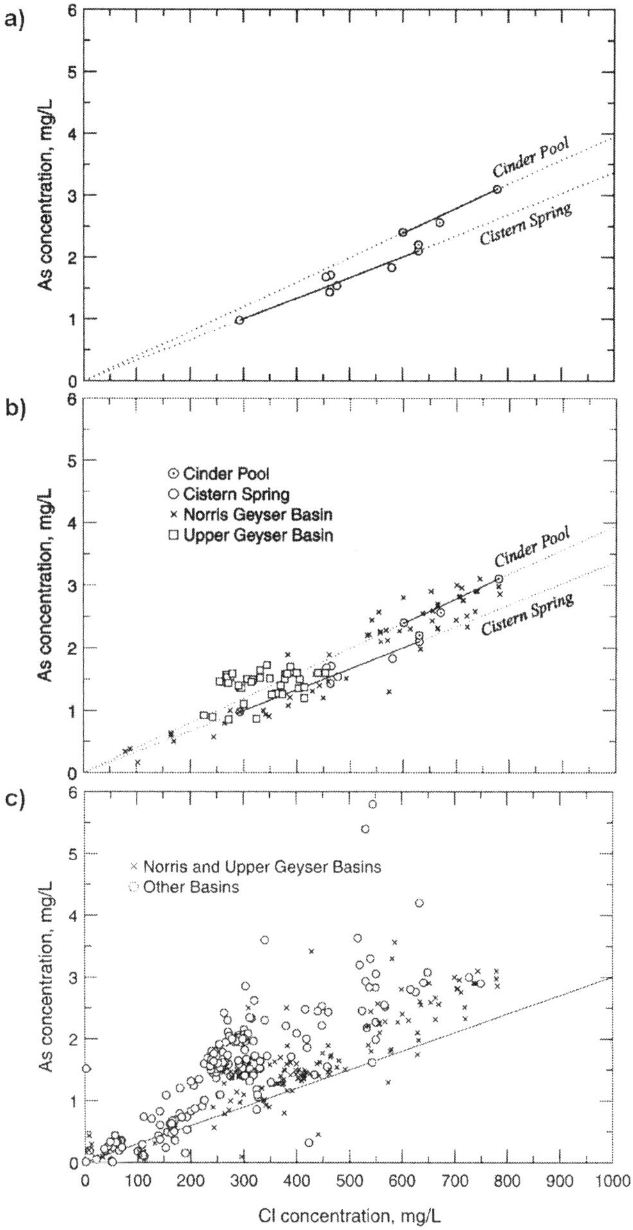

Figure 3. Variations in As and Cl in: a) Cistern Spring and Cinder pool in Yellowstone National Park. b) Springs of Norris and Upper Geyser Basin, together with the linear trends defined by seasonal variations in Cistern Spring and Cinder Pool (solid lines). c) A more general survey of Yellowstone National Park thermal waters, where "x" represents samples mainly from Upper Geyser Basin and Norris Geyser Basin (Nordstrom et al., 2001).

3. SPECIATION OF ARSENIC IN GEOTHERMAL FLUIDS

3.1 Dissolved As

The chemical behaviour of As in sulfide-rich fluids has been the subject of considerable, and ongoing, debate as there are several major obstacles to the accurate prediction of As speciation. In part, this reflects the lack of a complete thermodynamic database for As species. Stability constants for ion pairs and complexes between the As oxyanions and polyvalent cations are few and limited in applicability (Nordstrom, 2000). Also, polymerisation of arsenite, arsenate and thioarsenite complexes in high As or high sulphur solutions appears likely, but is largely unconfirmed. These reactions could significantly complicate As speciation. However, reliable thermodynamic data are available for some As species, and these can be used to interpret As behaviour in many geothermal systems and surface water environments, as long as the limitations are recognized.

3.1.1 Low sulfide, reduced fluids

Heinrich and Eadington (1986) and Ballantyne and Moore (1988) noted that, from thermodynamic considerations, As was more likely to be transported in most hydrothermal solutions as arsenious acid ($H_3As^{III}O_3^{o}$), rather than as a thio complex. H_3AsO_3 is considered to be the product of both As oxide (As_4O_6) and orpiment (As_2S_3) dissolution in reduced fluids at acid to neutral pH, over a wide range of temperature conditions (Eary, 1992; Ivakin et al., 1979; Mironova et al., 1984; Pokrovski et al., 1996). Orpiment solubility, as given by the following reaction:

$$0.5As_2S_3 + 3H_2O = H_3AsO_3 + 1.5H_2S$$

increases with temperature, to at least 300°C (Pokrovski et al., 1996; Webster, 1990; see Figure 4a). The solubility of orpiment is independent of pH under acidic, or acid-neutral pH conditions, but increases with further increase in pH. Note that the orpiment solubilities shown in Figure 4 assume a constant sulfide concentration, and do not take into account additional sulfide released from the orpiment. Once this additional contribution becomes significant, orpiment solubility will decrease (refer Section 3.1.2). Polymerisation has been observed using Raman spectroscopy (Pokrovski et al., 1996), and polymer configuration and stability has been calculated (Tossell, 1997). However, polymerisation is predicted to occur only at very high As concentrations (>1mol/kg; Pokrovski et al., 1996) and these conditions are unlikely to exist in geothermal fluids .

3.1.2 High sulfide, reduced fluids

In sulfide-rich hydrothermal solutions, orpiment solubility as H_3AsO_3 is inhibited but solubility as thioarsenite complexes increases. The thioarsenites are a group of covalently bonded arsenic-sulfide complexes that may or may not include oxygen. Reactions between arsenate ions and dissolved sulfide, for example, can result in the successive replacement of oxygen with sulfide (Pauling, 1970). However, the exact nature of the thio-complexes remains controversial (Helz et al., 1995; Wood et al., 1998). It was evident 40 yrs ago that a simple AsS_2^- stoichiometry and structure was likely to be unrealistic (Angeli and Souchay, 1960). A trimeric configuration ($As_3S_6^{-3}$) was proposed at that time and has since attracted support (Eary, 1992; Helz et al., 1995; Spycher and Reed, 1989a; Vorobeva et al., 1977; Webster, 1990). Dimer species have also been proposed (Mironova and Zotov, 1980; Mironova et al., 1990), but there has been little recent support for other earlier proposed species such as AsS_3^{-3}, $As_4S_7^{-2}$, AsS_4^{-5}, and $As_2S_5^{-4}$. Helz et al. (1995) suggest that the monomer may occur in undersaturated solutions, while the trimer is formed in the saturated solutions present in most experimental solubility studies and As-precipitating geothermal systems. Wood et al. (1998) provided Raman spectroscopic evidence for the simultaneous existence of more than one species.

Assuming a trimer structure occurs in a saturated geothermal fluid, orpiment solubility via the following reaction:

$$1.5As_2S_3 + 1.5HS^- + 0.5H^+ = H_2As_3S_6^-$$

increases with pH and sulfide concentration (Fig. 4b), but is not greatly affected by temperature changes below 200°C (Spycher and Reed, 1989a; Webster, 1990; Weissberg et al., 1966).

3.1.3 Redox state

The oxidation of arsenious acid to arsenate ($H_2As^VO_4^-$ or $HAs^VO_4^{-2}$) is thermodynamically favoured when the redox conditions in low sulfide fluids become sufficiently oxidizing. Oxidation occurs when a rising geothermal

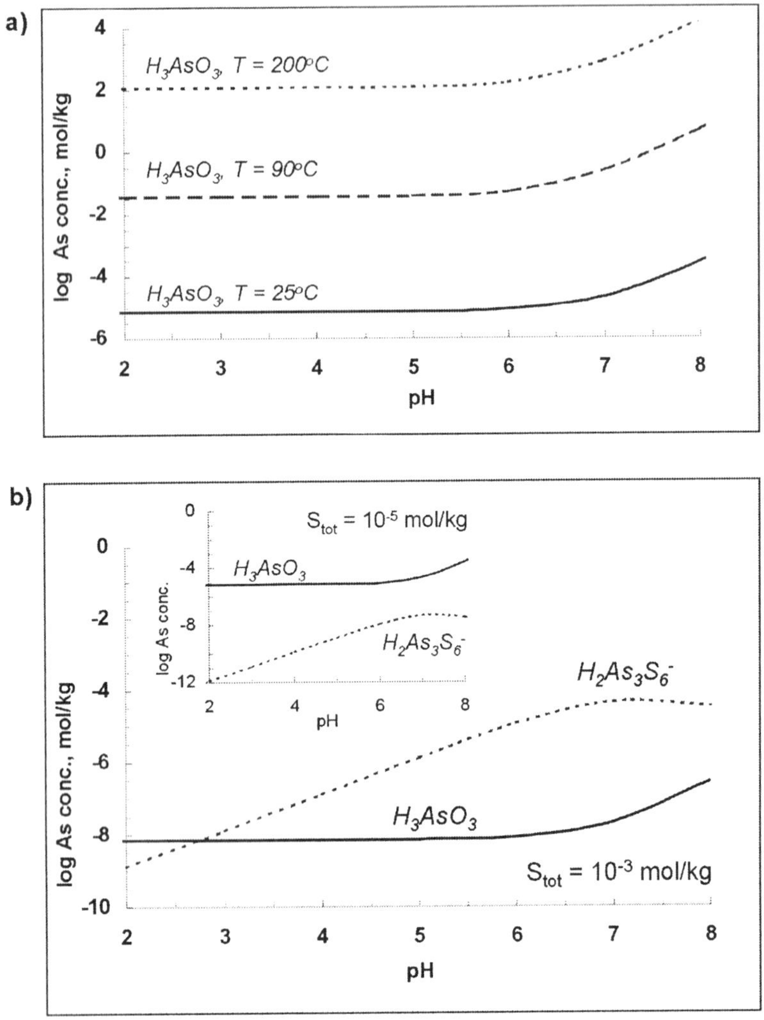

Figure 4. a) Orpiment solubility as a function of pH and temperature in low sulfide fluids, using stability constants for H_3AsO_3 from Webster (1990) and Pokrovski et al (1996). B) Orpiment solubility as a function of pH at 25°C, in a high sulfide fluids using stability constants for $H_2As_3S_6^-$ from Webster (1990).

fluid is exposed to atmospheric oxygen or mixed with another oxidizing fluid, such as a shallow ground water. Arsenate ions appear to be rapidly formed *in situ* in hot springs and their surface drainage systems. Of the many geothermal features in Yellowstone National Park, As^{III} predominates in most hot springs and geysers discharging directly from the reservoir (Ball et al., 1998; Nordstrom et al., 2001). The geothermal reservoir fluids are reduced with respect to the As^{III}/As^{V} redox couple as expected. Oxidation to As^{V}, however, occurs very rapidly in hot spring overflows and receiving streams and rivers. A few hot springs at Yellowstone contain mostly As^{V}, which is likely to be related to fluid flow rates and/or mixing with oxygenated ground waters. Very few Yellowstone geothermal waters have been found with similar proportions of As^{III} and As^{V}, unless they are overflow drainages in the process of oxidizing.

These results are similar to As^{III}/As^{V} distributions reported for hot springs on the island of Dominica (Carribean), in the Valles Caldera (New Mexico) and from the Massive Central in France (Criaud and Fouillac, 1989). In each case, hot springs formed from reservoir fluids contained mainly As^{III}, whereas acid-sulfate and bicarbonate features were more enriched in As^{V}. It is worth noting here that reported As^{III} concentrations may be underestimated if the sample has not been appropriately preserved prior to analysis. On site sample filtration through a 0.45μm (or finer) membrane to remove bacteria and cold storage in an air-tight container are recommended (Hall et al., 1999). Acidification with HNO_3 or HCl to pH 1.5- 2.0 is often recommended, as this inhibits oxidation of As^{III} by chemical oxidants such as Fe^{III}. However, in some waters, acidification has been observed to change As speciation slightly, reducing As^{V} to As^{III} (Hall et al., 1999, Webster unpublished data.) Arsenic oxidation in the stream draining Champagne Pool at Waiotapu geothermal field in the Taupo Volcanic Zone, and in an open, concrete drain conducting geothermal waste water across the Wairakei geothermal field suggests a short half-life of $\leq$5 minutes for *in situ* oxidation (Finlayson and Webster, 1989). More importantly, however, geothermal water collected immediately after steam separation, and before entering the drain, does *not* show a similar rate of As oxidation. Instead there is no appreciable As^{III} oxidation in the wastewater (collected directly into a sterile glass bottle), even if bubbled with air for several hours.

Such rapid As^{III}/As^{V} oxidation kinetics appear likely to be catalysed by microbial activity, although the mechanisms by which bacteria could facilitate arsenite oxidation at temperatures >50°C are not well studied. However, as bacteria are clearly implicated in As^{III} oxidation in geothermal systems, progress is being made in the identification of the thermophilic bacteria that can oxidise As^{III} (e.g., Gihring et al., 2001; Langner et al., 2001; Mielke et al., 2000). The ability of H_2S and thiosulphate (S_2O_3) to reduce As^{V} to As^{III} is well known (e.g., Cherry et al., 1979; DeKonnick, 1909; Forbes et al., 1922; Rochette et al., 2000), hence geothermal waters containing sulfide or thiosulphate will preserve arsenic as As^{III}, until the reduced sulfur is oxidised or volatilised (Nordstrom et al., 2001).

Further evidence of a microbiological influence on As chemistry may exist in the recent, intriguing observation of slight diurnal variations in the (total) As concentrations of various hot springs at the Waiotapu geothermal field (James Pope, pers. comm).

3.2 Arsenic deposition from geothermal fluids

Arsenic enrichment in geothermal systems occurs predominantly near the surface, along with other "epithermal" elements such as Sb, Au and Hg (White, 1981). In contrast, base metals such as Ag, Cu, Pb, Zn will be deposited at greater depth. A typical example of this type of zoning occurs in the Ohaaki geothermal system (formerly known as Broadlands) in the Taupo Volcanic Zone of New Zealand (Simmons and Browne, 2000). Here, As concentrations in drill core from BR 16 increased from 1.2-1.9 mg/kg at 1.5 km depth, to 19-22 mg/kg near the surface (Ewers and Keays, 1977). At depth, As was mainly concentrated in the pyrite, as has been noted at other fields (Ballantyne and Moore, 1988). Arsenic minerals such as arsenopyrite (FeAsS) appear to be uncommon in the rocks of the geothermal reservoir itself. However, a range of As minerals is precipitated from geothermal surface features such as hot springs (refer Section 3.2).

It has been suggested that the observed metal zoning may simply reflect different precipitation mechanisms for the base metals complexed by chloride and the epithermal metals complexed by sulfide (Drummond and Ohmoto, 1985; Weissberg, 1969; White, 1981). If transported as a thio-complex, As would be expected to precipitate in response to decreasing temperature and boiling (H_2S loss) in permeable zones (Ewers, 1977). However, as noted above, As is more probably transported as H_3AsO_3 in hydrothermal solutions (Ballantyne and Moore, 1988; Heinrich and Eadington, 1986; Spycher and Reed, 1989a). Spycher and Reed (1989b) argue that metal zonation is due to a three-step process of fluid boiling, gas-phase transport and acid reactions with metal-bearing waters. Base metal precipitation occurs in response to boiling and the pH increase which occurs when CO_2 gas is driven off. During boiling, As remains soluble as an oxyanion under the higher pH and lower sulfide concentrations present in the fluid. Precipitation of orpiment then occurs in response to the acidification of hot spring waters with acid-sulfate waters (Spycher and Reed, 1989b), subaerial cooling of the fluid or increased H_2S concentrations (Webster, 1990).

This scenario is consistent with the relationships discussed in Section 3.1.1 and shown in Figure 4, as rapid oxidation of As^{III} or H_2S would reduce the concentrations of these components and thereby increase orpiment solubility. Note that only if the fluids boiled dry would As be taken into the gas phase. Vapour deposition of realgar could occur when the temperature decreases below 130°C (Spycher and Reed, 1989b).

3.3 Arsenic in hot spring deposits and scales

The As contents of coloured precipitates forming at the periphery of hot springs, hot pool and geysers can be very high (Table 2). Analyses of red precipitates around the periphery of Champagne Pool in the Waiotapu geothermal field (Fig. 5a), for example, revealed an amorphous deposit containing up to 2wt% As, as well as sulfides of Hg, Sb, Tl and ore-grade Au and Ag concentrations (Weissberg, 1969). Similar red As-rich precipitates have been reported from Steamboat Springs in Colorado (White, 1967), Roosevelt Hot Springs in Utah (J. Moore, pers. comm.), and Beppu hot pool in Japan (Koga, 1961) .

The mineralogy of these coloured precipitates can be difficult to determine, given the amorphous nature of most recent deposits. Useful As:S ratios are obscured by other sulphide-bearing minerals and sometimes by native sulphur if there is a steam component to the hot spring. Nor is colour always a definitive indicator. For example, amorphous realgar is a bright red colour, but so is stibnite (Sb_2S_3) and cinnabar (HgS). Orpiment is a yellow-green colour, reminiscent of native sulfur. However, in the absence of high levels of antimony and mercury, As–rich precipitates in hot springs are commonly yellow orpiment (Fig. 5b). Thermodynamic calculations provide support for orpiment precipitation in hot spring environments. Hot springs on the Waiotapu and Rotokawa geothermal fields in New Zealand, for example, are saturated or slightly over-saturated with respect to orpiment (Webster, 1990). Also, orpiment precipitation has been observed where small H_2S-rich steam vents occur in the bed of hot spring-fed streams at these fields.

Realgar does not appear to be actively depositing in hot springs at Waiotapu but does occur as coatings and veins in the altered rocks of the geothermal field. This supports the hypothesis of different deposition mechanisms for realgar and orpiment, as proposed by Spycher and Reed (1989b). It has also been proposed that a transient As disulfide solid phase occurs in hot spring environments (Noguchi and Nakagawa, 1969), but it has not been isolated. Scorodite ($FeAsO_4.2H_2O$) has been reported from the Osorezan geothermal area in Japan (Aoki and Yui, 1981). Native As has been reported from low temperature hydrothermal deposits (eg., Sergeyeva et al., 1969) but is rare. Other As sulfides, such as dimorphite (As_4S_3) and duranusite (As_4S), have not been reported from geothermal systems.

a)

b)

Figure 5. Arsenic mineral precipitates at the Waiotapu geothermal field, Taupo Volcanic Zone, NZ: a) An As- and Sb-rich sulphide precipitate in Champagne Pool. b) Yellow orpiment in a small hot spring.

The scales which form in pipes and drains of a developed field can also be rich in As (Table 2). Scales rich in As and Au deposit on back pressure plates at the Ohaaki geothermal field, for example, due to the sudden drop in pressure and consequent boiling of the fluid (Simmons and Browne, 2000). The subaerial scale forming in the Wairakei borefield drain demonstrates a close positive correlation between As and Fe content, with As deposition mainly due to adsorption on Fe-oxide (Finlayson and Webster, 1989).

Table 2. The As content (mg/kg or wt%) of various hot spring and scale deposits

Field	As, mg/kg or wt%	Reference
Hot Spring Precipitates:		
Yellowstone Nat. Park, US	<2 mg/kg - >5%	White et al. (1992)
Steamboat Springs, US	50 – 700 mg/kg	White (1981)
Tamagawa, Japan	6 - 56%	Noguchi and Nakagawa (1969)
Waiotapu, NZ	1.0 - 25%	Weissberg (1969); Webster (1990)
Rotokawa, NZ	0.4 - 3.4%	Weissberg (1969);Krupp and Seward (1990)
Ohaaki, NZ	300-400 mg/kg	Weissberg 1969)
Kamchatka, Russia	0 - 0.67%	Karpov and Naboko (1990)
Power Station Scales:		
Wairakei waste water drain, NZ	1.6 mg/kg- 9.4%	Finlayson and Webster (1989); McKenzie et al. (2001)
Ohaaki back-pressure plate scale, NZ	10 mg/kg - 0.1%	Simmons and Browne (2000)

3.3.1 The role of bacteria in arsenic deposition

Thermophilic bacteria and cyanobacteria in hot spring biofilms facilitate the extracellular and intracellular formation of silica, silicates, carbonates and oxide minerals in sinter deposits (e.g., Konhauser and Ferris, 1996; Tazaki, 1995). Microbial activity can also influence the macrotexture of sinters, resulting in the formation of columnar and laminated stromatolites, coccoids and oncoids (Hinman and Linstrom, 1996; Jones et al., 1997; Renaut et al., 1996). Microbial biofacies identified in sinter deposits include cyanobacteria (eg. *Phormidium* and *Anabaena* sp., *Oscillatoria* sp., *Nostoc* sp., and *Calthrix* sp.; Tazaki, 1995; Jones et al., 1997). However, recent developments in molecular microbiology suggest that the biodiversity of hot spring microbial communities is considerably more complex than previously reported based on microscopic examination and cultivation (Ward et al., 1998).

An analysis of microbial mats in mudpools at the Rotokawa and Waimaungu geothermal fields (Taupo Volcanic Zone) indicated enrichment of As, along with Sb, P and Zn, in the organic matrix relative to the bulk samples (Hirner et al., 1990). However, this was not confirmed in a more recent investigation (McKenzie et al., 2001). Consequently the role of thermophilic bacteria in As precipitation, adsorption or other uptake into the sinter remains uncertain.

4. THE FATE OF ARSENIC FROM GEOTHERMAL SOURCES

The contamination of natural drainage systems by As is one of the most significant environmental effects of geothermal activity. Although a degree of soil contamination will occur in the vicinity of a geothermal field, this is a local effect and one which is generally regarded as acceptable. However, water flow through the surface and subsurface catchments has the potential to transport As beyond the boundary of the geothermal field. In the vicinity of geothermal systems, it is chemical contamination of surface waters, rather than ground waters, which is the most commonly detected and reported. In surface waters, As enters a cycle of chemical and biochemical reactions (review by Ferguson and Gavis, 1972).

4.1 Surface waters

Hot springs, geysers and steam features all produce an excess of fluid at the surface; fluid that usually drains unimpeded into the nearest catchment system. Even after orpiment precipitation, many hot springs fluids will still contain >1 mg/kg As (Table 1) because orpiment is a relatively soluble salt. Consequently, water to be used for drinking, stock watering, irrigation or simply to support aquatic life, may have unacceptably high As concentrations downstream of a geothermal system. The WHO (1993) drinking water limit is 0.01 mg/L, compared to higher limits for the protection of aquatic life of 0.19 mg/L (USEPA, 1986) and for stockwatering and irrigation of 0.2 and 0.1mg/L respectively (USEPA, 1972). Consequently, the use of a water body for a drinking supply is the use most likely to be compromised by As contamination. Arsenic contamination of water for drinking or domestic use has therefore received considerable attention (e.g., Kneebone and Hering, 2000; McLaren and Kim, 1995; Webster, 1999; Wilke and Hering, 1998).

Arsenic speciation is important in surface waters because acute toxicity of As^{III} is greater than that of As^{V} or the organic forms of As (e.g., Jain and Ali, (2000). For human chronic toxicity the redox form of arsenic may not matter because arsenic is largely reduced to As(III) and methylated (National Research Council, 1999). Another issue is the processes by which As interacts with sediments, organic and biotic substrates in freshwaters, as this affects both As concentration and speciation. Studies of As in the rivers draining the two geothermal systems specifically featured in this review: Yellowstone National Park and the Taupo Volcanic Zone, have shown that As is principally transported in dissolved form (Nimick et al., 1998; Webster-Brown, 2000). "Dissolved" in this case is defined as passing through a 0.45μm filter membrane. On a macro-scale, As also appears to behave conservatively downstream of these geothermal systems, with little change in As mass flux down the river (Nimick et al., 1998) or in lakes and estuaries (Timperley and Huser, 1996; Webster-Brown, 2000). Studies of geothermal-derived As in other freshwater systems support this observation (e.g., Deely and Sheppard, 1996; Kneebone and Hering, 2000).

4.1.1 Dissolved Arsenic Speciation

Assuming thermodynamic equilibria prevail, As released into an oxygenated surface water as H_3AsO_3 should undergo immediate oxidation to arsenate ions: $H_2AsO_4^-$, or at pH > 6.9, $HAsO_4^{-2}$. Arsenate certainly predominates in the Waikato River (Aggett and Aspell, 1978; McLaren and Kim, 1995) and in streams and rivers receiving hot spring discharge from Yellowstone National Park (Ball et al., 1998; Nimick et al., 1998; Nordstrom et al., 2001). This is also the case in rivers draining the Mt. Apo geothermal field in the Philippines (Webster, 1999), and in Hot Creek which drains geothermal activity in the Owens River catchment, Sierra Nevada, US (Wilke and Hering, 1998).

A half-life of ~20 mins for As^{III} oxidation to As^{V} was calculated for As in Hot Creek and attributed to microbial catalysis (Wilke and Hering, 1998). Langner et al. (2001) derived a similar half-life for As oxidation in an acid hot spring in Yellowstone National Park. Gihring et al. (2001) found *Thermus aquaticus* and *Thermus thermophilus* in a high pH (8.8) hot spring in Yellowstone Park that effectively oxidize arsenite at about the same rate as in acid hot springs. Tantalus Creek, which drains most of the thermal waters from Norris Geyser Basin and empties them into the Gibbon River in Yellowstone National Park, contains about 1.7 mg/L arsenic of which at least 97% is As^{V} (USGS, unpublished data). These results attest to the efficiency of microbial arsenic oxidation in thermal overflow waters.

In river waters, as in geothermal features, the role of bacteria in the oxidation process may be very important. Wilkie and Hering (1998) noted that As oxidation in Hot Creek did not occur if the river water was filtered under sterile conditions or after an antibiotic had been added. They surmised that bacteria attached to submerged plants were mediating As^{III} oxidation. Bacteria capable of As^{III} oxidation have been isolated, and some of the most common are *Pseudomonas* sp. (Cullen and Reimer, 1989). Other bacteria genera that have been found to oxidize As^{III} include *Xanthomonas* and *Acromobacter* (Ehrlich, 1996). As^{V} may be predominant, but it is not exclusive. In the Waikato River, for example, high concentrations of As^{III} do appear periodically. Sudden, short-lived pulses of As^{III}, and associated higher concentrations of total As, have been attributed to the annual, but unpredictable, overturn of the deep, stratified hydroelectric lakes (Aggett and Kreigman, 1988). Interstitial sediment waters in the anoxic depths of the lakes are rich in As^{III}, and when thermally-driven overturn occurs, these waters are released downstream. However, a broader seasonal variation in As in the Waikato has also been identified (McLaren and Kim, 1995; see Fig. 6) and attributed to seasonal activity of arsenate-reducing bacteria. Various bacteria and strains capable of As^{V} reduction have been identified (e.g., Macy et al., 2000; Stolz and Oremland, 1999; Switzer et al., 1998) including *Anabaena oscillaroides* which has been isolated from the Waikato River (Freeman, 1985).

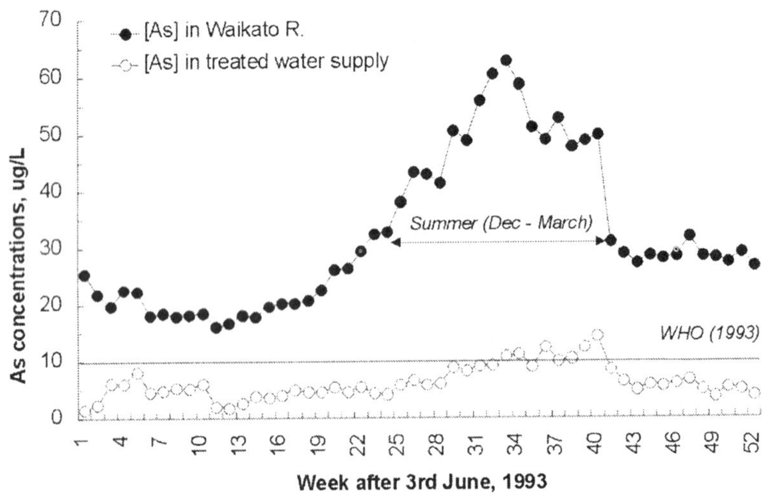

Figure 6. Seasonal variation in As concentrations in the Waikato River, and after passage through a conventional (alum floc) water treatment plant in Hamilton city, New Zealand. Residual As in the reticulated drinking water supply is indicative of As^{III} concentrations in the river water, as As^{III} is not removed alum floc treatment (modified from McLaren and *Kim (1995); data courtesy of Nick Kim).*

The third group of As species in natural freshwaters are the methylated arsenic species. Arsenic can be methylated by bacteria, algae and fungi to form gaseous mono-, di- and tri-methyl arsine ($(CH_3)_nAs$) (Baker et al., 1983; Cullen and Reimer, 1989; Maeda et al., 1987), which dissolve in water forming the methylarsenic oxyacids, monomethylarsonic acid (MMAA) and dimethyldarsenic acid (DMAA). Organic As complexes are estimated to make up <20% of the As in the Madison River draining Yellowstone National Park (Nimick et al., 1998), which is consistent with an estimate of 10-20% of total As for freshwater environments in general (Cullen and Reimer, 1989). DMAA has been reported as the dominant form of As in summer in a eutrophic lake in Japan (Sohrin et al., 1997), although the source of this As is unclear. Unlike methyl-mercury compounds, methylated arsenic does not appear to biomagnify in the food chain (Isensee et al., 1973; Maeda et al., 1992).

4.1.2 Transport & removal from the water column

A small portion of the As released into a river from a geothermal system will become part of the biochemical cycle, interacting with plants, biota, suspended material and bed sediments. In the Waikato River, for example, 7-8% of the As entering the river is being adsorbed onto the sediments, and sediment As concentrations as high as 1000 mg/kg have previously been reported from the deeper hydroelectric lakes (Aggett and Aspell, 1980). Suspended sediment As concentrations provide a better indication of the direct association between As and sediments. Concentrations of 0-60 mg/kg As in suspended sediments are reported from the Missouri River draining Yellowstone National Park (Nimick et al., 1998), while higher concentrations of 61-1790 mg/kg are found in the Waikato River draining Taupo Volcanic Zone (Webster-Brown et al., 2000). The higher As concentrations in suspended sediments of the Waikato River are a function of their proximity to geothermal activity. Where the Waikato River passes through the area of geothermal activity (Fig. 1b), As concentrations in the silica-rich suspended sediment are limited only by the adsorption capacity of the suspended sediment, which is ~1500-1800 mg/kg As (Webster-Brown, 2000).

A strong association between As and Fe-oxides in river and lake sediments has been reported (eg., Aggett and Roberts, 1986; Belzile and Tessier, 1990; Brannon and Patrick, 1987). This association is attributed to the adsorption of arsenate onto the Fe-oxide coatings on sediment particles. Arsenite ion, although also adsorbed, is not as rapidly or completely adsorbed onto Fe-oxide as arsenate (Belzile and Tessier, 1990; Pierce and Moore, 1982; Swedlund and Webster, 1999). This observation has important implications for the re-release of adsorbed As under anoxic conditions. Microbial reduction of arsenate can occur even when As is bound to Fe-oxide (Langner and Inskeep, 2000), and reduction or methylation of sedimentary arsenate under anoxic conditions is the principal mechanism for re-release of As into the water column (Aggett and

Kreigman, 1988; Brannon and Patrick, 1987; Dowdle et al., 1996). The seasonal variation in total As concentration in the Waikato River (Fig. 6) has been attributed to the seasonal activity of As^V-reducing bacteria (McLaren and Kim, 1995), as there is no evidence that seasonal changes in water temperature affects As^V adsorption onto sediments (Webster-Brown, 2000).

The life-cycle of aquatic plants may also affect the concentrations of As in the water column (Kuwabara et al., 1990). The uptake of As by aquatic macrophytes (plants) has been reported in rivers contaminated by geothermal waters. There is often a significant degree of As enrichment relative to the water column. Concentration factors of up to 20,000 times, depending on the plant species, have been reported for the Waikato River where the plants are estimated to take up 3-4% of the As available (Aggett and Aspell, 1980; Reay, 1972). Plant uptake of As is generally considered to be due to the chemical and behavioural similarity of arsenate and phosphate ions, the latter being an important nutrient for plant growth. High phosphate concentrations can inhibit As uptake by plants, as well as As adsorption on Fe- oxide surfaces (Kuwabara et al., 1990; Reuther, 1992).

Finally, As interacts with macrobiota either directly or via the food chain. As well as the issue of toxicity, there is the potential problem of As accumulation in animal flesh. Freshwater snails in the Waikato River, for example, were found to have 20 times the As content of snails in a control site (Golding et al., 1997). However, there has been little evidence that animals further up the food chain, such as trout, accumulate As from the Waikato River (Robinson et al., 1995).

4.2 Ground waters

Although As-rich ground waters appear to occur with alarming frequency (e.g., Del Razo et al., 1990; Frost et al., 1993; other chapters in this volume) other contributions in this volume), there are relatively few examples where the As is clearly related to geothermal activity. Geothermal fluids are themselves effectively deeply circulating ground waters. The thermally driven ascension to the surface limits opportunities of lateral spreading of these As-rich fluids at depth. Near the surface, however, contamination of shallow aquifer systems can and does occur.

High As concentrations in ground waters of the upper Missouri River catchment have been attributed to As in the thermal waters of Yellowstone National Park (Nimick, 1994; Welch et al., 2000). Arsenic-enriched ground waters of southwest Idaho and in the Intermontane Plateaus province are also considered to be influenced by thermal waters (Welch et al., 2000). Geothermal contamination of ground water has not yet been reported from the major geothermal fields of the Taupo Volcanic Zone, but it has been observed in a shallow ground water outflow to a wetland area at the southern end of Lake Taupo (Chague-Goff et al., 1999).

Elsewhere As contamination of drinking water wells in Taiwan, and the consequent toxic effects on the human population (Chen et al., 1994), is considered likely to be due to mixing with geothermal fluids. High pH ground waters of the Vulcano Porto field on Vulcano Island, near Sicily, have As concentrations ranging 0.6 to 134 μg/L (Aiuppa et al., 2000).

Ground water contamination is a concern for geothermal developers because there are several potential pathways for aquifer contamination. These include:

i) Unintentional reinjection of spent geothermal waste water into an aquifer. This will normally occur only if the integrity of the casing around a well is breached.

ii) Seepage from poorly-lined, or unlined, holding ponds for the retention of geothermal fluids and from pipelines, as reported from Los Azufres in Mexico for example (Birkle and Merkel, 1998).

iii) Burial of As-rich sludge from an As-removal waste treatment process or scale from drains and pipes (e.g., Peralta et al., 1996).

The speciation and solubility controls on As in ground water are similar to the abiotic processes influencing As behaviour in surface water. Arsenic may occur as both As^{III} and As^{V}, with solubility controlled by adsorption onto sediment surfaces (Welch et al., 2000). Under the higher rock:water ratio present in aquifers, clays and other silicate surfaces, as well as Fe- and Mn-oxides, may act to regulate As concentrations through adsorption under appropriate pH conditions (e.g., Goldberg and Glaubig, 1988a; Manning and Goldberg, 1997). It should be noted that high sulfide concentrations may accompany As of geothermal origin when it enters a ground water aquifer, and this will limit As solubility through the formation of orpiment.

5. ARSENIC REMOVAL FROM GEOTHERMAL FLUIDS

Clearly it is preferable to avoid the release of As from geothermal systems into surface or ground water systems. In developed geothermal fields, reinjection of waste water into the field is commonly used to avoid environmental pollution and maintain reservoir pressure. In many countries, including the US, reinjection is required by law. However, where this is not the case, reinjection is viewed as an expensive option and one which can reduce field productivity if the cooler waters find their way back into the reservoir too quickly (Brown and Bacon, 2000). Effective removal of As, and other contaminants of the "geothermal suite" from waste water prior to discharge, would enable the bore waters to be discharged or re-used without environmental damage. It may also allow valuable elements or minerals to be extracted, offsetting the cost of treatment, which may still be economically viable when compared to the costs of reinjection (Brown and Bacon, 2000).

Various methods for the removal of As from geothermal waste waters have been investigated at theoretical, laboratory, pilot plant and full plant scales. These include adsorption onto Fe-oxide floc and subsequent separation by dissolved air flotation (De Carlo and Thomas, 1985; Shannon et al., 1982) and co-precipitation with lime to form an As-rich calcium silicate (Rothbaum and Anderton, 1976). In both cases, effective removal was achieved only after oxidation of As^{III} to As^{V}. For Fe-oxide floc treatment, competitive adsorption of silica inhibits As adsorption, particularly that of As^{III} (Swedlund and Webster, 1999), suggesting that prior removal of silica would help optimise As removal efficiency. The use of ion selective chelating resins for As^{III} removal from geothermal waters has also been successfully tried (Egawa et al., 1985).

Arsenic contamination of surface or ground waters is, however, rarely a "point source" event, given the complex plumbing of most geothermal systems and the variety of opportunities this presents for thermal and ambient fluid mixing. While As removal prior to a point source waste water discharge will help to minimise impacts on aquatic ecosystems, particularly As accumulation in aquatic plants and sediments, it is unlikely to reduce As concentrations to a drinking water standard in regions of natural geothermal activity. Consequently, the ability of conventional water treatment plants to remove As from drinking water supplies continues to be important. Both alum- and Fe-oxide-based treatment systems effectively remove As^{V} if optimised to favour As adsorption and particulate removal, but alum-based systems are less effective in their removal of As^{III} (Hering et al., 1997; McNeill and Edwards, 1995). Figure 6 shows As removal from the Waikato River using a conventional alum-floc treatment station at Hamilton City, New Zealand (McLaren and Kim, 1995). Effective removal of As is achieved, except during the summer when As^{III} is also present in the water column. Reverse osmosis and nanofiltration membranes have greater potential for efficient As^{III} removal, as they do not discriminate between As^{V} and As^{III} (Waypa et al., 1997).

ACKNOWLEDGEMENTS

The authors would like to thank Jim Ball, James Pope, Donald Thomas and Nick Kim for allowing us to use their data, and Kevin Brown and Joe Moore for reading and providing a constructive review of this manuscript.

Chapter 5

Role of Large Scale Fluid-Flow in Subsurface Arsenic Enrichment

M.B. Goldhaber[1], R.C. Lee[1], J.R. Hatch[1], J.C. Pashin[2], J. Treworgy[3]

[1] *U.S. Geological Survey, Denver CO*

[2] *Geological Survey of Alabama, Tuscaloosa AL*

[3] *Earth Science Department, Principia College, Elsayh Il*

This chapter deals with the geologic controls on the distribution of arsenic in rocks. Specifically, it focuses on rare but geographically extensive paleohydrologic events that produced widespread arsenic enrichments in the earth's crust in the form of elevated contents of arsenic-rich (arsenian) pyrite. We summarize evidence documenting the existence of ancient large-scale hydrothermal fluid migration events in the central and eastern United States and discuss impacts on the arsenic content of aquifer rocks through which the fluid migrated. There are two specific geologic settings discussed. One is the midcontinent region, and the other is the Appalachian region (with a focus on coal in the Appalachian Basin). Evidence for a hydrothermal fluid flow event in the midcontinent comes largely from studies on the genesis of large zinc-lead deposits (the so-called Mississippi Valley-type or MVT ores) of the region that are hosted in Paleozoic carbonate rocks. These studies demonstrate that ore-fluids for MVT deposits (brines derived from adjacent sedimentary basins) were driven towards the craton by gravity flow during a late stage of the Ouachita orogeny about 270 million years ago. These warm brines migrated laterally for hundreds of kilometers. Arsenian pyrite was deposited along these fluid pathways. The same aquifers that were pathways for ore fluid migration are exploited today as drinking water aquifers, raising the possibility that given conditions favorable for release of arsenic from arsenian pyrite, drinking water arsenic concentrations might become elevated. Further east, Paleozoic sediments of the Appalachian basin were aquifers for westward migrating fluids during a late phase of the Appalachian orogeny (also about 270 million years ago). Coal beds were particularly favorable sites for deposition of arsenian pyrite. In marked contrast to the midcontinent flow event, the fluids involved in the Appalachian region were likely derived from deep-seated metamorphic processes and were not as saline as those in the midcontinent. We argue that in the southernmost Appalachians of Alabama, the metamorphic fluids impacting coal were related in time and composition to metamorphic fluids that formed arsenic-bearing gold deposits in the metamorphic rocks to the east. Pennsylvanian bituminous coals of Alabama locally contain abundant arsenian pyrite, which also is enriched to a lesser extent in copper, selenium, molybdenum, mercury and antimony. Mining of Alabama coal has resulted in elevated arsenic content in stream sediments, and in non-potable (saline) waters produced during coalbed methane recovery, but arsenic enrichment in drinking water supplies is apparently localized to individual water wells. Invasion of metamorphic fluids into coal-bearing Pennsylvanian rocks was widespread throughout the Appalachian basin, and arsenian pyrite occurrences are also widespread but sporadic. Large fluid flow systems are not restricted to the examples described in detail in this chapter, and additional areas impacted by such systems are noted.

1. INTRODUCTION

Environmental impacts of arsenic can arise from either anthropogenic or naturally occurring enrichments. In either case, the major human health impacts from arsenic arise from elevated drinking water (typically groundwater) concentrations (Anonymous, 1999). These elevated groundwater concentrations in turn are due to mobilization of arsenic from rocks and sediments into groundwater (Welch et al., 2000). Examples of such groundwater enrichments are numerous, and many are described in this volume. Given that release of arsenic from rocks governs human health impacts on drinking water, understanding controls on the abundance and form of arsenic in rocks is an important first step in predicting where health problems may arise.

The underlying hypothesis of this paper is that large-scale fluid migration can lead to widespread and potentially environmentally significant arsenic enrichments in rocks. Recognition that giant fluid-flow systems have existed in the past has come largely from the study of ore deposits. The hydrothermal fluid-flow systems described below represent the 'far field' (upstream and downstream) effects of ore-forming processes. Where such hydrothermal fluids are focused into small areas (and other conditions are met), ore deposits may form. When the fluids migrate laterally in aquifers feeding the ore districts, trace amounts of the ore constituents may precipitate. If arsenic is one of these ore constituents, it can accumulate in the aquifer rocks over geographically large areas, on the order of thousands of square kilometers or more. The processes described in this paper have produced elevated background levels of arsenic in rocks. Not all of the arsenic accumulations described are associated with elevated concentrations in groundwater or human health problems, but they have that potential given appropriate geochemical conditions for arsenic solubility in the groundwater.

It is critical to recognize that the same aquifers that functioned as transport paths for the ore fluids in the geologic past, can (and commonly do) serve as drinking-water aquifers in the modern environment. An important example of this coincidence of paleoaquifers with modern drinking-water aquifers occurs in the U.S. midcontinent (areas 1 and 2 in Fig. 1), the focus of the first part of this paper. In this part we describe the subcontinental-scale brine migration associated with the formation of the Mississippi Valley Type (MVT) lead-zinc ores in this region. Some local occurrences of elevated arsenic in groundwater of the U.S. midcontinent are almost certainly related to this association. Considerable research effort has gone into documenting in this region the timing, geographic extent, and impact of MVT ore formation on enrichments of various elements (Goldhaber et al., 1995; Leach and Rowan, 1986; Leach and Sangster, 1993; Viets and Leach, 1990). We draw on this work to form a conceptual basis of arsenic enrichment within the context of paleo-fluid migration events.

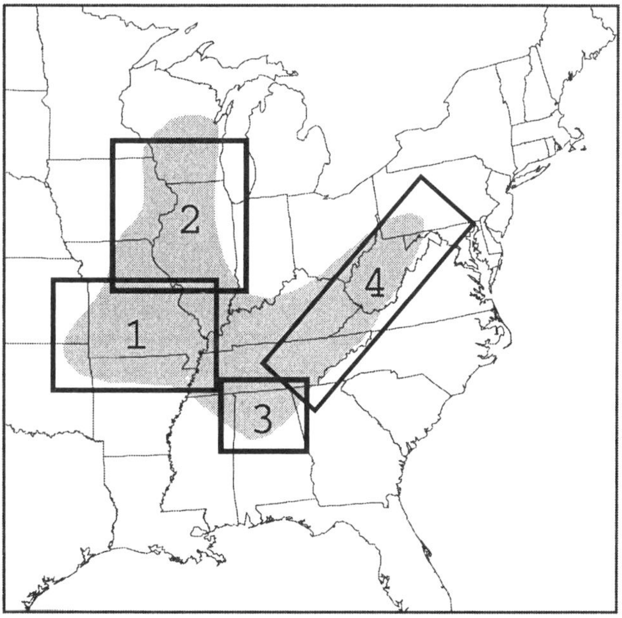

Figure 1. Map of the mid-continental and eastern United States showing regions in which rocks are impacted by large-scale fluid flow. The boxes show the four principal areas that are discussed in detail within this report: 1) The Ozarks region, 2) The Illinois Basin and upper mid-continent, 3) The Warrior basin, and 4) The central and northern Appalachian basin.

The second part of the paper addresses arsenic enrichment in coal of the Appalachian Basin and adjacent Appalachian Mountains in the eastern United States (areas 3 and 4 in Fig. 1). In this regional setting, elevated crustal arsenic concentrations in rocks are well documented, but processes associated with the original introduction of arsenic into coal and metamorphic rocks have not been systematically studied. However, the Appalachian region is very important because here both natural weathering and mining can potentially lead to dispersion of arsenic into the local environment.

Although we focus primarily on the mid-continental and eastern U.S., it should be noted that areas that have been affected by tectonically driven fluid-flow systems are found worldwide. Therefore, many of the concepts and characteristics of arsenic distribution developed in this paper are extendible to other settings, and examples are given at the end of the chapter.

2. LARGE-SCALE BRINE MIGRATION IN THE MIDCONTINENT AND EAST-CENTRAL REGIONS OF THE U.S.

2.1 Introduction

Recent studies to be summarized below have demonstrated that MVT zinc and lead ores of the southern midcontinent region of the U.S (Fig. 2), formed by long-distance migration of brine fluids from adjacent deep basins onto the continental craton. This fluid migration event is of interest not only in the context of ore formation but because, as documented below, trace mineralization deposited along migration paths of the ore fluids can have elevated arsenic contents. Below, we summarize evidence for the scale of this brine migration and describe its impact on arsenic concentrations in both rocks and waters of the region.

The discussion is divided into two parts: (1) the Ozark region which includes the Tri-State district, Southeast Missouri lead belts, Northern Arkansas district, Central Missouri district and (2) the east-central region which includes the Illinois-Kentucky fluorspar district, Upper Mississippi Valley Zn, Pb District, and central Tennessee Zn district. The reason for making this distinction is that the Ozark and east-central regions are differentiated on the basis of Pb isotope systematics (Goldhaber et al., 1995) and ore paragenesis (Hayes et al., 1997). These distinctions imply distinct or largely distinct fluid flow paths.

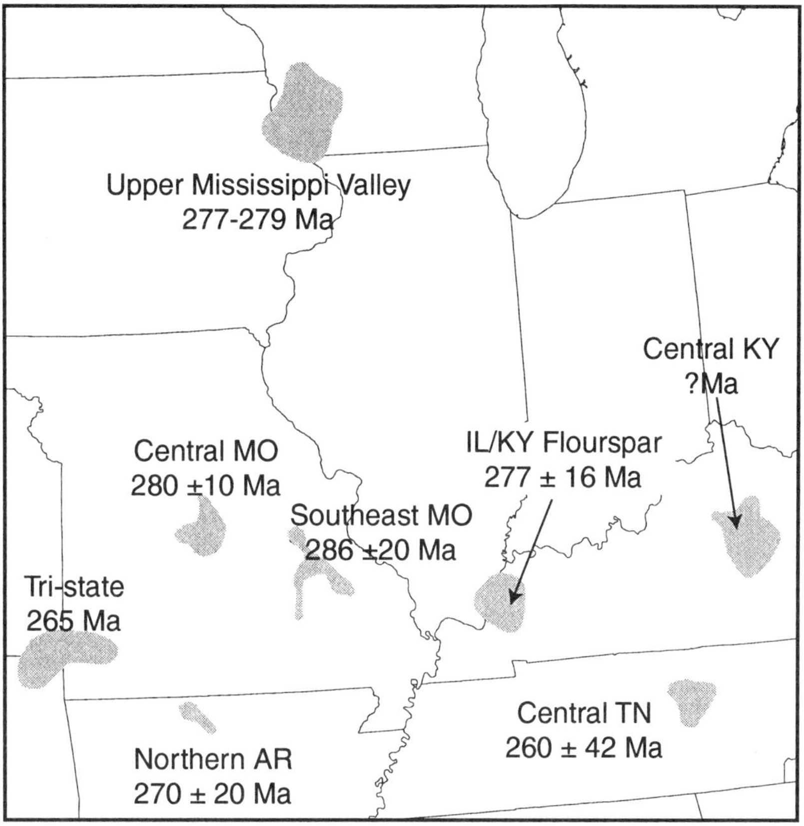

Figure 2. Locations and ages of major Mississippi Valley-type (MVT) ore deposit districts of the mid-continental United States. The ages are taken from Brannon et al.(1996); Chesley et al.(1994); Symons et al. (1997).

2.2 The Ozark Region

The sedimentary section of the U.S. southern midcontinent consists in large part of Paleozoic platform carbonate rocks interbedded with shale and subordinate sandstone. As documented below, a large volume of the Paleozoic carbonate rocks in this region is enriched in arsenic, as well as lead, zinc, copper, cobalt, molybdenum, and nickel compared to surrounding carbonate strata (Erickson et al., 1981; Lee and Goldhaber, 2001). In plan view this trace-mineralized area extends over thousands of square kilometers. The widespread metal enrichment process was related to the formation of much more spatially localized ore districts. These districts consist of economic concentrations of the ore elements zinc and lead, and to a lesser extent copper, cobalt, and barium that occur in the platform carbonate rocks. Rocks from Late Cambrian through Mississippian age host the MVT ore districts several of which are world-class in size (Fig. 2). For example, the lead, zinc, (copper, cobalt) ores of the southeast Missouri lead belts cover an area of approximately 2,000 square kilometers, with total metal production exceeding 20 million metric tones (Ohle, 1997). Collectively, the MVT ore deposits of southeast Missouri form the largest known concentration of lead in the crust of the Earth. Similarly, the Tri-State district of Missouri, Oklahoma, and Kansas covers an area of approximately 1,800 square kilometers and was one of the world's great zinc producing districts.

Despite the extent of ore-grade mineralization and size of MVT districts, the original mineralizing brine fluids are thought to have contained only low concentrations of metals--generally in the ppm to at most hundreds of ppm range (Anderson, 1983; Barnes, 1983). Because the brine fluids contained low metal contents, the formation of very large MVT ore bodies requires that enormous volumes of mineralizing fluids passed through the rocks of the region (Ohle, 1990). This requirement, coupled with the position of many deposits along the margins of large sedimentary basins, and the compositions of the ore fluids that are similar to those of sedimentary basin brines, has led to the conclusion that the elemental enrichments formed from fluids that were derived from adjacent sedimentary basins (Hanor, 1997).

It is now established that the genesis of MVT ores in the Ozark region was closely related in time (Brannon et al., 1997; Symons et al., 1997). Fig. 2 shows the ages of ore formation as determined by a variety of techniques. The ages all fall in the late Paleozoic (range 251-286 Ma), and most span an even narrower range. Late Paleozoic time represents the terminal phase of assembly of the supercontinent Pangea (Frazier and Schwimmer, 1987; Scotese et al., 1979). This was a time of massive orogeny that resulted in formation of the Appalachian and Ouachita mountain belts. The prevailing hypothesis for the origin of MVT ores involves hydrothermal fluids that were driven topographically i.e., due to recharge in these growing mountain chains and gravity flow through the adjoining sedimentary basins, and onto the shallow carbonate platform areas (Bethke, 1986; Bethke and Marshak, 1990; Garven et al., 1993; Sverjensky and Garven, 1992).

One of the necessary conditions for evaluating this gravity (topographically)-driven fluid-flow hypothesis is that it is physically credible. Computer modeling simulations incorporating the physics of fluid-flow through sedimentary rocks have shown that a large-scale, gravity-driven fluid-flow system is fully capable of transporting the heat and mass required to form the large MVT ore districts of the midcontinent region (Appold and Garven, 1999; Bethke, 1986; Garven, 1985; Garven et al., 1993). Modeling studies have also demonstrated that a variety of flow directions related to growth of both the Appalachian and Ouachita mountain chains are plausible (Garven et al., 1993).

The computer modeling results are reinforced by petrographic studies that demonstrate widespread precipitation of hydrothermal dolomite and sulfide minerals by migrating ore fluids. Hydrothermal vug-filling dolomite is known to have formed over much of the Ozark region (Gregg and Shelton, 1989; Leach et al., 1991; Shelton et al., 1993). Fluid inclusion studies have shown that this dolomite typically precipitated at temperatures >100°C from brine fluids whose salinity was very similar to that of fluids that formed the ore deposits (Gregg et al., 1992; Rowan and Leach, 1989; Shelton et al., 1993). The hydrothermal dolomite can be paragenetically correlated using zoning patterns (microstratigraphy) as revealed by cathodoluminescence microscopy. This zoning is recognized over most of southern Missouri, northern Arkansas, and parts of Kansas and Oklahoma (He et al., 1997; Rowan, 1986; Voss and Hagni, 1985) and can be traced directly into the ore districts showing that ore-forming fluids deposited the hydrothermal dolomite (Rowan, 1986). Sites where this distinctive dolomite zoning has been observed are shown on Fig. 3.

The minerals pyrite and to a lesser extent marcasite (both have the composition FeS_2 but differ in crystal structure) are typically intergrown with, and fill vugs in the regionally distributed hydrothermal dolomite. Although much less common, sphalerite (ZnS) and galena (PbS) are also locally present as fillings of vuggy porosity in dolomite. An example of pyrite filling a vug in hydrothermal dolomite is shown in Fig. 4.

The hydrothermal dolomite and associated iron sulfide minerals both formed from hydrothermal ore fluids. This conclusion is documented both by the temperature and salinity measurements on the dolomite noted above, and by similarity in the isotope compositions of trace Pb contained in the widespread epigenetic vug-filling FeS_2 and in the hydrothermal dolomite to that of ore lead in the MVT ore districts (Goldhaber et al., 1995). Studies on the gases dissolved in dolomite-hosted fluid inclusions (Leach et al., 1991) combined with computer-based, geochemical reaction-path modeling leads to the conclusion that the dolomite and paragenetically associated Fe, Zn, and Pb sulfides precipitated due to outgassing of CO_2 (causing pH to increase) from low pH brines as these fluids migrated up dip in aquifer rocks (Leach et al., 1991; Plumlee et al., 1994). The same dissolved gas studies establish that hydrogen sulfide was a component of the hydrothermal fluids over much of the Ozark region. This is important because depending upon specific conditions of pH, temperature, and total sulfide concentration,

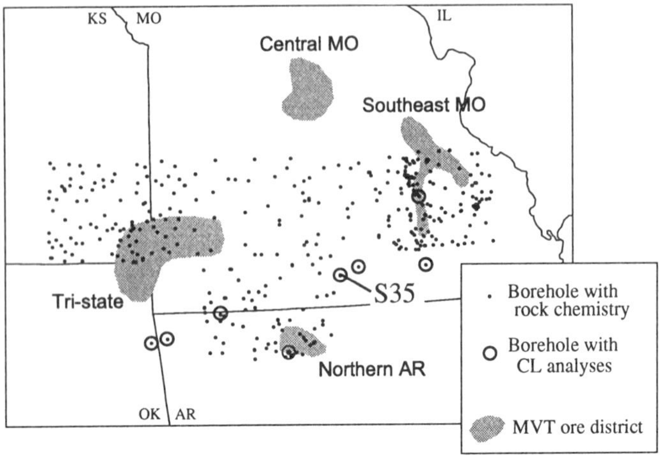

Figure 3. Locations of a subset of the major MVT ore-deposit districts of the Ozark region. Locations of boreholes studied during the USGS-CUSMAP program that were chemically analyzed are shown as black dots. These cores were composited over ten foot intervals and insoluble residues prepared from the composites. These residues were analyzed by emission spectroscopy. Locations of investigations of dolomite zonation were performed, in addition to chemical analyses, are circled (see text for references to these studies). The location of core S-35, discussed further in the text and Figs. 4 and 5 is labeled.

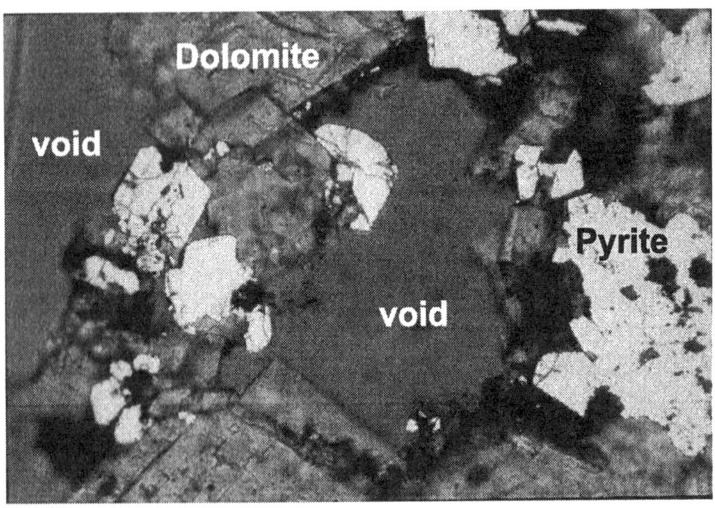

Figure 4. Pyrite filling vuggy porosity in core S-35 in the U.S. midcontinent. See Fig. 3 for the location of this core. Pyrite (bright phase) occurs as both intergrowths and intergrowths with hydrothermal dolomite. Both are growing into a void in the rock.

arsenic forms soluble complexes with dissolved sulfide (Spycher and Reed, 1989; Webster, 1990), and such complexes could have facilitated arsenic leaching and transport by the fluids.

Iron sulfide minerals are, on a regional scale, the dominant residence for elevated concentrations of metals such as Pb, Zn, Cu, Co, Ni, Mo, and As in aquifer rocks of the Ozark region. This mineralogical residence has been established by chemically analyzing pyrite separated from midcontinent rocks (Erickson et al., 1981; Goldhaber et al., 1995). The suite of metals enriched in the iron sulfide phases includes Pb, Zn , As, Cu, Co, Ni, Mo, and Ag. Here we focus on arsenic. Fig. 5 is from a polished section of rock from a depth of 225.2 m in core S-35 from Douglas County, Missouri (location shown on Fig. 3). This figure is a "dot map" illustrating the relative abundance of different elements by the density of dots. Note the covariance of the Fe, S, and As concentrations. The data for core S-35 are typical of others in the southern midcontinent region for which chemical analyses exist (M. Goldhaber, unpublished data).

In contrast to the extensive data collected on trace-mineralized rocks between ore districts, less information is available on the arsenic contents of the ore deposits. Because iron sulfide minerals are typically gangue phases, they are commonly ignored in trace element studies of ore deposits, which tend to focus on ore minerals such as sphalerite and galena. Therefore, there have not been extensive analyses for arsenic in iron sulfides in many of the MVT ore districts or trace-mineralized areas. However, a few studies on trace element contents in iron sulfides from the Ozark region have been performed. Wu et al. (1996) analyzed 80 pyrite and marcasite samples from the Viburnum Trend of the Southeast Missouri Lead District and found arsenic in concentrations of 2 to 900 ppm. Bhati and Hagni (1980) also analyzed iron sulfide minerals from this area, but did not publish results for arsenic. Hagni (1993) described the relatively rare occurrence of nickel-arsenic-sulfide ores from the Magmont-West ore deposit of the Viburnum Trend. Leach et al. (1995) list arsenic as a trace constituent in ores from the Northern Arkansas and Southeast Missouri MVT ore districts, but without abundances specified.

2.3 Stratigraphic and Geographic Distribution Of Arsenic In The Ozark Region

Erickson and colleagues (Erickson et al., 1988; Erickson et al., 1991; Erickson et al., 1978) extensively studied the geographic distribution of trace elements within the Ozark portion of the U.S. midcontinent. These studies were part of a USGS program to evaluate potential for undiscovered mineral

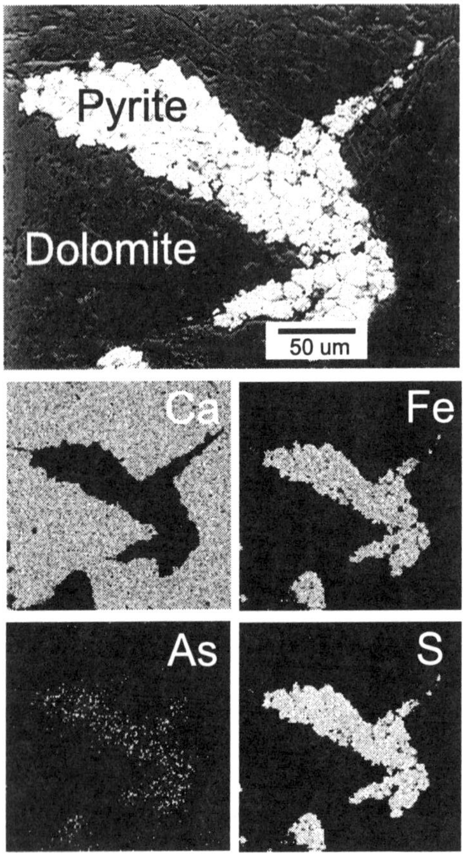

Figure 5. Scanning-electron microscope (SEM) images of pyrite within a dolomite host rock. Sample is from borehole S-35 (see Fig. 3 for location). A fifty micrometer scale bar is shown for reference. The uppermost panel is a backscattered electron image. The bright area is pyrite and the dull areas are dolomite. The remaining 4 images are dot maps showing the relative abundance of specific elements. The brighter the image, the higher is the relative abundance of the element. Note the correlation of arsenic (As) with pyrite as indicated by the Fe and S dot maps, and the anticorrelation with dolomite (Ca dot map).

resources of the Ozark region by identifying MVT ore-fluid pathways (Erickson et al., 1983). The primary approach used in these studies was to isolate and chemically analyze the acid (6N HCl)-insoluble residues of carbonate rock samples from boreholes. MVT sulfides, and therefore MVT-related metals, tend to be insoluble in HCl and are thus concentrated in this rock fraction (Erickson et al., 1981). The rock samples from which the acid-insoluble residues are derived were obtained from boreholes distributed throughout the Ozarks (Fig. 3). Individual rock samples represent

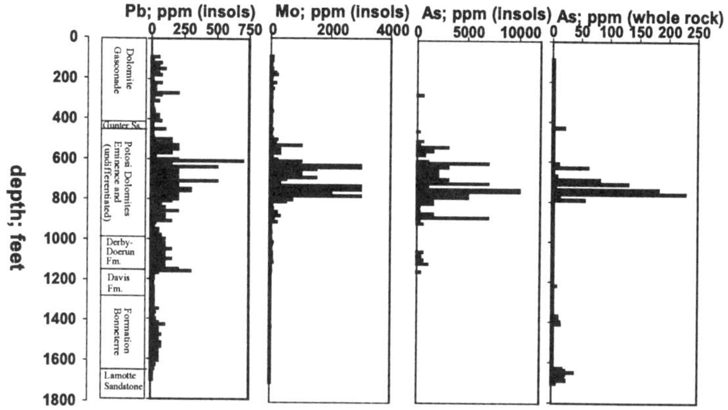

Figure 6. This figure shows the depth distribution of Pb, Mo, and As in whole rock and insoluble-residue samples of rocks from borehole S-35 in the Ozark region. See Fig. 3 for the location of core S-35. The Pb (insols), Mo (insols), and As (insols) data are from emission spectroscopy analyses of insoluble-residues from the core. The Pb and Mo data are shown because these elements often correlate with As in midcontinent region rocks. The As (whole rock) column is data from whole rock analyses of the same depth intervals by Atomic Absorption Spectroscopy. The overall core interval spans the upper Cambrian and lower Ordovician eras. The purpose of this plot is to show that insoluble residue data reflect, and even provide an enhanced distribution of As in rock column. The enhancement is because As is localized in the sulfide fraction that is concentrated in the insoluble residues.

composites of 10-foot borehole intervals. Because of the USGS mineral assessment effort, insoluble residue concentrations are extensively available for midcontinent rocks, whereas whole rock data are not.

In order to understand the relevance of the insoluble residue concentrations to the overall arsenic contents in midcontinent rocks, it is useful to compare data for the insoluble residues and the whole rocks. An example of this comparison for a suite of elements including arsenic is shown in Fig. 6. This figure plots depth versus arsenic content in whole rock samples, and arsenic, molybdenum, and lead contents in the acid-insoluble residue fraction of core S35 (see Fig. 3 for sample location). The detection limit for arsenic in the whole rock samples was rather high at ~4 ppm. Nevertheless, it is evident that several depth intervals, particularly those within the Ordovician Eminence and Potosi Dolomites are significantly enriched in arsenic relative to surrounding intervals. Arsenic is concentrated in acid-insoluble residues compared to whole rock samples by roughly one order of magnitude. The depth range of elevated arsenic values in the insoluble residues and whole rocks are positively correlated. It thus appears from this comparison, and from several others that could be presented, that geochemical analyses of insoluble residues provide a data set from which general trends in rock arsenic distribution can be evaluated.

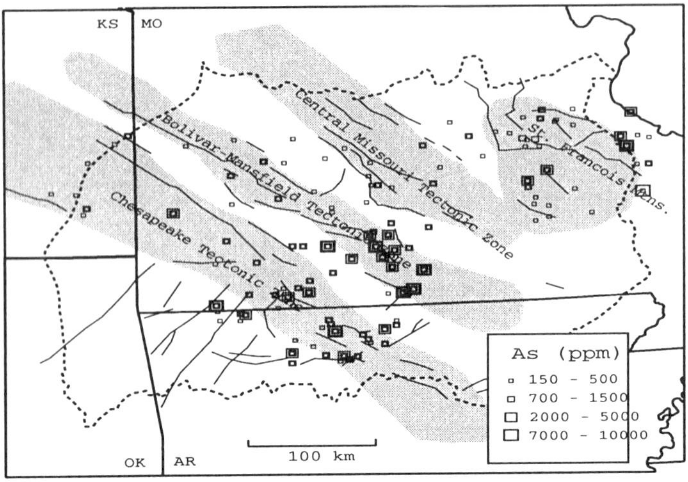

Figure 7. Plan-view distribution of arsenic in acid-insoluble residues of borehole rock samples as a series of squares whose size is proportional to the concentration of As in the sample. The plot only shows sites at which arsenic was detected in concentrations ≥150ppm. Multiple values at individual sites are shown as concentric symbols. Note the concentration of elevated values in the southern Ozarks and Data is from Lee and Goldhaber (2001). Major tectonic zones are shown for reference - Lee (2000) identified correlations between the faults and fractures of these tectonic zones and enrichments in MVT-related metals such as Zn and Pb although these correlations are not evident from the arsenic plot.

The distribution of arsenic in insoluble residue samples is shown in map view in Fig. 7 taken from Lee and Goldhaber (2001). This plot projects data from multiple depths and stratigraphic horizons to the surface. Parts of southern Missouri and northern Arkansas have consistently elevated rock arsenic concentrations. Fig. 8 shows the same data as in Fig. 7, but displayed in three dimensions. Samples having over 700 ppm arsenic in the insoluble residue are shown as larger blocks, and samples having less than this limit are shown as small blocks. Arsenic is enriched in the lower part of the Paleozoic section. Data not presented here show that arsenic abundance tends to correlate geographically and stratigraphically with elevated levels of lead and molybdenum (Lee, 2000). The arsenic enrichment is primarily within the Upper Cambrian through Lower Ordovician dolomites of the Ozark aquifer, which is the primary source of drinking water in the Ozark region (Imes and Emmett, 1994).

2.4 East-Central Region and Illinois Basin Region

As in the Ozark region, ore districts at either end of the Illinois Basin formed during the late Paleozoic (see Fig. 2 and associated references) where fluid-flow presumably was gravity driven as well (Bethke, 1986; Rowan and Goldhaber, 1996). Paleohydrologic modeling has shown that hydrothermal ore fluids migrating from south to north through the Illinois Basin were responsible not only for ore genesis at either end of the basin, but also partly responsible for the observed patterns of coal maturation (Rowan et al., 2001) and sediment diagenesis (Fishman, 1997; Pitman and Spoeetl, 1996). Furthermore, the pattern of fluorine distribution in rocks of the Illinois Basin is indicative of fluorine transport by hydrothermal fluids from the area of the Illinois-Kentucky Fluorspar district northward into the Illinois Basin (Goldhaber et al., 1994; Rowan et al., 1996).

Trace mineralization including sporadic arsenic enrichment is present within the Illinois Basin, particularly in the southern portion and along its western edge (Erickson et al., 1987). Follow on studies failed to detect elevated trace-element contents in rocks of the central Illinois Basin (Goldhaber, unpublished data). However, data from further north suggests enrichment in elements related to MVT deposits including arsenic extending into southeast Minnesota, based on chemical analysis of acid insoluble residues from that area (Lively et al., 1997).

3. FLUID FOCUSING RESULTING IN MORE INTENSE AREAS OF TRACE ELEMENT LOCALIZATION

The regional fluid-flow described above produced widespread diffuse trace mineralization over large areas of the U.S. Midcontinent. In addition to this diffuse mineralization, specific paleohydrologic processes can lead to somewhat more intense element enrichment, ranging from economic ore deposits down to sub-economic but nonetheless significant concentrations of ore and gangue minerals. Because arsenic can accompany MVT mineralization, these more intensely mineralized areas are important to identify because they may be particularly susceptible to environmental impacts.

Paleohydrologic controls resulting in this more intense style of mineralization were identified during a multi-year cooperative research program (CUSMAP for Conterminous U.S. Mineral Assessment Program) among the USGS, state geological surveys, and universities in the region (Pratt and Goldhaber, 1990). Favorable paleohydrologic settings occur where migrating ore fluids are focused into a volume of reactive rock or where the juxtaposition of aquifers allows two ore fluids of different composition to mix. Pratt (1990) has summarized these paleohydrologic conditions, which are illustrated schematically in Fig. 9. This figure shows a stratigraphic section composed of sandstone, shale, limestone, and dolomite beds whose vertical arrangement is roughly patterned after that of the U.S. midcontinent. Fluid-flow paths are shown as arrows and resulting areas of ore formation as capital letters. On a regional scale, these flow paths are predominantly lateral as described above. However, particular combinations of aquifers and aquitards allow local vertical fluid-flow. Shale and dense finely crystalline limestone are the major aquitards. Sandstone, vuggy dolomite, and occasionally porous limestone are the principle aquifers.

A specific rock type may be reactive towards migrating MVT brines for a variety of reasons. Chemical disequilibrium with hydrothermal fluids is one cause, by which migrating brines are in chemical equilibrium with dolomite but may be out of equilibrium with limestone (Plumlee et al., 1995). Thus, limestone reacts with the ore fluids resulting in geochemical changes such as a pH increase that promote ore and gangue mineral deposition. Likewise, coal is also a reactive rock type capable of fixing MVT ore and gangue minerals as it potentially contains a reservoir of reactive reduced sulfur capable of precipitating metals (Hatch et al., 1976; Whelan et al., 1988).

Regions where migrating brines rise and intersect a reactive rock type or mix with geochemically dissimilar fluids are favorable sites for mineralization. Geologic features favorable for vertical fluid-flow are: (1) faults (2) shale 'windows' and (3) aquifer pinchouts. Faults having a large vertical extent are major conduits for MVT ore fluids. This is illustrated in Fig. 9 by a fault rising from the basement through the section. Mineralization is shown as occurring in area 'A' where the rising fluids pass through a limestone bed; faults penetrating coal beds would also be mineralized in this model. Shale windows are areas where shale aquitards pinch out or are absent due to facies changes or erosional unconformities. In the absence of such aquitards, fluids might rise towards reactive sites. A striking example of a shale window is the world-class Tri-State mining district (Fig. 2) that formed in a shale window of large geographic extent (Pratt et al., 1993). Aquifer pinchouts such as the onlap of the lowermost sandstone against a basement high (leading to ore area 'C') or the transition of the upper sandstone to shale leading to ore area 'C′' (Fig. 9) result in fluids being forced upwards, potentially into a reactive rock type such as limestone, or into a fluid mixing zone (area C). A substantial amount of mineralization in the southeast Missouri lead belt was associated with pinchouts of the basal sandstone (Lamotte Sandstone) against basement highs (Ohle, 1990), and was located where these rising fluids entered a mixing zone (Goldhaber et al., 1995).

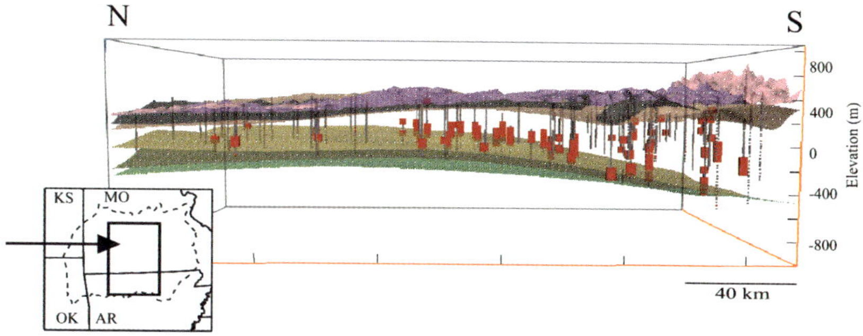

Figure 8. 3D cross-sectional view of the central portion of the Ozark Uplift. Inset box shows the areal extent of the 3D model and view direction. View is looking to the east, down the axial crest of the Ozark uplift. Surfaces of the model are the tops of the regional aquifer units of the Ozark aquifer system. Vertical exaggeration is approximately 40X. The vertical lines within the model are the locations of CUSMAP boreholes (also shown in Fig. 3). The enlarged red blocks are chemical analyses of insoluble residue samples that contain greater than 700 ppm arsenic (approximately equal to 70 ppm on a whole rock basis), in contrast to the small black boxes that represent samples with less than 700 ppm.

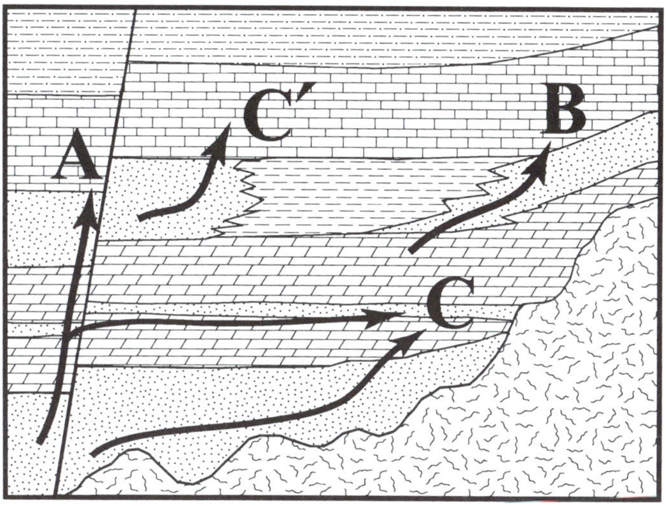

Figure 9. Conceptual diagram of the important paleohydrologic controls on the occurrence and distribution of MVT mineralization, and arsenic enrichment. This figure is simplified from one originally constructed by T. Hayes (written communication, 1989). Sandstone is shown as a dot pattern, limestone as a rectangular pattern, dolomite as a trapezoidal pattern, and shale as the dash-dot pattern. The arrows show pathways of fluid migration, and the letters show the locations of ore formation discussed in the text.

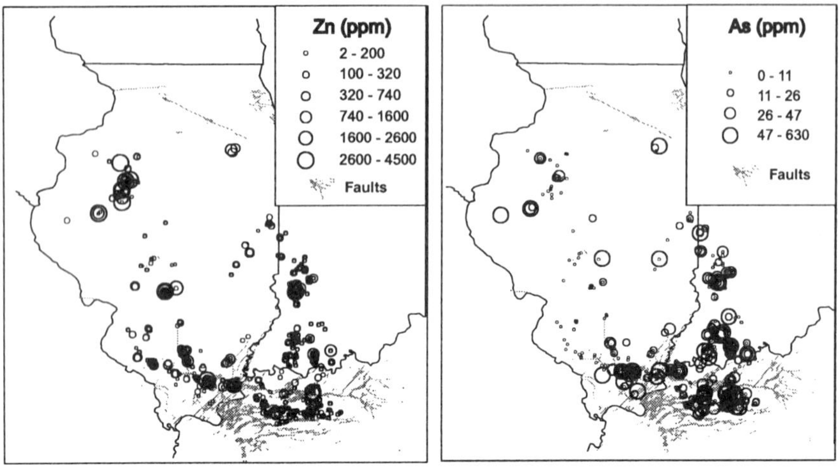

Figure 10. Distribution of zinc content (left panel) and arsenic content (right panel) in coals from the Illinois basin. Data sources are given in the text. The size of the symbol is proportional to the abundance of the element. The breakpoints for the element distributions were chosen using the default 'natural breaks' algorithm in the computer program 'Arcview'. Faults are shown as light lines. Note the concentration of zinc and arsenic associated with the intense faulting in the southern Illinois Basin, and the localization of zinc and arsenic in the northwestern portion of the basin.

We have chosen to illustrate arsenic enrichment resulting from a subset of these paleohydrologic conditions using geochemical data from coal beds of the Illinois Basin and adjacent portions of the western region of the interior coal province in Iowa and Missouri. These examples also serve to introduce coal as an important lithologic residence of arsenic, a subject that is pursued further in a subsequent section of this paper.

Figure 10 shows the distribution of zinc and arsenic in coal samples of the Illinois basin. The data are shown as circles whose size is proportional to the element's content on a whole coal basis. The data shown were compiled from Gluskoter (1977), Oman et al. (1992) and Bragg et al. (1991) and are listed in Affolter and Hatch (2002). Zinc in Illinois Basin coal ranges up to 10,000 ppm and arsenic up to 950 ppm. In addition to zinc and arsenic, data from the sources listed above shows that the elements lead, barium, cadmium, cobalt, copper, molybdenum, and nickel are also significantly enriched over average values for U.S. coal.

Major faults of the Illinois Basin area are shown in Fig. 10. The southern portion of the basin is intensely faulted; northeast-trending faults in southeast Illinois and northwest Kentucky are associated with the Illinois-Kentucky Fluorspar district. A number of these faults in Mississippian limestone beds are mineralized with the ore minerals fluorite (CaF_2), sphalerite, and galena, and with gangue minerals such as calcite and pyrite

(Heyl, 1983). Anomalously high zinc and arsenic contents in overlying Pennsylvanian coal beds also occur in the highly faulted area. These zinc and arsenic enrichments in the southern part of the Illinois Basin represent an example of fluid migration from deeper in the stratigraphic column into more shallowly buried coal beds (e.g. analogous to area 'A' in Fig. 9).

Sporadic high zinc and arsenic contents are localized towards the northwest limit of the coalfield in western Illinois. Detailed studies in western Illinois (Cobb, 1979; Hatch et al., 1976) show that zinc occurs as the mineral sphalerite and is hosted by cleats (vertical fractures) in the coal. The sphalerite is accompanied by pyrite kaolinite, quartz and calcite. Analysis of a sample of a mineralized cleat containing both pyrite and sphalerite supplied by Jim Cobb showed that arsenic was localized in the pyritic portion and not the sphalerite (M. Goldhaber, unpublished data). Because cleats only form after significant burial, the trace mineralization is considered to be epigenetic in origin (Cobb, 1979). Fluid inclusion studies on the sphalerite (Coveney et al., 1987) indicate that it formed at temperatures significantly above those expected for its present burial depth, from fluids with very high salinities similar to those of MVT ore fluids. Hatch et al. (1976) and Whelan et al. (1991) concluded that migrating hydrothermal fluids associated with MVT ore formation deposited the trace mineralization in coal.

Figure 11 shows a geologic cross section oriented roughly northwest southeast through the western shelf area of the Illinois basin. Coal beds are found in Pennsylvanian rocks. In the region of the zinc and arsenic anomaly, Mississippian strata are truncated beneath a sub-Pennsylvanian unconformity (labeled the sub-Absaroka unconformity in the figure) north of that area. This pinchout of the Mississippian section is of importance to understanding the incorporation of trace elements into the overlying Pennsylvanian coal beds. The Mississippian Burlington-Keokuk Limestone, in contrast to many other limestone units, likely had substantial permeability and could have transmitted hydrothermal fluids because the Burlington-Keokuk is known to contain trace mineralization within the Illinois Basin (Erickson et al., 1987) and host ore deposits (Heyl, 1983). Trace sphalerite in fact occurs in the Burlington-Keokuk Limestone near its pinchout (Erickson et al., 1987). It is thus likely that fluids were forced upwards into the overlying Pennsylvanian section, precipitating epigenetic minerals in coal cleats.

The effect of a shale window on the arsenic content of coal can be illustrated with data from the interior coal province (Fig. 12). Hatch (1983) pointed out that Pennsylvanian coal seams in southern Iowa and adjacent parts of Missouri, contain sphalerite. The mode of occurrence of this sphalerite is similar to that described in Illinois. Data in the USGS coal database (Bragg et al., 1997) documents the fact that these sphalerite-bearing

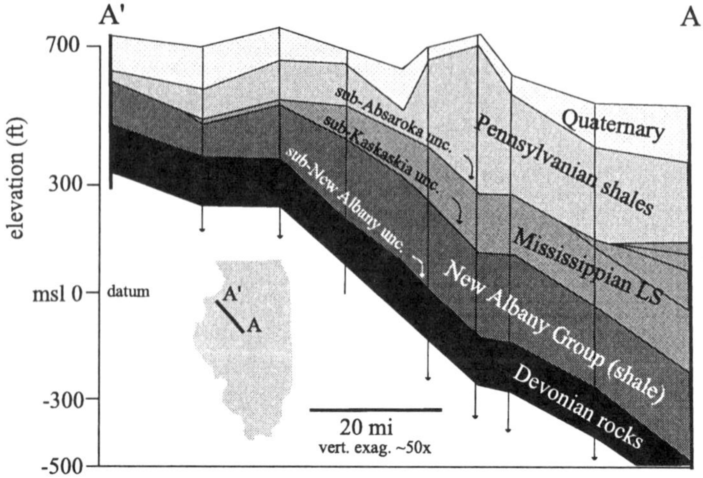

Figure 11. Generalized geologic cross-section of the Illinois basin along cross-section A-A' (see inset for the location of the cross-section). The locations of cores used a control points are shown as vertical lines. Note the pinchout of Mississippian limestone beneath the sub-Absaroka unconformity. Fluid flow in the Mississippian rocks would be confined between New Albany Group shales below, and Pennsylvanian (coal-bearing) shales above. Near the pinchout of the Mississippian rocks, fluids migrating in this part of the section would be forced upwards in the Pennsylvanian coal-bearing portion of the section. The extent of the zinc-arsenic enriched area in the northwestern part of the basin (see Fig. 10 for a map view) is shown for reference.

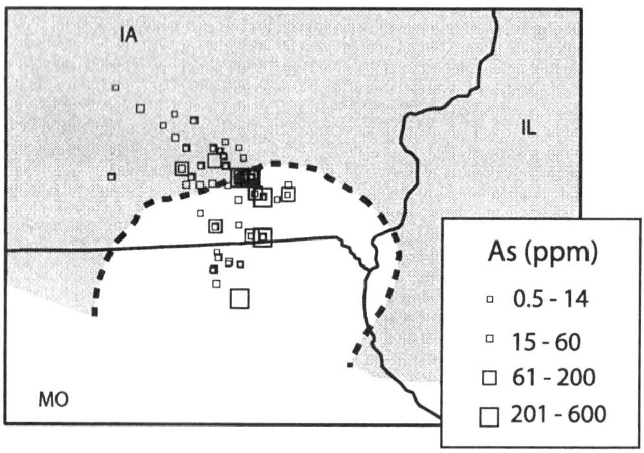

Figure 12. Arsenic concentrations in Pennsylvanian coals of southern Iowa and northern Missouri. The Shaded area shows the region in which the Ordovician Maquoketa shale is present. Dashed line shows the boundary of a window in this shale. Note the occurrence of elevated arsenic concentrations (>61 ppm). They are either within or near the edge of the shale window. The shale window evidently allowed metal-bearing solutions to migrate pwards into the overlying coal beds. The breakpoints for the arsenic distribution were calculated using the default algorithm ('natural breaks') for the program Arcview.

coal beds also may also be enriched in arsenic. The abundance of arsenic in coal samples from southern Iowa is shown together with lithologic information relevant to understanding the paleohydrologic controls on MVT fluid introduction into the coal beds. The coals having elevated arsenic (between 61-200 ppm) overlie, i.e. are up stratigraphic section from a window in the Ordovician Maquoketa shale (the non-shaded portion of the figure) (Pratt, 1992) or are adjacent to it.

The arsenian pyrite underlying the Fox River Valley (Schreiber et al., this volume) most likely represents another MVT type of deposit that is the source of arsenic-rich ground water.

4. ARSENIC IN COAL OF THE APPALACHIAN REGION

Arsenic is locally enriched to concentrations of one to two orders of magnitude above typical background concentrations in coal of the Appalachian Basin. We hypothesize that these enrichments result from interactions of coal with hydrothermal fluids mobilized by metamorphic processes during the late Paleozoic Alleghanian orogeny. Our interpretation (developed more fully below) is that these fluids represent the distal (western) part of a large metamorphic fluid-flow system that also was responsible for formation of gold deposits and gold occurrences further east in metamorphic rocks of northern Alabama and perhaps elsewhere in the Appalachian Piedmont. If true, the timing of the metamorphic hydrothermal system is similar to the topographically driven, basin brine derived system discussed above, which was also of late Paleozoic age. The responsible fluid flow system also may ultimately be tied to the same overall tectonic event, the assembly of the supercontinent Pangea during the late Paleozoic Appalachian and Ouachita orogenic events. However, the fluid source was at least in part metamorphic devolatilization reactions leading to low salinity fluids and not basin brines as in the midcontinent. The suite of elements enriched in the Appalachian basin sediments differs from that in the midcontinent, implying that the chemistry of the hydrothermal fluids may have been different, in part reflecting this difference in fluid source. This metamorphic fluid-flow event represents a second example of a fluid-flow system that is both of enormous extent and has profound implications for arsenic distributions in the crust of the United States.

We first present results from northern Alabama, where detailed studies have been conducted during the course of a joint USGS-Alabama Geological Survey project. The focus is than expanded to a larger portion of the Appalachian Basin.

4.1 Arsenic in Alabama Coal

The highest arsenic contents in U.S. coal are found in coal of the Warrior, Cahaba, and Coosa coalfields in Alabama (Goldhaber et al., 2000). The arithmetic mean content of arsenic in coal samples from the entire U.S. is 24 ppm with a standard deviation of ± 60 (n=7676) (Kolker and Finkelman, 1998). For comparison, the mean value for Alabama bituminous coal samples is 75 ppm. The striking enrichment in arsenic in Alabama coal can be seen in Fig. 13, which is a histogram contrasting arsenic abundance in Warrior (and Cahaba) basin coal samples with that in Illinois and Wyoming coals. The Alabama data has a long tail to elevated values with some samples exceeding 500 ppm on a whole coal basis.

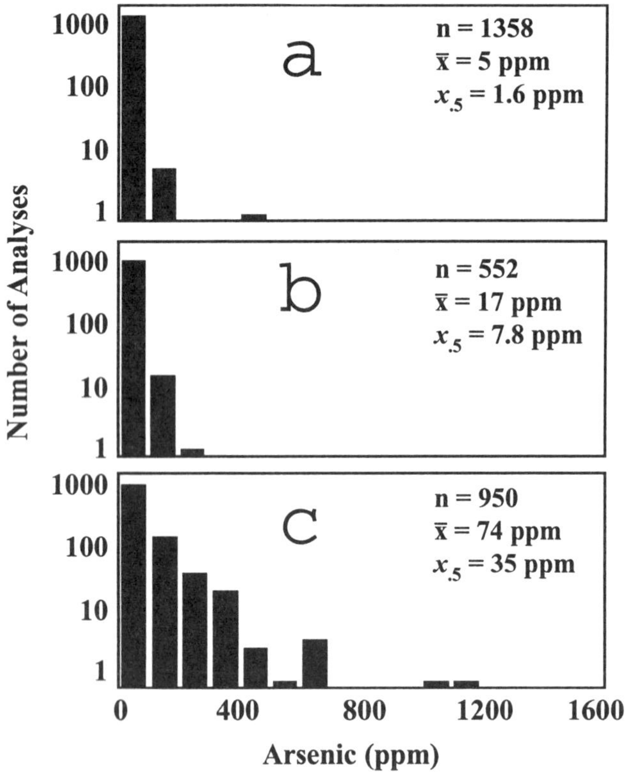

Figure 13. Histograms and descriptive statistics of arsenic abundance in U.S. coals. a) Coals from the Powder River basin of Montana and Wyoming. b) Coals of the mid-continental U.S. including Illinois, Indiana, and western Kentucky. c) Coals of the Black Warrior basin, Alabama. Also shown for each histogram are the number of samples (n), the mean (x) and median ($x_{.5}$)

4.1.1 Geologic Background

The Warrior basin, a foreland basin at the south end of the Appalachian Plateau constitutes a major structural element in northern Alabama. Other structural elements in this region are the unexposed Ouachita fold-thrust belt and central uplift southwest of the Warrior basin, and a succession of thrust belts in the southern Appalachians east of the basin that make up the Valley and Ridge province and Northern Piedmont terrains (Fig. 14). The Warrior basin formed by convergence of the Ouachita and Appalachian thrust belts. It has been described (Thomas, 1989) as a southwest-dipping regional homocline, consisting of a cratonal cover sequence that dips gently under the two orogenic fronts. The youngest rocks filling the basin are sandstones, mudstones and coal of the Pennsylvanian Pottsville Formation, which represents a clastic wedge of materials that was shed principally from the adjacent Ouachita orogenic high.

Directly east of the Warrior basin is the fold-thrust belt of the Valley and Ridge Province. The two tectonic blocks are in contact across the Opossum Valley thrust fault (and other faults along that strike zone), the westernmost of numerous thrust splays off the master Alleghanian decollement that reach the surface. The province is characterized by alternating fold ridges and valleys in a stratigraphic sequence spanning Cambrian to Pennsylvanian time. The Cahaba and Coosa coalfields occur in Pennsylvanian rocks of the Pottsville Formation within this province.

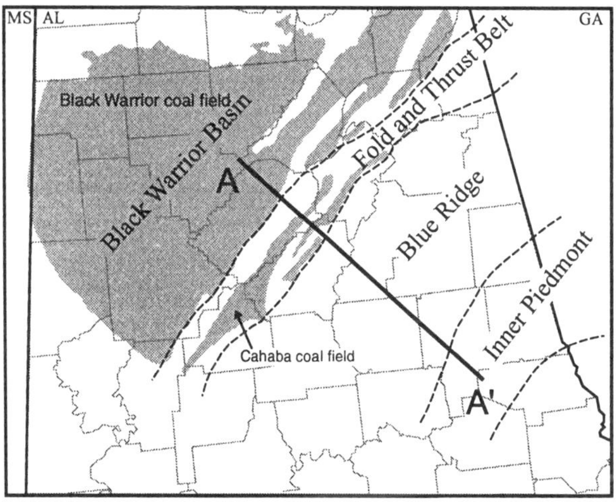

Figure 14. Physiographic provinces of Alabama. Shaded areas are the coalfields of Alabama. The line A-A' shows the location of the line of section in Fig. 22

4.1.2 Distribution of Arsenic in Alabama Coal

The geographic distribution of arsenic within the Warrior and Cahaba field coal beds is plotted in Fig. 15, which is a plan view map of arsenic concentrations. Coal beds occur over a range of depths in the Pottsville Formation and Fig. 15 combines data from all of these beds. This display of the data overemphasizes high values because lower concentration samples from the same geographic area or even the same coal-face may be obscured by overlying samples with higher concentrations. Nonetheless, it is evident that there are significant areas of the Warrior and Cahaba basins where the coal has been enriched in arsenic. These high arsenic areas tend to coincide with higher sulfur coal (Winston, 1990). The three dimensional distribution of arsenic in the Warrior basin is depicted in Fig. 16, which shows contoured values for major coal-producing groups (see Pashin et al., 1995 for a discussion of these groups) recognized in the basin. Elevated arsenic contents are particularly noticeable in northern Tuscaloosa and eastern Fayette Counties. Arsenic is nowhere present throughout the vertical extent of the coal-bearing section, but in places it is nearly so. The areas having elevated arsenic are accompanied by significant but less-striking enrichments (compared to average U. S. coal) in Mo, Sb, Se, Cu, and Hg (Goldhaber et al., 2000) and Au (see below).

4.1.3 Residence of Arsenic in Alabama Coal

As in the midcontinent, arsenic in Alabama coal resides in the mineral pyrite (Kolker and Finkelman, 1998). This is demonstrated in Fig. 17 and Fig. 18, which document arsenic localization in pyrite. A wavelength-dispersive elemental map obtained using an electron microprobe (Fig. 17) shows arsenic zonation within a mass of pyrite that contains a series of arsenic zones. The pyrite occurs along a microfracture in coal. Individual spot analyses indicate that the bright areas in this map (red-yellow) have up to 4.5-weight percent arsenic, and the dull areas (blue) have a few tenths of a weight percent. Figure 18 shows a group of early diagenetic pyrite framboids in filled by later pyrite cement. Elevated arsenic concentrations are found only in the cement. Spot analyses of As-bearing pyrite (termed arsenian pyrite) from a number of Alabama coal samples using a Laser Ablation- Inductively Coupled Plasma, Mass Spectrometer confirm that pyrite is also the residence of the elevated contents of Mo, Sb, Se, Hg and Au (M. Goldhaber and I. Ridley, unpublished data). Gold contents up to nearly 200 ppb were measured in pyrite nodules (ash basis) (Goldhaber and Hatch; unpublished data). Because high sulfur coal is avoided during mining and coarse pyrite is removed from coal by a cleaning process prior to

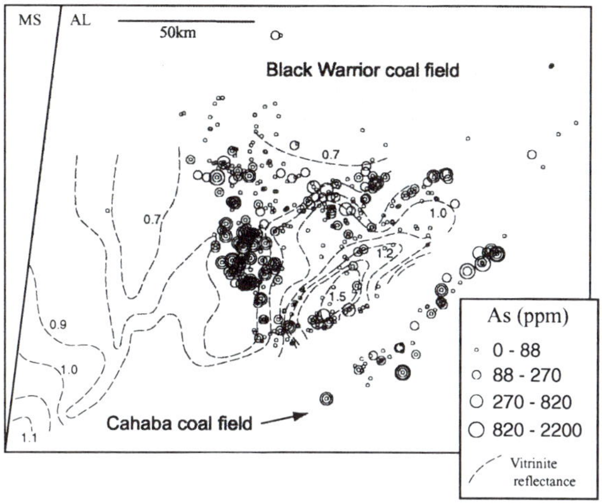

Figure 15. Arsenic concentrations in coals of the Black Warrior basin, Alabama. Shaded areas show the extent of coalfields of Alabama. The proportionally sized symbols show the content of arsenic in coal samples analyzed by the USGS. Some symbols that plot inside each other are from the same core or coal face. The break points for the symbol groups were calculated using the 'natural breaks' algorithm in the computer program 'Arcview'. Dashed lines are contours of vitrinite reflectance data from coals. We are presently evaluating whether the lack of correspondence of high arsenic content with high vitrinite reflectance values is an artifact due to low sample density or reflects an actual trend in the data.

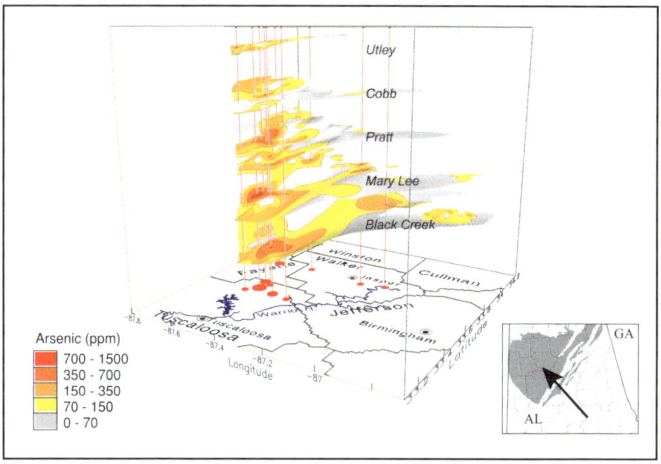

Figure 16. 3-D visualization of arsenic concentrations in the major coal groups of Black Warrior basin, Alabama. Inset box shows view direction of the visualization. Surfaces show the contoured concentrations of arsenic for the Utley, Cobb, Pratt, Mary Lee, and Black Creek coal groups. Vertical lines are drawn through areas with exceptionally high arsenic concentrations; the downward projection of these lines shows the plan-view location of high arsenic concentrations.

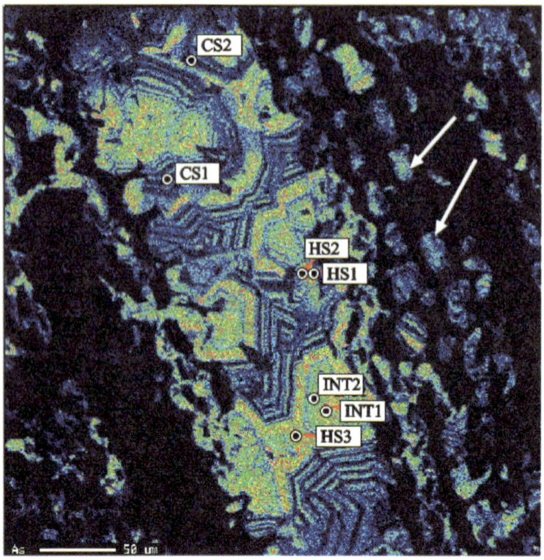

Figure 17. Wavelength-dispersive electron microprobe image of arsenic concentration in within a coal sample from the Black Warrior basin of Alabama. Brighter colors show areas with higher arsenic concentrations and reveal the oscillatory zonation of arsenic in pyrite. Black is coal. White arrows indicate cell lumens in the host coal that are filled with epigenetic pyrite. The labeled points are microprobe analysis sites (values in weight percent): HS1=4.38% As, HS2=4.45% As, HS3=3.97% As, CS1=0.40% As, CS2=0.33% As.

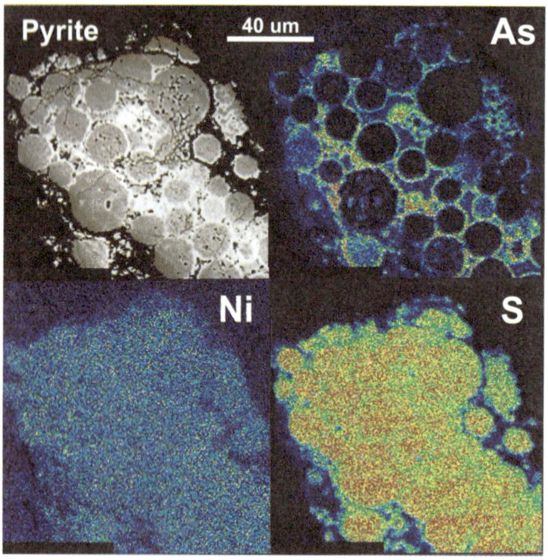

Figure 18. Wavelength-dispersive electron microprobe image of a cluster of pyrite framboids within coal from the Black Warrior basin Alabama. Bright colored areas show the presence of arsenic, nickel, and sulfur. Arsenic is concentrated in epigenetic overgrowths and cement surrounding the framboids, while the diagenetic framboid interiors themselves contain little or no arsenic. This indicates that the arsenic was added after earliest diagenesis.

its utilization, coal combusted in power plants will have considerably less arsenic than the whole coal concentrations documented here. On the other hand, the waste products from the cleaning process may be enriched in arsenic.

4.1.4 Timing of Arsenic Introduction into Coal

Observations on the mode of occurrence of arsenic in coal samples, such as those shown in Figs. 17 and 18 support a late diagenetic or epigenetic origin for the arsenic. This conclusion is reinforced by our field observations. We have found that pyrite recovered from fault zones is commonly arsenian, whereas pyrite from unfaulted beds at distances of tens of meters may not be. For example, a pyrite nodule from a faulted coal bed in the abandoned Kellerman coal mine (Pashin et al., 1999) in southeastern Tuscaloosa County contains 2200 ppm arsenic. Pyrite samples collected from coal within about 5.0 m from the faults were likewise arsenic enriched. Coal samples collected further from the fault contained only background levels of arsenic. Arsenic was observed associated with pyrite near small fault zones in at least two other mines. We have also recognized arsenian pyrite on cleats (vertical fractures) in coal. This observation supports an epigenetic origin as cleat formation requires that coalification proceed to a point where fracturing occurs. This requires considerable burial (Pashin et al., 1999).

Also supporting a post-depositional origin for arsenic are geochemical data on shales and sandstones adjacent to As-rich coal beds (J. Hatch and M. Goldhaber, unpublished data). These rocks are low in arsenic and associated elements. The mean arsenic content of sandstones adjacent to the coal beds is 9 ppm (n=20) and the median 5.5 ppm. The mean arsenic content for adjacent non-coaly shales is 20 ppm (n=29), and the median arsenic content is 12 ppm. For comparison, the average arsenic content of sandstones is 1 ppm and shales is 13 ppm (Turekian and Wedepohl, 1961). If as is typical arsenic is transported to the depositional site adsorbed on particle surfaces, it is likely that the adjacent rocks (particularly shales) would be enriched, reflecting the sedimentary erosional source of the arsenic.

In contrast to the evidence for a late diagenetic timing of arsenic addition, one observation may indicate relatively early introduction of arsenic in the depositional history of the coal beds. As illustrated in Fig. 17 arsenian pyrite locally fills cell lumens (the cavities bounded by the cell wall) of coalified plant debris. These cell lumens, if unfilled, are crushed during burial. Therefore, unless the arsenian (>~0.1 wt. % arsenic) pyrite is a replacement of earlier mineral phase, the presence of arsenian pyrite in uncrushed lumens argues for a relatively early formation.

4.1.5 Arsenic in Gold Occurrences and Ore Deposits

In order test for possible genetic linkage between processes introducing arsenic into coal and gold deposits, it is first necessary to present some background on the latter. Gold occurrences and deposits in northern Alabama are restricted to metamorphic rocks east of the Valley and Ridge Province (Fig. 19). The eastern boundary of this province is the Talladega fault, which forms the leading edge of the overthrust metamorphic complex that makes up the northern Piedmont of Alabama. The Piedmont terrain is separated into two units by the Hollins Line fault, the Talladega slate belt to the northwest and the Ashland-Weedowee belt to the southeast. A major Alleghanian strike-slip fault, the Brevard zone marks the southeastern limit of the northern Piedmont. The terms Blue Ridge and northern Piedmont are equivalent in this region.

Gold occurrences and deposits are known from both the Talladega slate belt and the Ashland-Weedowee belt, occurring in 11 separate districts (Guthrie and Lesher, 1989). In addition, gold occurrences are found along the Brevard fault zone (Guthrie and Lesher, 1989). The gold in all of the districts is associated with quartz-sulfide veins of slightly varying mineralogy. Some quartz veins contain only pyrite, but in most districts other sulfides such as pyrrhotite, chalcopyrite, and arsenopyrite are volumetrically significant. At all of the gold deposits, gold is closely associated with the sulfide minerals that are in structural loci related to late metamorphic and deformational features. Such observations have led recent investigators (Guthrie and Lesher, 1989; Stowell et al., 1989; Stowell et al., 1996) to conclude that the present configuration and distribution of the gold deposits reflect of fluid movement during metamorphism and deformation associated with the late Paleozoic Alleghanian orogeny.

The Alabama gold deposits are classified as 'mesothermal', or 'orogenic' (Groves et al., 1998). The term orogenic will be used here. This class of deposits is closely associated with the formation of metamorphic terrines ranging in all from Archean to Recent (Goldfarb et al., 2001; Kerrich and Cassidy, 1994). They are universally associated with orogeny. Groves et al. (1998) concluded that orogenic gold deposits form by regional hydrothermal fluid migration inherent to tectonism along convergent margins. They argued that deposits associated with orogenesis represent a continuum from gold-arsenic ores formed at great depth in metamorphic zone such as the Alabama deposits (Stowell et al., 1996), to the distal and shallow end of such systems that are characterized by low-temperature, mercury-rich deposits.

The timing of ore genesis for orogenic gold deposits relative to orogenic events is of relevance here. It is increasingly recognized that fluid migration occurred during a change in plate motion from dominantly compressive to

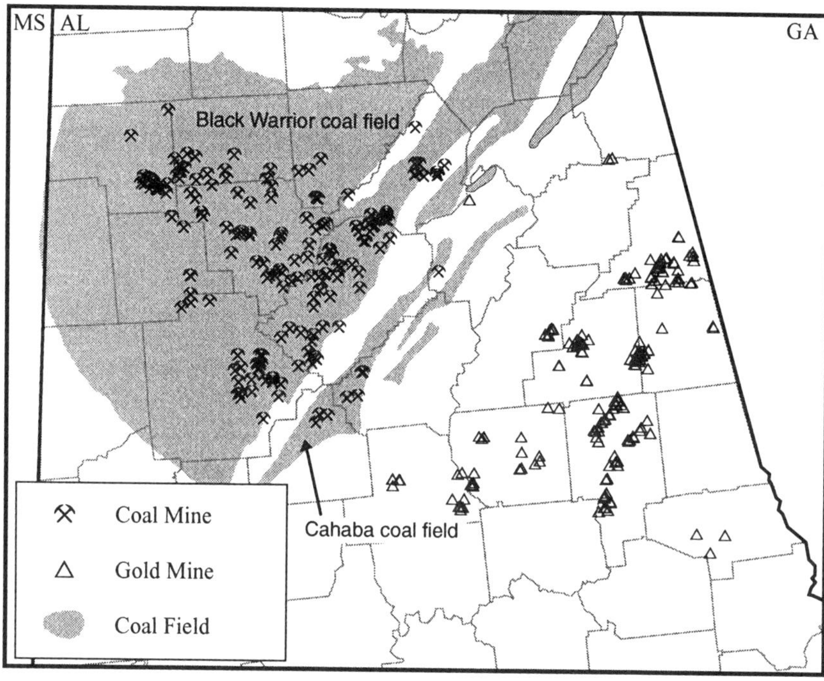

Figure 19. Locations of coal mines and gold mines (exclusive of placer deposits) in Alabama

strike-slip that reactivated--and thus increased permeability in earlier-formed structures in the orogenic zone (Miller et al., 1994). Goldfarb et al. (2001) argue that a modern analog occurs along the San Andreas fault, where hot springs that discharge deep crustal fluids (White, 1967) and form mercury vein systems (Rytuba, 1995) represent the shallow facies of a deep metamorphic hydrothermal system.

4.1.6 Evidence Supporting A Genetic Link Between Metamorphic Fluid And Arsenic Enrichment In Alabama Coal

Evidence presented above suggests that the arsenic and related elements present in Alabama coal were added to the coal after deposition and subsequent burial. Additional evidence that supports a relationship between the observed anomalous arsenic contents in coal and metamorphic fluids related to Alleghanian tectonism include (1) geochemical similarity of the elemental suite in nearby coal and ore deposits, (2) geographic distribution of As-rich coal, (3) evidence for thermal alteration of coal by migrating fluids (4) independent evidence for alteration of rocks of the fold and thrust belt during the Alleghanian orogeny, (5) permissive timing constraints and (6) availability of fluid-flow pathways from the east to west into the coal-bearing sequence.

The element suite (As, Cu, Mo, Sb, Hg, and Au) that is enriched in both coal and gold occurrences and deposits is similar with arsenic prominent in both. Arsenic distribution in the coal was discussed above. In the gold districts, the only known residence of arsenic is the mineral arsenopyrite ($FeAsS_2$). Guthrie and Lesher (Guthrie and Lesher, 1989) list arsenopyrite as occurring in gold-bearing quartz veins in five of the eleven Alabama gold districts. Mines in two other districts may have high arsenic values (up to 1000 ppm) in chemical analyses of grab samples (Lesher et al., 1989). Gold prospects along the Brevard fault zone contain up to five weight percent arsenopyrite, and soil samples near the prospects may contain over 1000 ppm arsenic (Cook and Thomson, 1995). In the quartz-sulfide veins, arsenopyrite is dominantly associated with pyrrhotite and pyrite, plus varying but lesser percentages of chalcopyrite ($CuFeS_2$) and, in some deposits, nickel-antimony sulfide. Elevated mercury and to a lesser extent molybdenum contents are reported in grab samples from a number of mines in the Alabama Piedmont (Lesher et al., 1989), and mercury ± antimony mineralization is characteristic of the distal (low temperature and shallow depth) ends of orogenic gold-forming systems (Groves et al., 1998). The significance of this similarity in element enrichment is that it may reflect similar compositions of the introduced fluids. For orogenic gold deposits, these fluids have been characterized by fluid inclusion studies as having are near-neutral pH, low to moderate salinity, and H_2O-CO_2 $\pm CH_4$ compositions, (Cartwright and Oliver, 2000), which transported gold as a reduced sulfur complex (Seward, 1991).

The presence of elevated gold contents in the arsenian pyrite in coal in amounts up to 200 ppb (ash basis) is consistent with deposition from an epigenetic hydrothermal fluid that likewise contained reduced sulfur. In fact, the suite of elements concentrated in Alabama coal, is one that tends to form mobile complexes in the presence of aqueous sulfide species. Conversely, the lack of significant enrichment in base metals such as zinc and lead in Alabama coal is an indication that the arsenic-bearing fluid had only moderate levels of chloride, because high-chloride fluids are capable of leaching, complexing, and transporting base metals (Goldfarb et al., 2001; Kerrich and Cassidy, 1994). In the Illinois Basin coal described earlier (Fig. 10), zinc is the dominant trace element added, and the hydrothermal fluid was brine with chloride as the dominant anion. However, some orogenic gold deposits were formed by saline ore fluids. These differ from the gold-only type in being enriched in base metals including copper (Goldfarb et al., 2001). The presence of significant amounts of copper (in the mineral chalcopyrite) in many of the gold occurrences in Alabama, and a minor enrichment in copper in arsenian pyrite from coal beds (Goldhaber et al., 2000), may indicate at least some chloride enrichment in the fluids associated with introduction of arsenic.

The distribution of elevated arsenic concentrations in Alabama coal basins bears on the transport path of the arsenic. If arsenic was carried in metamorphic fluids that were driven westward by tectonic forces, then the fold belt, which lies between the Warrior basin and the presumed source of the fluids in metamorphic rocks should have been along this flow path. Significantly, chemical analyses of coal from the Cahaba coalfield, which is located in the fold belt, confirm high arsenic concentrations in these coal beds (Fig. 15).

A compelling line of evidence that coal in the Warrior basin has been affected by westward-migrating fluids is the variation of vitrinite reflectance in the coal. Vitrinite reflectance of the Mary Lee coal is shown in Fig. 15. An area having elevated vitrinite reflectance (higher coal rank) occurs in an elliptical pattern along the eastern edge of the coal basin. Typically, an increase in vitrinite reflectance is attributed to increased temperature associated with increased burial depth. That isn't the case in the Warrior Basin. There is little relationship between the structure, both regional and local and the observed vitrinite reflection pattern (Pashin et al., 1999). The sedimentary section thickens systematically toward the southwest away from the anomaly. This relationship is particularly striking in the elliptical high-rank area. Winston (1994) and Carroll et al. (1995) hypothesized that the vitrinite reflectance anomaly records advective heat transport from westward-migrating fluids that entered the Warrior basin along faults.

Available timing constraints are consistent with the large-scale fluid flow model presented here. These constraints are shown schematically in Fig. 20, a plot of temperature versus time, upon which are superimposed some benchmarks. One benchmark is a tentative time-temperature path for the thermal evolution of the Inner Piedmont area of Alabama during the Alleghanian orogeny (Steltenpohl et al., 1995). This path is based on 'cooling ages' of hornblende, muscovite, and apatite as determined by Ar-Ar geochronology and fission track analyses. These cooling ages record the time at which the first two minerals passed through a temperature at which they retained argon, and at which apatite retained fission tracks produced by radioactive decay. The age of the Pottsville Formation is also a key benchmark because introduction of the arsenic must postdate Pottsville sedimentation. Alleghanian tectonism was active during Pottsville deposition. Pashin has presented evidence for synsedimentary growth of the Sequatchie anticline within the Warrior basin (Pashin, 1994, 1998). Furthermore, despite fossil evidence for contemporaneous deposition of coal beds in the Warrior and Cahaba fields, individual coal groups cannot be

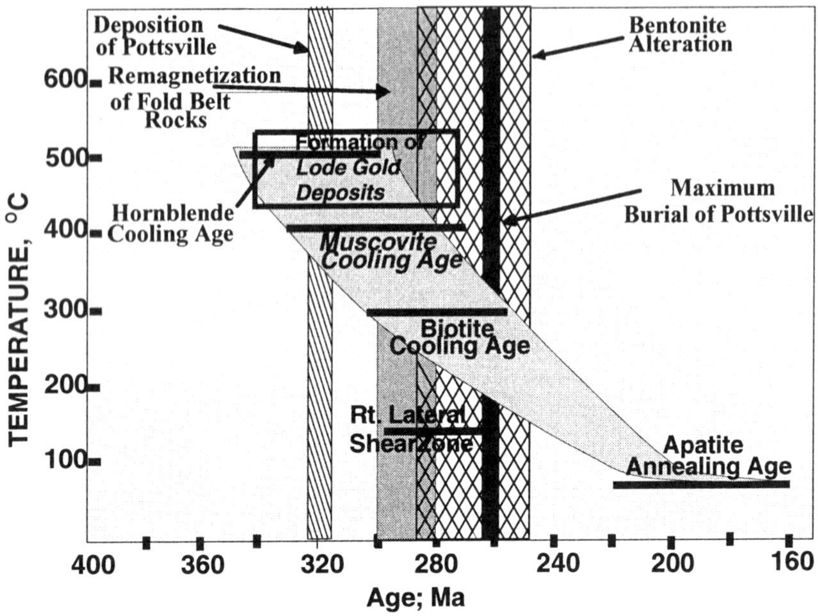

Figure 20. This figure shows the potential timing constraints on fluid-flow in northern Alabama. The data on Hornblende, Muscovite and Biotite cooling ages and apatite annealing age define a time temperature path for the Alabama Appalachians. This path is shown as a light gray band. The timing of deposition as well as maximum burial of the Pottsville Formation is shown as two vertical bars. Gold deposit formation time/temperature relations are represented as a box. The most likely time for introduction of arsenic into the Alabama bituminous coals is towards the end of gold deposit formation, during a change in tectonic stress regime (i.e. the right lateral shear zone), and coincides in part with alteration of Ordovician bentonite by K-rich fluids and remagnetization of fold belt rocks also by migrating fluids. The references for the components of this figure are given in the text.

correlated across the Birmingham anticlinorium (Pashin et al., 1995). This observation suggests that folding involving this structure became active during upper Pottsville deposition.

The timing of gold mineralization is shown Fig. 20 as a box that encompasses the range of measured ages of ore formation and their temperatures as determined by radiometric dating and fluid inclusion studies (Stowell et al., 1996). Ore genesis evidently spanned the period from peak metamorphism to somewhat past the peak. Pottsville deposition occurred during the initial phase of gold mineralization. However, Pottsville deposition predates the earliest alteration (illitization) of Ordovician bentonite beds in the fold-thrust belt. Potassium-argon age dating has shown that hydrothermal fluids altered these beds in the late Paleozoic (Elliott and Aronson, 1987), during a fluid-flow event in a crustal block (the fold-thrust belt) known to contain arsenic-rich coal (see Fig. 15). Similarly, paleomagnetic studies demonstrate that Ordovician and Mississippian rocks in the fold-thrust belt of Alabama, and immediately adjacent portions of Tennessee in the same crustal block were remagnitized by migrating fluids in the late Paleozoic (Miller and Kent, 1988). In addition, timing of the transition from compression-related folding and thrusting, to 'transpression' (i.e. initiation of right-lateral motion on the Brevard fault zone has been dated (Goldberg and Steltenpohl, 1990; Steltenpohl et al., 1995). This transition to a component of strike-slip motion is important because, as noted above, a change in plate motion is considered to be linked to orogenic gold mineralization elsewhere (Miller et al., 1994).

The timing data presented in Fig. 20 indicate that fluid migration through the fold-thrust belt (as documented by Ar-Ar ages and paleomagnetic studies) overlaps both the timing of right-lateral shear on the Brevard fault zone, and at least a portion of the gold mineralization stage. All of these events probably occurred prior to the time of maximum burial of the Pottsville Formation. The relations illustrated in Fig. 20 are best reconciled by a model involving the introduction of arsenic-bearing fluids into the coal beds some 10-15 million years after initial deposition of the Pottsville Formation (at which time the Pottsville would have been buried to depths of 600-1200 meters, continuing (most likely episodically) for some time thereafter, and ceasing before the Pottsville reached its maximum burial depth of between 3,000 and 3700 meters (see Carroll et al. 1995 for burial history curves).

4.2 Mechanisms of Fluid Generation and Transport

The fluid flow system we hypothesize is in accord with evidence worldwide that high water fluxes take place through metamorphic belts (Manning and Ingebritsen, 1999; Rumble III, 1994). The driving force for metamorphic fluid flow is discussed in Manning and Ingebritsen (1999) and consists of the difference between the hydrostatic and lithostatic pressure gradients. More general information on fluid evolution during metamorphism may be found in Yardley (1997) and references therein.

The hypothesized fluid-flow pathways are along major faults. Orogenic gold deposits are nearly always localized within large faults or fault zones (Kerrich and Cassidy, 1994), and those in Alabama are no exception (Lesher et al., 1989). Some potential fault-related fluid pathways are shown in a schematic cross section (Fig. 21). Major fault zones exist that could have provided ingress of westward migrating fluids from deeper and hotter parts of the section into the Warrior basin. The sedimentary rocks and thrust faults of the Valley and Ridge dip as a wedge to an unknown distance, but at least several kilometers eastward beneath the overriding thrust complex of the Blue Ridge (Hatcher et al., 1989). This juxtaposition of metamorphic and fold belt terrains provides a potential fault connection for fluid transport between metamorphic fluid sources in the Blue Ridge to the east and Cahaba and Warrior basins to the west. Channelized fluid-flow along major fault zones in regional metamorphic terrains at scales greater than a kilometer is well established (Oliver, 1996).

In summary, the evidence presented above is best interpreted as indicating that the arsenic enrichments found in Alabama bituminous coal was introduced following deposition and some (perhaps considerable) burial by the distal portion of a regional westward-directed metamorphic fluid-flow system.

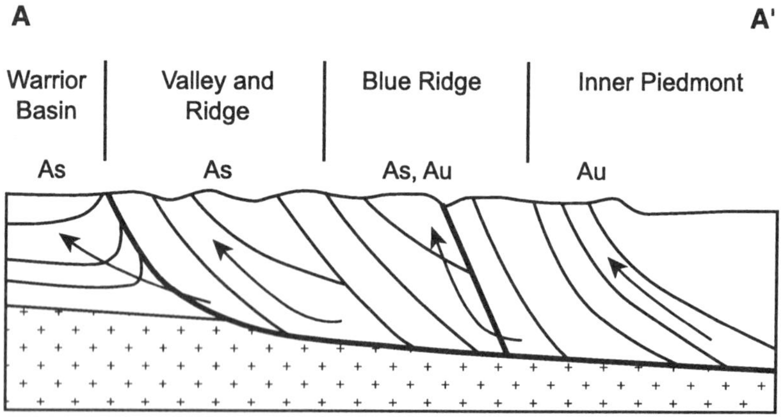

Figure 21. Schematic cross-section through the physiographic provinces of Alabama. The location of the section is shown in Fig. 14. Arrows illustrate the hypothesized directions of fluid flow along major fault zones. Sites with arsenic and/or gold enrichment near the presently exposed surface are noted.

4.3 Arsenic in Coal of the Central and Southern Appalachian Basin

4.3.1 Occurrence of Arsenic in Appalachian Basin Coal

There are a number of anomalously high arsenic contents (>100 ppm) elsewhere in the Appalachian Basin. Figure 22 is a plot of arsenic content in all Appalachian Basin coal samples from all horizons from the USGS coal quality database (Bragg et al., 1997). Sporadic high arsenic contents are present in Kentucky, West Virginia, Ohio, and Pennsylvania. This plot can be misleading because it visually overemphasizes high values due to the size of the largest dots, which may obscure underlying smaller ones. Nonetheless, it does document the presence of anomalous arsenic levels in the central and northern Appalachian coalfields.

An alternative depiction of the distribution of arsenic in Appalachian coal is shown in Fig. 23, which is a series of histograms of arsenic contents in coal by state. The high arsenic 'tails' for Appalachian basin coals of Pennsylvania, Ohio, West Virginia, and Kentucky are not as prominent as those for the Warrior basin of Alabama. But it is evident that a significant number of arsenic analyses are greater than 100 ppm outside of Alabama.

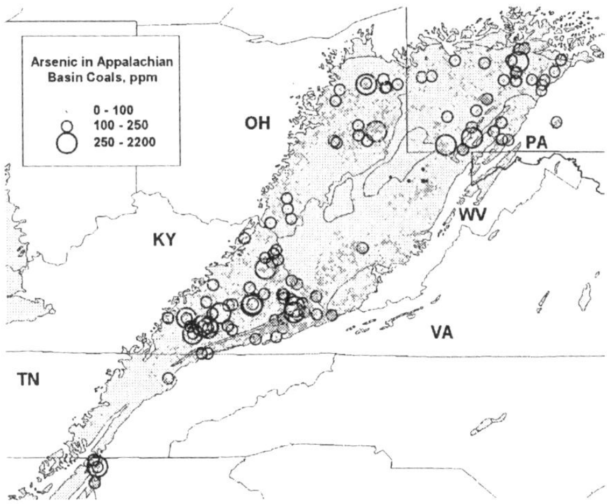

Figure 22. Arsenic in Appalachian Basin coal samples from the USGS Coalqual database. The Appalachian Basin is shown as light gray. Arsenic content is expressed by proportionally sized symbols. Note the occurrence of coal with elevated arsenic content (>100 ppm).

Evidence exists that the mode of occurrence of arsenic in central Appalachian coal beds is similar to that described above for Alabama. Hower et al. (1995) reported local concentrations of arsenic in pyritic samples from Kentucky. Arsenic contents measured using an electron microprobe were 1.9 wt% in epigenetic pyrite and marcasite. They also documented arsenian pyrite filling cell lumens with 3.5 wt% As. Field examination of pyrite accumulations in the eastern Kentucky coalfield using a portable X-ray fluorescence spectrometer showed arsenic present at levels in the 30-500 ppm range, and one sample with 1800 ppm (M. Goldhaber, M. Tuttle and J. Ruppert, unpublished data). In the mines we visited, this macroscopic pyrite was relatively rare. Wiese et al. (1990) report arsenic concentrated in pyrite and marcasite in Ohio coal, though they did not quantify the amounts. These mineralogical associations and arsenic concentrations are similar to those reported in Alabama.

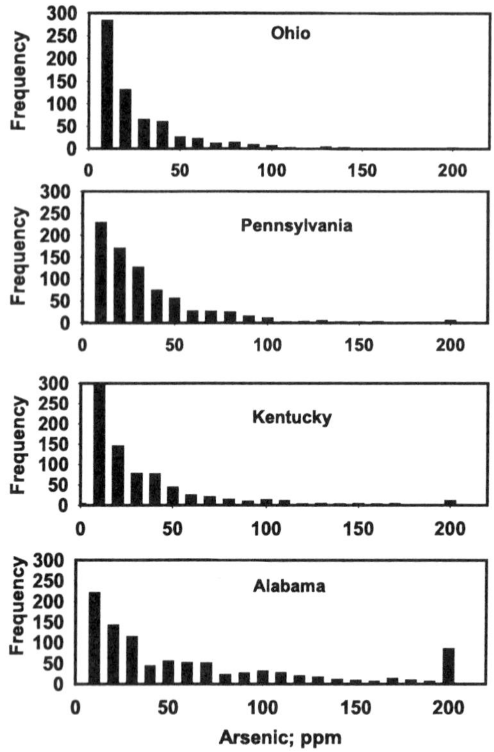

Figure 23. Stacked histograms showing the abundance of arsenic in Appalachian Basin coal samples from the USGS Coal Quality database (Bragg et al., 1997) by state. Note the 'tails' to elevated values present in these distributions.

4.3.2 Evidence for Alleghanian Fluid-flow In The Central And Southern Appalachians

The same lines of evidence cited above for fluid migration in the fold-thrust belt and Warrior basin of Alabama are valid for the southern and central Appalachians. This evidence is reviewed in several references (Evans and Battles, 1999; Montanez, 1994; Schedl et al., 1992), and is summarized briefly here.

Vitrinite reflectance anomalies, occurring as 'tongues' of elevated values extending from the leading edge of the Valley and Ridge province into the Appalachian Plateau (analogous to that shown in Fig. 15, are known from Kentucky, West Virginia, and Pennsylvania (Evans and Battles, 1999; Zhang and Davis, 1993). These authors concluded that burial alone cannot explain the vitrinite reflectance anomalies, and suggested that additional heating provided by hydrothermal fluids is required. Lewis and Hower (1978) concluded that warm fluids introduced during tectonic events might have played at least as important a role as that of stratigraphic and tectonic burial, in producing vitrinite reflectance trends in coal beds of the Valley and Ridge province in Virginia. Hower et al. (1990) summarized data on the development of high grade (anthracite) coal in the Appalachian region, many of which occur in the eastern portion of the fold-thrust belt. They concluded that circulation of hydrothermal fluids was involved in maturation of the coal, and was responsible for precipitating minerals including sulfides along coal cleats.

Diagenetic remagnetization of Paleozoic sedimentary rocks in the Valley and Ridge province is a widespread phenomenon (Lu et al., 1990; McCabe et al., 1989). McCabe et al. were able to link this event to Alleghanian deformation by showing through fold tests that the remagnetization was coeval with Alleghanian folding in some cases. Alleghanian K-Ar and Ar-Ar age dates in Ordovician bentonite beds are widespread throughout the Appalachian fold and thrust belt (Elliott and Aronson, 1987; Hay et al., 1988) and similar ages are recognized in authigenic K-feldspar (Hearn et al., 1987; Hearn and Sutter, 1985). In rocks of the Lower Ordovician Knox Group of Tennessee, authigenic magnetite responsible for the remagnetization and authigenic K-feldspar are in fact cogenetic (Suk et al., 1990). Similarly, late diagenetic dolomite in upper Knox Group rocks has been linked to late Alleghanian fluid-flow, having formed at approximately the same time as authigenic K-feldspar (Montanez, 1994).

Evans and Battles (1994) evaluated the nature of fluids that migrated through the Valley and Ridge Province. This region lies in the critical position between the presumed source of metamorphic fluids to the east and Appalachian Basin to the west. They studied fluid inclusions in tectonically-formed carbonate and quartz veins within Ordovician through Mississippian rocks of the Valley and Ridge Province in West Virginia. Their study identified a stratigraphically controlled series of fluid-flow systems of Alleghanian age, and recognized a three component mixture of fluids: (1) a saline *in situ* fluid (basinal brine) with high salinities generally > 20 wt% equivalent NaCl at temperatures of 70 to 120 °C, (2) Migrating meteoric water having temperatures similar to those of the basinal brine but lower salinities and (3) a warm migrating fluid with temperatures of 150-220 °C and salinities dominantly in the range of 5-15 wt% equivalent NaCl. The study by Evans and Battles (1994)is particularly important because it documents the passage of warm migrating fluids through the Valley and Ridge Province during the Alleghanian orogeny. Their study demonstrates significant compositional variability in migrating fluids, and suggests multiple sources for the fluids. It is possible that fluids similar to both MVT ore fluids and metamorphic fluids passed through the system at various times and/or at differing stratigraphic intervals.

4.3.3 Orogenic Gold Deposits In The Central And Southern Appalachians

Orogenic gold deposits and occurrences analogous to those in Alabama occur elsewhere in the central and southern Appalachian Mountains. Some of these deposits such as those in the Carolina Slate belt formed prior to the Alleghanian orogeny (Stein et al., 1997), but others are likely to be late Paleozoic in age. For example, the Dahlonega belt in Georgia, is along strike from the Alabama deposits (German, 1989), and may have formed in a similar fashion. Thus, conditions favorable for transport and deposition of gold may have occurred elsewhere during active tectonism along the Appalachian chain.

5. ENVIRONMENTAL IMPACT OF ARSENIC IN COAL AND METAMORPHIC ROCKS.

Given the local enrichments of arsenic locally present in pyrite of Appalachian Basin coal, environmental impacts are possible. Dispersal of arsenic from coal to stream sediments and soils may occur by both natural weathering and mining disturbance. Because arsenic in coal is commonly associated with iron sulfide minerals, much of the arsenic will be avoided by mining practices that favor selection of low sulfur (low pyrite) coal or that remove pyrite from the coal during coal preparation. In addition, disposal of the resulting coal or wash waste could present some concern because arsenic may be concentrated in these materials.

Some dispersion of arsenic into the local environment from historic coal mining areas can be demonstrated. Arsenic concentrations in stream sediments of the Warrior coalfield of Alabama are significantly elevated compared to areas outside the coal-mining area, and to typical U.S. streams (Goldhaber et al., 2001). The highest stream sediment arsenic contents measured in Alabama of (180 ppm) occur immediately adjacent to waste piles from old abandoned coal mines (Goldhaber et al., 2001). Surface oxidized portions of these waste piles likewise contain elevated values of rock arsenic ranging up to 470 ppm. Taken together, these data implicate old abandoned coal mines as a source of arsenic in stream sediments. There is presently no evidence that above-background arsenic concentrations in stream sediments and coal mine waste piles are impacting the surrounding ecosystems. In contrast to the abandoned mines, we have not detected release of arsenic to stream sediments from operating strip mines (Goldhaber unpublished data).

Concentrations of arsenic in stream waters of the Warrior coalfield typically are not commonly above the drinking water standard of 50 μg/l (ppb) arsenic (O'Neil et al., 1993). Data from the USGS Storet database provided by Brian Astor of the USGS Water Resources Division records thirty values greater or equal to 10 ppb arsenic in a dataset of over 2000 analyses. The highest contents are 90 ppb (3 samples). Our studies of streams with elevated arsenic in sediments likewise have not detected stream water arsenic contents above 2 ppb (Goldhaber et al., 2001), although within this narrow range of 0-2 ppb, the highest values do occur in acidified reaches of streams. Astor (1999) compared stream water chemistry in unmined and strip-mined drainages in the Warrior basin. He found significant differences in major ion chemistry between mined and unmined drainages and also found that on average the trace elements arsenic, copper and molybdenum were slightly elevated in locations with active mines. But even in mined drainages the concentrations of arsenic were low (≤ 2 ppb). More generally, however, high arsenic contents in the range of 100 to 10,000 ppb are known to occur in waters draining a variety of types of hard rock mines (Williams, 2001)

In contrast to stream water, groundwater in the Warrior coalfield may have elevated arsenic concentrations. Waters produced during methane recovery from coal (n=28) were found to have a mean arsenic content of 25 ppb with a maximum of 475 ppb (O'Neil et al., 1993). These production waters are typically saline and not potable. However, the presence of elevated arsenic in the deep groundwater samples indicates that arsenic may be mobilized from the coal into solution. Shallow drinking-water wells in the Warrior coalfield are lower in arsenic than the production waters, having a mean of 2 ppb and a maximum of 44 ppb (n=35)(O'Neil et al., 1993). The highest of these arsenic contents exceed the present drinking water standard and indicate the potential for at least isolated arsenic contamination.

6. ADDITIONAL EXAMPLES OF LARGE FLUID FLOW SYSTEMS

Although we have focused on two specific large fluid-flow systems in this paper, there are other examples. The world class Silesia-Cracow zinc-lead MVT district of southern Poland formed in a structural setting analogous to the ores of the U.S. midcontinent, and likely formed by a large scale gravity-driven fluid flow system (Leach et al., 1997). Interestingly, high concentrations of arsenic in the Polish ores (up to 2000 ppm) occur in sphalerite (Leach et al., 1997).

The Carlin-type gold (arsenic, antimony, mercury) deposits in the Basin and Range Province of the southwestern United States contain abundant arsenian pyrite as well as lesser amounts of the arsenic sulfides orpiment (As_2S_3) and realgar (AsS) (Hofstra and Cline, 2000; Stenger et al., 1998). The mineralization in northern Nevada is associated with significant enrichment of arsenic in stream sediments of the region (Kotlyar et al., 1998), and groundwater (Welch et al., 1988: Welch and Lico, 1998). The deposits may form from metamorphic fluids with a subordinate input of magmatic fluid (Hofstra and Cline, 2000), or alternatively, by large-scale circulation of meteoric fluids over distances of hundreds of kilometers (Berger et al., 1998; Ilchik and Barton, 1997). In either case, the potential exists for large blocks of the crust to have been enriched in arsenic by ore fluids.

Another example is to be found in the South Wales coalfield of the British Isles. This coalfield is in a foreland basin, and thus occupies the same structural-tectonic setting as the Appalachian Basin. There are other analogies to the Appalachian Basin as well. There is good evidence that hot fluids migrated through this coalfield leading to thermal alteration of the coal (Gayer et al., 1991). Furthermore, the hydrothermal alteration has resulted in enrichment in gold (Gayer, 1994) and arsenic (Gayer et al., 1999). Additional examples of coal alternation by hydrothermal fluids leading to substantially increased arsenic contents are to be found in Bulgaria (Eskenazy, 1995), Turkey (Karayigit et al., 2000), and China (Belkin and Anonymous, 1998). Gold-arsenic mineralizing fluids have altered intramontaine coalfields in the Massif Central, France (Copard et al., 2000).

ACKNOWLEDGEMENTS

We would like to dedicate this paper to Terry Offield who provided significant insights into the geology of Alabama, but who passed away during an early stage of the study. We would like to thank John Slack, Hal Gluskoter, and Elizebeth Rowan for reviewing an earlier version of the manuscript. Greg Meeker assisted with electron microprobe studies, and Sharon Fay Diehl and Isabelle Brownfield assisted with scanning electron microscope studies.

Chapter 6

Arsenic in ground water used for drinking water in the United States

Sarah J. Ryker
U.S. Geological Survey
Present address: Carnegie Mellon University

The U.S. Environmental Protection Agency recently established a new maximum contaminant level of 10 micrograms per liter for arsenic in drinking water in the United States. Ground water is the primary source of drinking water for half the population of the United States. Several national assessments have found that high arsenic concentrations (above 10 micrograms per liter) are widespread in drinking-water aquifers in the western United States, the Great Lakes region, and New England. Moderate to high concentrations were identified in ground water in parts of the central and southern United States. This chapter summarizes national trends in the use of ground water as drinking water, and national estimates of arsenic occurrence in potable ground water. The chapter also briefly describes several studies on arsenic in specific settings and water-use scenarios; these studies illustrate by example the potential power of a regional approach to understanding and managing arsenic in drinking water.

1. INTRODUCTION

Arsenic above 10 micrograms per liter is found in potable ground-water resources in many parts of the world (Nriagu, 1994; WHO, 1993). In the United States, widespread high concentrations of arsenic in ground water (>10 μg/L) have been documented in drinking-water aquifers in the West, the Great Lakes region, and New England. Moderate to high concentrations have been found in parts of the central and southern United States. In contrast, widespread high concentrations of arsenic are relatively uncommon in surface water (Federal Register, 2001; Lockwood et al., 2001; Welch et al., 2000). A few notable exceptions occur where ground water makes a significant contribution to surface-water flow; where surface water has an unusually high pH; or in drainages from geothermal areas (Welch et al.,

2000). While somewhat unusual, these situations can have implications for drinking-water quality, particularly in the western United States (Frey and Edwards, 1997; Lockwood et al., 2001). However, this chapter focuses on arsenic in potable ground water as the area of greatest concern for most of the United States.

2. GROUND WATER AS DRINKING WATER

Half the population of the United States relies on ground water for drinking water (Figure 1). In 2000, more than 100 million people (over 35 percent of the population) were served by public water supply systems using ground water, and more than 40 million people (over 15 percent of the population) used self-supplied ground water, predominantly from private domestic wells (Solley et al., 1998; U.S. Environmental Protection Agency, 2001).

Many large public water supply systems use substantial amounts of surface water, while smaller or intermittent-use public systems and domestic systems are generally much more dependent on ground water. Public water systems serving at least 25 people are divided into three main classes. Community water systems (CWS) provide water to the same population year-round. As of 2000, over 80 percent of CWS rely on ground water; thirty-three percent of the total population served by CWS receives water from these ground-water-dependent systems. Non-transient non-community water systems (NTNCWS; e.g. schools, factories, office buildings, and hospitals) provide water to the same population for at least six months per year; 87 percent of this population relies on ground water. Transient non-community water systems (TNCWS; e.g. gas stations and campgrounds) provide water to transient populations; 98 percent of the water provided by these systems comes from ground water (U.S. Environmental Protection Agency, 2001). About 99 percent of the rural population drink self-supplied ground water, generally from private domestic wells serving individual households or clusters of up to 10 homes (Carlson, 1999; Solley et al., 1998).

Water-use estimates from several different sources indicate that ground water is increasingly important for drinking water in the United States. Many areas of population growth are increasingly making use of ground water for drinking water (Carlson, 1999; Perry and Mackun, 2001; Solley et al., 1988; Solley et al., 1998; U.S. Bureau of the Census, 1999). In 1950, twenty-six percent of water withdrawals for public supply came from ground (Dziegielewski et al., 2002; Solley et al., 1988; Solley et al., 1998). Some of

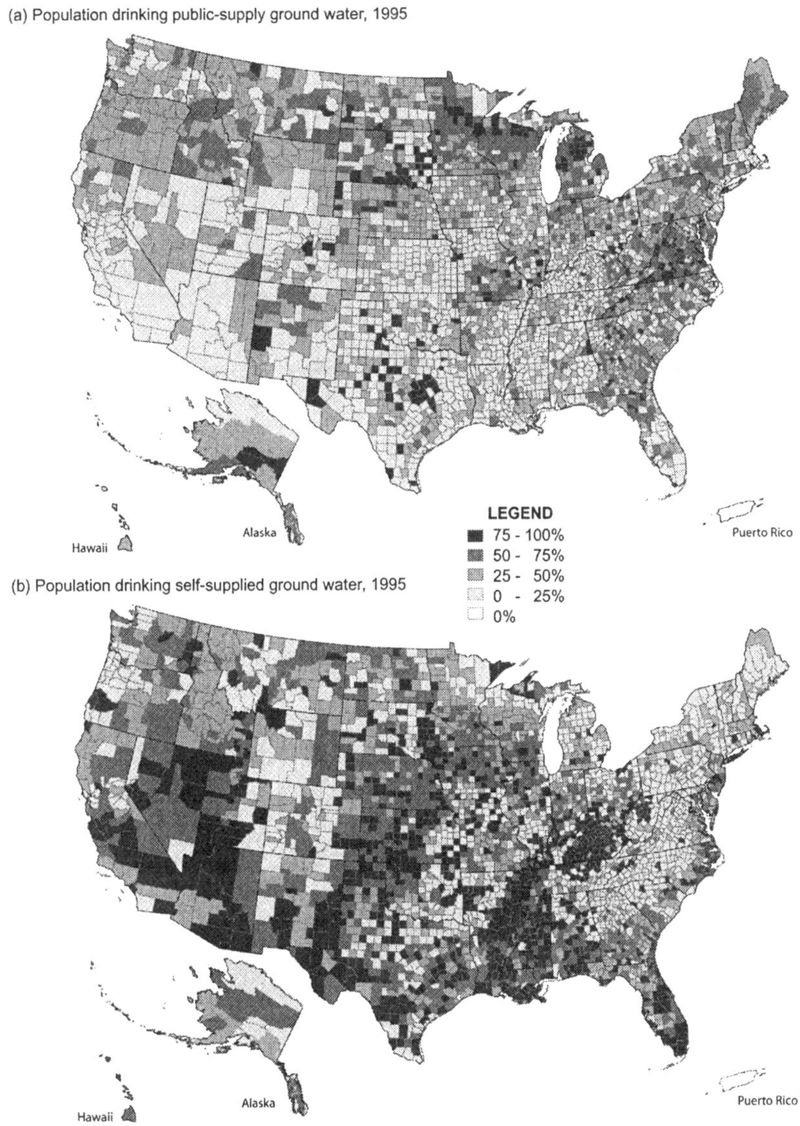

Figure 1. Population drinking ground water from (a) public supply systems and (b) private domestic wells, expressed as a proportion of the total population by county, 1995 (Solley et al., 1998).

this change corresponds with population shifts toward the southern and western United States, areas already dependent on water. By 1995, thirty-seven percent of water withdrawals for public supply came from ground water ground water; from 1960 to 1995, the greatest increases in public

supply withdrawals were in Nevada, Arizona, Florida, Georgia, California, and Texas (Dziegielewski et al., 2002). Some of the increase in publicly supplied ground water may also be explained by the expansion of rural water systems; the rate of conversions from private wells to piped water increased more than threefold during the 1980s and 1990s (Carlson, 1999; Dziegielewski et al., 2002; U.S. Bureau of the Census, 1999).
Geographic patterns have also been observed in use of self-supplied ground water. In 1995, over half of all ground water withdrawn by domestic wells came from 13 states: Florida, Michigan, Pennsylvania, North Carolina, New York, Ohio, Texas, Illinois, Washington, Virginia, California, Indiana, and Georgia (Dziegielewski et al., 2002; Solley et al., 1998). In densely populated regions such as New England, population growth appears to be largely in rural areas and outlying suburbs, with a corresponding increase in domestic use of ground water (Solley et al., 1988; Solley et al., 1998; U.S. Bureau of the Census, 1999). Nationally, between 1970 and 1990 the number of people supplied by private wells increased by 36 percent (U.S. Bureau of the Census, 1999), and during the 1990s alone, domestic well drilling increased by about 35 percent (Carlson, 1999). Carlson (1999) has also noted an apparent increase in small water systems serving clusters of 3-10 households; these systems are too small to be regulated as community water systems and so may be considered private domestic wells.

3. NATIONAL ESTIMATES OF ARSENIC OCCURRENCE IN DRINKING WATER

Arsenic is found in all types of ground-water supplies. Several recent estimates have been made of the number of public supply systems likely to exceed various concentrations of arsenic. Four primary national-, state-, and multi-state-scale datasets have been used as a basis for these estimates. The USEPA maintains the federal Safe Drinking Water Information System (SDWIS), which in 1999 included 44,087 ground-water public supply systems in fifty states. In addition, the USEPA has compiled a dataset of compliance monitoring samples of finished water from both ground-water and surface-water public supply systems in 25 states. These samples are generally of finished water, which may have been treated for a variety of contaminants before delivery to the consumer (Federal Register, 2001), although Frey and Edwards (1997) found that as of 1995, 65 percent of ground-water public supply systems had implemented no treatment expected to reduce arsenic. In 1984, the National Inorganics and Radionuclides Survey (NIRS) analyzed finished water from 270 public water supply wells in 49 states (all but Hawaii) and Puerto Rico (Longtin, 1988). In 1995, the

National Arsenic Occurrence Survey (NAOS) sampled raw water from approximately 270 public supply wells in all 50 states, and predicted arsenic concentrations in finished water based on known treatments implemented by the systems (Frey and Edwards, 1997). Frey and Edwards (1997) also provided national estimates based on reexamining the NIRS data and a dataset of finished water from 60 large systems (serving >10,000 people) sampled by the Metropolitan Water District of Southern California (Davis et al., 1994). In 1973 to 1997, the USGS analyzed for arsenic in 18,850 raw-water samples from wells in 49 states (all but Hawaii) and Puerto Rico; this dataset included 2,262 samples from public supply wells (Focazio et al., 2000). Several older national surveys of arsenic have been described in documents prepared for the USEPA (ISSI Consulting Group et al., 2000).

The most common approach to creating national estimates has been to summarize or model percentiles of arsenic concentration found in these existing datasets. The resultant rates of exceedance are then applied either to all ground-water public supply systems in the areas and aquifers represented in the dataset (Focazio et al., 2000), or to the entire population of public supply systems in the United States (Federal Register, 2001). Lockwood et al. (2001), Zhang et al. (2001), Gurian et al. (2001a), and others developed regression models based on additional explanatory variables and accounting for intra-source and intra-system variability. Several of the datasets used in these various efforts, and the resultant national estimates, are summarized in Table 1.

Also described in Table 1 and Figure 2 is a dataset of over 31,000 ambient (raw) samples collected from individual ground-water sources by the USGS and by state agencies. These raw-water samples provide a consistent national overview of where widespread high arsenic concentrations have been identified in ground water, but do not necessarily represent exactly what the population is drinking in any given region. Ninety-nine percent of self-supplied drinking water comes from individual domestic wells (Solley et al., 1998). At a national scale, therefore, raw-water samples are generally representative of the water consumed by domestic well owners. Interpreting raw-water data is more complicated for public water supply systems, which may take water from a single source (e.g. an individual well or surface-water intake), or may blend water from multiple sources. Multiple-source systems may include both surface water and ground water sources, and may change their sources over time—both seasonally, in response to short-term climate fluctuations, and over periods of years, in response to evolving demand for water. Based on the makeup of the raw-water dataset and on comparisons between USGS public-supply data, USGS data from all types of wells, and NAOS public-supply data (Focazio et al., 2000), the raw-water samples appear to be representative of arsenic concentrations in institutional supply wells, and in the untreated ground water used by public supply systems in many settings. For larger public supply systems, raw-water samples may serve primarily to highlight regions where blending or treatment of existing water sources, and careful siting of future water sources, may be indicated.

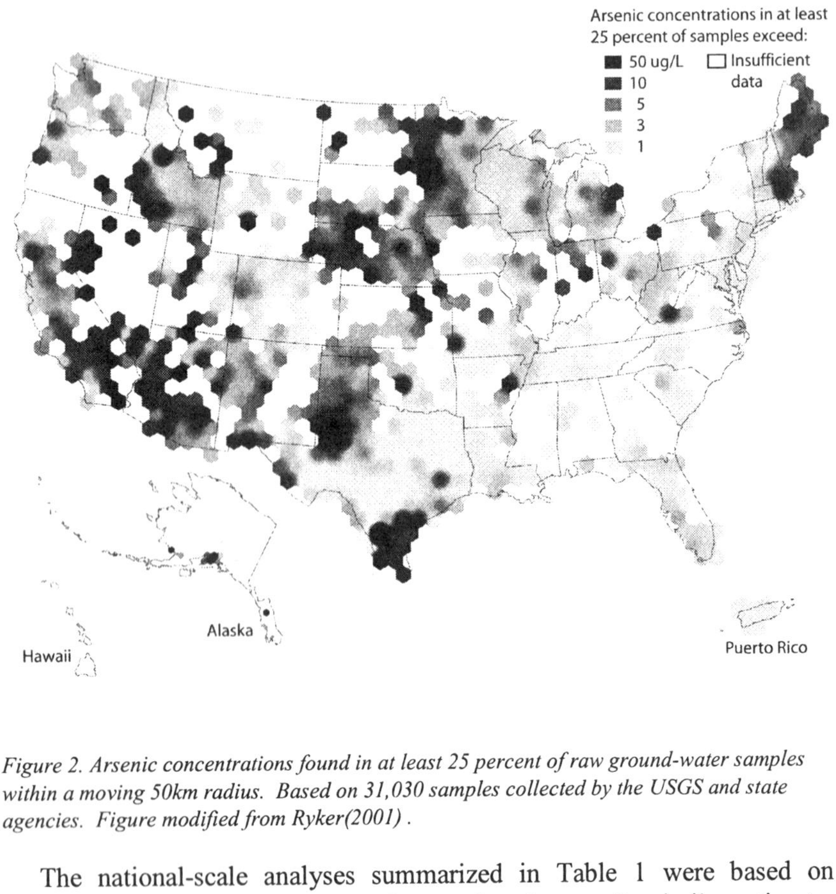

Figure 2. Arsenic concentrations found in at least 25 percent of raw ground-water samples within a moving 50km radius. Based on 31,030 samples collected by the USGS and state agencies. Figure modified from Ryker(2001).

The national-scale analyses summarized in Table 1 were based on disparate datasets and approaches, but produced generally similar estimates of arsenic concentration. Many of the differences between the estimates may result from differences in the sampling and measurement techniques represented by the underlying datasets. For example, the NIRS samples and the compliance monitoring samples compiled by the USEPA were not filtered, while the NAOS, USGS, and state environmental monitoring samples were generally filtered through a 0.45 μm pore-size filter. In addition, laboratory measurements of arsenic tend to have relatively high analytical variability. For ambient ground-water samples analyzed by GF/AA methods, the USGS has observed analytical variability up to ±14%

Table 1. National estimates of arsenic concentrations in public-supply and private domestic systems using ground water. [µg/L, micrograms per liter; NR, not reported.]

Type of supply ***(Population served)***	**Systems** ***(states)*** **in dataset**	**Type of water**	**Percentage of systems with arsenic exceeding concentrations (µg/L) of... 2**	**3**	**5**	**10**	**20**
USEPA (Federal Register, 2001) estimates of national distributions, based on state compliance monitoring data[1]							
Public—CWS *(all)*	15,980 *(25)*	Raw and finished	27.2	19.9	12.1	5.3	2.0
Public—NTNCWS *(all)*	4,380 *(17)*		NR	24.3	15.7	5.4	2.1
Frey and Edwards (1997) estimates from NAOS data[2]							
Public *(>10,000)*	} 270 *(50)*	Finished	28.8	NR	15.4	6.7	NR
Public *(1,000 - 10,000)*			23.5	NR	12.7	5.1	NR
Frey and Edwards (1997) estimates from NIRS and MWDSC data[2]							
Public *(>10,000)*	970 *(49)*	Finished	16.3	NR	9.1	2.5	NR
Public *(1,000 - 10,000)*	80	Finished	25.4	NR	9.3	3.6	NR
Summary of data from the National Inorganics and Radionuclides Survey[3]							
Public—CWS *(>1,000)*	330 *(49)*	Finished	17.4	11.9	6.9	2.9	1.1
Gurian et al. (2001a,b) estimates from single-sample NAOS, USGS, and SDWIS data[4]							
Public—CWS *(all)*	*(49)*	Raw	28	20	13	6	3
Public—CWS *(all)*	*(49)*	Finished	20	15	9	4	2
Focazio et al. (2000) estimates for 1,528 counties, based on USGS ambient data[5]							
All public supply wells	18,850 *(49)*	Raw	25.0	NR	13.6	7.6	3.1
Summary of data from USGS and state ambient monitoring programs[6]							
All wells	31,030 *(49)*	Raw	33.1	26.2	18.9	10.7	5.0
Public supply wells *(all)*	4,390 *(49)*	Raw	29.2	24.0	17.2	8.9	4.0
Domestic wells	7,530 *(49)*	Raw	33.7	26.4	19.6	11.5	5.1

[1] *Mean expected arsenic concentration for all systems. Information on specific sources is provided for 7 of the 25 states. (Federal Register, 2001).*
[2] *(Frey and Edwards, 1997)*
[3] *(Longtin, 1988)*
[4] *Mean expected concentration for all systems, based on public-supply data from both USGS and NAOS (Gurian et al., 2001a; Gurian et al., 2001b; Lockwood et al., 2001)*
[5] *Expected concentrations for ground-water systems only (Focazio et al., 2000)*
[6] *Ryker (2001)*

in analysis of identical spike samples; variability appears lower for ICP/MS methods (Garbarino, 1999; Hinkle and Polette, 1999; Long and Farrar, 1995). The current arsenic rule (Federal Register, 2001) permits a number of analytical techniques for compliance monitoring samples, and establishes ±30% as acceptable for future compliance monitoring data. The ambient ground-water samples collected by the USGS were quantified down to 1 μg/L, while the ambient ground-water samples collected by state agencies were quantified down to between 0.6 μg/L and 3 μg/L; approximately 35 percent of the ambient data are reported only as less than some reporting limit. Similarly, the NAOS samples were analyzed by ICP/MS; approximately 36 percent of the NAOS data were reported as "<0.5 μg/L." The state compliance-monitoring samples compiled by the USEPA were analyzed by a variety of techniques authorized under 40 CFR 141.23, with reporting limits typically between 3 μg/L and 10 μg/L. In the NIRS data, the reporting limit is 5 μg/L.

Other fundamental differences affect the estimates in Table 1. The exceedance rates estimated by the USEPA (Federal Register, 2001) and by Lockwood et al. (2001) are for whole public-supply systems, which may include multiple sources of water. All other estimates in Table 1 are exceedance rates for individual ground-water sources. In a model of blended water from four states' compliance monitoring data, Zhang et al. (2001) demonstrated that system-based exceedance estimates are likely to be slightly lower than source-based exceedance estimates. Further complicating comparisons, only the estimates by Lockwood et al. (2001), Gurian et al. (2001b), and an earlier analysis performed for the USEPA (Federal Register, 2001), incorporated intra-well and intra-system variability. The final USEPA estimates are based on the mean concentration for all sources within each system, while all other estimates in Table 1 used the single most recent value available for each well.

3.1 Regional variations

These national summaries provide a general overview of where arsenic may be an issue in drinking water. However, national analysis can be confounded by regional variations in explanatory factors including water-use patterns, well construction, aquifer characteristics and local geochemistry, and temporal variability. Such discrepancies can make national estimates difficult to apply locally. Estimates on a regional or state-by-state basis have been published by the USEPA (Federal Register, 2001), Focazio et al. (2000), Frey and Edwards (1997), and Lockwood et al. (2001). These more detailed analyses, along with additional information on local aquifer systems and water sources (e.g. specific aquifers), provide more refined descriptions of arsenic concentrations in drinking water.

3.2 Variation by type and size of water supply

Several of the national analyses described in Table 1 are also available by public-supply system size (population served); these system-size-based estimates have been important in cost-benefit analysis for the new arsenic rule (Federal Register, 2001). However, consistent relationships between system size and raw-water arsenic concentration are not well established on a national basis. The USEPA (Federal Register, 2001) found no consistent differences in mean arsenic concentration by system size, while Frey and Edwards (1997) found higher arsenic concentrations in public supply systems serving over 10,000 people than in systems serving 1,000 to 10,000 people. Lockwood et al. (2001) examined both absolute population-served estimates and categories of system size, and found that size-based differences in concentration were not nationally consistent but were greatest among large systems in the western United States.

System size may represent a combination of factors such as well capacity (pumping rate), ground-water contributing area, usage patterns, well drilling and maintenance practices, and the system's resources in selecting and treating water sources. Such factors have been shown to influence vulnerability to anthropogenic contaminants, but the relationships between these factors and natural contaminants are highly dependent on local conditions (Franke et al., 1997; Hinkle and Polette, 1999). For example, Bruce and Oelsner (2001) report that in the Ogallala aquifer of the central United States, larger-diameter, higher-pumpage wells tend to take in water from a larger contributing area, resulting in higher concentrations of anthropogenic contaminants; but that increased pumping does not appear to affect concentrations of natural contaminants such as arsenic. In parts of Minnesota, Wisconsin, and Ohio, however, larger-diameter public supply wells appear to alter geochemical conditions in the vicinity of the well, creating pockets of more highly oxygenated water with correspondingly lower concentrations of iron, manganese, and arsenic (Minnesota Pollution Control Agency, 1998; Ohio Environmental Protection Agency, 2001b; Riewe et al., 2000).

3.3 Variation by aquifer

In eastern New England, analysis of arsenic occurrence by aquifer type finds that arsenic concentrations are typically lower in glacial aquifers than in bedrock aquifers (Ayotte et al., 1999). Arsenic concentrations are also generally lower in public supply wells than in domestic wells, and lower in higher-pumpage wells than in smaller systems. However, Ayotte et al.

(1999) do not conclude that large public supply wells are inherently less likely to take in high-arsenic ground water. Ayotte et al. (1999) examined water use patterns by aquifer and found that aquifer type was the dominant factor, rather than well type. In eastern New England most public supply wells draw water from higher-yielding, lower-arsenic sand-and-gravel glacial materials, while domestic wells are typically cased through the glacial materials and draw their water from low-yielding, high-arsenic bedrock aquifers. A small but growing number of public supply wells (particularly in small systems) do tap bedrock aquifers; in terms of arsenic concentration, these bedrock public supply wells are more similar to bedrock domestic wells than to glacial-deposit public supply wells (Ayotte, Montgomery, et al., in draft, 2002; Ayotte et al., 1999). In this type of setting, arsenic occurrence estimates based on a mixture of data from different aquifers may overestimate the amount of arsenic in wells tapping glacial aquifers, and underestimate the amount of arsenic in wells tapping bedrock aquifers.

Water system types and sources are similarly differentiated in parts of the western United States. On a percentage basis, for example, both Oregon and Nevada are highly urbanized, with at least 70 percent of each state's population concentrated in urban centers that rely primarily on surface water (Bevans et al., 1998; Wentz et al., 1998). However, in terms of absolute numbers of people, both states have substantial rural populations that rely almost entirely on ground water. In Oregon, a total of 1.4 million people rely on ground water for their drinking water; rural domestic wells account for over 70 percent of this ground-water usage (Solley et al., 1998). High concentrations of arsenic (up to 2,000 ug/L) are found in wells in areas of Oregon where bedrock is exposed or near the land surface, and where wells tap several areally extensive volcanic formations (Hinkle and Polette, 1999).

Ground water in parts of Nevada contains some of the highest known arsenic concentrations in natural ground-water systems in the United States (Welch, 1994; Welch et al., 1997; Welch et al., 1988). In the past, arsenic in drinking water was of greatest concern in small communities where rural domestic wells tapped high-arsenic aquifers. However, since 1990 Nevada has seen greater population growth than any other state (Perry and Mackun, 2001). Accompanying this growth, Nevada communities have greatly increased ground-water withdrawals for both public water supply and rural domestic supply (Dziegielewski et al., 2002; Solley et al., 1988; Solley et al., 1998). Thus, arsenic may become a concern for larger population centers.

In parts of Minnesota, Illinois and Ohio, in contrast, domestic and public-supply wells commonly tap the same aquifer systems. In these states, arsenic is typically higher in the heavily used glacial aquifers than in bedrock aquifers. More than 70 percent of Minnesotans rely on ground water for drinking water, primarily from glacial aquifers in western and central Minnesota and from bedrock sources in southeastern Minnesota and the Twin Cities metropolitan area (Minnesota Pollution Control Agency, 1998). Within the Lower Illinois River Basin, close to 75 percent of the population relies on glacial deposits for drinking water (nearly half the population via ground-water community water systems, and an additional 25 percent via domestic wells) (Warner, 1998). In Ohio, close to half the population relies on ground water for drinking water (30 percent from public supplies, and 20 percent from domestic wells) (Ohio Environmental Protection Agency, 2001b; Solley et al., 1998). Glacial deposits of sand and gravel are used for public supply in metropolitan areas of southwestern Ohio (e.g. Dayton, Columbus, and Cincinnati). Within Illinois and Ohio's glacial aquifers, arsenic is especially high at greater depths; near buried valley deposits; and east of the Illinois River (Ohio Environmental Protection Agency, 2000, 2001a; Slattery et al., 2000; Warner, 1998, 2001). In all three states, arsenic concentration appears to depend on a combination of well type and construction, the aquifer tapped, and the position of the well within the aquifer.

3.4 Variation over time

Temporal variability can complicate comparisons of data from different regions and time periods. The geochemistry of the ground water passing through the aquifer, and the mineralogic composition of the surrounding aquifer materials, play important roles in determining concentrations of arsenic in ground water (Hem, 1985; Rogers, 1989). In general the chemistry of ground-water systems is slow to change; over time periods of weeks to decades, natural processes alone are unlikely to cause significant, widespread upward or downward trends in arsenic in ground water. (Exceptions occur near the surface of the water table where water levels can fluctuate seasonally or even daily, or in regions where seasonal variations in discharge strongly affect the water table.) Supporting this, several regional studies in North America have found that arsenic concentrations in ground water are relatively stable over periods of weeks to years (Alberta Health and Wellness, 2000; Ayotte et al., 2001; Focazio et al., 2000; Grantham and Jones, 1977; Hinkle and Polette, 1999; Karagas et al., 2001; Ryan et al., 2000). However, human influences result in significant exceptions in some regions of the United States.

Anthropogenic influences over arsenic in ground water may act at a variety of scales. As discussed above, the construction and usage patterns of a well may have a highly localized effect on arsenic concentrations in the vicinity of the well. This well-scale influence is highly dependent on local mineralogy and water chemistry, and can be obscured by pumping patterns that may or may not relate to the region's seasonal hydrologic cycle (Chen, 2001). For example, irrigation wells and vacation-home domestic wells may go unused for months at a time, while public supply systems may increase ground-water pumping when reservoirs are low.

Similar mechanisms may act at a regional scale—widespread human alteration of the natural ground-water system may alter ground-water geochemistry across a large area. For example, in several arid agricultural regions of the western United States, decades of intensive irrigation have either raised (Bauer and Vaccaro, 1990; Welch, 1998) or lowered (Cordy et al., 1998; Dennehy et al., 2002) the regional water table. A similar potential influence is the increasing use of ground water in many areas of population growth (see figure 1).

Wisconsin's use of ground water has increased steadily since 1950; between 1985 and 1995, the state's population grew by 24 percent while ground-water withdrawals for public supply doubled (Solley et al., 1988; Solley et al., 1998). Approximately 70 percent of Wisconsin residents now rely on ground water; this figure includes about 800,000 domestic wells (Wisconsin Department of Natural Resources, 1999). Two major aquifers, the St. Peter Sandstone and the overlying Platteville/Galena Dolomite, supply most domestic wells and many public supply systems in the state's more densely populated areas (Burkel and Stoll, 1999). Use of these aquifers is expected to increase through the year 2015; about 1,000 new wells are drilled every year in Winnebago and Outagamie Counties (Wisconsin Department of Natural Resources, 1997). These areas' increased consumption of ground water has decreased recharge to the underlying drinking-water aquifers (Axness et al., 2002; Steuer and Hunt, 2001; Wisconsin Department of Natural Resources, 1997). Recent studies indicate that the falling water table may have oxidized shallow aquifer materials, mobilizing arsenic from sulfide minerals into ground water (Burkel and Stoll, 1999; Riewe et al., 2000). Some arsenic is also found in water from deeper Cambrian sandstone aquifers, which are commonly used by public supply systems but are less impacted by changing water-use patterns (Burkel and Stoll, 1999; Riewe et al., 2000).

Anthropogenic influences over arsenic concentration may also act at a local scale, affecting an area larger than a well's immediate contributing area but smaller than the regional ground-water flow system. These local influences may provide an anthropogenic source of arsenic, or may alter local geochemistry to mobilize arsenic from aquifer sediments into ground water. In agricultural areas, historical use of arsenical pesticides and defoliants may have increased arsenic concentrations in soils (Aurelius, 1988; Chormann, 1985; D'Angelo et al., 1996; Frost et al., 1993; Peryea, 1998). Studies in the central United States have not documented effects on arsenic concentrations in ground water; it may be that this arsenic remains bound up in surficial soils (Welch et al., 2000). In other settings, agricultural practices such as the application of phosphate fertilizers may mobilize arsenic from soils into shallow ground water (Barringer et al., 2001; Peryea and Kammereck, 1997). However, there is little evidence that such mechanisms commonly influence water quality beyond the field scale (Welch et al., 2000).

Other local influences may include point sources such as Superfund sites, which in parts of the northeast appear to both provide an anthropogenic source of arsenic, and also mobilize naturally occurring arsenic into ground water (Welch et al., 2000). Mining operations and geothermal influences may similarly mobilize arsenic into ground water and create high variability in concentration within a small area (Mok et al., 1988; Paschke et al., 2001). Fortunately, such sites are not commonly used as drinking-water sources.

4. FURTHER RESEARCH

National-scale summaries of existing arsenic data have been useful in identifying areas with widespread high arsenic in drinking water, and in cost-benefit analysis for the new maximum contaminant levels for arsenic (Federal Register, 2001). In some areas of the United States, additional regional- and local-scale research has helped develop practical guidance on avoiding or reducing arsenic in ground water. For example, in Minnesota and Wisconsin regional studies have identified areas with high-arsenic ground water and high usage of ground water as drinking water, while geochemical studies have provided more detailed understandings of local sources of arsenic and mechanisms of release of arsenic into ground water (Burkel and Stoll, 1999; Cannon et al., 1998; Kanivetsky, 2000; Minnesota Department of Natural Resources, 2000; Minnesota Pollution Control Agency, 1998; Pelczar, 1996; Riewe et al., 2000; Schreiber et al., 2000; Wisconsin Department of Natural Resources, 1997). In 1996 the Wisconsin Department of Natural Resources designated an Arsenic Advisory Area in parts of two counties near the Lower Fox River Valley. Within the Advisory Area, although most wells yield low-arsenic water, extremely high concentrations (up to 15,000 μg/L) have been detected in domestic wells, monitoring wells, and public supply wells. While it is not possible to predict exactly which wells in the region will yield high-arsenic water, the state has established guidelines for well construction within the Advisory Area, based on what is understood about the distribution of natural sources of arsenic and mechanisms of release into ground water (Burkel and Stoll, 1999; Schreiber et al., 2000).

Regional studies also seem indicated in addressing arsenic in private domestic wells. Domestic wells are typically owned by individual households, and are not regulated or monitored under the Safe Drinking Water Act (U.S. Environmental Protection Agency, 2002). However, domestic water quality is an increasingly important issue in the United States. In 1995, ninety-nine percent of rural drinking water came from self-supplied ground water (Solley et al., 1998; U.S. Bureau of the Census, 1999), and this type of domestic ground-water use is increasing; about 366,000 new domestic wells are drilled each year (Carlson, 1999). In addition, the population served by private wells may be more vulnerable to water contamination than the population served by public water supplies. Many private well owners are unaware of contamination issues, or are unable to finance either treatment of an existing water source or new drilling to replace a source. The lack of national regulation means that there is no complete national database of domestic water quality data. The raw-water dataset compiled by the USGS (see Table 1) is large, but does not provide comprehensive coverage for all parts of the United States. Local efforts, incorporating information from state and local agencies, are critical to assess arsenic exposure in rural populations, and to identify areas of low-arsenic ground water for future use.

5. DATA SOURCES

Maine Department of Health

Massachusetts Department of Environmental Protection

Minnesota Pollution Control Agency, 1998, Ground Water Monitoring and Assessment Program, http://wwww.pca.state.mn.us/water/groundwater/gwmap/gwbaseline.html

New Hampshire Department of Environmental Services

Rhode Island Department of Health

Texas Water Development Board, http://www.twdb.state.tx.us/data/waterwells/well_info.html

U.S. Geological Survey, National Water Information System, http://water.usgs.gov/nwis/

U.S. Environmental Protection Agency, Safe Drinking Water Information System (federal version), http://www.epa.gov/safewater/sdwisfed/sdwis.htm

Wisconsin Department of Natural Resources, Drinking Water System, http://www.dnr.state.wi.us/org/water/dwg/DWS.htm

Chapter 7

Arsenic in groundwater – south and east Asia

Pauline L Smedley
British Geological Survey, Maclean Building, Wallingford, Oxfordshire, OX10 8BB, UK

Naturally-occurring high arsenic concentrations are a recognised problem in groundwater from many regions of eastern Asia, including parts of Bangladesh, West Bengal (India), China, Vietnam, Taiwan, Thailand and Nepal. A small number of tubewells with high arsenic concentrations have also been found in Pakistan, Myanmar and Cambodia, although sampling in these areas is so far limited and the scale of potential problems is uncertain. Chronic arsenic-related health problems are recognised in some, though not all, of these regions. The high-As groundwaters in eastern Asia typically occur under strongly reducing conditions in young, Quaternary, aquifers in alluvial and deltaic sediments. Under such conditions, arsenic is commonly accompanied by high concentrations of iron, manganese and ammonium and low concentrations of nitrate and sulphate. Arsenic speciation is variable but the element occurs principally as arsenite. High alkalinity values and phosphorus concentrations are also often a feature of the high-As groundwaters. Although high concentrations of dissolved humic matter are also found in some groundwaters from Inner Mongolia (China) and Taiwan, they are not observed in all regions. A notable exception to these conditions is that of the Ron Phibun area of Thailand. Here, localised arsenic contamination of surface and shallow groundwaters is due to the oxidation of arsenopyrite in an area impacted by former tin mining operations. These groundwaters tend to be more oxidising with relatively low iron concentrations, higher sulphate and with arsenic present more typically as arsenate, although high manganese concentrations occur in some. High concentrations of aluminium, cadmium and zinc are also observed in the worst-affected waters from the area. Much has been learned about the scale, distribution and causes of arsenic contamination in Asian aquifers over the last few years. However, still many uncertainties remain over the areas affected and the precise mechanisms involved. Doubtless, other areas with contaminated groundwater will be discovered in future, but an appreciation of the aquifers most at risk and the common indicators of likely arsenic contamination should help to identify new areas more quickly and promote more rigorous water testing during well installation programmes.

1. INTRODUCTION

The recent revelation of the serious incidence of chronic arsenic poisoning in Bangladesh and West Bengal from the long-term use of groundwater bearing naturally high arsenic concentrations has come as a shock to water providers, users and scientists alike. However, investigations particularly over the last few years have shown that the Bengal Basin is not unique. Many documented cases of arsenic problems in aquifers have now been identified in various parts of the world. A large number of these are in Asia and certainly Asia has the worst cases of arsenic-related health problems known in terms of patients affected and populations exposed. Despite this concentration of cases, the occurrences of high-arsenic groundwaters are constrained to a restricted range of geological, hydrogeological and geochemical conditions in aquifers, rather than being a 'general' problem. Recent investigations of a number of affected aquifers in Asia have led to a better understanding of the controlling conditions under which As problems occur. However, much remains unknown about both the scale and precise causes of the groundwater contamination. Much of the available information for Asian aquifers is found within unpublished reports by various agencies and hence is difficult to access and is typically not reviewed. This chapter provides a summary of the information currently available on recognised problem areas in Asia and an overview of current understanding of the principal hydrogeochemical controls on arsenic mobilisation. It is likely that other areas with high groundwater-arsenic concentrations will be discovered in the coming years. These are likely to be on a smaller scale than those already identified because of the lessons learned from recognised areas. However, uncertainties still remain in some remote or politically unstable regions where lack of access makes assessments of potential problems difficult. For most areas, targeted screening of 'at risk' aquifers, together with health surveillance given an improved knowledge of the symptoms of chronic arsenic poisoning, will both aid in the rapid identification of new areas.

The chapter discusses the incidence of As contamination in major aquifers in Asia that are typically used for potable supply by large numbers of people. Although As contamination is also documented in areas affected by sulphide mining or mineralisation (e.g. Malaysia, Breward and Williams, 1994; Korea, Woo and Choi, 2001), geothermal areas (e.g. Japan, Yokoyama et al., 1993) and occasionally as a result of industrial contamination (Japan, Tsuda et al., 1992), these tend not to be used directly for potable supply or are of localised extent and have not such a major and widespread impact on human health. They are therefore not discussed in this chapter. The Ron Phibun case of southern Thailand is a notable, though local, exception.

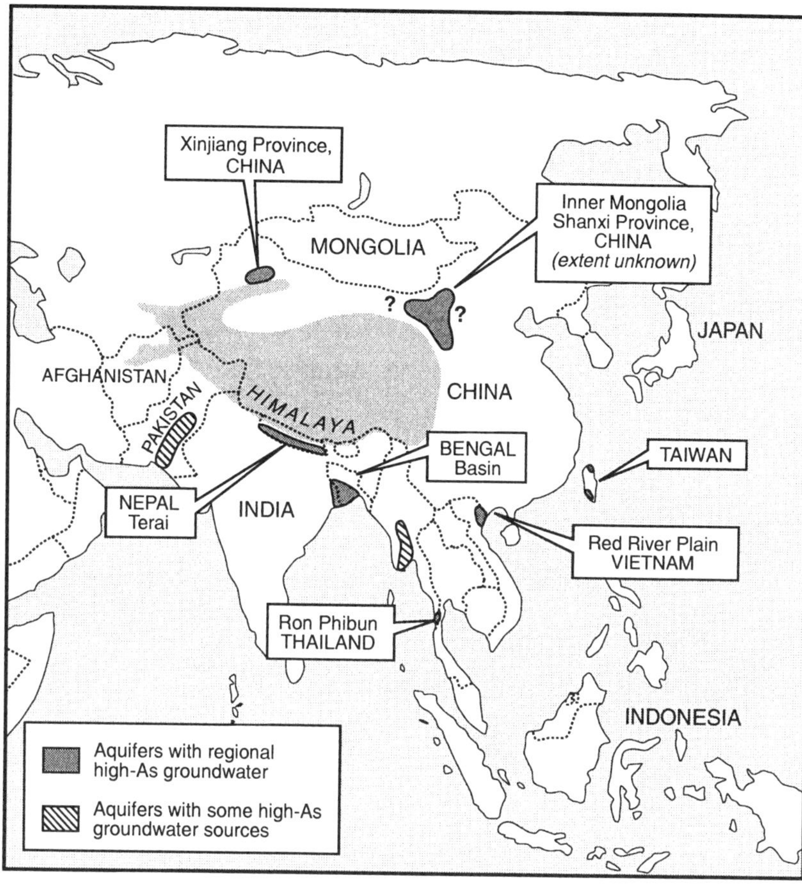

Figure 1. Map of south and east Asia showing the distributions of recognised aquifers containing high-arsenic groundwaters. The shaded area represents the Himalaya and associated uplands.

The World Health Organisation (WHO) guideline value for arsenic in drinking water was reduced provisionally to 10 μg l^{-1} in 1993 and other western countries have made moves to reduce national standards similarly. Despite this, national standards for As in most Asian countries (except Japan) remain at the pre-1993 WHO guideline value of 50 μg l^{-1}. This is borne partly out of analytical constraints, partly out of confusion over the appropriate value to use, and in some countries out of difficulties with compliance. 'High-arsenic' groundwaters discussed in this chapter are defined as those with arsenic concentrations greater than the 50 μg l^{-1} value (Fig. 1; Table 1).

Table 1. Summary of the distribution, nature and scale of arsenic contamination in recognised high-arsenic aquifers (>50 μg l^{-1}) in eastern Asia

Location	Areal extent (km^2)	Population at risk*	As concentration range (μg l^{-1})	Aquifer type
Bangladesh	150,000	35 million	<1–2300	Quaternary alluvial and deltaic sediment
West Bengal	23,000	5 million	<10–3200	Quaternary alluvial and deltaic sediment
Inner Mongolia, Xinjiang and Shanxi Provinces, China	68,000	5.6 million	40–4400	Quaternary alluvial and lacustrine sediment
Taiwan	6000	? 10,000 (now alternative supply)	10–1800	Quaternary? deltaic sediment
Vietnam	1000 recognised	>1 million	1–3100	Quaternary alluvial sediment
Nepal	30,000	550,000	<10–340	Quaternary alluvial sediment
Thailand	100	15,000	1–5000	Former tin mining area, disseminated sulphide ores in granite and alluvial/colluvial sediments

**Estimated to be drinking water with As concentrations >50 μg l^{-1}. Data sources: Kuo (1968); Williams et al. (1996); CGWB (1999); PHED/UNICEF (1999); Berg et al. (2001) ; Chitrakar and Neku (2001); Tandukar (2001); Sun, pers comm. (2001).*

2. DOCUMENTED PROBLEM AREAS

2.1 Bangladesh

Of the regions of the world with groundwater arsenic problems, Bangladesh is by far the worst case identified, with some 35 million people thought to be drinking groundwater containing As at concentrations greater than 50 μg l^{-1} (Table 1) and some 57 millions drinking water with more than 10 μg l^{-1} (Gaus et al., 2001). The large scale of the problem reflects the large area of affected aquifers, the high dependence of Bangladeshis on groundwater for potable supply and the large population accumulated in the fertile lowlands of the Bengal Basin.

The affected aquifers are Quaternary alluvial and deltaic sediments associated with the Ganges-Brahmaputra-Meghna river system which cover a major part of Bangladesh. Groundwater from these aquifers contains As at concentrations up to around 2300 μg l^{-1} (Smedley et al., 2001b). Several surveys of the groundwater have shown a highly variable distribution of As, both laterally and with depth, although highest concentrations are generally found in the south-east of the country (Fig. 2) and concentrations appear highest in shallow tubewells with depths in the range 15–30 m (see Kinniburgh et al., this volume, for a summary of the British Geological Survey and Department of Public Health Engineering, BGS and DPHE, 2001 groundwater survey of Bangladesh). Although often not detected in groundwater surveys with the density of that carried out by the BGS and DPHE survey, some areas with low overall As concentrations have localised 'hotspots' with high As concentrations. That of the Chapai Nawabganj area of western Bangladesh is a notable example (Fig. 2), where the median concentration in groundwater from Holocene sediments was found to be 3.9 $\mu g\ l^{-1}$ but with extremes up to 2340 $\mu g\ l^{-1}$ concentrated in a small area of around 5 x 3 km. Table 2 shows the frequency distribution of As concentrations in sampled groundwaters from the three study areas shown in Figure 2.

Table 2. Frequency distribution of arsenic concentrations in groundwater from Quaternary alluvial/deltaic aquifers in three study areas of Bangladesh (well depths respectively <50 m Chapai Nawabganj; <100 m Faridpur; <150 m Lakshmipur; data source BGS and DPHE, 2001)

Study area	*Number of samples (%)*			*Total samples*
	<10 μg^{-1}	*10–50 $\mu g\ l^{-1}$*	*>50 $\mu g\ l^{-1}$*	
Chapai Nawabganj	40 (58)	13 (19)	16 (23)	69
Faridpur	16 (27)	18 (31)	25 (42)	59
Lakshmipur	3 (5)	9 (15)	47 (80)	59

Groundwater from aquifers in older (Pleistocene) sediments such as those of the Barind and Madhupur Tracts (north-central Bangladesh) generally have low As concentrations (<10 $\mu g\ l^{-1}$, often significantly less). Low concentrations are also generally found in groundwaters from the coarser sediments of the Tista Fan of northern Bangladesh and the large majority of analysed groundwaters from deep aquifers (>150 m) which are exploited in north-east Bangladesh (Sylhet) and the southern coastal area (Kinniburgh et al., this volume).

Summary statistics for groundwater chemical compositions from the three study areas of Chapai Nawabganj, Faridpur and Lakshmipur are given in Table 3. Groundwaters from the Quaternary (mainly Holocene) alluvial/deltaic sediments are predominantly strongly reducing. Groundwater from the three study areas has typically high concentrations of Fe (up to a maximum of 25 $mg\ l^{-1}$; 90th percentile 11.3 $mg\ l^{-1}$), Mn (up to 4.4 $mg\ l^{-1}$; 90th percentile 1.5 $mg\ l^{-1}$) and NH_4-N (up to 18 $mg\ l^{-1}$; 90th percentile

6.4 mg l^{-1}) and with low concentrations of NO_3-N and SO_4 (each commonly <1 mg l^{-1}). Dissolved oxygen is generally not detected, while dissolved organic carbon concentrations reach up to 14 mg l^{-1} (90th percentile up to 8.7 mg l^{-1}; Table 3). Dissolved As occurs dominantly as As(III) (Smedley et al., 2001b). Concentrations of HCO_3 and P are also characteristically high, analysed samples having maxima of 1140 mg l^{-1} and 5.0 mg l^{-1} respectively. Halcrow and DHV (1995) found groundwater conditions to be generally less reducing in the aquifers of northern Bangladesh where sediment grain size is predominantly coarser. This is likely to be a contributory factor in determining the distributions of As in the groundwaters nationally.

Available data for As in sediments from Bangladesh, including those bearing high-As groundwaters, show the concentrations of As to be similar to, or even less than, those for average sediments which are typically in the range 5–10 mg kg^{-1} (Smedley and Kinniburgh, 2002). Pearce et al. (2001) found concentrations in the range 0.4–10.3 mg kg^{-1} (median 2.1 mg kg^{-1}) for 21 sediment samples from the three study areas given in Figure 2 (Chapai Nawabganj, Faridpur, Lakshmipur). Foster et al. (2000) found As concentrations in the range 1–16 mg kg^{-1} in a sediment profile from Brahmanbaria in north-east Bangladesh, although a concentration of 264 mg kg^{-1} was found in an iron-rich clay layer. Datta and Subramanian (1997) also reported As concentrations in the range 1.2–5.9 mg kg^{-1} in modern river sediments from the Ganges, Brahmaputra and Meghna river systems. The concentration ranges suggest that it is the geochemical conditions in the aquifers that maintain the high dissolved As concentrations rather than the presence of high-As source rocks.

Several Bengali workers have proposed that the As contamination of groundwater in Bangladesh is due to the oxidation of pyrite in the aquifers, brought about by overabstraction of groundwater over the last few decades as a result of increased irrigation demands (e.g. Das et al., 1996; Mandal et al., 1996). While this is a plausible mechanism for As release, the strongly reducing groundwater chemistry with low dissolved SO_4 concentrations and the lack of evidence for significant seasonal groundwater drawdown in the worst-affected areas of Bangladesh make the pyrite oxidation hypothesis unconvincing. Although sulphide minerals have been identified, albeit rarely, in some alluvial sediments from Bangladesh, this does not provide evidence for pyrite oxidation as an important mechanism of As release. Indeed, sulphide minerals are an expected product of sulphate reduction under the strongly reducing conditions. Alternative mechanisms, including reductive dissolution of iron oxides (Nickson et al., 1998) and coupled reduction and desorption of As from iron oxides (BGS and DPHE, 2001) have been proposed as the dominant mechanisms driving the As mobilisation under the reducing aquifer conditions in Bangladesh. Phyllosilicate minerals (notably chlorite, biotite and Al hydroxide) have been suggested to play an additional role in the cycling of As in the Bangladesh aquifers (Breit et al., 2001).

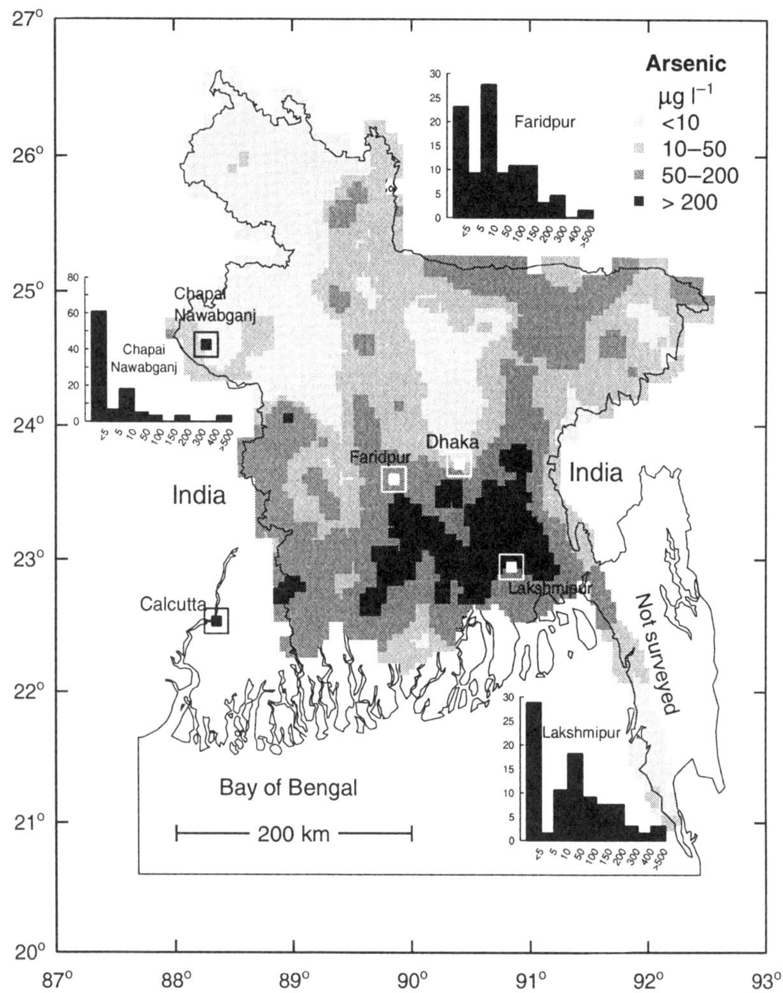

Figure 2. Map of arsenic distributions in groundwater from Bangladesh. The map shows a smoothed distribution determined by disjunctive kriging on 3207 samples of groundwater (<150 m depth, Gaus et al., 2001). Insets are frequency histograms of the concentrations of arsenic ($\mu g\ l^{-1}$) in groundwater from three selected study areas analysed in more detail.

One issue of considerable uncertainty is the variation in groundwater As concentrations with time. Temporal variations on all scales, from hours to decades, can cause problems with compliance, require monitoring and can affect the overall long-term viability of groundwater use nationally. Variations can be particularly critical for wells close to the accepted standard for drinking water (50 $\mu g\ l^{-1}$ in Bangladesh). It is a common perception that the problems encountered in the Bengal Basin have become progressively worse with time. This is partly borne out of the hypothesis that overabstraction of groundwater is responsible but may also be related to the increased testing of wells in recent years (and hence increased numbers of wells found exceeding 50 $\mu g\ l^{-1}$) as well as to increased numbers of identified patients. Few data have so far been produced to analyse reliably the temporal trends in As concentration in Bangladesh. Data given by (Smedley et al., 2001a) for a period up to 1 year showed minor variations but no convincing seasonal trends in As concentrations in given wells. Clearly, more well testing is needed to assess the temporal variations and the future prognosis of groundwater quality.

2.2 West Bengal, India

Endemic arsenic problems were first recognised in West Bengal in 1983 (Chakraborty and Saha, 1987) though were not appreciated internationally until the mid 1990s. Today, it is estimated that more than 5 million people in West Bengal are drinking water with arsenic concentrations greater than 50 $\mu g\ l^{-1}$ (Table 1).

Holocene alluvial and deltaic sediments similar to those of large parts of Bangladesh occur in the eastern part of West Bengal. Here, they form the western margins of the Bengal Basin. High As concentrations have been identified in groundwaters from some tubewells in up to eight districts of West Bengal, the five worst-affected being Malda, Murshidabad, Nadia, 24 North Parganas and 24 South Parganas (Fig. 3). These cover around 23,000 km^2, to the east of the Bhagirathi-Hoogli river system. Arsenic concentrations have been found in the range <10–3200 $\mu g\ l^{-1}$ (Table 1; CGWB, 1999). Groundwaters from the laterite upland in the western part of West Bengal, as well as the Barind and Ilambazar Formations, the valley margin fan west of the Bhagirathi river and the lower delta plain and delta front are reported to have low groundwater As concentrations (PHED, 1991).

Table 3. Summary of the chemical compositions of groundwater from Quaternary (mainly Holocene) aquifers in three study areas in Bangladesh (after BGS and DPHE, 2001)

		Lakshmipur shallow groundwater (<150 m)				*Faridpur shallow groundwater (<100 m)*				*Chapai Nawabganj shallow groundwater (<50 m)*			
	Units	*10th percentile*	*90th percentile*	*Median*	*n*	*10th percentile*	*90th percentile*	*Median*	*n*	*10th percentile*	*90th percentile*	*Median*	*n*
pH		6.86	7.37	7.18	59	6.61	7.09	6.94	59	6.74	7.15	6.99	65
Eh	mV	27	120	82	59	49.1	165	101	59	-18	158	60	68
DO	mg l^{-1}	<0.1	0.1	<0.1	59	<0.1	<0.1	<0.1	59	<0.1	0.42	<0.1	69
SEC	μS cm^{-1}	442	2240	886	59	533	1010	749	59	492	1040	726	69
Ca	mg l^{-1}	25.9	102	47.9	59	56.0	156	101	59	72.1	148	104	69
Mg	mg l^{-1}	23.0	84.9	42.6	59	21.5	43.7	31	59	13.5	39.2	24.6	69
Na	mg l^{-1}	17.6	478	79.2	59	9.78	53.7	18.6	59	8.66	59.8	18.4	69
K	mg l^{-1}	5.44	22.5	11	59	3.45	6.80	4.7	59	1.23	6.53	4	69
HCO_3	mg l^{-1}	277	729	463	59	380	690	536	59	298	659	464	69
Cl	mg l^{-1}	6.28	767	67.7	59	2.6	41.3	9.3	59	2.40	33.5	6.9	68
SO_4	mg l^{-1}	<0.2	40	0.7	59	<0.2	13.5	<0.2	59	<0.2	45.9	7.4	69
NO_3-N	mg l^{-1}	<0.5	2.2	<0.5	59	<0.3	<0.3	<0.3	59	<0.5	<0.5	<0.5	69
NO_2-N	μg l^{-1}	<4	2290	<4	59	<4	7	<4	59	<4	5.4	<4	69
NH_4-N	mg l^{-1}	<0.06	6.35	1.84	59	<0.06	5.90	1.02	59	<0.06	0.99	0.02	69
DOC	mg l^{-1}	1.3	8.7	3.1	57	0.37	6.5	1.9	48	0.44	2.2	1.3	68
P	mg l^{-1}	0.23	2.2	0.68	59	0.13	2.7	1.5	59	0.04	0.52	<0.2	69
As(III)	μg l^{-1}	3.5	213	7	59	0.31	105	9.2	59	<1	46	1.0	69
As_T	μg l^{-1}	29	369	9	59	<6	205	39	59	<1	136	3.9	69
Fe	mg l^{-1}	0.344	6.79	1.71	59	0.28	11.3	5.57	59	0.043	3.39	0.455	69
Mn	mg l^{-1}	0.268	1.28	0.577	59	0.086	1.47	0.48	59	0.288	1.44	0.76	69
B	μg l^{-1}	52	520	56	59	18	71	32	59	12	67	25	69
F	μg l^{-1}	140	352	230	59	40	200	110	59	70	482	160	69
V	μg l^{-1}	<0.2	0.26	<0.2	59	<0.2	0.34	<0.2	59	<0.2	1.9	<0.2	69
Cu	μg l^{-1}	<1	1	<1	59	<1	<1	<1	59	<1	<1	<1	69
Mo	μg l^{-1}	1.0	8.1	2.95	59	0.24	2.6	1.3	59	0.25	2.4	0.82	69
U	μg l^{-1}	<0.01	3.5	0.26	59	<0.01	7.2	0.013	59	0.066	7.8	1.6	69

The Quaternary sediment sequence is a complex mix of unconsolidated sands, silts and clays and the sediments contain variable amounts of organic matter, with peat fragments being a common occurrence (CGWB, 1999). Sediment thickness generally increases southwards. Sedimentation patterns vary significantly laterally, but sands generally predominate to a depth of 150–200 m in Nadia and Murshidabad while the proportion of clay increases progressively southwards into 24 North and South Parganas, as does the thickness of surface clay (Ray, 1997). The regional variations in sediment lithology have a strong influence on groundwater hydrogeology as the aquifers from the northern districts (Malda, Murshidabad, Nadia and part of 24 North Parganas) appear to be largely unconfined, whilst those from most of 24 North and South Parganas are confined (Ray, 1997).

The three-dimensional configuration of the aquifers is complex. A shallow 'first aquifer' has been described at 12–15 m depth, with an intermediate 'second aquifer' at 35–46 m and a deep 'third aquifer' at around 70–90 m depth (PHED, 1991). High arsenic concentrations are noted to be restricted to groundwater from the intermediate 'second aquifer'. Shallowest groundwaters ('first aquifer') appear to have low concentrations, presumably because many of the sources abstracting from this depth are open dug wells and are likely to contain groundwater which has oxidised by exposure to the atmosphere. Groundwaters from the deep aquifer are also reported to have low As concentrations, except where only a thin clay layer separates it from the overlying aquifer, such that a degree of hydraulic connection exists between them. CGWB (1999) noted that the depths of As-rich groundwaters vary in the different districts but that where high-As groundwaters exist, they are generally in the depth range of 10–80 m.

As with Bangladesh, the distribution of As concentrations in the groundwaters is known to be highly variable. Some particularly high concentrations (>200 $\mu g\ l^{-1}$) have been found in groundwaters from 24 South Parganas, along the international border of 24 North Parganas and in eastern Murshidabad (Acharyya, 1997; CSME, 1997).

As with many of the high-As groundwaters from Bangladesh, the groundwaters of eastern West Bengal are strongly reducing with characteristically high concentrations of Fe (up to 12 $mg\ l^{-1}$) and low concentrations of SO_4 and NO_3-N (each <1 $mg\ l^{-1}$). High alkalinity values (HCO_3 up to 700 $mg\ l^{-1}$, commonly >500 $mg\ l^{-1}$) and low F (<1 $mg\ l^{-1}$) are also reported (CGWB, 1999; PHED, 1991). pH values are near-neutral to slightly alkaline. Salinity increases further south in the near-coastal parts of the aquifers as a result of saline intrusion (Adyalkar et al., 1981).

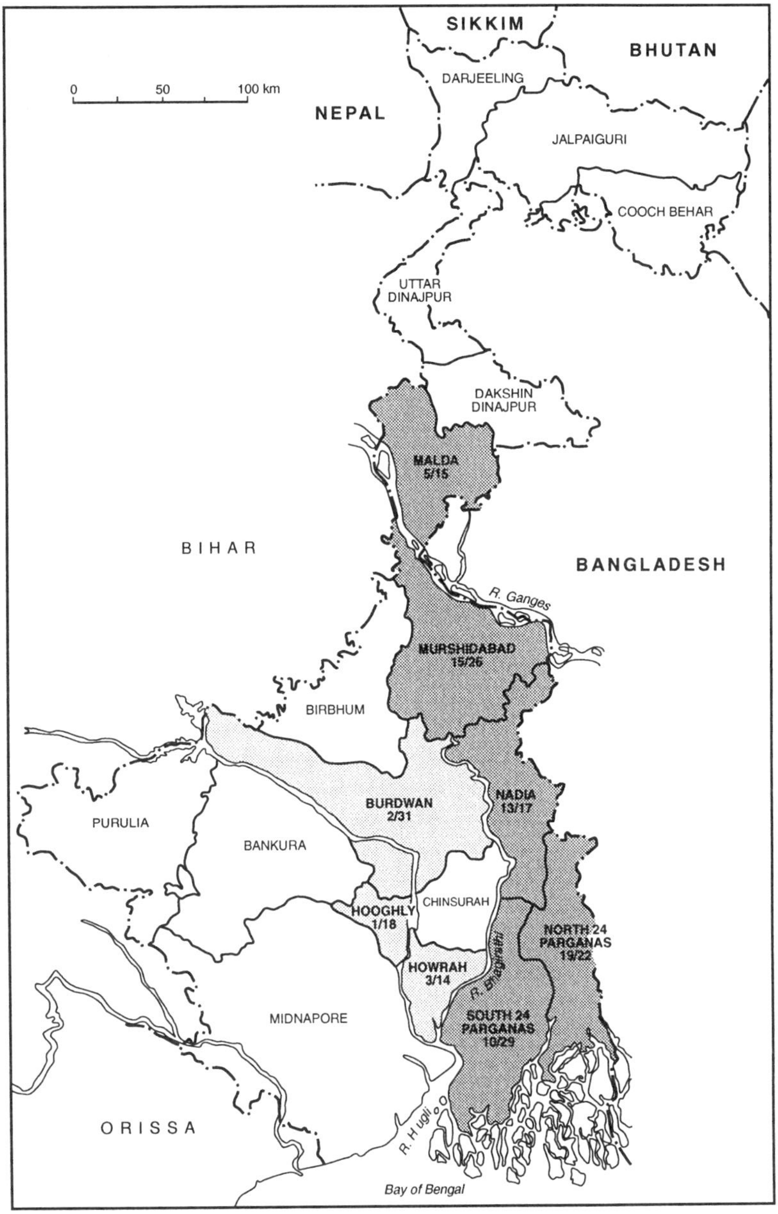

Figure 3. Map of West Bengal showing districts with groundwater containing high arsenic concentrations. Numbers refer to the number of blocks with arsenic concentrations above 50 μg l^{-1} and the number of blocks in the district. The darker shading shows the worst-affected districts (data from PHED/UNICEF, 1999).

Arsenic speciation studies have revealed a large range in the relative proportions of dissolved arsenate and arsenite (e.g. Acharyya, 1997; Das et al., 1995), although arsenite typically dominates. Nag et al. (1996) found a predominance of arsenate over arsenite, but this was in Burdwan district which is known not to be so badly affected by arsenic contamination and may not be representative of the worst-affected areas.

Few time-series data are documented for the groundwaters of West Bengal and, as with Bangladesh, there remains uncertainty over whether significant temporal variations in As concentration occur. CGWB (1999) concluded that groundwater As concentrations vary seasonally, with minima during the post-monsoon period, considered to be due to dilution of groundwater by monsoon recharge. However, the conclusion is apparently based on small sample sets (4–6 samples at any given location) collected over a short time interval (less than one year). Chatterjee et al. (1995) noted a variation of around 30% in time-series data from monitoring of groundwaters over the period of a year in their study of parts of 24 Parganas North and South, but detected no significant seasonal changes in the variation.

Monitoring of groundwater from deep tubewells has apparently shown some once low-As waters have experienced an increase in As concentration with time (e.g. Mandal et al., 1996). Few data are accessible to assess this assertion. Changes in As concentration with time in the deep aquifer may be for a number of reasons, including vertical leakage from the shallow aquifer above via inadequately sealed tubewells or by multiple screening in both shallow and deep sand horizons. If such increases are occurring in some deep tubewells, drawdown of contaminated shallow groundwater through thin or slightly permeable clay horizons between the aquifers may also be a factor.

The sediments of West Bengal appear to contain As in similar concentrations to those found in Bangladesh. Ghosh and De (1995) found As in the range 0.05–8.3 mg kg^{-1} (median 2.3 mg kg^{-1}) in sediments from 24 North Parganas. PHED (1991) found sediment concentrations of <0.1–93 mg kg^{-1}, although most were at the lower end of this range (median 3.2 mg kg^{-1}). Each of these studies showed the highest concentrations to be in fine-grained sediments (clays). Thin Fe- and As-rich coatings were observed on sand grains in some sediments (PHED, 1991). Although the mineral concerned was not identified, the coatings are likely to be iron oxide. Grains of pyrite have been found in some sediments from West Bengal, although it has not been observed as a common mineral. An investigation of sediments by CSME (1997) revealed no separate As mineral phases.

As with Bangladesh, the source of the As in the West Bengal groundwaters has been disputed for some time. Many workers (e.g. Das et al., 1995; Mandal et al., 1996) claim that oxidation of pyrite or arsenopyrite is responsible through recent overabstraction of groundwater. Although seasonal fluctuation in groundwater levels is of the order of 2–4 m in most of the five worst-affected districts of West Bengal, water-level monitoring since the early 1970s apparently shows little evidence for significant long-term changes in water level (CGWB, 1999). Groundwater abstraction is estimated to be typically less than 65% (up to 80% in a few blocks) of the annual renewable resource (CGWB, 1999; Ray, 1997) and groundwater is therefore not overexploited. Alternatively, release from iron oxides under strongly reducing conditions has been proposed as the dominant mechanism for the As release (Bhattacharya et al., 1997). Given the redox conditions in the aquifers of West Bengal and the many similarities with the Bangladesh aquifers, the iron-oxide reduction hypothesis appears more plausible as the dominant mechanism for As release, although the precise mechanisms of As mobilisation in the reducing groundwater environment are still poorly understood (Smedley and Kinniburgh, 2002).

2.3 Taiwan

Health problems experienced in Taiwan have been the subject of much research since their initial discovery in the early 1960s and have formed the basis of many epidemiological risk assessments over the last 30 years. Taiwan is the classic area for the identification of black-foot disease (e.g. Chen et al., 1985; Tseng et al., 1968) and other peripheral vascular disorders but cardiovascular disease, neurological problems, diabetes and internal cancers have also been described from the area. A clear dose-response relationship has been found for many of the disorders, including blackfoot disease and bladder cancer (e.g. Chen et al., 1985).

High-arsenic groundwaters have been recognised in two areas of Taiwan: the south-west coastal region (Kuo, 1968; Tseng et al., 1968) and the north-east coastal region (Hsu et al., 1997). Kuo (1968) observed As concentrations in groundwater samples from south-west Taiwan ranging between 10 $\mu g\ l^{-1}$ and 1800 $\mu g\ l^{-1}$ (mean 500 $\mu g\ l^{-1}$, n=126) with half the samples analysed having concentrations of 400–700 $\mu g\ l^{-1}$. An investigation by the Taiwan Provincial Institute of Environmental Sanitation found that 119 townships in the affected area had As concentrations in groundwater of >50 $\mu g\ l^{-1}$, with 58 townships having >350 $\mu g\ l^{-1}$ (Lo et al., 1977). In north-eastern Taiwan, Hsu et al. (1997) reported an average As concentration of 135 $\mu g\ l^{-1}$ for 377 groundwater samples.

In the south-west, the high As concentrations are found in deep (100–280 m) artesian well waters. The sediments from which these are abstracted are poorly documented, but appear to include deposits of black shale (Tseng et al., 1968). The groundwaters are likely to be strongly reducing as the As is found to be present largely as As(III) (Chen et al., 1994) and some of the groundwaters contain methane (Tseng et al., 1968) as well as humic substances. Groundwaters abstracted in north-eastern Taiwan are also reported to be artesian but more typically shallow, with a depth range of 16–40 m (Hsu et al., 1997). As found in Bangladesh and West Bengal, groundwater from shallow dug wells in Taiwan has low arsenic concentrations (Guo et al., 1994). This is probably a reflection of relatively oxidising conditions in the shallow parts of the aquifer immediately around the open wells.

2.4 China

The presence of endemic arsenicosis has been recognised in China since the 1980s and today the scale of the problem is known to be large. Arsenic problems related to drinking water have been identified in the Provinces of Xinjiang and Shanxi as well as in parts of Inner Mongolia (Fig. 1). Concentrations of arsenic up to 4,400 $\mu g\ l^{-1}$ have been found in groundwater from these affected areas. More recently, high-arsenic drinking water has also been identified in parts of Liaoning, Jilin and Ningxia Provinces in north-central and north-east China (Sun, pers. comm., 2001). However, the extent of these occurrences and the health consequences are not currently well-defined. As in many other countries, the Chinese standard for arsenic in drinking water is 50 $\mu g\ l^{-1}$. The population exposed to drinking water with concentrations in excess of this has been estimated as around 5.6 millions (Table 1) and the number of diagnosed arsenicosis patients currently around 20,000 (Sun et al., 2001). Today, the populations of Xinjiang Province are supplied with alternative low-As sources of drinking water. In other areas however, populations continue to rely on high-arsenic groundwater for potable supply. The majority of the affected patients are the rural poor.

2.4.1 Xinjiang Province

The first cases of arsenic-related health problems due to drinking water were recognised in Xinjiang Province of north-west China (Fig. 1). Artesian groundwater has been used for drinking in the region since the 1960s (Wang and Huang, 1994). Wang (1984) found As concentrations up to 1200 $\mu g\ l^{-1}$ in groundwaters from the province. Wang and Huang (1994) reported As concentrations of between 40 $\mu g\ l^{-1}$ and 750 $\mu g\ l^{-1}$ in deep artesian groundwater from the Dzungaria Basin on the north side of the Tianshan Mountains (from Aibi Lake in the west to Mamas River in the east, a stretch of ca. 250 km). Artesian groundwater from deep boreholes (up to 660 m) was found to have increasing As concentrations with increasing borehole depth. Shallow (non-artesian) groundwaters had observed As concentrations between <10 $\mu g\ l^{-1}$ and 68 $\mu g\ l^{-1}$. That in the saline Aibi Lake was reported as 175 $\mu g\ l^{-1}$, while local rivers had concentrations between 10 $\mu g\ l^{-1}$ and 30 $\mu g\ l^{-1}$.

Wang et al. (1997) reported As concentrations up to 880 $\mu g\ l^{-1}$ from tubewells from the Kuitan area of Xinjiang. A 1982 survey of 619 wells showed 102 with concentrations of As >50 $\mu g\ l^{-1}$. High fluoride concentrations were also noted in the groundwaters (up to 21.5 mg l^{-1}).

2.4.2 Inner Mongolia

In Inner Mongolia, concentrations of As in excess of 50 $\mu g\ l^{-1}$ have been identified in groundwaters from aquifers in the Ba Men region and Tumet Plain which includes the Huhhot Basin (Figure 1, Luo et al., 1997; Ma et al., 1999). These areas border the Yellow River Plain and include the towns of Boutou and Togto. In the region as a whole, around 300,000 residents are believed to be drinking water containing >50 $\mu g\ l^{-1}$ (Ma et al., 1999). Arsenic-related health problems from the use of groundwater for drinking were first recognised in the region in 1990 (Luo et al., 1997). The most common manifestations of disease are skin lesions (melanosis and keratosis) but an increased prevalence of cancer has also been noted. Ma et al. (1999) reported that 543 villages in Ba Men region and 81 villages in Tumet had tubewells with As concentrations >50 $\mu g\ l^{-1}$. Around 1500 cases of arsenic disease had been identified in the area by 1995.

The area of Ba Men with high-As groundwater appears to be around 300 x 20 km in extent and the sediments are Quaternary lacustrine deposits. Wells were mostly installed in the late 1970s and well depths are typically 10–35 m. Arsenic concentrations have been found in the range 50–1800 $\mu g\ l^{-1}$ (Ma et al., 1999) and around 30% of wells sampled had As concentrations >50 $\mu g\ l^{-1}$. The groundwaters are reducing with As being dominantly present as As(III) (Table 4). Some contain high F concentrations (average 1.8 mg l^{-1}; Ma et al., 1999).

The Huhhot Basin (area around 80 x 60 km) lies to the east of Ba Men area. The basin is surrounded on three sides by high mountains of the Da Qing and Man Han ranges and is itself infilled with a thick sequence (up to 1500 m) of poorly consolidated sediments, largely of Quaternary age (Smedley et al., in press). The Holocene sediments at the top of the sequence consist of arenaceous and often conglomeratic alluvial fan deposits along the basin margins and fine-grained lacustrine deposits in the central part of the basin.

Groundwater has been used for several decades for domestic supply and agriculture. Traditional sources of water were shallow hand-dug wells which were typically 10 m or less deep and tapped the shallowest groundwater. These have now generally been abandoned in favour of tubewells which abstract at shallow levels (typically <30 m) by hand pumps or in some cases by motorised pumps. Groundwater is also present within a distinct, deeper aquifer (typically >100 m depth). Tubewells tapping this deeper aquifer are often artesian in the central parts of the basin.

Table 4. Statistical summary of groundwater chemistry in both shallow and deep aquifers of the Huhhot Basin, Inner Mongolia (Smedley et al., in press)

	Units	*10th percentile*	*90th percentile*	*Median*	*n*
Well depth	M	6	196	40	73
DO	mg l^{-1}	<0.1	7.7	0.4	72
Eh	MV	15	351	154	72
pH		7.43	8.24	7.78	73
Ca	mg l^{-1}	11.5	98.6	50.1	73
Mg	mg l^{-1}	16.8	79.4	35.2	73
Na	mg l^{-1}	19.1	376	104	73
K	mg l^{-1}	1.0	6.2	1.9	73
HCO_3	mg l^{-1}	263	749	389	72
Cl	mg l^{-1}	13.0	263	74.3	73
SO_4	mg l^{-1}	0.8	144	34.6	73
NO_3-N	mg l^{-1}	<0.5	14.6	1.13	73
NO_2-N	µg l^{-1}	<3	287	6	73
NH_4-N	mg l^{-1}	<0.01	3.8	0.02	71
DOC	mg l^{-1}	1.1	10.9	3.6	72
P	mg l^{-1}	<0.1	1.6	0.18	73
B	mg l^{-1}	0.04	1.32	0.17	73
Fe	mg l^{-1}	<0.01	1.12	0.07	73
Mn	µg l^{-1}	1.0	378	26	73
Se	µg l^{-1}	<0.5	0.9	<0.5	73
As(III)	µg l^{-1}	<0.9	168	5.2	73
As_T	µg l^{-1}	<1	202	5.4	73
F	mg l^{-1}	0.22	2.95	0.53	73
V	µg l^{-1}	<0.5	3.2	0.9	73
Cu	µg l^{-1}	<1	1.9	<1	73
Mo	µg l^{-1}	0.3	10.4	1.9	73
U	µg l^{-1}	<0.01	12.2	1.24	73

The groundwaters display a strong redox change along the groundwater flow line from the basin margins to the central low-lying area. This reflects in part the changing lithology of shallow sediments from relatively coarse marginal alluvial facies to finer lacustrine facies in the central part of the basin. The change leads to the development of increasingly confined aquifer conditions downgradient. Both shallow and deeper groundwaters are aerobic (dissolved oxygen up to 9.8 mg l^{-1}) along the basin margins but become anaerobic along the flow line. The aerobic groundwaters contain NO_3-N (up to 34 mg l^{-1}, 90th percentile 14.6 mg l^{-1}), SO_4 (up to 1007 mg l^{-1}, 90th percentile 144 mg l^{-1}) and small but detectable concentrations of Se (up to 5 μg l^{-1}, 90th percentile 0.9 μg l^{-1}). These each diminish to below detection limits in the anaerobic sections of the aquifers. Detectable redox-sensitive elements in the anaerobic groundwaters include in particular, Fe (up to 4.5 mg l^{-1}, 90th percentile 1.1 mg l^{-1}), Mn (up to 1.3 mg l^{-1}, 90th percentile 378 μg l^{-1}) and NH_4-N (up to 18 mg l^{-1}, 90th percentile 3.8 mg l^{-1}; Table 4). In addition, concentrations of P and HCO_3 are high in many of the groundwaters, particularly in central parts of the basin. Maximum observed concentrations were 3.1 mg l^{-1} and 1150 mg l^{-1} respectively. As with the Ba Men region, many of the Huhhot groundwaters also have high concentrations of F (up to 6.8 mg l^{-1}). The F does not show any correlation with As however.

A feature worthy of special note in the Huhhot groundwaters is the relatively high concentrations of dissolved organic carbon (DOC). This reaches up to 15 mg l^{-1} in sampled groundwaters from the shallow aquifer but is especially high in groundwaters from the deep aquifer, up to 31 mg l^{-1} DOC (Smedley et al., in press). These deep groundwaters are often brown as a result of the abundance of humic acid. Whether or not the DOC has a critical influence on arsenic mobilisation is unclear. Nonetheless, the high concentration of dissolved organic matter in many of the groundwaters is believed to have been an important control on the generation of strongly reducing conditions in the low-lying parts of the aquifers and will have played a role at least indirectly. Although high concentrations of dissolved organic matter are not universally recognised in high-arsenic groundwaters across the world, they have been observed in groundwaters from Taiwan (Chen et al., 1995) and elsewhere. Indeed, they have been claimed to be a critical factor in the development of blackfoot disease in the populations of Taiwan (Lu et al., 1990).

Arsenic concentrations in the Huhhot Basin groundwaters range between <1 and 1480 μg l^{-1} in the shallow aquifer (≤100 m) and between <1 and 308 μg l^{-1} in the deep aquifer (>100 m). Of a total of 73 samples, summarised by Smedley et al. (2001c), 25% of shallow sources and 57% of deep sources have As concentrations in excess of 50 μg l^{-1} (Table 5). The regional distributions of As in the groundwaters from the shallow and deep aquifers are shown in Figure 4. Concentrations in the aerobic groundwaters from the basin margins are universally low. High concentrations are generally restricted to the low-lying part of the basin where groundwaters are strongly reducing and where sulphate reduction has occurred (Smedley et al., in press). The As is present predominantly as As(III) which typically constitutes some 60–90% of the total As. The redox characteristics of the Huhhot Basin groundwaters have many similarities with those of Bangladesh and it is logical to conclude that the main geochemical processes controlling arsenic mobilisation are similar in the two areas.

Table 5. Frequency distribution of arsenic concentrations in groundwater from the Huhhot Basin, Inner Mongolia (from Smedley et al., in press)

Well depth	Number of samples (%)			Total samples
	<10 $\mu g\ l^{-1}$	10–50 $\mu g\ l^{-1}$	>50 $\mu g\ l^{-1}$	
≤100 m	35 (59)	9 (15)	15 (25)	59
>100 m	6 (43)	0 (0)	8 (57)	14

Of a limited number of samples of dug-well water investigated, some are observed to have As concentrations in excess of 50 $\mu g\ l^{-1}$ (Smedley et al., 2001c). The affected wells are from the part of the aquifer with high concentrations in tubewell waters. This observation contrasts with the situation observed in other reducing groundwater environments such as Taiwan (Guo et al., 1994) and the Bengal Basin (Smedley et al., 2001b). The dug well waters of the Huhhot Basin appear to be reducing with dissolved Fe and Mn present, arsenic dominantly as As(III) and with low NO_3-N concentrations. They also have high DOC concentrations (up to 11.4 mg l^{-1}) and this may be a reason for the maintained reducing conditions and the stabilisation of arsenic in solution (Smedley et al., 2001c). High concentrations of sulphur in the dug well waters (measured SO_4 up to 254 mg l^{-1}) may be due to inputs from surface pollution, oxidation of sulphide minerals or evaporation. A smell of hydrogen sulphide in wells containing water with high As concentrations suggests that the sulphur present is dominantly in reduced form rather than as sulphate.

Few data are available for sediments in the Huhhot Basin. Concentrations from 12 analysed samples are reported in the range 3–29 mg kg^{-1} with a median of 7 mg kg^{-1} (Smedley et al., in press). These concentrations are typical of average values for sediments and are comparable to values observed in high-As aquifers from the Bengal Basin and elsewhere.

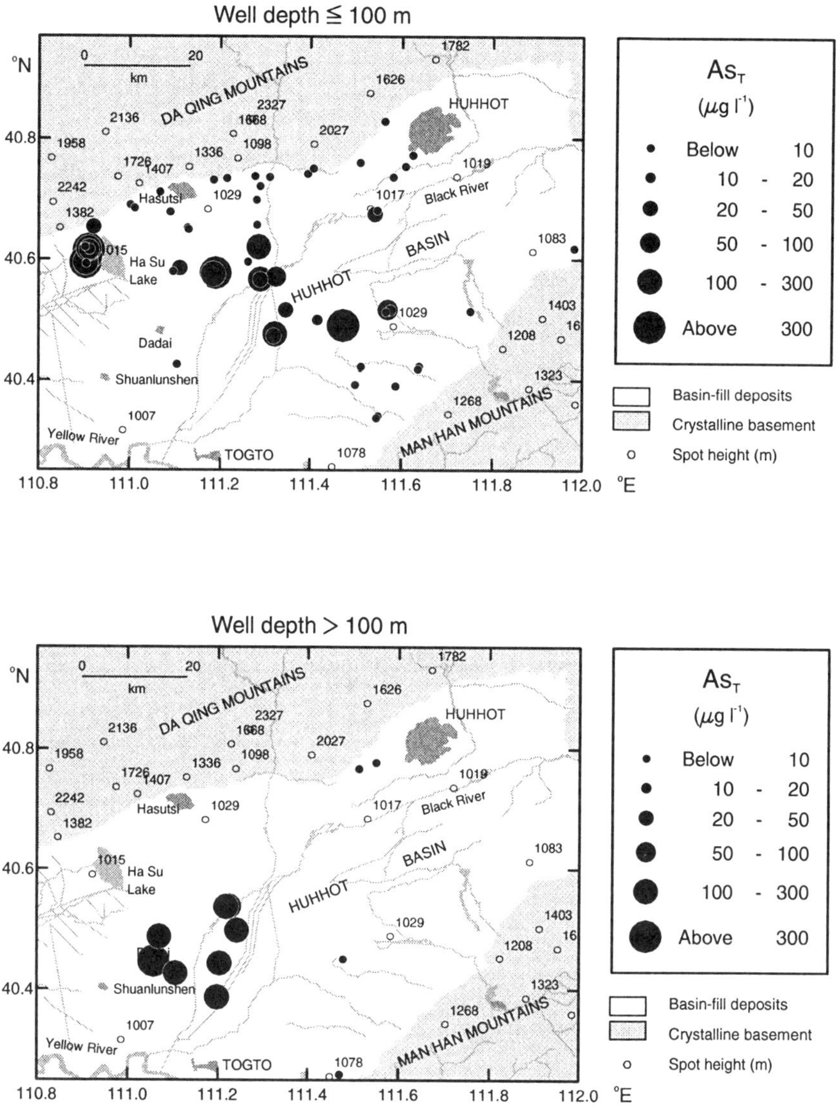

Figure 4. Distribution of arsenic in groundwater in the shallow and deep aquifers of the Huhhot Basin (from Smedley et al., 2001c).

2.5 Vietnam

Arsenic concentrations in the range 1–3050 μg l^{-1}, (average 159 μg l^{-1}; Berg et al., 2001) have been found in groundwater from the city of Hanoi and surrounding rural areas. Recent suggestions are that arsenic-related health problems are beginning to be identified among the exposed populations of Vietnam, although this is as yet unsubstantiated. In a surveyed area of some 1000 km^2 around Hanoi, Berg et al. (2001) found that the As concentrations were spatially variable, but generally higher to the south of the city on the southern margins of the Red River. Concentrations were high in groundwaters from both the shallow and deeper aquifers, but the extremely high values were found in the shallow groundwater from private tubewells. Groundwater from deeper tubewells had As concentrations up to 440 $\mu g\ l^{-1}$. The causes of the variations are not fully clear, but differences in sediment thickness, sediment age and hydraulic connection between layers are believed to be factors.

The groundwaters in the urban area are shown to be anaerobic with high concentrations of Fe (up to 56 $mg\ l^{-1}$) and Mn (up to 3.2 $mg\ l^{-1}$) and generally low concentrations of NO_3-N (typically <0.4 $mg\ l^{-1}$ but up to 7 $mg\ l^{-1}$). In the southern area worst affected by As, high NH_4-N concentrations, up to 20 $mg\ l^{-1}$, are also observed (Trafford et al., 1996).

Alkalinity values are high in some groundwater samples from the shallow aquifer in the Hanoi urban area (up to 528 $mg\ l^{-1}$) but are typically lower (ca. 200–350 $mg\ l^{-1}$) in the deeper aquifer. Concentrations of SO_4 are typically <10 $mg\ l^{-1}$ in both the shallow and deeper aquifers, though occasional high values occur. Concentrations of dissolved P are also occasionally high, reaching up to 1.4 $mg\ l^{-1}$ (as P; Trafford et al., 1996).

Sediments from the area have As concentrations in the range 0.6–33 $mg\ kg^{-1}$ (6–33 $mg\ kg^{-1}$ in brown-black clay horizons, 2–12 $mg\ kg^{-1}$ in grey clay and 0.6–5 $mg\ kg^{-1}$ in grey sand; Berg et al., 2001). Associations between sediment As and sediment Fe concentrations suggest that iron minerals are the dominant sources, though organic matter may contribute additional As to the groundwater. Since groundwater drawdown close to Hanoi city is pronounced, the source of the groundwater As contamination in the Red River aquifers is unclear. The presence of strongly reducing conditions in the parts of the aquifer with high dissolved As concentrations suggests that As release by iron-oxide reduction and desorption reactions (Kinniburgh et al., this volume) is most likely. However, partial oxidation of sulphide minerals during groundwater drawdown could be a contributory factor. Total sulphur concentrations of <0.1% in the Vietnamese sediments were suggested to be pointers that sulphide minerals are not a significant source of the groundwater As (Berg et al., 2001). Unfortunately, Trafford et al. (1996) did not measure As and Berg et al. (2001) did not measure SO_4. Hence, the relationship between the two, and the mechanisms involved in As release are not defined.

Monitoring of selected wells up to three times in the affected area indicated significant changes in As concentrations with time, with maxima typically occurring during September and December (dry season) and lower values during May (early monsoon period; Berg et al., 2001). Extreme variations (e.g. between 220 μg l^{-1} and 1.8 μg l^{-1}) were observed in some wells. If such temporal variations are typical each year, connectivity of the aquifers with the Red River may be a significant factor impacting the seasonal trends, with dilution by river water during monsoon intervals.

To date, high As concentrations have not been observed in the groundwaters from the Mekong Basin of southern Vietnam. This may be a reflection of the fact that the Vietnamese part of the Mekong Basin is the delta region with groundwater use being limited due to the presence of saline groundwater in the aquifer. Limited sampling for As is also a likely factor.

2.6 Nepal

Groundwater is abundant in the Quaternary alluvial sediments of the lowland Terai region of southern Nepal and is an important resource for domestic and agricultural use. The region is estimated to have around 200,000 tubewells which supply groundwater for some 11 million people (Chitrakar and Neku, 2001). Both shallow and deep aquifers occur throughout most of the Terai region. The shallow aquifer appears to be mostly unconfined and well-developed, although it is thin or absent in some areas (Upadhyay, 1993). The deep aquifer (precise depth uncertain) is artesian. Quaternary alluvium also infills several intermontaine basins in Nepal, most notably that of the Kathmandu Valley of central Nepal (ca. 500 km^2) where sediment thickness reaches in excess of 300 m (Khadka, 1993).

A number of surveys of groundwater quality in the Terai region have revealed the presence of arsenic at high concentrations in a number of samples from shallow tubewells (<50 m depth), though most analysed appear to have <10 μg l^{-1}. The Nepal Department of Water Supply and Sewerage (DWSS) has recently carried out a survey of some 4000 tubewells from the 20 Terai districts, mostly analysed by field test kits but with laboratory replication of some analyses (Chitrakar and Neku, 2001). Results from the survey indicated that 3% of the samples had As concentrations of >50 μg l^{-1} (Table 6). The worst-affected districts were found to be Rautahat, Nawalparasi, Parsa and Bara in the central area of the Terai. The highest observed concentration was 343 μg l^{-1} (Parsa District). From testing in 17 of the 20 Terai districts, the Nepal Red Cross Society (NRCS), also found 3% of groundwater sources sampled having concentrations above 50 μg l^{-1}, the highest observed concentration being 205 μg l^{-1}. The spatial distribution of the worst-affected areas was found to be similar to that reported by Chitrakar and Neku (2001). A Finnida survey found 12% of analysed samples exceeding 50 μg l^{-1} while a survey by Tandukar found 9% of samples exceeding this value (Table 6). The highest arsenic concentrations observed by Tandukar (2001) were around 120 μg l^{-1}, most of the high-arsenic samples being from the River Bagmati area. The high As concentrations occur in anaerobic groundwaters and are often associated with high concentrations of dissolved Fe (Tandukar, 2001). The number of samples with exceedances above 50 μg l^{-1} are generally small, but are nonetheless cause for further testing and remedial action.

Although the amount of As data so far reported is limited, preliminary surveying appears to indicate that deeper tubewells in the Terai have lower As concentrations (<10 μg l^{-1}). Data so far available from the Kathmandu Valley have also revealed no arsenic problems there, although more analysis is required to verify the groundwater quality in the region. At the time of writing, no arsenic-related health problems are known to have been identified in Nepal.

Table 6. Frequency distribution of arsenic concentrations in analysed groundwater samples from Nepal (from Chitrakar and Neku, 2001; Tandukar, 2001)

Agency	*Number of samples (%)*			*Total samples*
	<10 μg l^{-1}	*10–50 μg l^{-1}*	*>50 μg l^{-1}*	
DWSS	3479 (89.3)	289 (7.3)	128 (3.3)	3896
NRCS	2206 (79)	507 (18)	77 (3)	2790
Finnida	55 (71)	14 (18)	9 (12)	78
Tandukar	54 (61)	27 (30)	8 (9)	89

2.7 Thailand

In terms of documented health problems from drinking water, the worst identified case of As poisoning related to mining activity is that of Ron Phibun District in Nakhon Si Thammarat Province in peninsular Thailand (Fig. 1). Arsenic-related health problems were first recognised in the area in 1987. Over 1000 people have been diagnosed with As-related skin disorders, particularly in and close to Ron Phibun town (Williams, 1997). At the time of first recognition of the problems, some 15,000 people are thought to have been drinking water with >50 μg l^{-1} (Fordyce et al., 1995). The affected area lies within the South-East Asian Tin Belt which is related to the emplacement of Triassic granitoid intrusions. Primary Sn-W-As mineralisation and alluvial placer tin deposits have been mined in the district for over 100 years, although mining activities have now ceased. Arsenic is contained principally in arsenopyrite and its reaction products. Legacies of the mine operations include arsenopyrite-rich waste piles, waste from ore dressing plants and disseminated waste from small-scale panning by local villagers. Remediation measures include transportation of waste to local landfill.

Waste piles from former bedrock mining in the area are found to contain up to 30% As, the majority in secondary arsenate minerals, particularly scorodite (Williams et al., 1996). Alluvial soils also contain up to 5000 mg kg^{-1} As. In these, Fordyce et al. (1995) concluded that some 20% of the As was present in crystalline iron oxides, with the remainder assumed to be in sulphate mineral phases or other oxidised products.

High As concentrations found in both surface and shallow groundwaters from the area around the mining activity are thought to be caused by oxidation of arsenopyrite, exacerbated by the former mining activities and subsequent mobilisation in groundwater during post-mining groundwater rebound (Williams, 1997).

Surface waters draining the bedrock and alluvial mining areas are commonly acidic (pH<6) with SO_4 as the dominant anion (up to 142 mg l^{-1}) and with high concentrations of some trace metals, including Al (up to 10,500 μg l^{-1}), Cd (up to 250 μg l^{-1}) and Zn (up to 4200 μg l^{-1}; Williams et al., 1996). Strong positive correlations are observed between SO_4 and Cd, Al, Be, Zn and Cu, and SO_4 correlates negatively with pH (Fordyce et al., 1995). Concentrations of these trace metals diminish downstream of the mining area as a result of increased pH from buffering reactions with carbonate minerals. However, highest As concentrations (up to 580 μg l^{-1}) were found some 2–7 km downstream of the bedrock mining area (Fig. 5). As(V) is the dominant dissolved As species in the surface water samples (Williams et al., 1996).

A summary of groundwater quality in shallow and deep aquifers from the area is given in Table 7. Shallow groundwaters (<15 m) are from alluvial and colluvial deposits and deeper (>15 m) groundwaters are from an older carbonate aquifer. The shallow aquifer shows the greatest degree of contamination with As (Fig. 5), with concentrations reaching up to 5100 μg l^{-1} (90th percentile 1440 μg l^{-1}). In the shallow aquifer, 39% of samples (collected randomly) had As concentrations >50 μg l^{-1}, while in the deeper aquifer, 15% exceeded 50 μg l^{-1} (Table 8; Williams et al. 1996).

Conditions in the shallow aquifer are generally oxidising, with redox potentials being typically around 400 mV, NO_3-N up to 8.9 mg l^{-1}, (90th percentile 0.88 mg l^{-1}), low Fe concentrations (<0.4 mg l^{-1}) and dissolved As present dominantly as As(V), although Mn concentrations are relatively high in some samples (up to 8 mg l^{-1}; 90th percentile 0.93 mg l^{-1}; Table 7). Groundwaters from the deeper aquifer are typically more reducing with dissolved Fe concentrations up to 23 mg l^{-1} (90th percentile 11.8 mg l^{-1}), Mn up to 2.3 mg l^{-1} (90th percentile 0.61 mg l^{-1}) and with low NO_3-N concentrations (mostly <0.02 mg l^{-1}; Table 7).

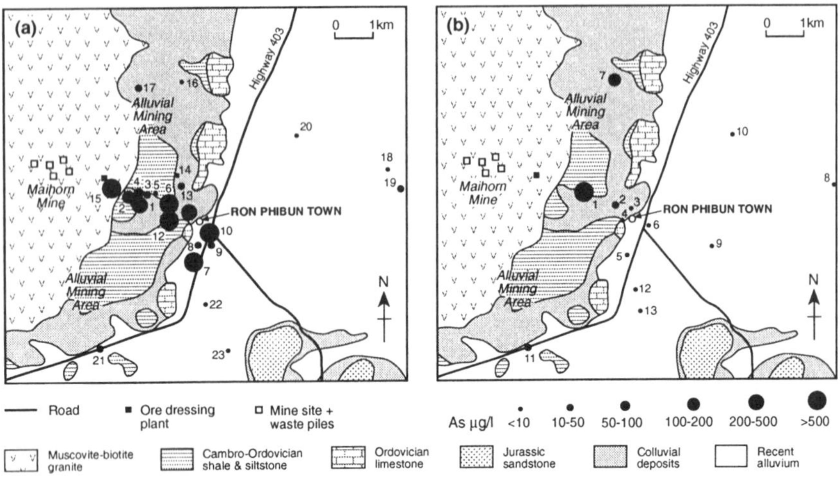

Figure 5. Simplified geology of the Ron Phibun Area, Thailand showing the distribution of arsenic in analysed groundwaters (from Williams et al., 1996). The distributions are (a) arsenic in groundwaters from shallow tubewells (<15 m depth), (b) arsenic in groundwaters from deeper tubewells (>15 m). Numbers refer to samples given in Williams et al. (1996)

The high-As groundwaters of the Ron Phibun area clearly differ in chemical composition from many other high-As groundwater provinces in Asia. In the worst-affected (shallow) aquifer, conditions are more oxidising than those prevalent in the worst-affected areas of the Bengal Basin for instance. The sample with the highest As concentration (5100 $\mu g\ l^{-1}$) has a SO_4 concentration of 40 $mg\ l^{-1}$. This value is much greater than those found in typical Bengal delta high-As groundwaters. In these, SO_4 concentrations are typically <1 $mg\ l^{-1}$, and the groundwaters show evidence of SO_4 reduction (Smedley et al., 2001b). The SO_4 concentrations in the Ron Phibun groundwaters are not as high as in many waters impacted by oxidation of sulphide minerals (e.g. Nordstrom et al., 2000; Schreiber et al., 2000), but this is not inconsistent with the generation of SO_4 by sulphide oxidation, since observed high dissolved As concentrations can be generated by oxidation of very small amounts of arsenopyrite. The groundwaters of the shallow aquifer also differ from those of the Bengal Basin and other high-As Asian aquifers by having much lower HCO_3 concentrations (10–386 $mg\ l^{-1}$) and being more acidic (pH 6.0–7.4). One feature worthy of note however, is the relatively high concentrations of dissolved P in many of the Ron Phibun groundwaters (up to 1.5 $mg\ l^{-1}$ in the shallow groundwater and up to 0.9 $mg\ l^{-1}$ in the deeper samples). In the groundwaters from the deeper aquifer, conditions appear more similar to those from other high-As aquifers in Asia and the retention of As in solution appears to be more of a function of the presence of reducing conditions, although leakage of high-As water from the overlying shallow aquifer is also a possibility.

Table 7. Summary statistics of groundwater chemistry in shallow (<15 m) alluvial and colluvial aquifers and deeper (>15 m) carbonate aquifers in Ron Phibun area, Thailand (data from Williams et al., 1996)

		Shallow aquifer (<15 m)				***Deeper aquifer (>15 m)***			
	Units	***10th percentile***	***90th percentile***	***Median***	***n***	***10th percentile***	***90th percentile***	***Median***	***n***
Temp	°C	25.8	27.6	26.7	23	26.64	28.8	27.3	13
pH		6.05	7.12	6.78	23	6.54	7.11	6.92	13
Eh	mV	331	429	377	23	243	391	331.9	13
SEC	$\mu S\ cm^{-1}$	108	778	332	23	560	952	650	13
Ca	$mg\ l^{-1}$	6.38	87.9	46.4	23	61.9	100.0	90.4	13
Mg	$mg\ l^{-1}$	0.494	6.16	2.37	23	4.8	15.9	6.6	13
Na	$mg\ l^{-1}$	4.90	37.4	14.9	23	9.46	96.2	15	13
K	$mg\ l^{-1}$	1.03	15.2	3.54	23	1.3	5.74	2.2	13
Cl	$mg\ l^{-1}$	3.94	51.5	19.5	23	4.1	41.2	7.9	13
SO_4	$mg\ l^{-1}$	2.51	36.4	11.3	23	0.61	18.1	2.17	13
HCO_3	$mg\ l^{-1}$	10.0	299	149	23	273	446	354	13
NO_3-N	$mg\ l^{-1}$	0.011	0.88	0.061	23	0.011	0.024	0.011	13
Si	$mg\ l^{-1}$	6.6	15	9.9	23	13	33	16	13
P	$mg\ l^{-1}$	0.017	0.62	0.066	23	0.071	0.49	0.15	13
Fe_T	$mg\ l^{-1}$	0.004	0.05	0.010	23	0.035	11.8	0.34	13
Fe(II)	$mg\ l^{-1}$	0.003	0.003	0.003	11	0.003	11.8	0.33	13
Mn	$mg\ l^{-1}$	0.016	0.93	0.19	23	0.018	0.61	0.10	13
As_T	$\mu g\ l^{-1}$	1.99	1440	41.4	23	1.56	111	3.55	13
As(III)	$\mu g\ l^{-1}$	17.8	90.1	49	8	47.7	52.9	50.3	2
Sr	$mg\ l^{-1}$	0.016	0.200	0.11	23	0.164	0.364	0.217	13
Ba	$mg\ l^{-1}$	0.013	0.080	0.034	23	0.016	0.196	0.043	13
B	$mg\ l^{-1}$	0.008	0.039	0.022	23	0.023	0.057	0.03	13
Zn	$\mu g\ l^{-1}$	9.1	200	18	23	82	4100	1000	13
Cd	$\mu g\ l^{-1}$	3.5	3.5	3.5	23				
Li	$mg\ l^{-1}$	0.001	0.005	0.001	23				
Al	$\mu g\ l^{-1}$	180	183	7	23				
Be	$\mu g\ l^{-1}$	0.10	0.68	0.1	23				

Table 8. Frequency distribution of arsenic concentrations in water from Ron Phibun area (from Williams et al., 1996)

Aquifer	*Number of samples (%)*			*Total samples*
	<10 $\mu g\ l^{-1}$	*10–50 $\mu g\ l^{-1}$*	*>50 $\mu g\ l^{-1}$*	
Surface water	1 (4)	2 (8)	20 (83)	24
Groundwater from shallow aquifer (<15 m)	7 (30)	7 (30)	9 (39)	23
Groundwater from deeper aquifer (>15 m)	9 (69)	2 (15)	2 (15)	13

2.8 Other surveyed areas

2.8.1 Pakistan

Quaternary sediments, mainly of alluvial and deltaic origin, occur over large parts of the Indus Plain of Pakistan. The sediments comprise several hundred metres of mainly fine-medium sand, silt and clay, although coarser sands and gravels occur in parts, especially on the margins of the plain abutting upland areas (WAPDA/EUAD, 1989). Aeolian sand deposits occur to the east of the Indus Plain (Thar and Cholistan desert areas) as well as over large parts of the Baluchistan Basin of western Pakistan. As a result of the arid climate of large parts of Pakistan, groundwater availability is severely limited in many areas and poses a major problem for water supply.

Of the aquifers of Pakistan, those of the Indus Plain (predominantly in Punjab and Sindh Provinces) are potentially most vulnerable to As contamination. The Indus sediments have some similarities with the As-affected aquifers of Bangladesh and West Bengal, being Quaternary alluvial-deltaic sediments derived from Himalayan source rocks. However, the region differs in having a more arid climate, greater prevalence of older Quaternary (Pleistocene) deposits and dominance of unconfined and aerobic aquifer conditions, with greater apparent connectivity between the river systems and the aquifers. Aerobic conditions are demonstrated by the presence of relatively high nitrate concentrations in many Indus groundwaters (Mahmood et al., 1998; Tasneem, 1999). Cook (1987) also found dissolved oxygen in a number of groundwater samples from the lower Indus Plain (Sindh Province). Hence, the aquifers appear to have different redox characteristics from those of the lower parts of the Bengal Basin. Under more aerobic conditions (at the near-neutral pH of the Indus groundwaters), As mobilisation in groundwater should be less favourable.

Additional contrasts between the groundwater characteristics of the Indus Plain and those of the Bengal Basin are found in the concentration ranges of HCO_3 and P. Although few HCO_3 data are available for groundwater in Pakistan, values obtained for both fresh and saline groundwater samples from the lower Indus of Sindh were in the range 180–250 mg l^{-1} (Cook 1987). In the saline groundwaters which predominate in Pakistan, Cl is often the dominant anion. From the limited amount of data available for P in the Indus groundwaters, concentrations appear to be generally low, although Rahim (1998) found up to 0.46 mg l^{-1} (as P) in groundwater from Rawalpindi area.

To date, only a limited amount of groundwater testing for As has been carried out in Pakistan. However, the Provincial Government of Punjab together with UNICEF began a testing programme in northern Punjab in 2000. Districts to be tested were selected on the basis of geology and available water-quality information. These included areas affected by coal mining and geothermal springs (Jhelum and Chakwal Districts), areas draining crystalline rock (Attock and Rawalpindi), areas with high-iron groundwaters (Sargodha) and one district from the main Indus alluvial aquifer (Gujrat). A total of 364 samples were analysed. The majority (90%) of samples had As concentrations less than 10 $\mu g\ l^{-1}$, although 6 samples (2%) had concentrations above 50 $\mu g\ l^{-1}$ (Table 9; Iqbal, 2001). Further well testing for As is on-going. No confirmed cases of arsenic-related disease have been found in Pakistan, although epidemiological investigations are also being undertaken in some areas. From the available data, the scale of arsenic contamination of Indus groundwaters appears to be relatively small although further results are needed to verify the scale of contamination. Under the arid conditions in Pakistan, high fluoride concentrations appear to be a more widespread water-related health problem.

Table 9. Frequency distribution of arsenic concentrations in groundwater samples from northern Punjab (from Iqbal, 2001)

District	*Number of samples (%)*			*Total samples*
	<10 $\mu g\ l^{-1}$	*10–50 $\mu g\ l^{-1}$*	*>50 $\mu g\ l^{-1}$*	
Gujrat	33 (87)	3 (8)	2 (5)	38
Jhelum	32 (86)	4 (11)	1 (3)	37
Chakwal	63 (88)	9 (12)	0 (0)	72
Sargodha	49 (83)	7 (12)	3 (5)	59
Attock	68 (92)	6 8)	0 (0)	74
Rawalpindi	81 (96)	3 (4)	0 (0)	84
Total	326 (90)	30 (8)	6 (2)	364

2.8.2 Myanmar

As elsewhere in Asia, traditional sources of water for domestic supply in Myanmar were dug wells, ponds, springs and rivers. However in the Quaternary aquifer of the Irrawaddy delta, many of these have been superseded since 1990 by the development of shallow tubewells. Little testing for As in groundwater has been carried out in tubewells from the alluvial aquifer. However a few reconnaissance surveys have been undertaken and As has been found in excess of 50 $\mu g\ l^{-1}$ in some. Save The Children reported from analysis of 145 shallow tubewells in Thabaung, Laymyethna and Henthada townships (Ayeyarwaddy Division) that 35% of samples exceeded 50 $\mu g\ l^{-1}$ (WRUD, 2001). The Water Resources Utilisation Department carried out a survey of groundwater in Sittway Township in the western coastal area, as well as Hinthada and Kyaungkone Townships close to the south coast (delta region) of Myanmar (WRUD, 2001). In Sittway Township, salinity problems occur in some groundwaters and surface waters and most tubewells are <15 m deep as a result. Merck field-test kits were used for the analysis of As. In Hinthada and Kyaunkone Townships, well depths are typically around 30–50 m deep, although some deeper tubewells (55–70 m) were also sampled. In the southern Townships of the delta area, high Fe concentrations are noted. The distribution of As concentrations in the sampled areas is given in Table 10, although it must be emphasised that this distribution is crudely defined due to the limited resolution and accuracy of the field kits. Exceedances above 50 $\mu g\ l^{-1}$ in shallow tubewells from Sittway, Hinthada and Kyaunkone Townships were around 10–13% (Table 10). One sample from the depth interval 56–70 m also exceeded 50 $\mu g\ l^{-1}$. As with a number of other affected aquifers, dug wells generally had As concentrations of <10 $\mu g\ l^{-1}$ (WRUD, 2001).

Table 10. Frequency distribution of arsenic concentrations in groundwaters from the alluvial aquifer of Myanmar (from WRUD, 2001)

Township	*Well type*	*Number of samples (%)*			*Total samples*
		<10 $\mu g\ l^{-1}$	*10–50 $\mu g\ l^{-1}$*	*>50 $\mu g\ l^{-1}$*	
Sittway	STW	17 (29.3)	35 (60.3)	6 (10.3)	58
	DW	22 (96)	1 (4)	0 (0)	23
Hinthada	STW	56 (68.3)	15 (18.3)	11 (13.3)	82
	DW	6 (75)	1 (12.5)	1 (12.5)	8
Kyaungkone	STW	48 (80)	5 (8)	7 (12)	60
	DW	21 (95)	1 (5)	0 (0)	22
	'DTW'	1 (33.3)	1 (33.3)	1 (33.3)	3

STW: shallow tubewell, DW: dug well, 'DTW': deep tubewell (55–70 m)

2.8.3 Cambodia

The Mekong River system has a substantial proportion of its length within Cambodia and tubewells provide a significant source of potable supply in the country. Little water testing for As has been carried out so far in Cambodia. However, an initial reconnaissance screening of around 100 tubewells from 13 provinces has been carried out with analysis including As, F, some trace metals and some pesticides. Approximately 9% of the samples analysed had As concentrations >10 $\mu g\ l^{-1}$, with observed concentrations in the range 10–500 $\mu g\ l^{-1}$. Exceedances above 10 $\mu g\ l^{-1}$ were found in 5 out of the 13 provinces investigated. The highest concentrations observed to date are from Kandal Province, close to Phnom Penh. Several districts in this province have a high percentage of wells with water containing As in excess of the WHO guideline value for As (Feldman and Rosenboom, 2000). Since this initial screening for As, field testing using portable kits has identified groundwater sources with concentrations above 10 $\mu g\ l^{-1}$ in two additional provinces. High Fe and Mn concentrations and low Eh values are also a common feature of the groundwaters throughout the lowland areas of Cambodia.

UNICEF and other organisations continue to support and carry out field testing using portable kits. However, there is a need for a systematic survey of As in groundwater across the Mekong Valley to determine the true scale of the problem in Cambodia. As the Mekong Valley also covers parts of Laos and Thailand, arsenic problems are also possible in the alluvial and deltaic parts of these countries. However, few data are so far available from these regions.

3. DISCUSSION AND CONCLUSIONS

In terms of the populations at risk from As exposure and the percentage of wells with high As concentrations, Bangladesh is clearly the worst affected region identified. However, serious problems are also recognised in parts of China as well as West Bengal (India) and a number of other countries have problems severe enough to require further testing and mitigation. The geochemistry of the aquifers is complex and many features of the processes involved in As mobilisation remain unclear. However, some common features exist between the recognised provinces. Excepting the special cases (such as Ron Phibun, Thailand) linked to arsenic contamination due to past mining activity, those incidences listed in this account are located in delta areas and alluvial plains where the aquifers are composed of poorly consolidated and geologically young (Quaternary) sediments. Many affected aquifers appear to be of Holocene age, i.e. just a few thousand years old. It is perhaps not a coincidence that many of the world's arsenic-affected aquifers are concentrated within Asia. This is a region of major recent orogenic activity and weathering rates of rock material in the Himalayan region are high. Thus, geologically young alluvial and deltaic plains are a relatively common feature of many of the lowland areas of Asia. This, together with the fact that large populations concentrate in lowland deltaic and river floodplains because of fertile soils and relative abundance and accessibility of water supplies has led to a loading of groundwater-arsenic problems in the Asian continent.

A common feature of arsenic-affected aquifers is the presence of generally slow-flowing groundwater. This may be for a number of reasons, such as low recharge under arid climatic conditions (e.g. Inner Mongolia), small hydraulic gradients (e.g. Bangladesh) or the presence of low-permeability strata. Slowly moving groundwater may be critical in the generation of high-arsenic groundwaters as the element can be mobilised (dissolved, desorbed) from fresh, reactive minerals in the sediments and not have been flushed from the aquifer by large volumes of flowing groundwater (Smedley and Kinniburgh, 2002). From available data for sediments in affected aquifers, the generation of high-arsenic groundwaters appears by and large not to be a function of unusually high concentrations in the host sediments. Indeed, the sediments typically have close to world-average As concentrations. Rather, the distribution of high-As groundwaters appears to be a function of the local (or regional) hydrogeological and geochemical environment. In the Asian aquifers at least, this is overwhelmingly related to the occurrence of strongly reducing conditions.

Reducing conditions are manifested by regionally high Fe and Mn and often high NH_4-N (Table 11), though these rarely correlate well with dissolved As concentrations, except on very local scale. Other features common to high-arsenic groundwater are high alkalinity (>500 mg l^{-1} as HCO_3) and high P concentrations (>0.5 mg l^{-1}). Derivation of these is likely to be largely by decomposition of fresh organic matter in the young alluvial and deltaic aquifers, although P may also be desorbed from iron oxides in a similar way to As (Kinniburgh et al., this volume). High concentrations of dissolved HCO_3 and P may also enhance As mobilisation by competing for available surface sorption sites on iron and other oxides (Manning and Goldberg, 1996). High concentrations of dissolved organic matter are also commonly observed in high-arsenic groundwaters, though they appear not to be a prerequisite for As mobilisation.

Table 11. Typical chemical characteristics of high-arsenic groundwaters occurring under reducing conditions in young alluvial and deltaic aquifers

Determinand	*Typical concentration (mg l^{-1} unless indicated)*
High Fe	>1
High Mn	≥0.5
High NH_4-N	>1
High HCO_3	>500
Often high P	≥0.5
Low SO_4	<5
As(III) often dominant	≥60% of the As_T

Additional problems of high F occur in some groundwaters from arid areas (notably northern China, Pakistan). This creates potential additional health problems from dental and skeletal fluorosis. However, the presence of one of these elements in the Asian aquifers is not necessarily an indication of the presence of the other since As is dependent on the presence of strongly reducing conditions, whilst F occurs where Na-HCO_3 groundwaters dominate and where Ca concentrations are comparatively low.

There is now broad agreement amongst scientists working in West Bengal and Bangladesh about the composition of the arsenic-affected groundwaters, their depth distribution and to some extent, about their evolution. The strongly reducing nature of the affected groundwaters is a key feature and many scientists now consider that a mechanism involving reduction of iron oxides has played a key role in creating the As problem. The 'pyrite oxidation' hypothesis favoured by some West Bengal scientists is no longer widely supported. However, many uncertainties remain over the precise mechanisms of As release from sediments, and the diagenetic changes that take place on sediment burial. Considerable uncertainties also remain about the extent to which the arsenic concentrations may change or have changed with time and more investigations are needed to establish whether temporal variations in As concentration in a given aquifer are significant. If so, this has serious implications for long-term mitigation strategies and groundwater management.

It is possible that other problem areas in Asia remain to be identified. Other deltas and alluvial plains are particularly at risk but probably only under strongly reducing conditions. In As-affected aquifers from some inland basins from other parts of the world, especially those containing recent volcanogenic material, high As concentrations have been found under oxidising conditions where high pH values ($\geq$8.5) occur (e.g. Robertson, 1989; Smedley et al., 2002). Such conditions have not been found in aquifers in Asia. However, it is possible that some so-far unsurveyed Asian aquifers could have increased As concentrations under such high-pH conditions. Quaternary arid inland basins such as occur in parts of Pakistan and northern China are possible areas. Random screening of tubewell water in unsurveyed high-risk areas is therefore a priority.

4. ACKNOWLEDGEMENTS

Acknowledgements are given to collaborators in various projects on arsenic carried out over the last few years, including David Kinniburgh for investigations in Bangladesh and West Bengal, Matin Ahmed and Ihtishamul Huq for work in Bangladesh and Luo Zhen-dong, Zhang Ge-you and Zhang Mei-yun for work in Inner Mongolia. Fiona Fordyce also kindly provided data for water samples from the Ron Phibun area. The manuscript has benefited from reviews by Alan Welch, Yan Zheng and Fiona Fordyce. Support has been provided for the various projects by the UK Department For International Development (DFID), WaterAid and UNICEF. This chapter is published with the permission of the Director, British Geological Survey.

Chapter 8

The scale and causes of the groundwater arsenic problem in Bangladesh

David G. Kinniburgh*, Pauline L. Smedley*, Jeff Davies*, Chris J. Milne*, Irina Gaus*, Janice M. Trafford*, Simon Burden**, S. M. Ihtishamul Huq+, Nasiruddin Ahmad+, Kazi Matin Ahmed$

** British Geological Survey, Wallingford*

*** British Geological Survey, Keyworth*

+ Department of Public Health Engineering, Dhaka

$ Department of Geology, University of Dhaka, Dhaka

Groundwater is now extensively used for drinking water in Bangladesh and present estimates indicate that there are some 6–11 million tubewells in Bangladesh. It is now apparent that approximately ¼ of these wells contain arsenic at concentrations exceeding the Bangladesh drinking water standard (50 $\mu g\ L^{-1}$). As many as 35 million people may be drinking arsenic-affected groundwater. We discuss a national survey of groundwater quality in Bangladesh that attempted to map the distribution and nature of affected wells. Other solutes measured included Na, K, Ca, Mg, Fe, Mn, P and SO_4. The worst-affected part of Bangladesh lies in the south-east of the country where the sediments are of Holocene age and where concentrations of arsenic frequently exceeded 200 $\mu g\ L^{-1}$. Where sampled, deep groundwaters (>150 m) were only rarely affected as were shallower groundwaters from older sediments including the aquifers underlying the Barind and Madhupur Tracts. Seven groundwater samples from the capital city of Dhaka also suggest that the city is not affected. The arsenic is undoubtedly of natural origin and the problem arises even though the sediments do not contain abnormal quantities of total arsenic. There is no evidence to suggest that the dissolved arsenic is derived from the oxidation of pyrite as some have suggested. Rather it appears that the high concentrations reflect a combination of factors: (i) young sediments undergoing rapid change from an oxidizing to a reducing environment following sediment burial; (ii) the release of arsenic by one or more mechanisms which are poorly understood at present but which probably involve the desorption and dissolution of arsenic from iron oxides which are quite abundant in many of the worst-affected sediments; (iii) the very low hydraulic gradients throughout much of Bangladesh mean that groundwater flow is very slow which, combined with the 'young' age of many of the sediments, means that the natural flushing of the shallow aquifer will be slow allowing any released arsenic to accumulate. The rapid rate of deposition of sediments in Bangladesh and the Bengal Basin means that the chance of a well intercepting arsenic-rich water is likely to be relatively high compared with smaller deltas and other alluvial environments where the sedimentation rate is much lower.

1.1 Background

Before 1960, most of the drinking water in Bangladesh was derived from surface water, often from numerous ponds, as well as from dug wells. However, as a result of poor bacteriological quality of the pond water and the fear of cholera and typhoid epidemics, new sources of safe drinking water were sought. Groundwater was seen as the obvious answer as it did not suffer from the microbiological contamination of surface water and the deltaic environment of most of Bangladesh was ideal for its exploitation both in terms of groundwater availability and the ease by which tubewells could be installed. The changeover to groundwater gathered pace in the 1980's in part driven by the UN International Decade for Water Supply and Sanitation. This change in the source of drinking water is believed to have been partly responsible for the reduction in infant mortality in Bangladesh since 1980.

At the same time, the introduction of high-yielding varieties of rice and the rapid increase in the use of groundwater for irrigation enabled Bangladesh to increase yields of rice. Irrigation enabled increased production during the latter part of the dry season (February–April). The use of groundwater for irrigation varies regionally but is greatest in the drier parts of north-western Bangladesh.

Groundwater is now the major source of both drinking water and irrigation water and 97% of the rural population derives its drinking water from tubewells. There has been no accurate country-wide survey of the distribution of tubewells but there are believed to be somewhere in the region of 6–11 million hand-pump tubewells, mostly privately owned.

This success in providing what was believed to be safe drinking water and irrigation water is now in jeopardy because of the finding that many of the tubewells supply water containing excessive arsenic. While the principal concern is the arsenic in drinking water because of its direct influence on human health, there is also concern about the use of arsenic-rich groundwater for irrigation and its consequent effect on the crop uptake of arsenic and its impact in the food chain. There is also concern about the long-term impact of arsenic from irrigation water on the soil quality and on the possible further release of arsenic to groundwater as a result of redox changes induced by paddy cultivation or as a result of displacement of arsenic by phosphate fertilizers.

There have been a number of surveys of arsenic in groundwater in Bangladesh. Most of the data have been collected by government departments, particularly the Department of Public Health Engineering (DPHE) and the Bangladesh Water Development Board (BWDB), but there have also been a number of studies carried out by consultancies in support of numerous drinking water supply and irrigation projects. There have also been a number of detailed surveys by universities (e.g. Burgess et al., 2000). Each of these surveys has had different objectives, regional coverages, sample densities and analytical suites.

Following reports of Bangladeshi patients visiting Calcutta, West Bengal for 'treatment', the DPHE began their arsenic investigations. The first detection of arsenic-rich groundwater in Bangladesh is usually attributed to DPHE in 1993. They observed high concentrations of arsenic at Nawabganj district, close to the West Bengal border. At that time, extensive areas of high-As groundwater were being investigated in West Bengal by the Government of West Bengal. Arsenic problems in groundwater elsewhere were usually associated with areas of mining activity or natural mineralization (often involving sulfide minerals) and with areas of hydrothermal activity. Few detailed geochemical studies of arsenic in major aquifers had been carried out at that time.

Most surveys of Bangladesh groundwater have not included accurate geo-referencing of sampling points making computerised mapping difficult. Nonetheless, the early surveys gave a broad indication of the major element groundwater quality in Bangladesh. The major cause of concern in these early water quality surveys concentrated on the often high iron content of many of the groundwaters and high salinity in the coastal zone. There was also concern about the poor bacteriological quality of drinking water from some hand-pumped tubewells.

After 1995, surveys for arsenic in groundwater in Bangladesh increased rapidly. However, most of these surveys were not sufficiently comprehensive in terms of the range of water quality parameters measured or in terms of sample density or coverage to give a reliable picture of groundwater quality on a national scale. The first survey to achieve a national coverage for arsenic was carried out by DPHE with UNICEF assistance in 1997 using field-test kits.

This chapter gives an account of the results from the DPHE/British Geological Survey (BGS) National Hydrochemical Survey (NHS) carried out during the dry seasons (March–June) of 1998 and 1999. This survey was based on laboratory measurements of arsenic and included a range of other elements which taken together broadly define the geochemical environment of the high-As groundwaters.

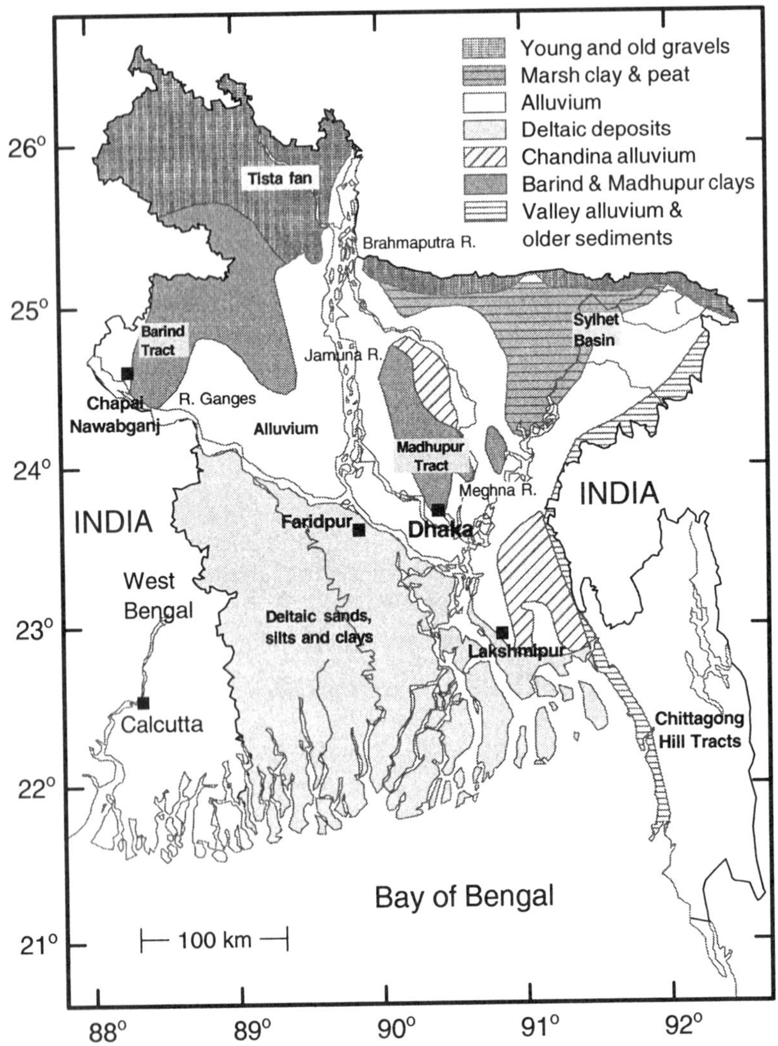

Figure 1. Map showing the distribution of major sediment types (after Alam et al., 1990)

1.2 Sedimentology, geology and hydrogeology

The Bengal Basin in Bangladesh contains a sequence of Cretaceous to Recent sediments which is up to 15 km thick and which occupies some 100,000 km^2 of lowland floodplain and delta. The combined deltas of the Ganges, Brahmaputra and Meghna (GBM) river systems lie within Bangladesh. The GBM system carries the greatest total sediment load of any river system in the world. The contribution to the overall sediment load is greatest from the Brahmaputra, followed respectively by the Ganges (Padma) and the Meghna. The surface geology is shown in the map by Alam et al. (1990) and is described in the accompanying text. It has also been reviewed in DPHE/BGS/MML (1999) and Davies (2001). A simplified description of the geology is given in Fig. 1. The present coastline appears to have remained largely unchanged for the last 3 ka BP (Goodbred and Kuehl, 2000).

Apart from the Pleistocene Tracts, the Chittagong Hill Tracts and the extreme north-west, most of Bangladesh lies no higher than 10 m above mean sea level. The depth to the water table varies seasonally from 1–3 m or less (extensive flooding) in the south to between 5–15 m in the north. Regional hydraulic gradients are therefore very low, typically about $1{:}10^4$. Locally hydraulic gradients tend to be greatest near rivers.

The geomorphology of the GBM system has developed in response to a series of glacio-eustatic sea-level cycles and long-term tectonic activity. Most of the surface sediments are of late Quaternary age but raised areas of Pleistocene sediments are exposed in the Barind Tract to the north-west and the Madhupur Tract in north-central Bangladesh (Fig. 1). The late Quaternary sediments have been deposited in two geomorphologically distinct environments, one to the north of the Ganges and lower Meghna rivers and the other to the south. The northerly sediments are fluviatile deposits in mountain front fan deltas and in the floodplains of major rivers. The southerly sediments have principally been deposited in an estuarine delta environment. At the time of the last glacial maximum some 20–18 ka BP, sea level was up to 130 m below the present-day level. This led to the weathering of exposed sediments and erosion created deeply-incised valleys which as a consequence of the later rise in sea level became filled with a variety of sediments ranging from coarse sands and gravels to fine silts.

The sandy, unconsolidated Pleistocene to Holocene fluvial and deltaic sediments that underlie much of Bangladesh form prolific aquifers. Well depths are up to 350 m but most of the wells for domestic water supply are in the depth range 15–70 m with the well screen usually placed in horizons of medium to fine sand. The exact ages of the shallow Holocene sediments, which provide the bulk of the water, are not known in detail but the available ^{14}C dates point to an age range of 3–10 ka BP (Umitsu, 1993). Across large parts of the southern coastal region, the shallow aquifer is saline and a deeper aquifer, normally found deeper than 150 m below ground level, is used for water supply. The deep aquifer is separated from the shallow aquifer by extensive silts and clays which form a confining layer. Further north, the nature of the deep aquifer has not been extensively explored and in places either may not exist or may not be effectively separated from the shallow aquifer.

The sediments of the Dupi Tila aquifer, which underlie the Pleistocene terraces and are also believed to underlie many of the younger sediments in other parts of the floodplain, are of Lower Pleistocene age. All sediments older than 21 ka BP will have undergone at least one cycle of sea-level change, and perhaps many. The large hydraulic gradients that existed at these times of low sea level (glacial maxima) and the different climatic conditions then existing are likely to have induced significantly greater groundwater flow through the aquifers.

1.3 Early groundwater-quality surveys

Davies (1994) reported groundwater quality in the Dhamrai area of north-central Bangladesh and included the Holocene aquifer in parts of the Jamuna valley and further east as well as the Dupi Tila aquifer of the Madhupur Tract. Differences were identified between groundwaters from the two aquifers, with higher HCO_3, Mn, Ca, Mg, K concentrations and some high P concentrations in the Jamuna valley. Arsenic was not measured in this survey.

Halcrow and DHV (1995) carried out a survey of baseline groundwater quality across Bangladesh, concentrating particularly on heavily-cropped areas of the country and the effects of pesticide and fertilizer use. Halcrow and DHV (1995) identified high concentrations of iron and ammonium in many samples, high phosphorus in some and uniformly low fluoride concentrations (typically 0.8 mg L^{-1}). Arsenic was not measured in this survey.

NRECA (1997) surveyed around 570 tubewells for As and Fe(II) as well as a subset for other determinands including dissolved oxygen, phosphate, sulfate and chloride. Sample sites were in clusters spaced about 25–50 km apart and spread around the country. This survey identified the highest average arsenic concentrations at Chandpur, Faridpur and Feni. Arsenic was also identified (>50 μg L^{-1}) in isolated pockets west of Rajshahi and in northern Bangladesh. Redox potentials were found to be highest (most oxidizing) in northern Bangladesh, especially in the Madhupur and Barind Tracts and the Tista Fan area. The distribution of iron was spatially very variable but salinity indicators and phosphate were generally higher in the coastal area. NRECA concluded that in general, high arsenic concentrations correlated with low redox potential, electrical conductivity greater than 700 μS cm^{-1}, low dissolved oxygen, total Fe greater than 10 mg L^{-1}, phosphate (as PO_4) greater than 4.5 mg L^{-1} and Cl greater than 25 mg L^{-1}. They found low SO_4 concentrations (typically 1–2 mg L^{-1} or less) in groundwater from almost all areas.

In 1997, the Bangladesh University of Engineering and Technology (BUET) analysed around 1200 samples from irrigation wells in north-eastern

Bangladesh for arsenic. They found about 33% of the wells exceeded the Bangladesh standard for arsenic (50 $\mu g\ L^{-1}$) and 60% exceeded the WHO guideline value (GV) of 10 $\mu g\ L^{-1}$. The worst-affected districts were found to be Moulvibazar and Sunamganj. This was the first indication that there was a significant problem in northern Bangladesh.

From 1997 onwards, numerous surveys have been undertaken including in February 1997 a detailed survey by BGS and DPHE of the geochemistry of well waters from the arsenic 'hot spot' of Chapai Nawabganj. Planning for a more extensive survey began in mid-1997 and led to the DPHE/BGS National Hydrochemical Survey (NHS) reported here. While we concentrate on arsenic, we also discuss other elements and properties of Bangladesh groundwaters since these provide the necessary background for understanding the evolution of the arsenic-rich groundwaters.

1.4 The DPHE/BGS National Hydrochemical Survey

1.4.1 Aims

In view of the patchiness of pre-1997 analyses of arsenic and other elements, the lack of accurate geo-referencing and the known limitations of field-test kit surveys, a national survey of arsenic and a range of other diagnostic elements in the groundwaters was carried out. The main aims of the survey were: (i) to establish the regional distribution of arsenic and other elements in the groundwaters based on existing wells, and (ii) to provide estimates of the percentage of wells exceeding various limits for arsenic and other elements.

It was not feasible to measure the normal set of field parameters (pH, redox potential, dissolved oxygen, temperature etc) at each of the sampled wells. Instead more detailed hydrochemical studies were undertaken in three Special Study Areas – the sadar upazilas* of Nawabganj, Fardipur and Lakshmipur districts (Fig. 1). Additional measurements in these areas included the above field parameters as well as a broad range of trace element, isotope and sediment analyses and a limited amount of time series sampling. In this Chapter we concentrate on the results of the national survey but include selected data from these three Special Study Areas where appropriate. The full set of data and maps for all of these surveys can be found in BGS and DPHE (2001).

* An upazila, formerly known as a thana, is an administrative area smaller than a district and larger than a union.

1.4.2 Organization

Details of the planning and organization of the sampling are described by DPHE/BGS/MML (1999). The survey covered the whole of Bangladesh apart from the three districts of the Chittagong Hill Tracts (CHT).

The intention was to adopt a stratified random sampling scheme as closely as possible with the stratification being designed to get a reasonably uniform density of coverage. Accordingly well selection was made using the following strategy: (i) a 3 × 3 grid was pencilled on the upazila map to divide the upazila into approximately nine equal-area cells; (ii) at least one well was selected from each cell ensuring that there was at least 2 km between adjacent sampled wells – normally about 10–12 wells were initially selected by the DPHE field staff in this way; (iii) the final selection of the wells – where possible one per cell – was made by the sampling team leader while in the field. Preference was given to DPHE-constructed wells since details of their construction were better recorded than for the private wells. This was not believed to introduce any bias into the sampling. Most sampled wells were shallow tubewells but where possible, deep tubewells (>150 m) were also sampled in order to provide information about possible depth relationships.

Most of the wells sampled were fitted with a standard Bangladesh number 6 hand pump. Each well was purged prior to sampling by pumping approximately one stroke per foot of well depth. The water sample was filtered through a 0.2 µm membrane filter into a plastic 30 ml bottle and acidified to 1% with concentrated Analytical Grade nitric acid.

The samples were analysed for total arsenic by hydride generation-atomic fluorescence spectrometry (HG-AFS) in the BGS laboratories in the UK. The detection limit was generally 0.25 or 0.5 $\mu g\ L^{-1}$. Additional elements were measured by ICP-AES and in a few cases by ICP-MS. The samples were periodically interspersed with standard reference samples. The analytical data were combined with the site details and entered into a database.

1.4.3 Location and characteristics of the sampled wells

1.4.3.1 Distribution of sampled wells

A total of 3534 wells were sampled, amounting to a sample density of approximately one per 37 km^2 or an average site-to-site separation of about 6 km. The areas of low sample density are those where access was particularly difficult, for example the Sundarbans region in the south-west, the flooded *haor* regions around the Sylhet Basin to the north-east and around the Atrai Basin in the west. These regions have very low population densities for the same reasons. Sixty one of the sixty four districts of

Bangladesh were sampled, the remaining districts being in the CHTs. On average, there were 58 samples per district. Of the 496 upazilas in Bangladesh, 433 were sampled, giving an average of about 8 samples per upazila. A total of 327 samples were taken from deep tubewells (>150 m).

1.4.3.2 Depth of sampled wells

A map of the sampled wells and their depth distribution (Fig. 2) shows that there is a distinct geographical distribution in well depths. This reflects the minimum depth needed to obtain water of acceptable yield and perceived quality (i.e. salinity). The wells in the southern coastal region either need to be very deep (>150 m) as in the Barisal-Patuakhali region (west of the Meghna) or very shallow, as in the Lakshmipur-Noakhali region (east of the Meghna), in order to avoid salinity problems.

Relatively deep wells are also found in the Sunamganj-Sylhet region in north-eastern Bangladesh where shallow aquifers are poor-yielding or non-existent. Very shallow wells are also found in north-western Bangladesh where there is little or no overlying silt or clay layer, so there is no necessity to go deeper. The cluster of deeper wells shown in central Bangladesh were those sampled in the city of Dhaka where extensive groundwater drawdown has occurred due to heavy abstraction. These tubewells abstract from a deeper pre-Holocene aquifer typically from depths of about 100 m.

In percentage terms, the distribution of well depths was: <15 m, 8%; 15–30 m, 33%; 30–60 m, 36%; 60–90 m, 9%; 90–150 m, 5%; 150–200, 1% and >200 m, 8%. The majority of wells were therefore 'shallow'. The median depth was 35 m; only 14% of wells were in the depth interval 60–150 m and 9% exceeded 150 m. In the following discussion, we have chosen 150 m as the cut-off between 'shallow' and 'deep' wells since this often serves to separate wells from the shallow and deep aquifers, where present.

1.4.3.3 Age of sampled wells

There has been considerable growth in the number of installed tubewells in recent years. The survey showed that the date of installation of the sampled wells was: before 1970, 1.9%; 1970–74, 2.5%; 1975–79, 6.0%; 1980–84, 8.1%; 1985–89, 13%; 1990–94, 27% and 1995 or later, 42%. Hence two-thirds of the sampled wells had been installed since 1990. There are regional variations in the age distribution with the greatest percentage of 'old' (pre-1980) wells sampled in the Khulna area and the smallest percentage in the Rajshahi area.

1.4.4 Overview of the distribution of arsenic

1.4.4.1 Regional pattern

The distribution of sampled wells and the arsenic concentration in the groundwaters can be seen from the point-source map (Fig. 3). Class limits were based on rounded quartile values. A very large range of arsenic concentrations was found. The minimum concentration was less than 0.25 μg L^{-1} while the maximum was 1670 μg L^{-1}, a range of four orders of magnitude. Only two samples (0.06%) exceeded 1000 μg L^{-1} and 850 samples (24%) fell below the detection limit. The median concentration was 4.0 μg L^{-1}; 42% of all samples exceeded 10 μg L^{-1}, the WHO guideline value for drinking water and 25% exceeded the Bangladesh standard, 50 μg L^{-1}. If only shallow wells (<150 m depth) are considered, the exceedances increase to 46% and 27%, respectively. 9% of the complete sample set exceeded 200 μg L^{-1} and 5.6% exceeded 300 μg L^{-1}. The average concentration was approximately 60 μg L^{-1} (calculated assuming values below the detection limit were half the detection limit). 249 or 58% of the 433 sampled upazilas contained at least one well exceeding the Bangladesh standard.

There is a distinct geographical distribution of high arsenic groundwater concentrations with the highest concentrations in the south and south-east and the lowest concentrations in the north and north-west. These regional trends are more clearly seen in the smoothed map produced by disjunctive kriging (Gaus et al., 2001) based on a 5 km grid (Fig. 4).

The district with the greatest percentage of wells exceeding the Bangladesh standard was found to be Chandpur in the south-east where 90% of the sampled wells exceeded this standard. The average arsenic concentration in Chandpur was 366 μg L^{-1}. The remaining 11 most-contaminated districts (Madaripur, Munshiganj, Gopalganj, Lakshmipur, Noakhali, Bagerhat, Shariatpur, Comilla, Faridpur, Satkhira and Meherpur) all had 60% or more of their sampled wells exceeding the Bangladesh standard and also had average As concentrations exceeding 100 μg L^{-1}.

These high-As districts contrast sharply with other districts where most of the wells contained <50 μg L^{-1}. However, in only one of the 61 sampled districts, Thakurgaon in the extreme north-west of Bangladesh, were there no sampled wells exceeding the WHO guideline value of 10 μg L^{-1}. Eight districts had no sampled wells exceeding the Bangladesh standard (Thakurgaon, Natore, Barguna, Jaipurhat, Lalmonirhat, Nilphamari, Panchagarh and Patuakhali). These were either southern coastal districts which are dominated by deep wells or north-western districts where contamination is regionally lower. Perhaps more significant is the contrast in

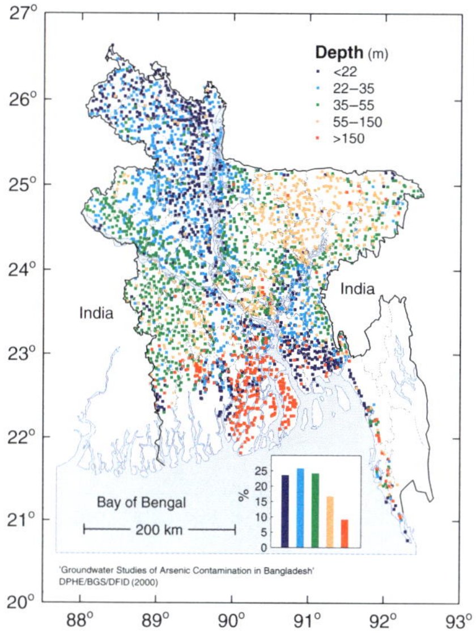

Figure 2. Distribution of depths of wells sampled in the National Hydrochemical Survey

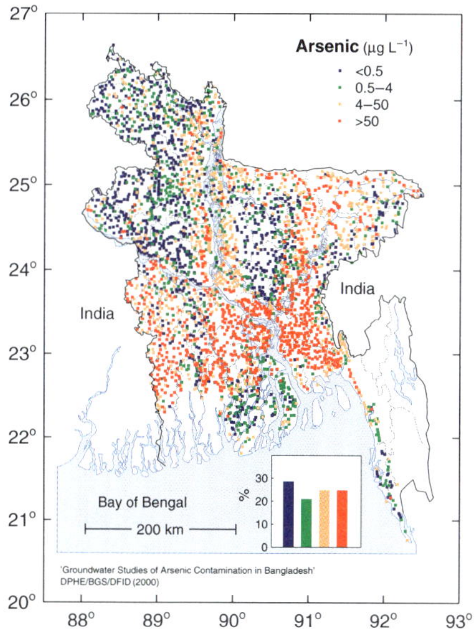

Figure 3. Point-source map showing the distribution of arsenic in Bangladesh well waters. Points in red are wells with greater than the Bangladesh standard for arsenic in drinking water (50 μg L⁻)

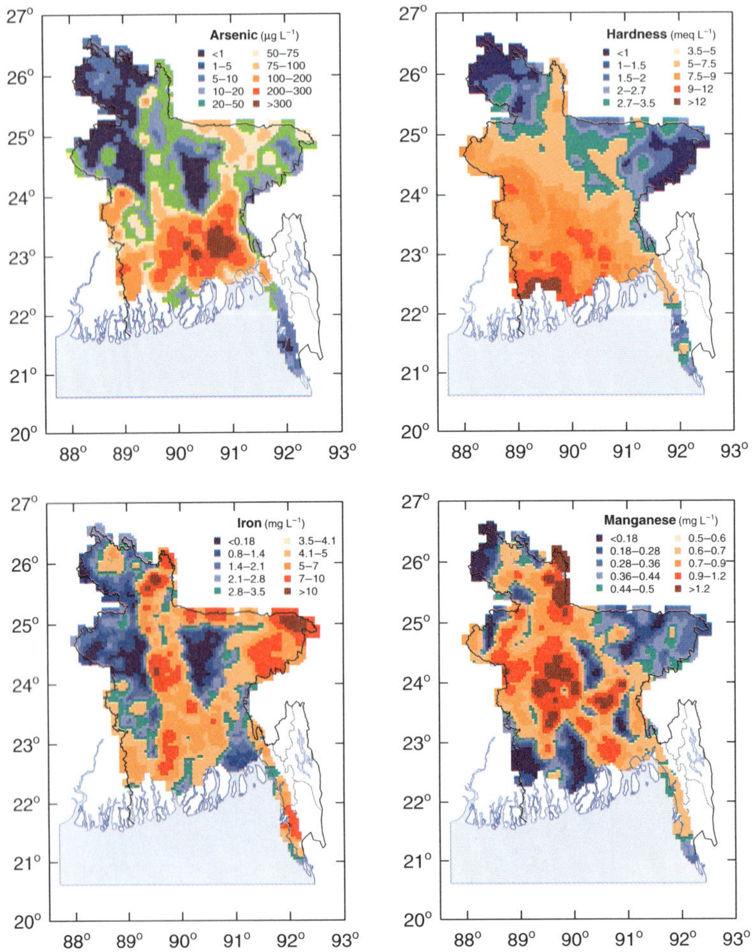

Figure 4. Smoothed maps showing the regional distribution of arsenic, hardness, iron and manganese in shallow groundwaters. The smoothing has been achieved by estimating the concentrations at the centre of a 5 km grid square using disjunctive kriging

average arsenic concentration which exceeds 100 µg L^{-1} in the worst-affected districts and is just a few µg L^{-1} in the least-affected districts.

Neither the raw data nor the smoothed maps show the complete picture. For example, a number of arsenic 'hot spots' have been detected in northern Bangladesh initially by identification of arsenic-affected patients followed by more detailed field testing. The arsenic maps give little indication of such areas since the areasare too small to be reliably identified at the sample density employed in the NHS. These hot spots include some of the worst-affected villages in Bangladesh, for example, Chapai Nawabganj (Nawabganj district) and Char Ruppur (Pabna district).

1.4.4.2 Variation of arsenic with well depth

One of the most important distinctions in arsenic concentrations found in the survey is between 'shallow' and 'deep wells' (Table 1 and compare Fig. 2 and Fig. 4). Wells deeper than 150–200 m show a sharp reduction in average As concentration, and in the percentage of wells that exceed the Bangladesh standard for arsenic (Table 1). The 'cut-off' depth is likely to depend on the nature of the aquifer and on the presence of intermediate clay layers and is therefore likely to vary with location. Relatively few wells in the depth range 90–150 m were sampled (most were from the Sylhet region) but it appears that many of these contained high concentrations of arsenic. Therefore, wells in this region may need to exceed 150 m before low-As water is encountered. Only 3 (1%) of the deep wells (greater than 150 m) exceeded 50 $\mu g\ L^{-1}$. A more detailed investigation of both the deep aquifer and the deeper part of the shallow aquifer is needed across Bangladesh. This should be linked to the geology and sedimentology as well as to details of the construction of the tubewells.

Table 1. Average arsenic concentration in sampled wells as a function of the depth interval

Depth interval (m)	Number of wells	% of wells	Average As concentration ($\mu g\ L^{-1}$)	% of wells with >50 $\mu g\ L^{-1}$
<15	287	8	56	25
15–30	1180	33	76	31
30–60	1258	36	56	26
60–90	317	9	33	21
90–150	165	5	45	35
150–200	32	1	7	1
>200	295	8	3	1

The average As concentration appears to fit a 'bell-shaped' depth distribution with the maximum concentration being found in the 15–30 m depth range. A broadly similar bell-shaped depth trend has been recognised in West Bengal where the aquifer has traditionally been divided into three units with the middle unit, Unit 2, being described as the 'arseniferous' unit (Bhattacharya et al., 1997; Chowdhury et al., 1999; PHED, 1991). The geographical distribution of shallow and deep wells in this survey is far from uniform (Fig. 2). Many of the wells deeper than 200 m were from the southern coastal region. Very few deep wells were found or sampled in the rest of Bangladesh. It is therefore not possible to extrapolate the results from the relatively few deep wells sampled to the other parts of Bangladesh which are not well represented in the NHS coverage.

Of the few deep wells investigated in Faridpur as part of more detailed investigations (Smedley et al., 2001), two wells contained arsenic concentrations exceeding 10 μg L^{-1}. This is in an area which does not have a thick, confining clay layer at intermediate depths. Moreover, there is reason to believe that in the worst of these cases, the borehole may have also been screened in the shallow aquifer. Occasional relatively high As concentrations have been found in groundwaters from 'deep' tubewells recently installed by DPHE in the As-affected areas. Of 170 'deep' tubewells recently drilled by DPHE to assess the water quality in the deep aquifer in various parts of Bangladesh, 95% were below the Bangladesh arsenic standard (unpublished data).

1.4.4.3 Variation of arsenic with surface geology

Depth is an important variable, but is just one of many – the geology and hydrogeology of the aquifers are also important. Each of the sampled wells was assigned to a geological unit based on the Geological Survey of Bangladesh (GSB) classification as given on the geological map of Bangladesh (Alam et al., 1990). This geological classification is only based on the surface geology, and in that respect can only be expected to have a direct influence on the behaviour of the underlying aquifer where there is a close relationship between surface geology and subsurface geology. As the depth of the well increases, this relationship is likely to become weaker. Nevertheless, with this caveat, it usefully distinguishes between regions having major geological differences.

Classification of sample sites by surface geological unit (Table 2) shows that the highest arsenic concentrations were found in the groundwaters from recent deltaic and alluvial deposits with more than half of the well waters having As >50 μg L^{-1} in each of the Estuarine deposits, Chandina alluvium, deltaic silt and deltaic sand units. These are also amongst the most abundant deposits. 30.5% of the shallow wells in recent alluvial and deltaic units contained As >50 μg L^{-1}.

On the other hand, the lowest As concentrations were consistently found beneath the older deposits including the Tertiary and Quaternary deposits and the Pleistocene Barind and Madhupur Clay. There were very few wells exceeding 50 μg L^{-1} associated with these deposits. Wells in the 'old gravelly sand' unit, and to a lesser extent the 'young gravelly sand' unit, also gave low As concentrations. These units form part of the Tista Fan in north-western Bangladesh where the sediments tend to be young, coarse, with low iron oxide concentrations and often lacking in a thick overlying layer of silt/clay.

Table 2. Arsenic concentrations in shallow well waters from the various surface geological units (ordered according to their average arsenic concentration)

Geological unit	GSB code*	No. of wells	% of wells in unit	Average As ($\mu g\ L^{-1}$)**	% of shallow wells with As>50 $\mu g\ L^{-1}$
Estuarine deposits	de	10	0.3	248	60.0
Chandina alluvium	ac	173	5.4	172	65.9
Deltaic silt	dsl	395	12.3	114	52.4
Deltaic sand	dsd	51	1.6	111	52.9
Tidal deltaic deposits	dt	205	6.4	108	48.8
Alluvial sand	asd	116	3.6	67.6	24.1
Alluvial silt & clay	asc	466	14.5	62.3	26.2
Marsh clay & peat	ppc	317	9.9	56.2	33.8
Alluvial silt	asl	597	18.6	43.0	21.4
Valley alluvium	ava	75	2.3	25.0	9.3
Beach and dune sand	csd	19	0.6	23.2	15.8
Young gravel	afy	326	10.2	17.4	8.3
Barail	Tba	1	<0.1	14.3	<0.1
Dupi Tila	QTdt	12	0.4	6.4	8.3
Dihing & Dupi Tila undiff.	QTdd	38	1.2	6.2	2.6
Old gravel	afo	111	3.5	2.2	1.8
Boka Bil	Tbb	5	0.2	1.8	<0.1
Tipam Sandstone	Tt	13	0.4	1.4	<0.1
Girujan clay	QTg	4	0.1	0.9	<0.1
Barind clay	rb	205	6.4	0.7	<0.1
Madhupur clay	rm	67	2.1	0.6	<0.1
Mangrove swamp	dsw	1	<0.1	<0.5	<0.1
Dihing	QTdi	1	<0.1	<0.5	<0.1
All		3208	100	60	27.4

**GSB= Geological Survey of Bangladesh*

***For values less than the detection limit (DL), averages were estimated by substituting ½DL*

Groundwater from the Chandina alluvium has amongst the highest As concentrations. There are two major areas mapped as Chandina alluvium and these may differ in age and diagenetic history. The larger area is near Comilla in the south-east while there is a smaller area near Mymensingh north-east of the Madhupur Tract (Fig. 1). The chemical compositions of the well waters from these two areas are quite distinct – the southerly block contained high concentrations of arsenic (average = 210 $\mu g\ L^{-1}$, n = 116) while, with one exception, the northerly Chandina alluvium gave very low groundwater As concentrations (average As <1 $\mu g\ L^{-1}$, n = 29). The groundwaters from the northerly area also contained relatively little dissolved Fe by Bangladesh standards (average = 0.4 mg L^{-1}). It is therefore likely that the wells in the northern part of the Chandina alluvium derive their water from the older underlying sediments of the Madhupur Tract. This illustrates one of the limitations of using surface geology alone to classify the wells.

Overall, the most significant observation is that high arsenic concentrations are confined to the groundwaters from the most recent (Holocene) alluvial and deltaic sediments – conversely, groundwaters from the older sediments are normally essentially 'arsenic-free'.

1.4.5 Variation of groundwater arsenic concentrations with time and depth

1.4.5.1 Piezometer monitoring

There is a paucity of systematic information about the variation of arsenic concentration with time or with depth for a given well. Five separate shallow piezometers (10 m, 20 m, 30 m, 40 m and 50 m depth) and one deep piezometer (150 m) were therefore installed in each of the three Special Study Areas. Water levels were monitored and water samples collected from these piezometers at approximately two-weekly intervals for ten months. The variation of water levels in the Lakshmipur piezometers is shown in Fig. 5.

The water table is shallow and varies systematically between 1–3 m below ground level (Fig. 5). It is nearest the surface during the monsoon season (July–October). This high water table and limited seasonal fluctuation is typical of southern Bangladesh. The five shallow piezometers have a common piezometric head and therefore appear to be in hydraulic continuity with one another. However, the deep (150 m) well has a different head indicating that the shallow and deep aquifers are either separated by a clay layer of low permeability or that pumping from the deep aquifer is resulting in a vertical hydraulic gradient. The borehole log shows that the lithology of the top 60 m is dominated by grey silts with occasional bands of fine to medium sand but there is a layer of clay at 75–80 m which could act at as a confining layer. The most productive aquifers are found between 100–130 m but these contain saline groundwaters.

1.4.5.2 Piezometer and deep tubewell water quality

The variation in arsenic and sulfate concentrations between June 1999 and March 2000 in the piezometers and the adjacent deep tubewell at Lakshmipur is shown in Fig. 6. There is some variation with time but in most cases there is no convincing trend. There are indications of a decreasing trend in arsenic with time at 30 m and 50 m, and of sulfate at 40 m but the reasons for these are not clear. They may reflect a recovery period following drilling. The 50 m piezometer gave a low yield which may make the results for this depth unrepresentative. The variation with depth is

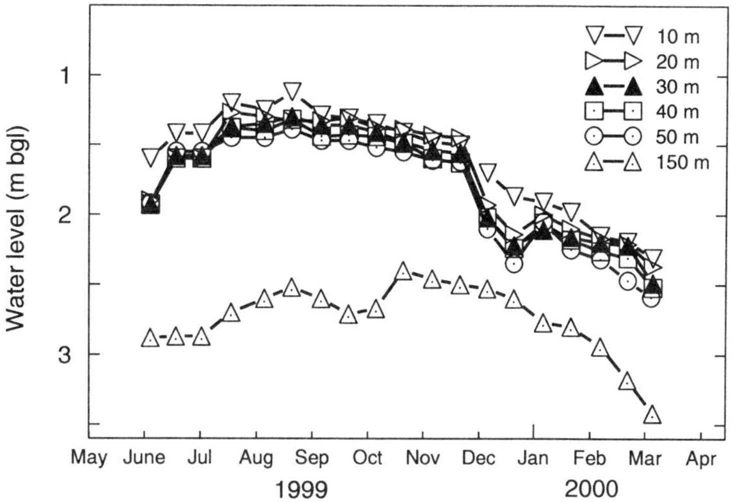

Figure 5. Seasonal variation in the depth to the water table in the Lakshmipur piezometers

much more distinct. Arsenic increases rapidly from 100–150 μg L^{-1} at 10 m to 400–600 μg L^{-1} at 20 m. Thereafter there are no clear trends with depth down to 50 m but there is a large decrease to <10 μg L^{-1} between 50 and 150 m. A nearby deep hand-pump tubewell (275 m depth) was also monitored and this also gave consistently low As concentrations. It is not clear exactly where in the interval 50–150 m the decline in concentration takes place.

The sulfate concentrations are in most cases <5 mg L^{-1} between 10–50 m and at 275 m but the concentration is considerably greater at 150 m. This follows the well-established observation that water at intermediate depths in the southern coastal region is saline and is why there are few wells in this depth interval in this region. The water at 150 m is strongly saline (Cl = 7000–8000 mg L^{-1}) and the high sulfate concentration represents residual seawater sulfate that has not been reduced. Approximately half of the seawater sulfate remains. The water at 275 m is fresh (Cl concentration approximately 6 mg L^{-1}) and has a low sulfate concentration.

1.4.5.3 Dug wells

Historically hand-dug wells have provided drinking water in those parts of Bangladesh where other sources of water have been unreliable. This is particularly true in the drier north-east region. Three dug wells in the Chapai Nawabganj Special Study Area were monitored at approximately fortnightly intervals for a year. These gave consistently low As concentrations (usually < 10 μg L^{-1}) even though this was an area where the shallow tubewells

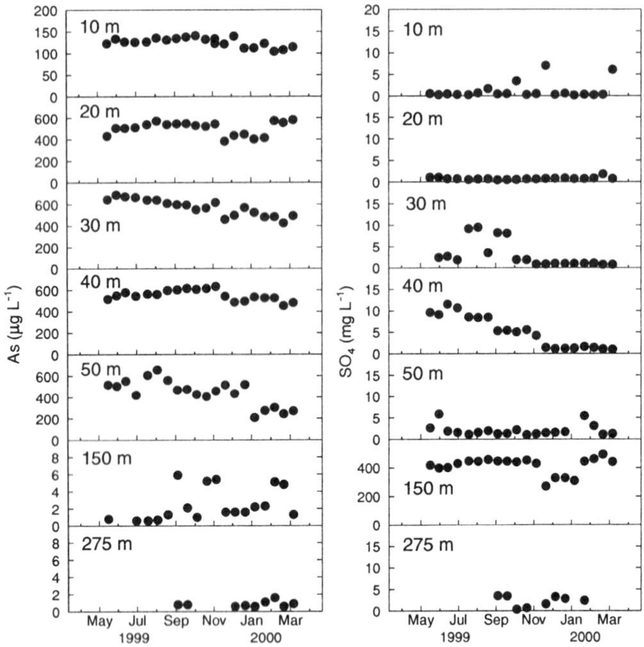

Figure 6. Variation of arsenic and sulfate concentrations with depth and time in Lakshmipur. All results are for piezometers except those for the 275 m depth which refers to a tubewell. Note the differences in the concentration scales

contained large concentrations of As (Smedley et al., 2001). It is believed that this is because the open conditions characteristic of a dug well lead to the shallow groundwater being not so strongly reducing as in an adjacent tubewell which in turn (for whatever reason) favors sorption of As to the iron oxides in the sediment. The shallowest horizons are also likely to be the most actively recharged with fresh rainwater which would also lead to somewhat lower As concentrations. Therefore our limited data indicates that dug wells can provide low-As water in areas where the tubewells contain excessive concentrations of arsenic.

1.4.5.4 Variation in arsenic concentration in groundwater with the age of the well

The data for the shallow wells (<150 m) were divided into six classes with respect to arsenic concentrations. Deep wells were excluded because most of their arsenic concentrations were very low. The percentage of sampled wells in each class is shown in Table 3. Many of the high-As wells have been constructed since 1990.

There is a trend for water from the oldest wells to contain larger As concentrations than water from the younger wells. It is however not possible to deduce from this that the shallow wells have become more contaminated with time. There could be other correlated variables that may account for the trend. For example, proportionately more sampled wells have been drilled in recent years in the Rajshahi Division than further south and this is a generally low-arsenic area. Long-term monitoring of individual wells is required to establish reliable time trends.

Table 3. Percentage of all shallow wells in given arsenic and 'Year constructed' classes

Year constructed (number)	% of shallow wells in given arsenic concentration (μg L^{-1}) class						% in class exceeding Bangladesh standard	% in class exceeding WHO guideline value
	<10	10–50	50–100	100–200	200–300	>300	%>50 μg L^{-1}	%>10 μg L^{-1}
Before 1970 (48)	25	19	21	21	4	10	56	75
1970–74 (83)	30	23	17	10	8	12	47	70
1975–79 (205)	42	20	13	8	10	6	37	58
1980–84 (268)	49	20	14	8	5	5	32	51
1985–89 (408)	49	19	10	8	5	8	31	51
1990–94 (866)	54	18	9	9	4	7	29	47
Since 1995 (1296)	61	18	8	6	3	4	21	39
All years (3174)	54	18	10	8	4	6	28	46

1.4.5.5 Spatial statistics for arsenic

The spatial dependence of the arsenic data for the shallow tubewells was analysed using both classical and geostatistical methods (Gaus et al., 2001). An analysis of variance (ANOVA) using log-transformed arsenic data and stratified by districts demonstrated that the between-district variance was more than 40 times greater than the within-district variance, confirming the distinct regional patterns shown by the map. More detailed spatial analysis was undertaken by computing the variogram for arsenic (Fig.7). It was suspected that the variogram for high-As areas (districts with a mean arsenic contamination exceeding 50 μg L^{-1}) might be different from the less-contaminated districts. However, this was found not to be the case (Gaus et al., 2001). A single variogram could therefore be applied to the whole country.

The data fitted a spherical model with the following parameters:

$$\gamma(h)=c_0 + c(1-e^{-h/a})$$

where $\gamma(h)$ is the semi-variance for a separation of h (in km), $c_0 = 0.5335$, $c = 0.724$ and $a = 48.1$ km. The semi-variances including c_0 and c have units of $(\log \mu g\ L^{-1})^2$. The variogram is smooth and has a range in excess of 125 km. This reflects the scale of variation of the major geological features. The 'nugget' variance, c_0, is large and accounts for about 40% of the total variance in log terms. This nugget variance represents the variation that occurs at separations of less than about 2 km, the minimum intersample distance, and includes sampling and analytical errors. The sampling errors include short-term fluctuations in well water chemistry. These results demonstrate the importance of arsenic variation at the village scale. The density of sampling in the NHS was insufficient to provide any information about the nature of the variation in arsenic concentrations at the village scale.

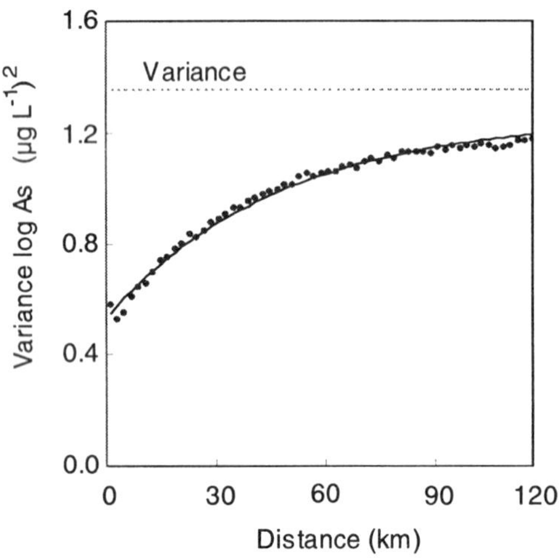

Figure 7. Variogram (based on log-transformed data) of arsenic concentrations in shallow tubewells. The line is the variogram fitted to a spherical model.

Additional geostatistical analysis, particularly disjunctive kriging, was undertaken in order to calculate the probablility that particular wells would exceed given concentrations of arsenic and other solutes. This analysis was also used to calculate the smoothed maps shown in Fig. 4 and is discussed in more detail in Gaus et al. (2001).

Using the 1991 Bangladesh census statistics, the population exposed to >50 $\mu g\ L^{-1}$ in its drinking water was estimated at either 28 million people or 35 million people depending on the method of overlaying the arsenic and population maps (by upazila or by disjunctive kriging to a 5 km grid, respectively). The corresponding figures for the 10 μg L^{-1} WHO guideline value were 46 million and 57 million, respectively.

1.4.6 Other water-quality parameters

A broad range of parameters (mostly metals) was measured in the water samples. These are of interest not only in terms of potential health impacts (e.g. Mn) and acceptability (e.g. Fe) but also in providing a better understanding of the chemical evolution of the groundwaters and therefore possible reasons for the arsenic concentrations. A summary of the average concentrations of a range of solutes in the shallow wells, subdivided according to major geological groupings is given in Table 4. 'Deltaic deposits', for example, includes the deltaic sand (dsd), deltaic silt (dsl), estuarine deposits (de), tidal deltaic deposits (dt) and tidal muds (tm) units (the codes in parentheses refer to the codes used in the GSB map of Alam et al. (1990) and given in Table 2).

There are few strong inter-element correlations within these data. The highest correlations for the shallow groundwaters were between Na and B (linear data, r^2=0.59, n=3203) and amongst the alkaline earth metals (r^2 up to 0.71). Arsenic was not strongly correlated with any of the solutes given in Table 4 – the highest correlation was between arsenic and the alkaline earth metals and with phosphate but these were weak in all cases (r^2<0.1).

1.4.6.1 Iron

The concentrations of iron and manganese were high in most of the groundwaters (Fig. 4). This reflects the prevalence of reducing conditions in the aquifers. The distribution of Fe shows some similarity to that of As, e.g. both elements tend to be higher in the younger (Holocene) sediments than in the older sediments, but the overall correlation is weak (r^2=<0.05, n=3530). The median Fe concentration in all sampled wells was 1.1 mg L^{-1} and the maximum was 61 mg L^{-1}. The median concentration of Fe in high-As (>50 μg L^{-1}) shallow groundwaters was 4 mg L^{-1}. In common with As, the NHS highlights the large difference in Fe concentrations between the shallow tubewells (median Fe concentration, 1.38 mg L^{-1}) and the deep tubewells (median Fe concentration, 0.17 mg L^{-1}).

Particularly high Fe concentrations are found in the Jamuna valley and in north-eastern Bangladesh. The broad north-south band of high-Fe groundwaters through central Bangladesh appears to follow the course of the palaeo-Brahmaputra valley (Davies, 2001). High-Fe groundwaters are also found along the present Meghna valley but interestingly not along the course of the Ganges. The lowest overall Fe concentrations are found in groundwaters from the Barind and Madhupur Tracts and in the coarse

Table 4. Average depth and average concentration of a range of solutes in the shallow well (<150 m) waters classified according to geology (all concentrations in mg L^{-1}).

Surface geology	n	Average													
		Depth (m)	Na	K	Ca	Mg	Fe	Mn	Si	SO_4	P	As	B	Sr	Ba
Deltaic deposits	659	39	150	6.8	101	37	3.4	0.55	18.9	7.9	1.11	0.114	0.130	0.418	0.164
Alluvial deposits	1179	39	39	4.0	53	19	4.0	0.84	21.5	6.9	0.67	0.053	0.060	0.220	0.080
Chandina alluvium	173	37	123	8.7	39	31	2.2	0.42	21.8	2.3	1.39	0.172	0.154	0.257	0.041
Beach & dune sand	19	26	204	12.8	15	21	1.3	0.31	17.3	8.6	2.47	0.023	0.392	0.145	0.021
Marsh clay & peat	317	57	126	4.1	50	21	3.4	0.41	20.2	8.2	1.25	0.056	0.116	0.239	0.095
Gravels	436	27	15	4.4	18	6	4.6	0.53	19.7	4.4	0.33	0.014	0.033	0.082	0.041
Valley alluvium	74	41	96	7.2	16	18	6.7	0.48	20.9	7.8	0.84	0.025	0.121	0.160	0.071
Barind clay	205	30	24	1.4	38	11	0.8	0.34	25.7	3.5	0.09	0.001	0.030	0.164	0.033
Madhupur clay	67	53	20	1.5	25	8	0.2	0.19	30.6	1.2	0.13	0.001	0.033	0.149	0.046
Older Quat./Tert.	73	41	24	3.2	10	6	5.8	0.33	19.4	4.1	0.37	0.005	0.034	0.095	0.056
All	3202	39	72	4.8	53	21	3.6	0.60	21.0	6.3	0.77	0.061	0.081	0.236	0.087

sediments from Tista Fan in the north-west. The Dupi Tila aquifers of the Barind and Madhupur Tracts occur in older (Plio-Pleistocene) sediments with longer histories of groundwater flow and sediment diagenesis. The sediments of the Barind and the Madhupur Tracts are commonly brown or yellowish brown (Alam et al., 1990) and reflect past episodes of oxidative weathering. The iron oxides in these sediments may therefore be less readily reduced (being more crystalline) than those associated with the younger and predominantly grey Holocene deposits. A band of relatively low-iron waters also follows the north-south trending Gorai-Bhairab feature in south-western Bangladesh.

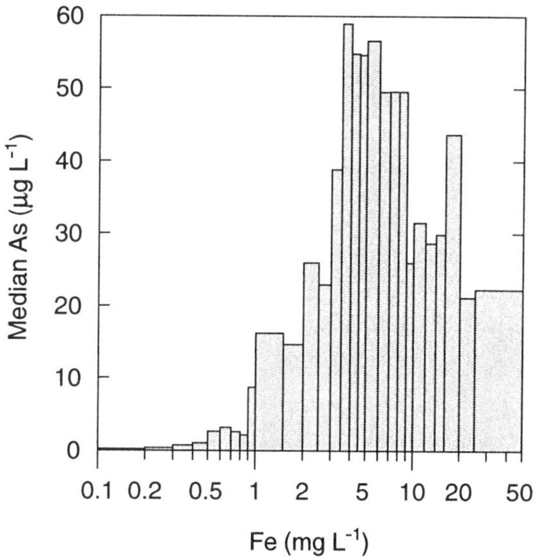

Figure 8. Variation of median arsenic concentration for well waters within a given iron concentration class interval

The highest As concentrations are found in well waters with Fe concentrations in the range 3–8 mg L^{-1} (Fig. 8) rather than in those with the highest Fe concentrations. This is because some of the areas with relatively Fe-rich groundwaters have groundwaters with low concentrations of As, e.g. the Jamuna valley and north-eastern Bangladesh. It would be misleading to attempt to predict high As groundwaters on the basis of their iron contents alone. On the other hand, low-Fe waters (Fe <0.7 mg L^{-1}) tend to have low As concentrations. Therefore the presence of a high Fe concentration is a definite risk factor indicating a possible high-As groundwater but is not in itself sufficient to make reliable predictions about the As concentration of a given well. On the regional scale at least, there are other factors involved.

Iron is released by iron oxide reduction and from the weathering of other Fe-containing minerals such as biotite and chlorite. Some Fe-reducing bacteria can use Fe(III) oxides directly to oxidize organic matter (without using O_2(aq) as the terminal electron acceptor). Bangladesh sediments contain up to 25% by weight of micaceous minerals and biotite frequently makes up a major component of this (Khan et al., 1998). The dissolution of iron oxides is a microbiologically-mediated reaction and the bioavailability of iron oxides depends to some extent on their structure and crystallinity. Relatively freshly precipitated iron oxides, especially of the very fine-grained hydrous ferric oxide (HFO) type, tend to be the most readily bioavailable and are probably responsible for the very high Fe concentrations found in many Bangladesh groundwaters. HFO is particularly abundant in soils and sediments that are, or have been, subject to changing redox conditions. All other things being equal, dissolved Fe concentrations can be expected to be greatest in young sediments.

1.4.6.2 Manganese

Like iron, manganese is also released microbiologically by the reductive dissolution of metal oxides, in this case manganese(IV) oxides. Manganese concentrations in Bangladesh groundwaters are generally high (Fig. 4), particularly along the line of the present-day Jamuna valley and its southern extension. They are lower in Sylhet to the north-east where iron concentrations are notably high. The differences between iron and manganese distributions are interesting and probably related in part to their different positions in the redox sequence – as the redox potential is lowered (the environment becomes more reducing), oxygen and nitrate will be reduced and then Mn(IV) followed by Fe(III) and SO_4^{2-}. At the near-neutral pHs of most Bangladesh groundwaters, Mn(II) is much more slowly oxidized and precipitated than Fe(II). The largest manganese concentrations might therefore be expected to be found in groundwaters that are less strongly reducing than those with high iron and arsenic concentrations. These redox differences, especially after repeated redox cycles and the movement of water, can also lead to a separation in the abundance of secondary iron and manganese oxides in the sediments and may later influence the scale of any re-dissolution.

Therefore the higher groundwater manganese concentrations in the Jamuna valley may reflect the redox status of these groundwaters. These are believed to be less reducing those found in the more distal parts of the delta. However, the high concentrations may also reflect a higher source term – intrinsically larger sediment manganese oxide concentrations.

The Bangladesh standard for manganese in drinking water, based on its health effects, is 0.1 mg L^{-1}. Some 74% of groundwater samples collected in the NHS exceeded this value and 35% exceeded the provisional WHO guideline value of 0.5 mg L^{-1}.

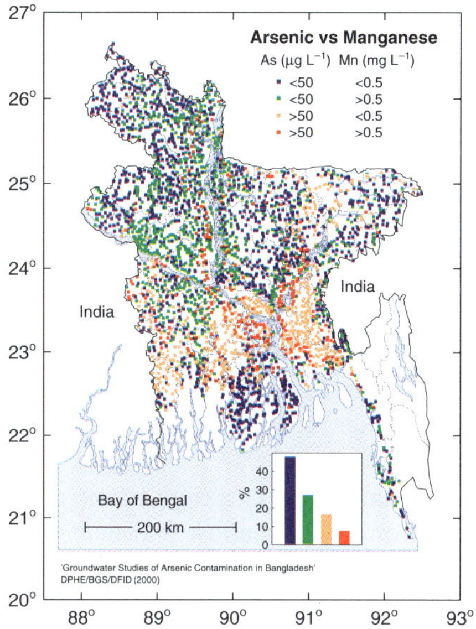

Figure 9. Map showing the joint variation of arsenic and manganese concentrations in well waters

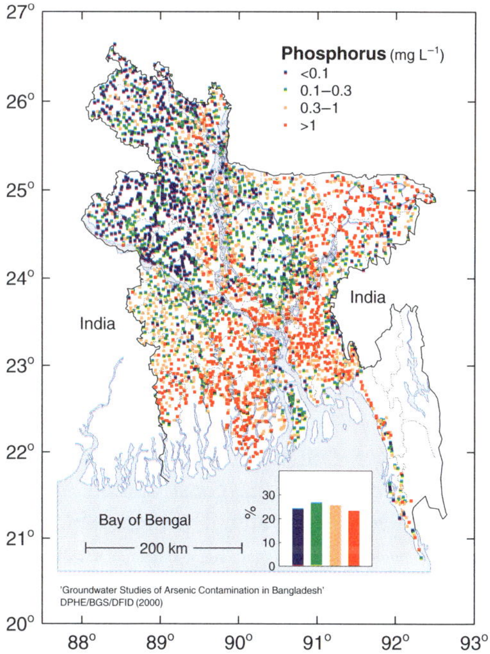

Figure 10. Point source map of the phosphorus concentrations in groundwater sampled from wells during the NHS

Arsenic and manganese are the two elements for which the health-related water quality standards are most commonly exceeded in Bangladesh. Some groundwaters exceed both of these standards while others pass one standard but fail the other (Fig. 9). The correlation is important for assessing the overall distribution of perceived risk posed by these two constituents. The NHS survey has shown that many groundwaters from north-western Bangladesh in particular have low arsenic but high manganese concentrations. Overall, the survey found that 8% of samples exceeded both 50 $\mu g\ L^{-1}$ of As and 0.5 mg L^{-1} of Mn, while 48% of samples were below both of these criteria. Groundwaters from the Barind and Madhupur Tracts (including Dhaka), from the deep aquifer in the southern coastal region and Sylhet and from the shallow but coarser sediments of north-western Bangladesh all tended to comply on both counts.

As was found with iron, there is a significant difference between the number of exceedances for Mn in shallow and deep wells. 39% of shallow wells exceeded the WHO guideline value for Mn and 67% exceeded one or other of the As and Mn WHO guideline values. The figures for the deep wells (>150 m) are 2% and 7%, respectively. The magnitude of the WHO guideline value exceedances for As was greater than for Mn: of the shallow wells exceeding the respective WHO guideline values, the ratio of average well water concentration to guideline value was 13 for As and 2.5 for Mn. This is important when considering the possible health impacts of the exceedances.

1.4.6.3 Phosphorus

The sampled groundwaters had relatively high P concentrations. A median P concentration of 0.29 mg L^{-1} was observed for the shallow groundwaters and 0.32 mg L^{-1} for the deep groundwaters. The maximum P concentration in the shallow groundwaters was 19 mg L^{-1} and in deep groundwaters was 6 mg L^{-1}. The median concentration of P in high-As (>50 $\mu g\ L^{-1}$) shallow groundwaters was 1.1 mg L^{-1}. The highest P concentrations were found in groundwaters from south and eastern Bangladesh and especially from along the Chittagong coast (Fig. 10). These high P concentrations are often associated with saline conditions which account for their prevalence near the coast and in areas recently (in the geological sense) inundated by seawater. The high median P concentration in the deep groundwaters reflects the generally high concentrations found in the southern coastal area. This contrasts with that of arsenic where the deep groundwaters sampled had almost invariably low arsenic concentrations. In other parts of the country and for shallow groundwaters, the P map shows some similarity to the As map (Fig. 3) although the overall correlation is poor (linear data, r^2=0.085, n=3204).

1.4.6.4 Sulfate

Sulfate concentrations in groundwaters in Bangladesh are normally very low (for analytical reasons, the 'sulfate' referred to in this section is actually 'total sulphur', expressed as sulfate). Concentrations from the NHS range between <0.2 mg L^{-1} and 753 mg L^{-1} in the shallow groundwaters and between <0.2 mg L^{-1} and 96 mg L^{-1} in the deep groundwaters. Although the maximum values are high, the median values in both the shallow and deep groundwaters are below 1 mg L^{-1} which is low for groundwaters and reflects the reducing nature of most groundwaters in Bangladesh.

The concentrations of sulfate were generally lowest in the south-western and southern parts of Bangladesh as well as in the Sylhet region in the north-east. Low concentrations of sulfate can be found under strongly reducing conditions as a result of sulfate reduction. The deep groundwaters from the southern coastal region had mostly low sulfate concentrations (<4 mg L^{-1}).

The low sulfate concentrations found in Bangladesh, even where Na concentrations were high, suggest that some bacterial sulfate reduction has occurred. Sediments from Bangladesh are generally Fe-rich which provides the Fe necessary for iron sulfide formation. Such sulfate reduction is supported by limited $\delta^{34}S$ isotopic data from the Special Study Areas and the low SO_4/Cl ratios relative to seawater in the more saline groundwaters of Lakshmipur upazila, indicating sulfate loss from solution (Smedley et al., 2001). Sulfate reduction appears to have been an important process in both the shallow and deep aquifers. Authigenic pyrite including framboidal pyrite, although rare, has been observed by SEM in some sediments from Bangladesh (Pearce et al., 2001).

Sulfate concentrations were typically higher (>4 mg L^{-1}) in the north, particularly in groundwaters from the Tista Fan, the Jamuna valley and the Rajshahi–Pabna area (Ganges and Atrai Floodplains) and from parts of the Barind Tract (although Madhupur Tract groundwaters appear to have low sulfate concentrations, typically <1 mg L^{-1}). Relatively high sulfate concentrations can also occur in areas affected by residual seawater (south-eastern Bangladesh and the Sylhet Basin; Kinniburgh, 2001). The high-SO_4 groundwaters of northern Bangladesh are thought to be more oxidizing than those from the more southerly parts of the delta, a conclusion supported by measurements of redox potential (NRECA, 1997; Smedley et al., 2001). High sulfate concentrations may be derived from residual seawater, from surface pollution (many of the groundwaters from the Jamuna valley in particular are abstracted from shallow depths) or from the oxidation of sulfide minerals (e.g. pyrite) in the zone of water table fluctuation. These processes are difficult to distinguish given the available geochemical data. If the relatively high sulfate concentrations are derived by oxidation of pyrite, then this does not appear to be a mechanism for the release of arsenic since the high-sulfate groundwaters typically contain low As concentrations.

1.4.6.5 Magnesium, calcium, strontium and barium

As noted above, the groundwater chemistry can provide additional information about the likely composition of the aquifer sediments. This is particularly true for the alkaline earth elements (Mg, Ca and Sr in particular) since their concentrations are often controlled by the dissolution and precipitation of carbonate minerals.

The regional distribution of 'hardness' (i.e. Ca plus Mg) is shown in Fig. 4. This probably reflects the distribution of free carbonates in the sediments with areas of hard water reflecting their presence and low hardness areas reflecting their absence. The most probable carbonate minerals are calcite ($CaCO_3$) and dolomite ($CaMg(CO_3)_2$). The absence of pH and bicarbonate data from the National Hydrochemical Survey dataset means that saturation indices for these minerals could not be calculated. However, data from the three Special Study Areas (Smedley et al., 2001) suggest that there is both calcite and dolomite saturation, even slight supersaturation.

Barium correlates only weakly with these cations. Mg^{2+} and Sr^{2+} readily substitute for Ca^{2+} in calcite whereas Ba^{2+} does so much less readily because of its substantially larger ionic radius. It appears from the NHS that the Holocene sediments derived from the River Ganges (south-western and south-central Bangladesh) probably contain free calcium-magnesium carbonates. Soils developed from the sediments in this region are also defined as carbonate-rich (Brammer and Brinkman, 1977). In contrast, the Holocene sediments of the Tista Fan (north-west) and the north-east *haor* region of the Sylhet Basin, as well as the older Quaternary and Tertiary sediments of the Chittagong area, probably do not contain free carbonates. Concentrations of Ca in these latter groundwaters are less than 15 mg L^{-1}. This is also true for the deep groundwaters from the southern coastal region.

These alkaline earth elements are also derived from other minerals, including aluminosilicates, which reduces the overall correlations. Where there has been a substantial inundation of seawater in the past, some Mg may remain from this source: for every 8.3 mg L^{-1} of Na that is derived from residual seawater, there will be 1 mg L^{-1} Mg along with small amounts of Ca (0.3 mg L^{-1}), Sr (6 μg L^{-1}) and Ba (2 μg L^{-1}). The effects of ion exchange can alter these simple relationships.

1.4.6.6 Other elements

Sodium, potassium and boron are indicators of groundwater salinity and reflect relict seawater influences either by marine inundation of low-lying areas or saline intrusion of near-coastal aquifers. Potassium can also be derived from mineral reactions (weathering of clays, ion exchange) and so although the distribution of K bears some similarity to that of Na and B, differences also occur (Kinniburgh and Smedley, 2001). Concentrations of Na, K and B are in general greater in the deep groundwaters sampled in the

NHS. This reflects the large proportion of deep groundwaters sampled from the southern coastal region and the Sylhet Basin.

In addition to the major and minor elements discussed above, a selection of trace elements was also measured by ICP-MS in 16–18 samples from the National Hydrochemical Survey. The samples were selected on the basis of their high Fe and Mn concentrations determined previously by ICP-AES. The data are therefore biased in terms of metal-rich groundwaters, but give an indication of water quality in these groundwaters. Statistical summaries are given in Table 5. The results indicate that even in these samples, concentrations of most analysed trace metals are below recommended WHO guideline values for drinking water. Exceptions were most commonly found for U for which the provisional WHO guideline value is 2 $\mu g\ L^{-1}$. The low concentrations of 'heavy metal' elements such as Cu and Pb probably reflect control by active sulfide mineral precipitation.

1.5 Other extensive national hydrochemical surveys

Two other large-scale surveys for arsenic have been undertaken since arsenic became an issue in Bangladesh. Starting in 1993 and particularly since 1997, DPHE has undertaken a large number of arsenic analyses using both field-test kits and laboratory analyses. The results of these, as of 1998, were summarised in DPHE/BGS/MML (1999). The 1997 DPHE/UNICEF survey based on nearly 23,000 samples analysed with field-test kits was the most extensive, and was the first to show the regional pattern for the whole of Bangladesh. Since the survey only tested whether there was detectable arsenic or not (so-called 'yes/no' results), it could only show the percentage of tested wells in which As was detected, not the magnitude of the contamination (the detection limit of the early field-test kits was probably in the range 50–200 μg As L^{-1}). Nevertheless the patterns revealed were broadly similar to those shown by the NHS. The greatest difference arises from the fact that the field-test kits were rather insensitive at the 50 $\mu g\ L^{-1}$ level and so the DPHE/UNICEF survey did not provide reliable resolution for areas in which most wells contained As in the 10–50 $\mu g\ L^{-1}$ range. This included extensive areas of alluvial sediments in northern Bangladesh especially along the Jamuna valley.

The School of Environmental Sciences, Jadavpur University, Calcutta and Dhaka Community Hospital (SOES/DCH) have also carried out, and are continuing to carry out, extensive surveys of arsenic in wells across Bangladesh (e.g. DCH, 1997; SOES/DCH, 2000). Their surveys have identified severe arsenic contamination in many areas, including northern Bangladesh, and have so far reported contamination (>50 $\mu g\ L^{-1}$) in 43 out of 64 districts in Bangladesh. The affected populations were estimated to be 51 million drinking water with >10 $\mu g\ L^{-1}$ and 25 million drinking water

Table 5. Summary of trace element data for selected metal-rich waters from the National Hydrochemical Survey

Element	Minimum	Median	Maximum
		μg L^{-1}	
Al	4	8	27
Be	<0.05	<0.05	<0.05
Cd	<0.02	0.035	0.51
Ce	<0.005	0.0215	0.587
Co	0.4	1.32	34.6
Cr	<0.5	<0.5	1.4
Cs	<0.05	<0.05	0.19
Cu	<1	<1	8
Li	2	2.8	25
Mo	<0.1	1.9	9.4
Ni	2.4	3.6	132
Pb	0.09	0.3	10.8
Rb	<0.1	0.45	10.2
Sb	<0.02	0.03	0.16
Sn	<0.1	<0.1	0.6
Tb	<0.005	<0.005	0.01
Tl	<0.01	<0.01	<0.01
U	0.03	2.365	11.6
V	<0.2	1.45	4.2
Y	0.017	0.062	0.32
Yb	<0.008	<0.008	0.029
Zn	3	9.5	94

with >50 μg L^{-1} As, slightly lower than our estimates from the NHS. They have found that 12% of the 'deep' (100–300 m) tubewells from areas overlain by 'Flood Plain and Deltaic' sediments had arsenic concentrations exceeding 50 μg L^{-1}. Tubewells greater than 300 m depth mostly had As concentrations of less than 50 μg L^{-1} but 22% contained arsenic between 10 and 50 μg L^{-1}. Tubewells less than 300 m deep from the 'Flood Plain and Deltaic Plain (including coastal region)' (n = 371) were mostly (except for four wells) below 50 μg L^{-1}, but about 22% of tubewells had arsenic concentrations in the range 10–50 μg L^{-1}. While we have not seen a map of the arsenic distribution from these surveys, it is believed to be broadly similar to that shown by the NHS.

1.6 Dhaka deep tubewells

Dhaka is a city of some nine million people and derives its water supply from more than 200 tubewells with depths of approximately 100 m. Groundwater samples were collected from seven public supply wells within Dhaka city.

These wells derive their water from the Plio-Pleistocene Dupi Tila aquifer which lies beneath the Madhupur Clay. The screened interval of the wells is unknown. The chemical analyses (Table 6) indicate that the groundwater is of low salinity (Na <50 mg L^{-1}; B <0.1 mg L^{-1} (data not shown); SO_4 0.6–35 mg L^{-1}). Concentrations of most trace elements are low to very low. In particular, As concentrations are all <0.5 μg L^{-1} and P concentrations are less than or equal to 0.1 mg L^{-1} (data not shown). The low As concentrations are believed to have been confirmed by other more extensive, but unpublished, surveys of Dhaka city water. Dissolved iron concentrations are also low (<0.25 mg L^{-1}). Of the measured elements which have WHO health-based guideline values, only Mn has any exceedances: one sample had a Mn concentration of 0.674 mg L^{-1} which just exceeds the provisional Guideline value of 0.5 mg L^{-1}.

1.7 Mineralogy and sediment chemistry

Sediments from Bangladesh vary in texture from coarse sands and gravels to clays and are frequently found as a series of upwardly-fining sequences. In some places, the texture can vary on a centimetre depth scale. The sands and silts are dominated by iron-coated quartz particles, feldspars, micas and lithic fragments with small amounts of calcite and heavy minerals (Khan et al., 1998; Pearce et al., 2001; Rahman and Hossain, 1999). The clays are dominated by a smectite-illite-chlorite clay mineral assemblage with minor amounts of kaolinite and contain variable amounts of fine-grained calcite. The sediments contain variable amounts of organic matter, often less than 0.5% (Pearce et al., 2001), although concentrations are sometimes much greater and occasionally include peat horizons.

Early work by the Public Health Engineering Department, Government of West Bengal (PHED, 1991) reported that 'arsenic is present as adsorbed ions on clay particles/quartz particles and also as primary arsenic mineral(s)'. X-ray diffraction of the 'arseniferous horizons' was reported to have shown traces of arsenopyrite. The reported As contents of sediments from Kochua, West Bengal were extremely low, <0.2 mg kg^{-1} in all cases, and below the detection limit (probably about 0.1 mg kg^{-1}) in the sandy horizons.

More recent studies have identified pyrite in the sediments (but not arsenopyrite). Chowdhury et al. (1997) found selected grains having As concentrations in excess of 2000 mg kg^{-1}. However, these studies do not reflect overall sediment concentrations and these high As concentrations were for sediments from depths exceeding 100 m where As concentrations in the groundwater are generally low.

Other studies have commented on the paucity of pyrite. For example, Khan et al. (1998) noted the 'lack of any significant of pyrite/arsenopyrite'

Table 6. Chemistry of deep groundwater from public supply wells across Dhaka city

	Units	RIP7501	RIP7502	RIP7503	RIP7504	RIP7505	RIP7506	RIP7507
Latitude	deg	23.7497	23.7260	23.7606	23.8017	23.7985	23.7405	23.7149
Longitude	deg	90.3888	90.3852	90.3637	90.3597	90.4066	90.4072	90.4287
Location		Well S170A, Green Rd	Azimpur colony 7, Dhanmondi	Muham-madpur 8	BIBM compound, Mirpur	Banani Pump 5, Gulshan	Circuit House, Ramma	Syedabad, Demra
Well depth	m	130	140	130	118	160	154	156
Ca	mg L^{-1}	31.4	50.7	24.3	15.4	15.4	16.5	59.2
Mg	mg L^{-1}	15.4	16.6	8.82	5.23	5.02	6.78	22.6
Na	mg L^{-1}	23	43	22	16	18	18	41
K	mg L^{-1}	2	5.6	1.8	2.3	1.7	1.7	2.5
SO4	mg L^{-1}	11.5	31.6	2.8	0.6	1.2	6.4	34.6
As_T	μg L^{-1}	<0.5	<0.5	<0.5	<0.5	<0.5	<0.5	<0.5
Fe	mg L^{-1}	0.171	0.021	0.232	0.248	0.024	0.097	0.041
Mn	mg L^{-1}	0.674	0.027	0.017	0.066	0.021	0.058	0.223
Si	mg L^{-1}	38.8	25.4	35.7	40	37.9	38.9	37.2
Ba	μg L^{-1}	39	52	41	18	11	13	23
Sr	μg L^{-1}	189	482	179	118	110	107	377

Table 6 (cont'd).

	Units	RIP7501	RIP7502	RIP7503	RIP7504	RIP7505	RIP7506	RIP7507
P	mg L^{-1}	<0.1	<0.1	<0.1	0.1	0.1	<0.1	0.1
Al	µg L^{-1}	3	5	23	5	4	4	3
Li	µg L^{-1}	15.3	6.0	5.8	9.9	8.4	9.4	9.5
Be	µg L^{-1}	0.02	<0.01	0.01	0.05	0.01	0.03	0.02
Cd	µg L^{-1}	0.09	0.04	0.06	0.03	0.03	0.02	0.04
Co	µg L^{-1}	1.53	0.3	0.17	0.2	0.09	0.21	0.65
Cr	µg L^{-1}	<0.5	2.4	0.7	8.6	3.1	3.2	1
Zn	µg L^{-1}	13	20	5	97	13	6	6
Cu	µg L^{-1}	1	1	<1	2	2	1	11
Ni	µg L^{-1}	2.3	2.1	0.9	1.1	0.6	1	1.8
Pb	µg L^{-1}	0.21	0.28	0.37	0.23	0.2	0.17	0.42
Sb	µg L^{-1}	<0.01	0.01	0.02	0.06	<0.01	0.01	0.02
U	µg L^{-1}	0.3	0.42	0.16	0.02	0.03	0.04	0.97
La	µg L^{-1}	0.006	0.012	0.013	<0.005	0.015	0.006	0.007
Ce	µg L^{-1}	0.015	0.022	0.02	0.007	0.023	0.008	0.009
V	µg L^{-1}	1.8	1.2	1.5	1.1	1.5	1.3	1.2

in sediments from a high-As region of Bangladesh but did note the abundance of iron-rich coatings on many of the sand particles. Our observations (DPHE/BGS/MML, 1999; Pearce et al., 2001) are that authigenic pyrite can be found in the sediments but that it is rare. We found that the total As content of a selection of sediments from high-As groundwater areas ranged from 0.4 to 10.3 mg As kg^{-1} (n = 21) and tended to be greatest in the clay-rich horizons even at depth. This is in the normal range for sediments (Smedley and Kinniburgh, 2002)

Selective dissolution of the sediments with acid ammonium oxalate, a reagent that is efficient at dissolving 'amorphous' iron oxides of the HFO-type as well as magnetite and probably other reduced iron oxides, extracted by dissolution and desorption up to nearly 6 mg kg^{-1} of arsenic, i.e., approximately half of the arsenic present. There was a strong correlation between the amount of Fe dissolved and the As extracted but there were also strong correlations between the extracted Fe and many other elements including Mg, Al and P. Oxalate extracts of sandy sediments from 13 sites across Bangladesh showed that there was a strong correlation between the average amounts of As_{ox} and Fe_{ox} extracted (Fig. 11). Multiple linear regression of the oxalate extract data, retaining the top two predictors, gave

$$As_{ox} = 3.9\ \%Fe_{ox} + 0.00045\ Mg_{ox} \qquad (R^2=0.990)$$

where the oxalate-extractable As and Mg concentrations are in units of mg kg^{-1} and $\%Fe_{ox}$ is in percent by weight. Mg_{ox} is believed to be related to the clay content of the sediments. This suggests that iron oxides and clays are the principal sources of extractable As with about ¾ of the As being derived from iron oxides. Furthermore all of the sediments from areas with low-As groundwaters (including sediments from the Dupi Tila aquifer) contained less than 0.15% Fe_{ox} and As_{ox} less than 0.5 mg kg^{-1}.

It appears from this indirect evidence that high As groundwater areas in Bangladesh tend to have sandy sediments with a relatively high concentration of oxalate-extractable arsenic and that this may in turn be related to the relatively high concentration of oxalate-extractable iron oxides and clays. The nature of the iron oxides in sediments from Bangladesh is uncertain. Many of the sandy sediments in the arsenic-affected aquifers are grey, not brown, and it is likely that the oxides in these sediments are at least partially-reduced Fe(II)-Fe(III) oxides, possibly with a green rust structure.

Recent studies of sediments from Ramrail, Brahmanbaria in eastern Bangladesh by USGS/GSB (Foster et al., 2000) found 24 m of grey, reduced Holocene sediments overlying brown, oxidized sediments probably of Pleistocene age. The total As contents of the sediments were in the range 1–16 mg kg^{-1}. A 5–10 cm thick Fe-rich layer had an As content of 264 mg kg^{-1} and XANES spectra confirmed that most of this As was present as As(V). In

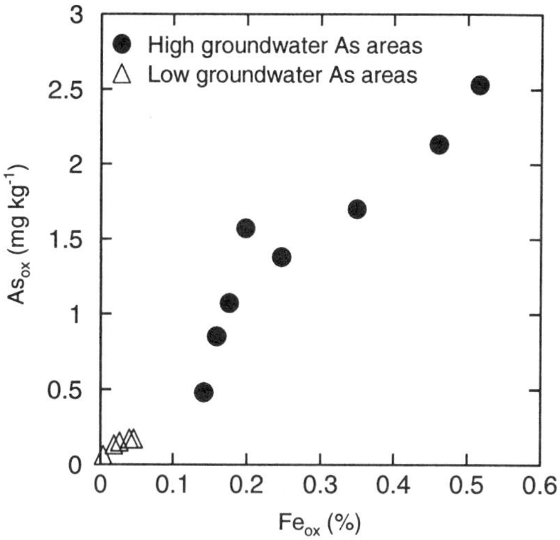

Figure 11. Relationship between average amounts of oxalate-extractable iron and arsenic from sandy sediments from a range of sites across Bangladesh. The sites are divided into those with known high and low arsenic groundwaters

the grey reduced sediments, most of the solid-phase As was present as As(III) and was associated with clays, micas or Al oxyhydroxides. In the brown sediments, most of the solid phase As was present as As(V) and was associated with iron oxides. The concentration of As in the groundwater was greater in the Holocene sediments than in the Pleistocene sediments but the highest concentration (760 $\mu g\ L^{-1}$) occurred close to the redox boundary.

1.8 Causes of the groundwater arsenic problem

1.8.1 Processes involved

There is much speculation about the reasons for the high arsenic concentrations in groundwaters in Bangladesh and elsewhere. It is useful in these discussions to consider three aspects of the problem: the *source* of the arsenic, its *mobilization* and the extent of its present and past *transport* or movement in the groundwater flow.

The total amount of arsenic present in sediments from Bangladesh, typically <1–10 mg kg^{-1}, appears to be in the normal range for alluvial and deltaic sediments (Smedley and Kinniburgh, 2002). Therefore on a global scale, total concentrations of arsenic in sediments do not appear to be a good guide to the likelihood of their associated groundwaters developing high-As concentrations – there are many sediments with As in this concentration range that do not host high-As groundwaters. Drinking water guidelines for arsenic are also very small in relation to the amount of arsenic present in typical sediments. The dissolution or desorption of just 1 mg As kg^{-1} from an aquifer sediment to groundwater is sufficient to give an As concentration in the groundwater of some 6 mg L^{-1} and so would exceed the WHO guideline value by more than two orders of magnitude. Groundwater is therefore very sensitive to any disturbance to the solid-solution equilibria controlling dissolved arsenic concentrations.

Given this, and since trace amounts of arsenic are associated with many minerals and organic matter, it is difficult to rule out any particular sources, even apparently minor sources, at the outset. A good quantitative understanding of the processes involved is required and at present, this is not available. Plausible sources of arsenic that have been suggested include: arsenopyrite, pyrite, iron oxides, biotite, chlorite, organic matter and even phosphate fertilizers. The possible sources and mechanisms of release can be reduced by considering ancillary 'evidence' including the concentrations of other solutes, both major and minor, and the nature of the sediments and the sediment-water interactions.

Hydrological factors and time are also likely to play a key role and these leave relatively little trace in present-day groundwater. For example, it is now apparent that high As-groundwaters are often associated with sediments in which groundwater flow is small and in which the cumulative flushing since burial has been minimal. The necessary low hydraulic gradients are found in flat, low-lying areas such as deltas, in closed inland basins and in deep 'dead end' aquifers. Younger sediments like those in Bangladesh are therefore particularly 'at risk' since the cumulative flushing that has taken place through them will be less than for similar, older sediments. This combination of geologically young sediments with hydrologically old groundwater appears to be an important factor in the development of many As-rich groundwaters.

1.8.2 The source of arsenic

Historically, high-As groundwaters have mostly been associated with mining and mineralized areas, and the affected areas have been quite localized (a few km^2). Areas of geothermal activity are also often associated with As-rich groundwaters and this can affect quite large areas (Smedley and Kinniburgh, 2002).

It is not surprising therefore that when high-As groundwaters were first discovered in West Bengal, and later in Bangladesh, there was a search for a possible As-rich source. For example, there has been speculation of high-As source minerals being derived from the Rajmahal Traps and the Rajmahal Basin of eastern Bihar (Acharyya et al., 1999; Khan, 1995). This persistent speculation delayed the realization that high-As groundwaters are possible without invoking unusual primary sources. High-As groundwaters are now being recognised in alluvial sediments and deltas around the world and even in lake and ocean sediments (Smedley and Kinniburgh, 2002). This points to a single source of widespread occurrence, or of many different sources. The evidence from Bangladesh is that it is primarily young (Holocene) sediments that are affected. In the Bengal Basin, where the rate of sedimentation has been very rapid and the volume of Holocene sediments is correspondingly large, many wells are developed (screened) in these young sediments with their potentially As-rich groundwaters. This may explain why the most extensive groundwater arsenic problem so far in deltaic environments is associated with the largest delta in the world.

Given this background, it is only possible at present to pose hypotheses and to test them against the available data. The two main hypotheses regarding the source and mobilization of As that have emerged are: (i) the *pyrite oxidation* hypothesis, and (ii) the *iron oxide reduction* hypothesis. A number of other hypotheses have also been proposed but so far, at least, with less widespread support. We discuss these below. A more detailed discussion is given in Kinniburgh (2001).

1.8.3 Pyrite oxidation hypothesis

This hypothesis was strongly advocated by West Bengal scientists in 1995 (Chatterjee et al., 1995; Chowdhury et al., 1997). It is based on the idea that arsenopyrite, or later As-rich pyrite, was initially present in the sediments and has been at least partially oxidized as a result of the recent seasonal lowering of the water table. This lowering has been attributed to the use of groundwater for irrigation and, by some, to the construction of the Farakka barrage (a controversial dam that was completed in 1975 across the River Ganges in West Bengal close to the West Bengal-Bangladesh border and which diverts River Ganges water to the Bhagirati-Hoogly River and ultimately to Calcutta). This hypothesis therefore supports the notion that the release of arsenic to the groundwater is a recent phenomenon induced by man's activities.

Certainly such a hypothesis is a possibility and needs to be considered. However, proponents of the hypothesis have offered little scientific evidence in support of it other than demonstrating the presence of pyrite in the sediments. On the contrary, the evidence is strongly against such a hypothesis for the following reasons:

- The presence of pyrite in at least some of the sediments is not in doubt, although the amounts are usually small and can be difficult to observe and to quantify. Arsenopyrite, if present, is certainly rare. Given the strongly reducing nature of the groundwaters in much of Bangladesh and, in many cases, their recent evolution from saline water containing abundant sulfate, pyrite formation by sulfate reduction is expected. Pyrite neoformation therefore seems more likely than pyrite dissolution. Limited sulphur isotopic data (Smedley et al., 2001) are consistent with sulfate reduction in the reducing sediments rather than oxidation. The morphology of the pyrite observed in the sediments also confirms this – it does not show signs of the corrosion that would be expected if oxidation were taking place. The formation of pyrite is more likely to scavenge arsenic from the porewater rather than to release it. However, the extensive areas of Bangladesh with high concentrations of arsenic in groundwater suggest that this process does not remove sufficient arsenic to generate low dissolved concentrations.

- Extensive oxidation of pyrite releases large amounts of sulfate and yet the evidence from the NHS and many other studies is that sulfate concentrations in high-As groundwaters in Bangladesh are usually low, often <1 mg L^{-1}. Where high sulfate concentrations are found, these are not correlated with high As concentrations. There is no convincing evidence for the formation of secondary sulfate minerals and indeed this is unlikely given the low sulfate concentrations found in most Bangladesh groundwaters.

- The groundwater As problem is greatest in the south-east of Bangladesh where the water table is highest and where there is only a small unsaturated zone (1–3 m at Lakshmipur). There is a much thicker unsaturated zone and a greater seasonal fluctuation of the water table in north-western Bangladesh, although the groundwater arsenic concentrations there are generally low. The greater depth to the water table and the greater zone of fluctuation in the north-west reflects hydrologic factors as well as the greater use of groundwater for irrigation. The pyrite oxidation hypothesis would predict high-As groundwaters in this area, not the low concentrations observed. The severely As-affected areas in south-east Bangladesh are too far away from the Farakka barrage to be significantly influenced by it. Many are part of the Meghna catchment, not the Ganges catchment.

- The pyrite oxidation hypothesis would predict that arsenic concentrations would be greatest in the shallowest groundwaters close to the water table since this is where the influence of oxidation would be greatest. The NHS results indicate that this is not the case: the Lakshmipur piezometers show that arsenic concentrations increase sharply between 10 and 20 m rather than decrease. This is also supported by the NHS data which shows that the very shallowest wells (<15 m) tend to have lower As concentrations than the deeper wells. Dug wells would also be expected to be the source of high As concentrations whereas the evidence is precisely the opposite.

- The hypothesis does not account for the high phosphate concentrations found in many of the As-affected groundwaters. While arsenic and phosphate may not be directly related, there is considerable evidence that there is at least some relationship between the two. Fertilizers have been suggested as this link but high phosphate concentrations (> 1 mg L^{-1}) are frequently found at considerable depths (>50 m) and do not show the pattern with depth expected if phosphate fertilizer was the principal source. Quite high phosphate concentrations (>0.5 mg L^{-1}) were also found in our piezometers in Chapai Nawabganj town. These are located far from an obvious fertilizer source.

- The situation in Bangladesh has some similarities with conditions at a site in the Atlantic Ocean shelf off the Amazon Basin, where in the absence of an unsaturated zone, the observed high pore water As concentrations cannot be related to pyrite oxidation (Sullivan and Aller, 1996).

1.8.4 The iron oxide reduction hypothesis

The iron oxide reduction hypothesis was used to explain the high As concentrations in some groundwaters from Ohio, USA by Matisoff et al. (1982) and later more generally by Korte (1991). It has been applied to the West Bengal and Bangladesh situations by Bhattacharya et al. (1997), Nickson et al. (1998) and DPHE/BGS/MML (1999).

Despite a common title, the hypothesis has many variations and subtleties and various authors have interpreted it in significantly different ways. For example, there is much confusion between the processes of 'dissolution' and 'desorption'. These are distinctly different processes although they may occur simultaneously in the same way that adsorption processes can be responsible for metal uptake during coprecipitation – 'reductive codissolution' is perhaps an appropriate description of the proposed release mechanism. Also, the role of organic matter and of competitive desorption by ions such as phosphate and bicarbonate are stressed to varying degrees.

The following factors lend support to some kind of iron oxide reduction hypothesis:

- Iron oxides have long been known to be efficient scavengers of arsenic – a property widely used for arsenic removal in water treatment plants. They are present to a varying extent in Bangladesh sediments and there is usually a good correlation between the arsenic and iron concentrations of sediments.

- There is strong circumstantial evidence that the release of arsenic is associated with the imposition of strongly reducing conditions, i.e. where sulfate reduction is taking place, both from the maps of arsenic and sulfate and from the As-SO_4-depth relationships.

- Reductive dissolution of iron oxides is a well-established process by which iron oxides can be dissolved. However, the detailed mechanism by which trace constituents are released is poorly understood. While the As–Fe correlation in Bangladesh groundwaters is often significant in the statistical sense, it is usually far from perfect and there is an example in Chapai Nawabganj (DPHE/BGS/MML, 1999), where more than 1 mg As L^{-1} was found in a groundwater with <0.2 mg Fe L^{-1}. This implies either that some of the iron has been reprecipitated or adsorbed (perhaps during transport) or that the arsenic was released without a congruent release of iron, i.e. that desorption was important. This is more widely supported by the relatively high As/Fe molar ratios in groundwater compared with that found in the sediments.

- Experimental results and model calculations show that reduction of strongly bound As(V) to As(III) is a plausible mechanism for the release of As from iron (and other) oxides under reducing conditions (Zobrist et al., 2000). As(III) is known to be the dominant species in most of the high-As groundwaters in Bangladesh (Smedley et al., 2001) and has been shown experimentally to be present in reduced Bangladesh sediments (Foster et al., 2000). The rate of this reductive desorption is unknown but is likely to be rapid on a geological timescale (here meaning 'thousands of years'). Recent results from the USGS/GSB study in Ramrail have shown that the dominant mechanism of arsenic sorption changes from iron oxide-As(V) in the oxidized sediment to clay-As(III) in the reduced sediments pointing to the lesser importance of iron oxides in the reduced sediments – a feature that is consistent with our interpretation of the iron oxide reduction hypothesis. The extent of reductive desorption is poorly understood quantitatively, particularly in the presence of high concentrations of other strongly bound anions such as phosphate and bicarbonate (Kinniburgh, 2001).

- Simultaneous release of phosphate and arsenic from a common source provides at least a partial explanation for the high phosphate concentrations found in many Bangladesh groundwaters and for the spatial correlation between As and P that is sometimes observed (Fig. 3 and Fig. 10). Gavrieli et al. (2000) came to a similar conclusion;

- Little is known about the diagenetic changes that iron(III) oxides undergo when they are buried, especially under strongly reducing conditions, but it is likely that some changes in mineral structure take place. This is indicated, for example, by the red/brown to tan/grey colour change that is observed across redox boundaries and observed in Bangladesh sediments. Any changes in bulk or surface coordination or in specific surface area would be likely to alter the overall adsorption isotherm of the bulk sediment (Kinniburgh, 2001).

- Changes in the nature of the overall adsorption isotherm for As (or of one of its competitors), for whatever reason, could lead to the release of arsenic to groundwater. We know that even small changes can be highly significant in view of the sensitivity of groundwater to small changes in adsorbed amounts.

1.8.5 Other hypotheses

Suggestions that the use of arsenic wood preservatives was in some way responsible for the groundwater arsenic problem were discounted by scientists early on (NRECA, 1997) although such arguments continued to appear in the Bangladesh press for some time. As the enormous scale of the problem became apparent, such local sources became increasingly less plausible.

The same can be said for the hypothesis that points to the leaching of phosphate fertilizers as the primary source of high P groundwaters and indirectly to As-rich groundwaters through the competitive desorption of arsenic, especially As(V), from the sediments. Such competitive reactions must be occurring given the known groundwater chemistry but there is no evidence that they are the primary driving force behind the As release. For example, there are As-rich groundwaters that are not P-rich (and *vice versa*), and as mentioned above, at a detailed level, the spatial pattern of high-P and high-As waters does not match that expected from a predominantly fertilizer source. This makes fertilizers an unlikely cause of the present-day arsenic problem although future phosphate leaching could exacerbate the problem in the shallowest groundwaters. Finally, as discussed below, the inferred timing of arsenic release does not implicate fertilizer leaching as a primary driving force for the As release.

1.8.6 The timing of the arsenic release and temporal changes in groundwater arsenic concentrations

Since arsenic was not detected in groundwater in Bangladesh until relatively recently, there is no direct information on when the arsenic was initially released to the groundwater. This is presently the source of considerable discussion amongst scientists and has implications for how the arsenic concentrations might change with time. It also has implications for the scope and likely efficacy of possible mitigation measures. The relatively recent appearance of arsenic-affected patients has been assumed by some to point to the recent release of arsenic (Acharyya et al., 2000; Chowdhury et al., 1997). However, the timing of the appearance of arsenic patients is only related to the timing of the arsenic release in so far as it confirms that the arsenic has been present in the groundwater for sufficiently long to have caused chronic health problems, i.e. more than six or so years. The recent rapid expansion in the use of tubewells for drinking water in Bangladesh means that, not surprisingly, the increase in the detection of the number of arsenic patients is also recent. Dug wells have a longer history of use than tubewells in Bangladesh but as we have indicated above (Section 1.4.5.3), these tend to have low As concentrations and are unlikely to have led to chronic arsenicosis.

The issue of changes in arsenic concentration with time is complex and there are many factors involved. First, there is the issue of when the arsenic was first mobilized, that is released from some non-labile mineral or organic phase in the sediment to a labile form such as dissolved or exchangeable arsenic. Secondly, there are questions about the subsequent movement of As in the aquifer both as a result of the movement of water under natural hydraulic gradients ('aquifer flushing') and as a result of gradients induced by pumping. This redistribution within the aquifer will lead to changes in As concentration at a particular well with time. Such chromatographic movement of As involves the continual sorption-desorption of the exchangeable arsenic and so it is in principle possible to have water and arsenic of different 'ages'.

There are two schools of thought about the time of the initial As mobilization: either (i) it is recent and has been induced by man's activities [there are proponents of this who support both the pyrite oxidation hypothesis and the iron oxide reduction hypothesis (Acharyya et al., 2000)], or (ii) it occurred much earlier and is therefore dominantly a 'natural' process. While we believe that an early release date, (ii) above, is the more likely, this is not to imply that man's recent activities have not had, or will not have, any impact on the extent of the groundwater arsenic problem. For example, recent changes in land use such as irrigation will not only alter the groundwater flow patterns but could also affect the boundary conditions for oxygen diffusion into the aquifer and so could also affect its redox status (Bhattacharya et al., 1997).

The most direct evidence concerning the age of the arsenic mobilization and release is probably from isotopic measurements. For example, in the Lakshmipur Special Study Area in SE Bangladesh, there are high-As groundwaters which contain essentially no tritium (Smedley et al., 2001). This points to a dominantly pre-1952 'age' for the groundwater and so does not support the concept that it was the introduction of recent oxygenated water into the aquifer as a result of water table lowering that led to the formation of high-As groundwaters. On the other hand, high-As groundwaters have been found in other parts of Bangladesh (e.g. Chapai Nawabganj in north-west Bangladesh) which contain significant tritium concentrations and point to a component of recent (post-1952) recharge.

Radiocarbon measurements for a small number of shallow groundwaters from our Special Studies Areas gave a ^{14}C enrichment generally in the range of 65-95% modern carbon, suggesting that the water contains a large component that is 'modern' (probably less than 1000 years old; Smedley et al., 2001). Radiocarbon activities in the few deeper (> 150 m) groundwaters measured indicated enrichments of 18-51% modern carbon with modelled groundwater ages in the range 2000–12,000 years. These deep groundwaters all contained low As concentrations.

These results indicate that the shallow groundwaters are predominantly modern in the ^{14}C sense but are not necessarily modern in the tritium sense, i.e. they are likely to be in the range 50–1000 years old. As indicated above, the initial As release must be older than this. Not surprisingly, the isotopic results from the three Special Study Areas suggest that the youngest waters are most likely to be found where the annual water table fluctuation is greatest, i.e. in the north west of Bangladesh rather than the south east. It is uncertain how many pore volumes of recharge water have passed through the sediments since the As was first mobilized but early calculations based on field and laboratory evidence suggest a sediment/water partition coefficient for the affected Bangladesh sediments on the order of 2–6 L kg^{-1} (Smedley and Kinniburgh, 2002). This indicates a low retardation coefficient for the As, probably of less than two. Studies of the variation of sediment and porewater chemistry over the scale of metres and tens of metres are needed to establish the nature of the sediment/water interactions and the extent of mixing.

It is clear that given the extremely heterogeneous nature of the sediments, groundwater flow may induce variations in arsenic concentration at the tubewell as a function of time. In general, the greater the volume of pumped groundwater, the greater the extent of mixing in the aquifer is likely to be. Therefore initially at least, present-day low-As wells may increase in As concentration and high-As wells may decrease. The evolution of As concentration for a particular tubewell will depend on many factors including the distribution of labile As in the aquifer in relation to the screened interval and the groundwater flow. The labile arsenic in the aquifer will gradually be removed in the pumped groundwater or discharged to the rivers and the sea. However this is unlikely to occur on a human timescale, or even for thousands of years.

Changes in the extent of contamination of deep wells are an important issue. While West Bengal scientists, e.g. CGWB (1999) have claimed that the results of repeat testing of boreholes, even deep tubewells, show a tendency for the arsenic concentration in individual tubewells to increase with time, the detailed data demonstrating that have not been published in an accessible form. From our experience, in at least one of the deep wells where this occurred, this appears to be due to downhole leakage from the shallow aquifer (RGAG & RGACG, 2000) and this may account for other cases too (BGS and DPHE, 2001). Deep wells in the southern coastal region of Bangladesh have been operating for more than a decade without being contaminated with saline water from the shallow aquifer. This suggests that, in this region at least, vertical leakage to the deep aquifer is minimal. The situation may be different in other parts of Bangladesh where there may not be such an extensive clay layer separating the aquifers.

The results of the relatively short-term monitoring, as described above, do not provide convincing evidence of any systematic changes within the aquifer over a 10-month period. Modelling of possible time trends in As concentrations in Bangladesh aquifers (Burgess et al., 2000) suggests that many shallow wells presently free of As will eventually become contaminated. However, such calculations depend critically on the nature of groundwater flow, the distribution of As in the aquifer sediments and the nature of the sediment/water interactions. A longer-term monitoring programme needs to be instigated at a few representative places. This monitoring should include all major solutes as well as arsenic and should consider all of the temporal scales involved (minutes to years and longer). It should also cover a range of pumping rates.

1.9 Conclusions

About one quarter of the tubewells sampled in the National Hydrochemical Survey of Bangladesh groundwaters gave groundwater exceeding the Bangladesh standard for arsenic (50 $\mu g\ L^{-1}$) and 42% exceeded the WHO guideline value. High arsenic concentrations are almost entirely restricted to groundwaters from shallow aquifers (<150 m) in alluvial and deltaic sediments largely of Holocene age. The maximum concentration found exceeded 1000 $\mu g\ L^{-1}$ and 5.6% of shallow wells (<150 m) sampled exceeded 300 $\mu g\ L^{-1}$, a concentration where the lifetime probablility of dying from various forms of cancer are likely to exceed 1:100 and may be 10 times greater than that (Smith et al., 2000). Aquifers in older (Plio-Pleistocene) aquifers including those from the Barind and Madhupur Tracts and from greater depths (>150 m) are usually not affected and frequently contain groundwater with As <0.5 $\mu g\ L^{-1}$. Most of the deep aquifers sampled were from the southern coastal region and this observation may not apply to the deep aquifer everywhere in Bangladesh. The deep wells in Sylhet appear to contain more As than those from the southern coastal regions, possibly as a result of multiple depth screening of Sylhet tubewells. Water from dug wells and to a lesser extent from the shallowest tubewells (<10 m) have relatively low As concentrations, perhaps as a result of slightly more oxidizing conditions and the greater degree of flushing and dilution with recent recharge.

Groundwater As concentrations show distinctive spatial trends, with the highest average concentrations in the south and south-east of the country. In the worst-affected district, Chandpur, 90% of the sampled wells were contaminated with As (>50 μg L^{-1}). Apart from the occasional 'hot spot' in the north, concentrations there are considerably lower and are frequently, but not always, <50 μg L^{-1}. Relatively high concentrations (10–50 μg L^{-1}) are found in the low-lying areas in the north-east and beneath the Holocene sediments of the Brahmaputra valley. Some high concentrations are also found in groundwaters from the Holocene sediments of the Jamuna valley. Of the shallow groundwaters from Holocene sediments, the north-west region (Tista Fan) has the most consistently low As concentrations.

The affected groundwaters are invariably reducing, usually with low sulfate concentrations, and often are relatively hard, having high Ca and Mg concentrations. They also tend to have a high concentration of phosphate and dissolved iron (Fe^{2+}). Many groundwaters contain high manganese concentrations although the distribution does not match that of arsenic closely. Groundwater from coastal areas is impacted by saline intrusion and suffers from high boron concentrations. Deep wells tend to have better quality, not only in terms of arsenic, but also in terms of manganese and iron. However, boron concentrations are commonly high in the deep coastal aquifer.

Analysis of seven deep public supply wells from Dhaka city showed that they all contained <0.5 μg L^{-1} As. These wells also had low concentrations of most other analysed trace elements. Concentrations of Cd, Cr, Pb, Ni, Sb and U were substantially below WHO health-based Guideline values. One of the Dhaka wells exceeded the WHO guideline value (0.5 mg L^{-1}) for Mn.

The spatial pattern of As concentrations in the Holocene aquifer is believed to reflect a complex interrelationship between the lithology of the aquifer, its redox characteristics and the history of groundwater flow within the aquifer. High As concentrations in the south and south-east reflect the strongly reducing conditions in the aquifers in this area, poor permeability of overlying sediments, higher proportions of finer-grained material and iron oxides and slow rates of groundwater movement resulting in low rates of aquifer flushing.

In the north-central Jamuna valley, many groundwaters have very high concentrations of Fe (often exceeding 10 mg L^{-1}) and Mn (up to 10 mg L^{-1}) relative to most of Bangladesh, and many have high SO_4 concentrations compared to the high-As groundwaters further south. The young sediments of the Jamuna valley are thought to be undergoing active reduction but have not achieved such strongly reducing conditions overall as in the high-As areas of southern Bangladesh. This may be for a number of reasons for this: less fine-grained sediment at the surface and hence less restriction of recharge and dissolved oxygen to the aquifers, shallow well depths and a relatively thick unsaturated zone compared to further south, and perhaps more active groundwater flow. Some of the higher SO_4 concentrations may be related to the lack of sulfate reduction or even to *in situ* sulfide oxidation. However, if sulfide oxidation has occurred, it apparently has not resulted in substantial As mobilization as As concentrations are relatively low in the high-SO_4 groundwaters. In contrast, the high-As groundwaters (>50 $\mu g\ L^{-1}$) from the Jamuna valley have very low SO_4 concentrations, typically 1 mg L^{-1} or less, and indicate that these are more strongly reducing. This association of high As with low SO_4, as elsewhere in Bangladesh, precludes sulfide oxidation as the dominant cause of arsenic mobilization, although groundwaters from some tubewells may have been affected by this process.

Low As concentrations in groundwaters from the Tista Fan deposits of northern Bangladesh occur in generally more oxidizing conditions with low overall Fe concentrations. These redox characteristics probably reflect the coarser grain size of the Tista Fan sediments and lack of impermeable sediment cover, as well as more active flushing of the aquifer in this region.

The regional distributions of Fe and Mn in the groundwaters also strongly reflect redox controls. Nationally, the Fe distribution shows a weak correlation with that of As. Despite the strong redox control of Fe and Mn, these two elements have differing regional distributions which probably reflect differing redox potentials at which the respective reactions take place. Many groundwaters exceeded the health-based provisional WHO guideline value for manganese.

Distributions of Na, B and K largely reflect seawater influences, with the greatest contributions from old seawater in southern Bangladesh (shallow and deep aquifers), the Sylhet Basin and the Atrai Floodplain north of the River Ganges. Potassium is also derived from mineral weathering reactions. Distributions of Ca, Mg, Sr and to some extent Ba, reflect the distributions of free carbonates in the aquifer sediments and soils. Concentrations are generally highest in south-west Bangladesh. Of the elements considered, As represents by far the worst recognized inorganic problem in Bangladesh groundwaters.

ACKNOWLEDGEMENTS

We thank all those who helped in the collection and analysis of the samples discussed in this Chapter. In particular, we thank Peter Ravenscroft for coordinating the Bangladesh input to the NHS Phase I, Ashfiquzzaman Aktar, Khairul Amin, Khairul Bashar, Abdul Malek, Abdul Noor, Md Rabiul Islam, Shahidul Islam, Md Ibrahim Khalil, Md Taibur Rahman, Abdus Salam and DPHE district staff for sampling and Linda Ault, Sally Bourliakas, Simon Burden, Jenny Cook, Kerry Dodd, Suzan Gordon and

John Thorns for chemical analyses, Balt Verhagen for the tritium analyses, Azharul Huq for access to the Dhaka wells and the people of Bangladesh, particularly from Mandari, for allowing us access to their wells. We also thank John Talbot for GIS support, Richard Webster for help with the geostatistical analyses, Willy Burgess and Ian Gale for constructive reviews and the UK Department for International Development (DFID) for funding and support. This Chapter is published with the permission of the Director of the British Geological Survey (NERC).

Chapter 9

Mechanisms of Arsenic Release to Ground Water from Naturally Occurring Sources, Eastern Wisconsin

ME Schreiber[1], MB Gotkowitz[2], JA Simo[3], and PG Freiberg[4]
[1]*Department of Geological Sciences, Virginia Tech, Blacksburg VA*
[2]*Wisconsin Geological and Natural History Survey, Madison WI*
[3]*Department of Geology and Geophysics, University of Wisconsin-Madison, Madison WI*
[4]*Redwood National Park, Orick CA*

Arsenic concentrations up to 12,000 µg/l have been measured in ground water from a sandstone aquifer in the Fox River valley in eastern Wisconsin, USA. In addition to a sulfide-bearing secondary cement horizon (SCH), which is present at the top of the aquifer, sulfide mineralization is also present throughout the aquifer. Within the SCH, arsenic occurs in pyrite and marcasite, and in iron hydroxides, but not as a separate arsenopyrite phase. Geologic, hydrogeologic, and geochemical data were used to characterize the arsenic source and the predominant geochemical process that controls its release to ground water. Several lines of evidence suggest that oxidation of sulfides is the cause of high (>100 µg/l) concentrations of arsenic in ground water, including 1) the presence of the arsenic-bearing sulfides in the aquifer; 2) water chemistry data that show a positive correlation between arsenic, iron, and sulfate and negative correlation between arsenic and pH; and 3) similar sulfur isotopic signatures in sulfides of the SCH and dissolved sulfate in ground water. We propose that atmospheric oxygen, introduced to the SCH through well boreholes, provides an oxidant to the system. This hypothesis is supported by the occurrence of high arsenic concentrations where water levels within the well intersect the SCH. However, the data do not unequivocally show sulfide oxidation to be the cause of the moderate (10-100 µg/l) and low (<10 µg/l) arsenic concentrations measured in ground water in the study area. The variability in thickness of the SCH and the concentration of arsenic within the sulfides, as well as the local availability of oxygen to the SCH, likely contribute to the spatial variability of ground water arsenic concentrations.

1. INTRODUCTION

Arsenic contamination of ground water has been attributed to several geochemical processes, including oxidation of arsenic-bearing sulfides, desorption of arsenic from oxides and hydroxides, reductive dissolution of arsenic-bearing oxides and hydroxides, release of arsenic from geothermal waters, and evaporative concentration (Welch et al., 2000), as well as leaching of arsenic from sulfides by carbonate (Kim et al., 2000). Due to the complexities of these processes and the potential occurrence of simultaneous processes, identification of mechanisms responsible for a particular case of arsenic contamination is often difficult. For example, there has been considerable controversy over the cause of arsenic contamination of ground water in Bangladesh. Some researchers have proposed sulfide oxidation as the predominant mechanism for arsenic release (Bagla and Kaiser, 1996; Dipankar et al., 1994; Mandal et al., 1996) while others contend that reduction of iron hydroxides is the cause of high dissolved arsenic concentrations in ground water (Bhattacharya et al., 1997; McArthur et al., 2001; Nickson et al., 2000).

Multiple lines of evidence are needed to distinguish mechanisms that release arsenic to ground water in a particular setting. To fully document the process of arsenic release, the lines-of-evidence approach should incorporate data on the: (1) geologic setting, to constrain the petrologic, stratigraphic and/or structural controls on the occurrence of arsenic-bearing units; (2) site mineralogy, to determine the mineralogic hosts of arsenic; (3) hydrogeologic setting, to delineate ground water flowpaths and to understand local water-rock interactions; (4) water chemistry, to examine the relationship between arsenic and key chemical parameters; (5) isotopes, to infer mechanism of release; and (6) subsurface microbiology, to evaluate the role of bacteria in the geochemical processes that control arsenic release. Accurate assessment of the mechanism of arsenic release will help to identify safe drinking water resources and to evaluate treatment methods for contaminated waters.

The lines-of-evidence approach was used to determine causes of elevated arsenic concentrations in the Fox River valley (FRV) of eastern Wisconsin. As part of this study, geologic and water level mapping were conducted to evaluate the regional source of the arsenic and the relationship of the source to the regional and local water levels. Quarry outcrops and rock cores were examined to determine the distribution of arsenic-bearing minerals and to observe stratigraphic control on mineralization. Ground water chemistry data were used to assess correlations between arsenic and other chemical parameters. The results of these analyses were then used to evaluate hypotheses of arsenic release to ground water.

1.1 Arsenic release via sulfide oxidation

Oxidation of arsenic-bearing sulfides has been implicated as the cause of arsenic contamination of ground water in regions of active and historic mining in many regions worldwide (e.g. Armienta et al., 1997; Ball and Nordstrom, 1985; Bottomley, 1984; Moore, 1994; Nicolli et al., 1989; Prior and Williams, 1996; Smedley et al., 1996). Sulfide oxidation has also been identified as the source of arsenic in ground water of non-mined areas (e.g. Gotkowitz et al., 2000; Peters et al., 1999). The rates of, and conditions leading to, sulfide oxidation have been extensively reviewed (e.g. Bierens de Haan, 1991; Goldhaber, 1983; Lowson, 1982; McKibben and Barnes, 1986; Moses et al., 1987; Nicholson et al., 1988; Nordstrom, 1982; Toran, 1987; Wiersma and Rimstidt, 1984). Pyrite oxidation can follow several reaction pathways. The first is the chemical oxidation of pyrite by oxygen:

$$2FeS_2 + 7O_2 + 2H_2O \Rightarrow 2Fe^{2+} + 4SO_4^{2-} + 4H^+ \qquad (1)$$

A second possible pathway is the oxidation of pyrite by nitrate:

$$5FeS_2 + 14NO_3^- + 4H^+ \Rightarrow 5Fe^{2+} + 10SO_4^{2-} + 7N_2 + 2H_2O \qquad (2)$$

Pyrite can also be oxidized by Fe(III):

$$FeS_2 + 14Fe^{3+} + 8H_2O \Rightarrow 15Fe^{2+} + 2SO_4^{2-} + 16H^+ \qquad (3)$$

Fe(III) is regenerated by oxidation of Fe(II):

$$4Fe^{2+} + O_2 + 4H^+ \Rightarrow 4Fe^{3+} + 2H_2O \qquad (4)$$

Under abiotic conditions, reaction (4) is slow (Singer and Stumm, 1970). However, under low pH conditions, acidophillic iron-oxidizing bacteria, including *Thiobacillus* and *Leptospirillum spp.*, mediate the oxidation of Fe(II) to Fe(III) (reaction (4)), allowing for reaction (3) to proceed several orders of magnitude more rapidly than under abiotic conditions (e.g. Nordstrom and Southam, 1997; Schrenk et al., 1998). It is important to note that at the near-neutral pH and low dissolved organic carbon (DOC) concentrations typically observed in drinking water aquifers, dissolved Fe(III) is controlled by the precipitation of iron hydroxides. Thus, in waters having near-neutral pH, the majority of Fe(III) will be in the solid or colloidal phase, which is important for arsenic release and transport for two reasons. First, at near-neutral pH, dissolved Fe(III) will not be present at high concentrations unless it is complexed to DOC and will consequently be a less important oxidant than dissolved oxygen (DO) or nitrate. Second, arsenic adsorbs strongly to iron hydroxides (e.g Dzombak and Morel, 1990; Fuller et al., 1993; Korte, 1991; Pierce and Moore, 1982). Thus, if iron hydroxides are present as coatings on aquifer materials, arsenic transport may be retarded via adsorption. However, if colloidal-size iron hydroxides are present, arsenic may undergo colloidal transport.

Because sulfide oxidation can occur via multiple pathways and can proceed under both biotic and/or abiotic conditions, a step-wise approach is useful for documenting its occurrence. A primary piece of evidence is the presence of arsenic-bearing sulfides. The second requirement of sulfide oxidation is the presence of an oxidant. Dissolved oxygen (reaction (1)) is the most ubiquitous oxidant of reduced compounds in the unsaturated zone and in shallow ground water. In areas where nitrogen has impacted ground water supplies, nitrate (reaction (2)) may act as a sulfide oxidant (e.g. Postma et al., 1991). At lower pH or in the presence of DOC, dissolved Fe(III) can be a significant oxidant, particularly in the presence of iron-oxidizing bacteria (reactions (3) and (4)).

The reaction by-products of sulfide oxidation are distinct; thus, evidence of sulfide oxidation can be found in the chemical signatures of ground water. In aquifers where sulfide oxidation occurs, ground water chemistry should show a positive correlation of arsenic with sulfate, iron, and trace metals contained in the sulfide minerals. Increases in total dissolved solids and specific conductance also result from sulfide oxidation, due to an increase of dissolved ions in the impacted waters. Ground water impacted by sulfide oxidation may reveal a negative correlation of arsenic with pH, provided that there is minimal buffering capacity provided by the host rocks.

The sulfur isotopic signature in sulfide minerals and in sulfate in ground water may also be used to evaluate the occurrence of sulfide oxidation. Although fractionation of sulfur isotopes can occur due to biological and redox cycling (e.g. Kaplan and Rittenberg, 1964), in most cases of sulfide oxidation, minimal fractionation between sulfide and sulfate is observed (Toran and Harris, 1989). This suggests that similar sulfur isotopic signatures in sulfide minerals found within the aquifer and in sulfate from ground water are indicative of sulfide oxidation.

Microbiological evidence can also be used to infer the occurrence of sulfide oxidation. The presence of acidophillic bacteria, such as *Thiobacillus* or *Leptospirillum,* in ground water or attached to aquifer surfaces strongly suggests bacterial involvement in sulfide oxidation reactions. More commonly found in ground water and borehole environments are *Gallionella, Siderocapsa,* and *Crenothrix*, neutrophillic iron-oxidizing bacteria. These bacteria may influence sulfide oxidation by mediating oxidation of Fe(II) to Fe(III) within the borehole (reaction (4)).

2. CASE STUDY: FOX RIVER VALLEY, NORTHEASTERN WISCONSIN

In the Fox River valley (FRV), which extends from Lake Winnebago to Green Bay in eastern Wisconsin, USA (Fig. 1a), high concentrations of arsenic (up to several mg/l) have been measured in drinking water wells completed in the Ordovician Sinnipee Group, St. Peter Sandstone, and Prairie du Chien Group (Burkel and Stoll, 1999; Gotkowitz et al., 2000; Pelczar, 1996; Simo et al., 1996). These formations provide a significant portion of the drinking water in eastern Wisconsin.

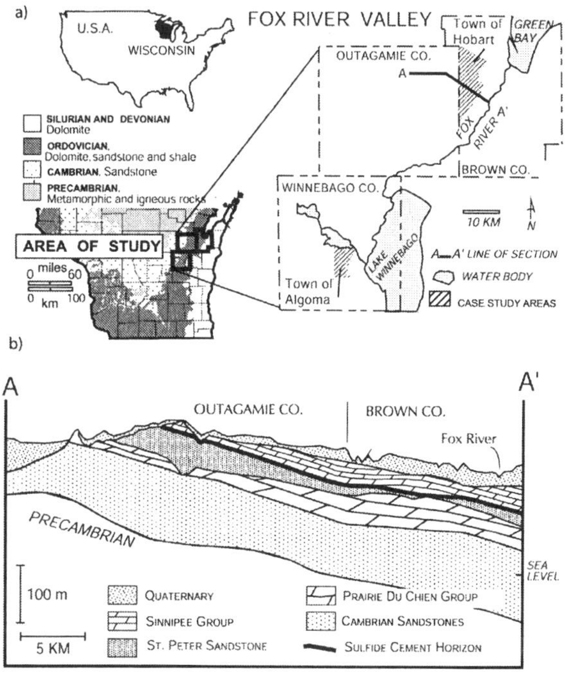

Figure 1. a) Location and bedrock geology of the Fox River valley, WI. b) Geologic section showing major units and the sulfide cement horizon (SCH). Modified from Batten and Bradbury (1996).

Arsenic contamination is of concern in the region due to the high concentrations and widespread occurrence of the problem (Fig. 2). According to data collected by the Wisconsin Department of Natural Resources (WDNR), approximately 21% of private water supply wells in Outagamie and Winnebago counties exceed 10 μg/l; about 4% of the wells exceed 50 μg/l. Elevated arsenic has also been observed in localized regions of Brown County. Within these broad areas, there is significant variability in arsenic concentrations. Wells located near each other with similar casing depths, static water levels and total depths, show vastly different levels of contamination. For example, in Algoma township (Fig. 1a), four similarly constructed wells located within a distance of 350 m had arsenic concentrations of 2, 36, 765 and 12,000 μg/l, respectively.

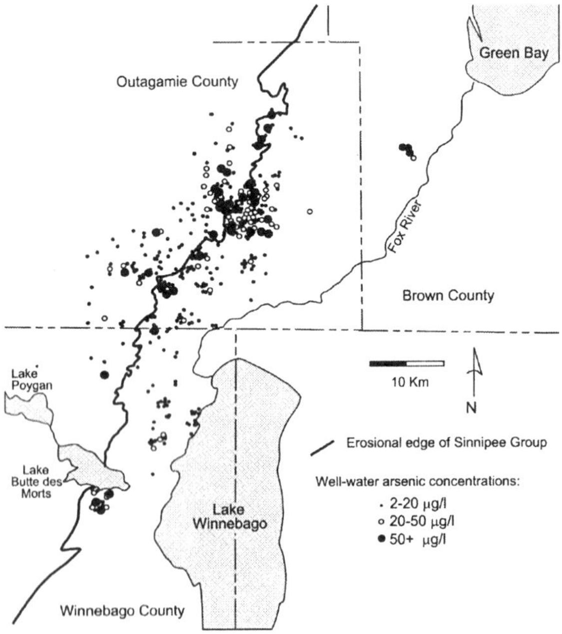

Figure 2. Distribution of arsenic concentrations in wells in the Fox River valley and the location of the erosional edge of the Sinnipee Group. Data from WDNR (1995, unpublished).

2.1 Conditions in the Study Area

2.1.1 Geologic Setting

A simplified stratigraphic section through the study area is presented in Fig. 1b. In the study area, Precambrian rocks are overlain by a thick succession of Cambrian sandstones. The sandstones are in turn overlain by the Ordovician Prairie du Chien Group, the St. Peter Sandstone, and the Sinnipee Group (Fig. 1b). The Prairie du Chien Group consists of dolomitized oolitic limestones, muddy facies, and stromatolites, with cyclic interbedded thin sandstones and shales. A major unconformity (Tippecanoe unconformity; Sloss (1963) separates the Prairie du Chien Group from the St. Peter Sandstone and the Sinnipee Group. This break is reflected by a paleokarstic topography (Mai and Dott, 1985). The St. Peter Sandstone, which fills the karst topography and pinches out against Prairie du Chien highs, contains a lower eolian unit and an upper marine unit. The overlying Sinnipee Group, consisting of dolostone with minor shales, represents the progressive marine flooding of the craton and drowning of the siliciclastic source in the Middle Ordovician (Choi, 1998). Typically, the Sinnipee Group rests upon the St. Peter Sandstone; in areas of karstic highs where the St. Peter is absent, it rests upon the Prairie du Chien. Quaternary glacially derived sediments overlie the Ordovician strata. These unlithified deposits are predominantly fine-grained till, a mixture of silty or sandy clay with some gravel. In the FRV, the thickness of glacial deposits ranges from 15 to 30 m (Batten and Bradbury, 1996).

Fig. 1b illustrates the stratigraphic position of a concentrated zone of sulfide mineralization, referred to as the sulfide cement horizon (SCH) (Simo et al., 1996). The SCH is laterally extensive across the study area and is present at the base of the Sinnipee Group and the top of the St. Peter Sandstone. In regions where the St. Peter Sandstone is absent, the SCH is found at the top of the Prairie du Chien. In addition to the SCH, sulfide mineralization occurs throughout the St. Peter as bands and nodules.

2.1.2 Hydrogeologic Setting

There are three primary aquifers in the Fox River valley (Fig. 3); an upper unlithified aquifer, a middle sandstone aquifer (herein called the St. Peter aquifer), and a deep sandstone aquifer (Conlon, 1998; Krohelski, 1986). The upper aquifer is unconfined and is made up of glacial sediments and weathered dolomite of the Sinnipee Group. The unweathered portion of the Sinnipee Group separates the upper aquifer from the St. Peter aquifer and is generally considered a confining unit. The St. Peter aquifer consists of the Prairie du Chien Group and the St. Peter Formation. The St. Lawrence-Tunnel City confining unit separates the St. Peter aquifer from the deep sandstone aquifer (Batten and Bradbury, 1996; Conlon, 1998; Krohelski, 1986). Residential and other low-volume supply wells are generally open to the Sinnipee and the upper portion of the St. Peter aquifer. Wells serving municipalities and large industries in the FRV are typically open to the St. Peter and the deep sandstone aquifers. Due to high rates of groundwater withdrawal, the potentiometric surface of the St. Peter and deep sandstone aquifers is declining at a rate of approximately 0.6 m/yr (Conlon, 1998).

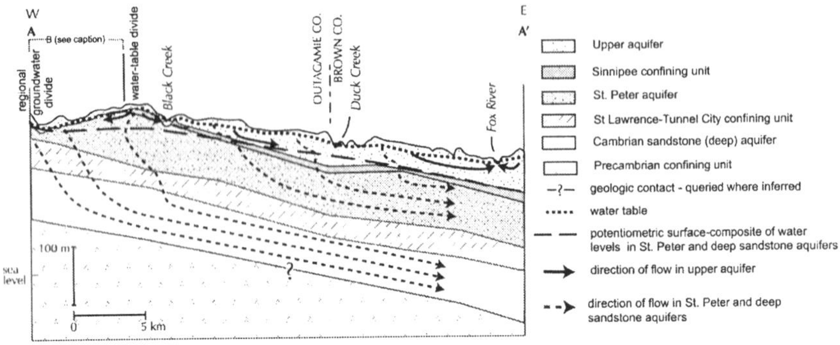

Figure 3. Cross-section showing major hydrogeologic units in the Fox River valley (see Fig. 1a for location of section A-A'). The regional recharge area for the St. Peter and deep sandstone aquifers occurs between the erosional edge of the Sinnipee and the regional ground water divide (denoted as "B" in figure). Modified from Batten and Bradbury (1996).

The St. Peter aquifer is regionally unconfined in the western portion of the FRV where the Sinnipee Group is absent, and is regionally confined in the eastern portions where the Sinnipee is present (Fig. 3). The general direction of ground water flow in the St. Peter and deep sandstone aquifers is from the regional ground water divide toward the regional discharge area (Lake Michigan) and pumping centers near the Fox River. Both the St. Peter and the deep sandstone aquifers receive recharge from an area between the erosional edge of the Sinnipee Group and the regional ground water divide.

Because residential wells are often open to both the St. Peter aquifer and the overlying confining unit, water levels do not provide an accurate measurement of hydraulic head in the aquifer. Therefore, the water levels measured in residential wells are termed "static water levels". This is distinguished from the "potentiometric surface", defined by water levels in wells open only to the confined St. Peter and deep sandstone aquifers.

3. METHODS

3.1 Rock and mineral analysis

Whole rock analysis was performed on 32 samples of the SCH to determine the concentration of arsenic and other elements. Samples were collected from the SCH at quarry locations within the study area and from one drill core from just outside the study area. All samples were crushed and pulverized prior to analysis. Measurement of 34 elements was performed by neutron activation, inductively coupled plasma spectrometry, or titrimetric methods by Bondar-Clegg, Inc. Mineral phases and elemental compositions of the SCH were determined by X-ray diffraction (XRD) and electron microprobe analysis (EMPA) using energy dispersive spectrometry (EDS). Transmission electron microscopy (TEM) was used to examine spatial variations in arsenic concentrations within individual crystals.

3.2 Groundwater chemistry

Geochemical data from wells in the Fox River valley region were provided by WDNR. All ground water samples were collected by WDNR personnel, and thus, field sampling procedures are assumed to largely conform to WDNR protocol (Lindorf et al., 1987). Procedures for sampling from water supply wells include sample collection from a cold-water tap situated prior to in-line treatment systems. Taps are flushed for 3 to 5 minutes in order to purge standing water in the pipes. Samples for cation and metals analysis are preserved with nitric acid. In all cases, arsenic was analyzed using ICP-AES or ICP-MS at laboratories certified by the State of Wisconsin. Detection limits ranged from 0.6 to 3 μg/l. Analyses of other water quality parameters also followed standard procedures and were conducted at certified laboratories. Because samples were typically collected for assessment of drinking water quality, samples were not filtered. Therefore, the metals concentrations reported represent total, rather than dissolved phase, constituents. Redox data are not routinely collected by WDNR. Thus, information on redox conditions of regional ground water was compiled from other sources (Batten and Bradbury, 1996; Pelczar, 1996; Saad, 1996; Siegel, 1989; Weaver and Bahr, 1991).

3.3 Other analyses

WDNR scientists collected one sample of the SCH and a groundwater sample from a highly impacted well for sulfur isotope analysis (J. Pelczar, personal communication, 1996). Sulfur isotope analysis was performed at the University of Waterloo Environmental Isotope Laboratory (UWEIL, 1994). Isotope data are reported as permil differences from the Canyon Diablo troilite.

One water sample, collected from the same highly impacted well, was analyzed by the Wisconsin State Laboratory of Hygiene for iron- and sulfur-oxidizing bacteria.

4. RESULTS AND DISCUSSION

4.1 Characterization of Sulfide Cementation

Pyrite and marcasite are the major minerals forming the sulfide cement, as identified by XRD and optical microscopy. These sulfides occur as both well-formed cubes and anhedral masses. Arsenic-rich areas (up to 1% by weight as estimated by EDS) occur in the pyrite and marcasite crystals as well as in iron hydroxides, but no separate arsenopyrite phase has been identified. Colloidal size (10-20 nm) iron hydroxide phases were identified using TEM. TEM-EDS analysis showed qualitative differences in arsenic, nickel, and zinc in the iron hydroxides on a nanometer scale.

The SCH contains four styles of mineralization, as shown in Fig. 4a: (1) a layer <0.5 to 30 cm thick, fairly continuous, consisting of sulfide-cemented quartz sandstone (uppermost St. Peter Sandstone, and lowest sandy-dolostone of the Sinnipee Group) (Fig. 4b); (2) a zone up to 3 m thick, in the St. Peter Sandstone (or the Prairie du Chien Group, if the St. Peter Sandstone is absent) with sparse to common sulfide nodules; (3) a zone 0.2-0.6 m thick consisting of thin (1-10 mm) bands of sulfide cement with sharp or diffuse boundaries that reflect mineralization across or within sedimentary structures; and (4) a zone of nodules filling moldic porosity (up to 8 cm in diameter) in the lowest 3 m of the Sinnipee Group and within the uppermost 1.5 m of the Prairie du Chien Group if the St. Peter is absent. The style of mineralization is variable at the scale of the study area as well as at smaller (e.g., quarry) scales.

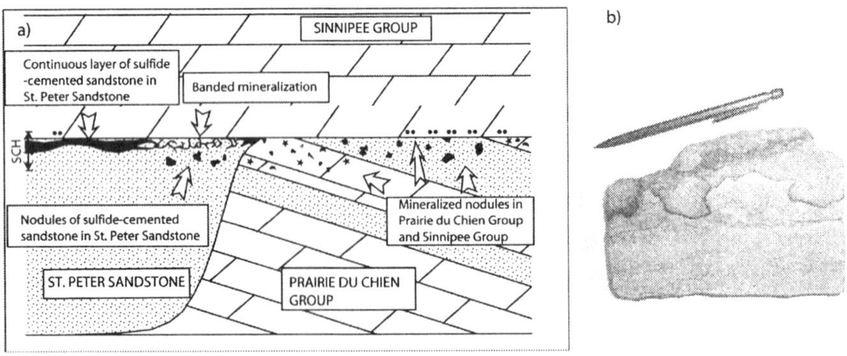

Figure 4. a) Generalized schematic of observed morphologies of the sulfide cement horizon (SCH). b) The continuous mineralization style of the SCH.

In addition to the concentrated horizon, sulfide mineralization occurs throughout the St. Peter sandstone in bands and nodules. Fig. 5 illustrates the variation in mineralization and arsenic content throughout a rock core. Most of the sandstone appears oxidized, is white-pink to red, and has whole rock arsenic concentrations less than 10 mg/kg. Zones of light and dark gray, reduced sandstone are also present in the core. While the gray sandstone is texturally similar to the pink sandstone, its arsenic content is on the order of 90 mg/kg. Pyrite nodules found within the oxidized sandstone have arsenic concentrations similar to the SCH, ranging from 325 to 585 mg/kg.

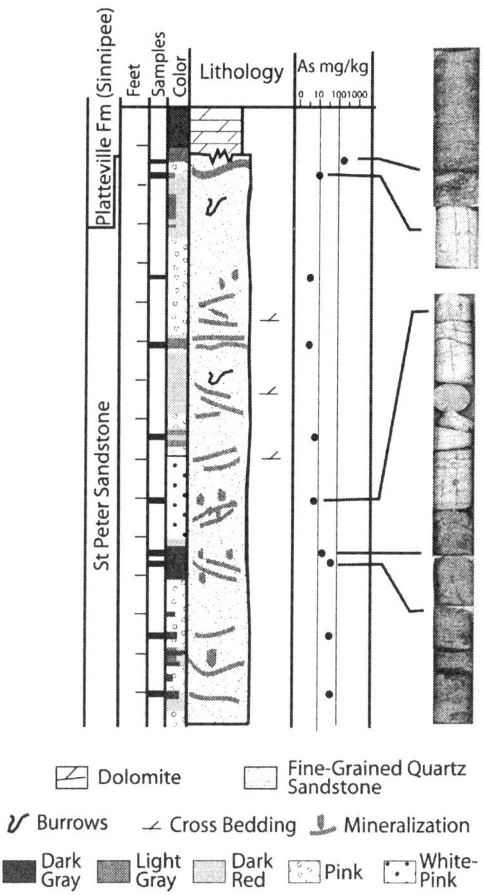

Figure 5. Lithologic section and description of a core collected in Algoma township (see Fig. 1a for location) and whole-rock arsenic concentrations in mg/kg at selected intervals within the core. Notice the color difference in the core and the relationship between color and arsenic concentrations. Also note the high arsenic at the location of the SCH and the overall lower (but still elevated) concentrations throughout the St. Peter section.

4.2 Ground water quality

Ground water samples were collected by WDNR for a variety of purposes (e.g., routine monitoring, studies designed to assess arsenic impacts, and studies focusing on other water quality issues). Thus, each well was not analysed for the same constituents. In addition, because some wells are open to more than one hydrogeologic unit, the results may reflect differences between aquifers, rather than between waters affected by reactions with the SCH. However, comparison of the data to studies of regional water quality aids in evaluating the water chemistry of arsenic-impacted wells with respect to background ground water.

For this study, we consider four types of water: non-impacted (arsenic below detection limit), low (greater than the detection limit and less than 10 µg/l As), moderate (10-100 µg/l As), and high (greater than 100 µg/l As). Typical of water quality data, the data set is censored and as such, the concentrations reported for most constituents are not normally distributed (Gibbons, 1994). Because of this limitation, box plots (Fig. 6) are used to evaluate and compare the water quality results. The distributions of concentrations were compared for iron, sulfate, pH, alkalinity, nickel, zinc, manganese and specific conductance (SC). The box plots should be viewed with consideration of sample size, which is indicated on Fig. 6.

The data set for non-impacted ground water in this study (Fig. 6, Type A) is generally in agreement with studies of regional ground water quality (Saad, 1996; Siegel, 1989; Weaver and Bahr, 1991). The notable exception to this is the elevated sulfate, which on the basis of the 73 wells included in the WDNR database, has an average concentration of 169 mg/l and ranges up to 1100 mg/l. Further examination of the data set shows that a majority of wells with elevated sulfate concentrations are municipal supply wells open to both the St. Peter and the deep sandstone aquifer. An area of relatively saline groundwater is present in the vicinity of Lake Winnebago (Ryling, 1961) and is a likely source of the elevated sulfate levels in these wells. The few shallow wells that exhibit elevated sulfate levels and no detectable arsenic are located in the eastern portion of Brown County, where the source of sulfate may be gypsum dissolution from the overlying shale confining unit (Saad, 1996; Weaver and Bahr, 1991).

4.2.1 Chemistry of arsenic-impacted ground water

As illustrated in Fig. 6, ground water with low (<10 µg/l) concentrations of arsenic (Type B) has similar chemistry to non-impacted ground water, with pH ranging from 7 to 8, iron concentrations less than 10 mg/l, and sulfate concentrations below 200 mg/l. Although alkalinity concentrations have a similar range as the non-impacted water, the low-arsenic wells display a higher median concentration. The box plots of manganese, zinc, and to a lesser extent, nickel, show higher maximum concentrations for the low arsenic wells than for the non-impacted wells. However, values for both data sets are within the same order of magnitude.

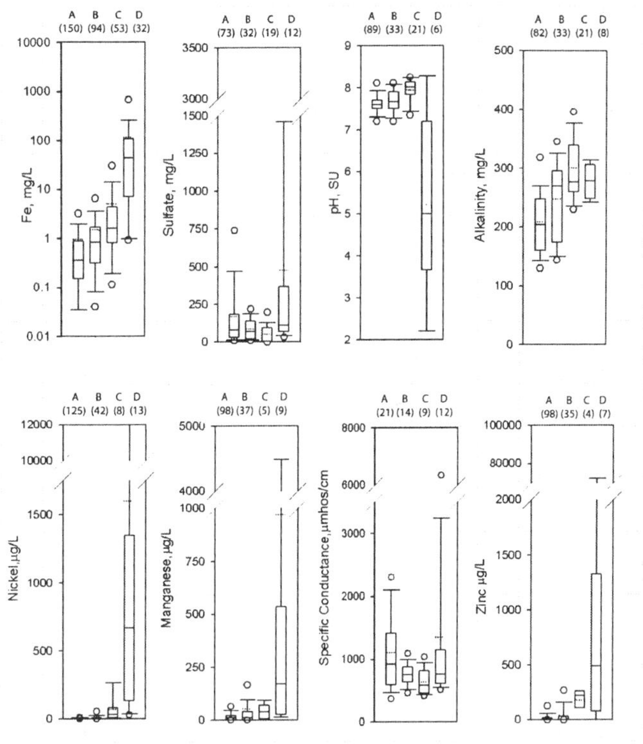

Figure 6. Box plots of concentrations of selected constituents. Plots show the 25th and 75th percentiles as end-lines. Mid-line of the box represents 50th percentile, and capped bars indicate 10th and 90th percentiles. Circles mark 5th and 95th percentiles. The mean value is shown as a dotted line. A, B, C, and D refer to four water types: non-impacted, low, moderate and high arsenic, respectively. Number of samples in parentheses. Data from WDNR (2001, unpublished).

In wells with moderate (10-100 µg/l) arsenic concentrations (Type C), iron, pH, alkalinity, as well as nickel, manganese and zinc, appear elevated above background. Although moderate-arsenic wells show different water quality from non-impacted wells, sulfate concentrations are within the range observed for non-impacted water.

In contrast, ground water with high (>100 µg/l) arsenic concentrations (Type D) exhibits a distinctly different chemical signature than non-impacted water. Sulfate is notably higher, reaching concentrations over 3000 mg/l. Iron concentrations, which range up to 1000 mg/l, are significantly higher than those measured in moderate, low, or non-impacted water. The box plots also illustrate the low pH associated with the high levels of arsenic, with values dropping to a minimum of 2.1. Levels of nickel, manganese and zinc in the high-arsenic wells are orders of magnitude greater than non-impacted wells. The increase in concentrations of ions increases the SC; values of SC range from 1000 to 7000 µmhos/cm. Alkalinity levels are similar to those in the low- and moderate-arsenic wells.

As illustrated in reactions (1) and (2), oxidation of pyrite by dissolved oxygen (DO) or nitrate releases sulfate and iron in a molar ratio of 2. Ground water samples with high arsenic exhibit a strong positive correlation between sulfate and iron; the slope of the trendline, which represents the molar ratio of sulfate to iron, is approximately 2 (Fig. 7b; Table 1). These data strongly support sulfide oxidation. In contrast, ground water samples with low and moderate arsenic do not exhibit a strong correlation (Fig. 7a; Table 1).

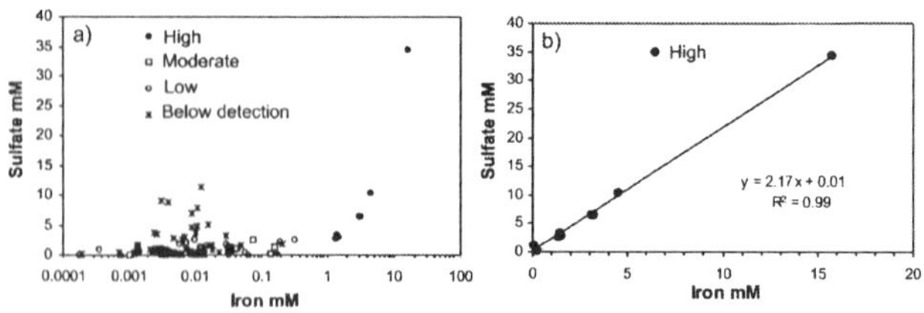

Figure 7. a) Molar relationship of sulfate to iron (log scale) for non-impacted water, and low, moderate, and high arsenic concentrations. b) Molar relationship of sulfate to iron (linear scale) for high arsenic samples only. Data from WDNR (2001, unpublished).

Table 1. Molar ratios of sulfate to iron in solution

Ground water type	*Molar ratio sulfate:iron*[1]
High arsenic	2
Moderate arsenic	26
Low arsenic	263
Arsenic below detection	358

[1] *Average values reported for sample sets shown in Fig. 6.*

Sulfur isotope data provide additional evidence that sulfide oxidation is occurring in the aquifer. The sulfur isotopic signature ($\delta^{34}S$) of pyrite from the SCH sample was -6.05 (-6.25 replicate) permil. This can be compared to the $\delta^{34}S$ of sulfate (-6.48 permil) in ground water to infer origin of the sulfate. The similar values of $\delta^{34}S$ for sulfide and sulfate indicate a lack of fractionation of sulfur isotopes between sulfide and sulfate, and are suggestive of sulfide oxidation. Because the ground water sample was collected from a well with an arsenic concentration above 100 μg/l, this inference can only be made to the highly impacted wells and cannot be extended to all arsenic-impacted ground water in the region.

Iron-oxidizing bacteria, including *Siderocapsa spp.* and *Crenothrix spp.*, were measured in elevated concentrations (> 500 per mL) in a highly impacted well (221 μg/l arsenic, pH 3.8). Although not a direct measure of their role in mediating and/or accelerating sulfide oxidation, the presence of these bacteria provides a potential for production of Fe(III), which can act as an oxidant of sulfides under low pH. Thus, in highly impacted wells, iron-oxidizing bacteria may accelerate sulfide oxidation.

4.3 Correlation of the SCH to elevated arsenic

Several types of information were used to assess stratigraphic constraints on arsenic concentrations in well water. Water level and well construction data were obtained from well records. Regional correlations of stratigraphic units were inferred from field mapping and drill core.

The relationship of wells to stratigraphic unit and arsenic concentrations was investigated by grouping wells according to the stratigraphic interval over which they are open (Cambrian sandstones, Prairie du Chien Group, St. Peter Sandstone, Quaternary deposits). The maximum arsenic concentration within each well group was noted, along with the percentage of wells in the group that had low, moderate, and high arsenic concentrations (Simo et al., 1996). This analysis did not demonstrate any obvious correlations between arsenic concentrations and stratigraphy. Arsenic concentrations range from below detection to 12,000 μg/l in wells that are open to the SCH. While most of the arsenic contamination in the FRV occurs in wells that are open to the SCH, arsenic is also present in wells that are cased through the SCH.

The relationship between the SCH and arsenic concentrations is clarified, however, if the static water levels in wells are considered. Fig. 8 shows the correlation between arsenic concentrations in well water, the relative location of the SCH, and the static water level measured in the individual wells. The highest arsenic concentrations (greater than 100 μg/l) occur in wells where the air-water interface in the well is within 15 m of the SCH, while lower concentrations (less than 100 μg/l) are found in wells with a variety of SCH-SWL configurations. It should be noted that not all wells that have a static water level near the SCH have a high concentration of arsenic.

This may be due to heterogeneity observed at local and regional scales of the thickness of the SCH and the associated mass of sulfides.

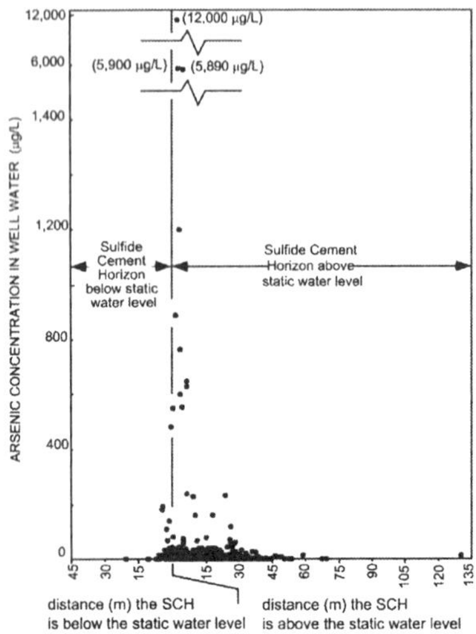

Figure 8. Relationship between arsenic concentrations in well water and the relative position of the SCH with static water levels. Arsenic data from WDNR (1995, unpublished).

4.4 Possible causes of sulfide oxidation

In addition to recognizing sulfide oxidation as the process by which arsenic is released to ground water, identification of the mechanism that causes sulfide oxidation is necessary to understand the controls on arsenic in well water. Two issues of concern in the Fox River valley case are the nature of the oxidant and its source.

As previously discussed, DO, nitrate, and Fe(III) can each serve as an oxidant of sulfides. In the St. Peter aquifer in the study area, nitrate is generally present at concentrations less than 0.5 mg/l as N (Saad, 1996) and the near-neutral pH and low DOC content of ground water precludes the possibility of significant concentrations of dissolved Fe(III) on a regional basis. On a local basis, some wells exhibit low pH, and in these wells Fe(III) may be an important oxidant. Given that DO is the most ubiquitous oxidant in the near-surface, it is likely the most significant oxidant of the SCH.

In studies of oxidation of sulfides in surface or near-surface environments, the source of oxygen, the atmosphere, is obvious. However, in a partially confined aquifer, DO is not necessarily present in significant concentrations. As illustrated in Fig. 9, several hypotheses have been posed regarding the source of DO to the SCH: (1) oxygen-rich water infiltrates to the SCH in the recharge area; (2) vertical leakage of oxygen-rich water through the Sinnipee confining unit initiates a regional oxidation of sulfides in the SCH; (3) partial dewatering of the St. Peter aquifer due to extensive ground water withdrawal exposes the SCH to oxygen; and (4) boreholes provide a direct conduit for atmospheric oxygen to reach the SCH in areas where the static water level is coincident with, or below, the SCH.

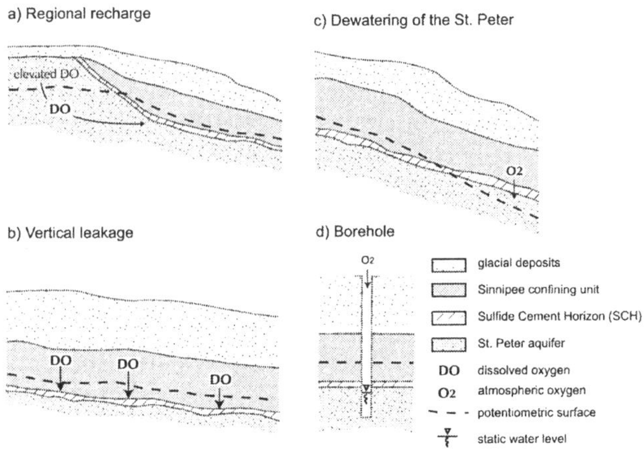

Figure 9. Schematic diagram showing different potential sources of oxygen to the SCH. a) regional recharge; b) vertical leakage; c) dewatering; and d) the borehole.

4.4.1 Regional Recharge

As oxygenated water recharges the St. Peter, it flows past, and can therefore interact with and oxidize, the SCH (Fig. 9a). Many wells impacted by arsenic occur within the recharge area (Fig. 2), where DO concentrations of up to 2.5 mg/l are measured in ground water.

However, DO-enriched recharge water is not the sole source of oxidant to the SCH across the FRV region. Between the erosional edge of the Sinnipee Group (the confining unit) and the Fox River, where the St. Peter aquifer is regionally confined, DO concentrations are less than 0.5 mg/l (Pelczar, 1996; Saad, 1996). Yet, high arsenic, in combination with high sulfate, iron, metals, and depressed pH, have been measured in ground water in this part of the study area. If DO in recharge water was the sole source of oxidant to the SCH in the FRV, arsenic concentrations should be independent of static water levels. As shown in Fig. 8, low and moderate arsenic concentrations are not correlated with the static water level, but high arsenic concentrations are. This suggests that recharge may supply DO to the SCH in the unconfined portion of the St. Peter, causing low- and moderate arsenic impacts in this area, while another source of oxidant causes high arsenic concentrations. This additional source appears to be related to the position of static water levels with respect to the SCH.

4.4.2 Vertical Leakage

Where the St. Peter aquifer is confined by the Sinnipee Group, hydraulic-head measurements indicate downward flow though the Sinnipee (Batten and Bradbury, 1996; Conlon, 1998). This flow may induce vertical leakage of DO from the upper unconfined aquifer through the Sinnipee and into the St. Peter aquifer (Fig. 9b).

This hypothesis can be qualitatively evaluated by examining patterns of arsenic contamination. Again, the finding that high arsenic levels correlate with the intersection of the static water level and the SCH (Fig. 8) negates the possibility that vertical leakage is the primary source of DO for these high concentrations, but it may contribute to the low and moderate impacts. Furthermore, most wells in the area where the St. Peter is confined, even those open to the SCH, do not appear to have significant arsenic contamination problems (Fig. 2), unless the static water level intersects the SCH (Fig. 8). Thus, although vertical leakage of oxygen-rich water may contribute to some limited oxidation, it is not a dominant source of DO for regional sulfide oxidation and high arsenic concentrations.

4.4.3 Dewatering

Dewatering of aquifers due to heavy ground water extraction can promote oxidation of sulfides. (Kinniburgh et al., 1994) report that local oxidation of sulfides within a confined sand aquifer was caused by aquifer dewatering and associated introduction of air at the well borehole. In some areas near the Fox River, municipal well pumping has caused the regional potentiometric surface of the St. Peter and deep sandstone aquifer to decline below the base of the Sinnipee confining unit (Fig. 9c). Thus, the lowering of heads into the St. Peter aquifer may expose the SCH to oxygen in this part of the study area.

4.4.4 Borehole

In areas where static water levels are at or below the SCH, the borehole provides a direct conduit for atmospheric oxygen to interact with water and the SCH (Fig. 9d). The spatial correlation of high arsenic concentrations with the intersection of the SCH and the static water level also supports this hypothesis as the cause of the more severe arsenic contamination. In wells open to both the Sinnipee and the St. Peter, the static water level is generally at the contact of the two units, because the Sinnipee Group has low hydraulic conductivity and is slowly draining to the more conductive St. Peter aquifer. With the exception of areas affected by significant ground water withdrawal, both the confining unit and the aquifer surrounding the wells are fully saturated. Thus, the borehole provides the only location where air, water, and sulfides can interact.

In this "borehole model" of SCH oxidation, the source of arsenic is local. Ground water from the aquifer formation that has not been in contact with oxygen at the borehole should not show the chemical signature of sulfide oxidation. Pelczar (1996) collected a ground water sample from a private well in Hobart (see Fig. 1a for location) prior to purging. The sample had a pH of 3.1 and an arsenic concentration of 790 μg/l. After several hours of pumping, the pH increased to 6.3 and the arsenic concentration decreased to 180 μg/l. The lower pH and higher arsenic concentrations in the standing water suggest that much of the sulfide oxidation occurs in close proximity to the well. As the well is pumped, fresh formation water, which is not as contaminated, enters the well.

Dissolved oxygen levels measured in well purge water also indicate that the borehole is a conduit for oxygen. Pelczar (1996) purged three nested monitoring wells installed in the confined portion of the St. Peter aquifer. DO concentrations in standing water in the boreholes ranged from 3 mg/l to 10 mg/l. During purging, DO concentrations decreased, stabilizing in each well at concentrations less than 0.1 mg/l. These results confirm that where the St. Peter aquifer is confined, borehole water contains significantly more DO than formation water.

In several wells, a lag time has occurred between well construction and initiation of sulfide oxidation, which further supports the hypothesis that the borehole provides an oxidant to the system. For example, a monitoring well located within the regional recharge area had an arsenic concentration of 34 μg/l and near-neutral pH (7.2) upon installation in 1992 (Batten and Bradbury, 1996). The well was sampled again in 2000, at which time arsenic was measured at 2700 μg/l, with a pH of 4.1 (Rick Stoll, WDNR, written communication, 2000). Although the recent sampling was conducted without purging, these results suggest that severe sulfide oxidation occurred subsequent to the installation of the well. If the source of DO was regional recharge or vertical leakage, the degree of oxidation should remain constant over time, independent of well installation and water levels.

Carbonate mineral dissolution may also play a role in the observed lag time between well installation and the initiation of sulfide oxidation. Sulfide oxidation in carbonate-hosted sulfide deposits may not produce low pH and high metals concentrations due to pH buffering from carbonate dissolution (Nicholson et al., 1988; Toran, 1987). In the FRV, the buffering capacity of bicarbonate in ground water may contribute to the observed lag period between installation of a well open to the SCH and development of low pH and/or high metals concentrations.

4.5 Conclusions: Hydrogeochemical Framework of Arsenic Contamination

Understanding the release of arsenic to ground water requires knowledge of the source of arsenic and the dominant geochemical processes that affect its fate. The Fox River valley provides a unique example of arsenic release within a confined aquifer system. Here, we use a lines-of-evidence approach to summarize the geologic, mineralogical, hydrogeologic and geochemical data used for evaluating mechanisms of arsenic release.

4.5.1 High (> 100 μg/l) arsenic concentrations

Ground water chemical and sulfur isotope data indicate that oxidation of sulfides in the SCH is the primary cause of the high arsenic concentrations in the St. Peter aquifer. These high concentrations are negatively correlated with pH and positively correlated with metals contained in the SCH (iron, nickel, cobalt, zinc, and manganese). The molar ratio of sulfate to iron of 2 in the high arsenic samples (Table 1) is indicative of sulfide oxidation. Similarity in the sulfur isotope signature of sulfate in groundwater to that of pyrite in the SCH suggests that the sulfate originated from the sulfide. In addition, the presence of iron oxyhydroxides in the SCH provides some evidence of *in-situ* weathering of sulfides.

The source of the oxidant in the severely impacted wells is most likely the borehole. The strong correlation between high arsenic and the interface of the SCH with the static water level in the well, in both confined and unconfined areas if the St. Peter aquifer, indicates that this interaction exerts significant control on arsenic release. Under confined conditions, the air-water-sulfide interface only occurs within the well. Field data also show that in the confined portion of the aquifer, the DO concentration in borehole water is significantly higher than in formation water. The borehole appears to promote oxidation of sulfides in the SCH in wells where the static water level is approximately coincident with the SCH.

Water chemistry patterns in high arsenic wells are not consistent with other geochemical mechanisms of arsenic release. The positive correlation of high arsenic to sulfate and trace metals, and the negative correlation to pH, are not suggestive of desorption or dissolution of iron hydroxides.

Desorption of arsenic is often initiated by higher pH or the presence of competing anions. Higher pH was not measured in wells with high arsenic (Fig. 6). In addition, the concentrations of competing anions, such as phosphate, are low in the ground water (WDNR, 2001, unpublished). Carbonate leaching of arsenic from sulfides has been proposed to cause high arsenic concentrations in the Marshall Sandstone of Michigan (Kim et al., 2000). The data from the FRV do not show a positive correlation between high arsenic and alkalinity, indicating that carbonate leaching does not cause the most severe arsenic contamination in the FRV.

4.5.2 Moderate (10-100 μg/l) and low (<10 μg/l) arsenic concentrations

Moderate and low arsenic concentrations are found in numerous wells in the FRV, irrespective of the presence of the SCH in the open interval of the well. These levels of arsenic in ground water do not correlate to the intersection of the SCH with the static water level (Fig. 8), indicating that the sulfide oxidation initiated by oxygen introduced at the borehole does not cause all arsenic occurrences. There is no single stratigraphic unit that controls the low to moderate concentrations; wells with open intervals in the Sinnipee Group, the St. Peter Sandstone, the Prairie du Chien Group or the Cambrian sandstones are all affected.

Although some data support sulfide oxidation as a cause of some of the low to moderate arsenic concentrations, the evidence for this is not conclusive. Many wells with these levels of arsenic are located within the recharge area, where DO concentrations of several mg/l are present in ground water. Ground water chemistry data for low and moderate impacts show slight differences in iron and nickel with respect to background ground water but do not show elevated sulfate or manganese. Water chemistry data were examined for evidence of other mechanisms of arsenic release. Reductive dissolution is expected to occur only under reducing conditions and thus, would not be likely in the recharge area. However, in areas where the St. Peter is regionally confined and DO is depleted, reduction of iron hydroxides may release arsenic to groundwater. This process would result in elevated iron and arsenic levels. Low arsenic wells do not show a strong correlation between iron and arsenic, but iron is elevated with respect to background in wells with moderate arsenic. These higher iron concentrations are not correlated with elevated sulfate or depressed pH, suggesting that another process may be controlling arsenic release for the moderate impacts.

Reductive dissolution of iron hydroxides is expected to be associated with increases in alkalinity as microorganisms couple the reduction of the solid phase iron with the oxidation of organic carbon. As shown in Fig. 6, ground water with moderate concentrations of arsenic does have higher alkalinity than the low arsenic or non-detect water. Although these data support the possibility of reductive dissolution as a mechanism for arsenic release, the evidence is not definitive.

Desorption of arsenic from iron hydroxides is another potential cause of arsenic release, although the data do not indicate that it plays a significant role in the region. However, pH is slightly elevated in wells with low and moderate arsenic as compared with background ground water. Further research is needed to evaluate desorption as a potential mechanism for arsenic release.

Similarly, more extensive data are needed to fully evaluate carbonate leaching as a potential cause of the low and moderate arsenic concentrations. While alkalinity appears elevated above background in low and moderate arsenic wells, this is not sufficient evidence to suggest carbonate leaching is the process controlling the release of arsenic. In the FRV, multiple lines of evidence suggest sulfide oxidation as the cause of high levels of arsenic in well water, however the available data do not conclusively identify a single process that controls the low to moderate arsenic concentrations found across the region.

ACKNOWLEDGMENTS

This research was supported by funding (# A349432) from the University of Wisconsin Water Resources Center and the Wisconsin Department of Natural Resources (WDNR) through the Wisconsin Groundwater Coordinating Council. The authors would like to thank the WDNR Lake Michigan District and Central Offices, including A. Weissbach, J. Pelczar, R. Stoll, M. Hronek, E. Heinen, K. O'Connor, and D. Johnson, for providing information, data, maps, and advice. Also thanked are K. Freiberg for helpful discussions and field assistance, J. Banfield for conducting the TEM analyses, and M. Diman and S. Hunt for help with drafting of figures. The authors would also like to thank A. Kolker and D. Krabbenhoft for their constructive comments on the manuscript.

Chapter 10

Arsenic in southeastern Michigan

Allan Kolker[1], S. K. Haack[2], W. F. Cannon[1], D. B. Westjohn[2], M.-J. Kim[3*], Jerome Nriagu[3], and L. G. Woodruff[4]

[1] *U.S. Geological Survey, Reston, VA*
[2] *U.S. Geological Survey, Lansing, MI*
[3] *University of Michigan, School of Public Health, Ann Arbor, MI,*
[4] *U.S. Geological Survey, Mounds View, MN.*
**present address: Korea Institute of Science and Technology, Seoul, South Korea*

Arsenic levels exceeding 10 μg/L are present in hundreds of private supply wells distributed over ten counties in eastern and southeastern Michigan. Most of these wells are completed in the Mississippian Marshall Sandstone, the principal bedrock aquifer in the region, or in Pleistocene glacial or Pennsylvanian bedrock aquifers. About 70% of ground water samples taken from more than 100 wells, have arsenic contents ≥ 10 μg/L, with a maximum value of 220 μg/L. Water samples and continuous cores were taken from two test wells. Arsenic content of core samples ranges from <5 to more than 300 ppm, with the highest values found for pyritic black shales. Authigenic cements in the Marshall Sandstone include patchy authigenic pyrite that locally contains arsenic-rich (up to 8.5 wt. % As) domains. Bulk arsenic contents of pyrite-bearing intervals, sampled in well cuttings, are as high as 1020 ppm. Arsenic-rich pyrite is likely the ultimate source of arsenic in eastern and southeastern Michigan ground water, but evidence for pyrite oxidation at depth in bedrock aquifers is generally lacking. Pyrite oxidation may occur or have occurred in tills derived from the Marshall Sandstone and Coldwater Shale, which were found to contain arsenic-rich (up to at least 0.7 wt. % As) iron oxyhydroxides. Plausible mechanisms for widespread arsenic mobilization in eastern and southeastern Michigan ground water include weathering of pyrite in tills, reductive dissolution of iron oxyhydroxides in tills, and potentially, pyrite oxidation in bedrock aquifers, due to drawdown in wells or lowering of water-table levels in response to Pleistocene glaciation.

1. INTRODUCTION

Naturally occurring arsenic in water supplies is one of the most widespread environmental/public health problems we face as a society. In eastern and southeastern Michigan, unpublished surveys of supply wells show maximum arsenic contents in excess of 300 µg/L. Arsenic levels exceeding the proposed U.S. drinking water standard of 10 µg/L, and approaching or exceeding the longtime standard of 50 µg/L, are present in numerous supply wells. To investigate the problem, the U.S. Geological Survey (USGS), in collaboration with the University of Michigan, sampled more than 100 wells and examined aquifer materials for possible sources of arsenic. The study area covers 10 counties, having a combined population of more 2 million people (Fig. 1).

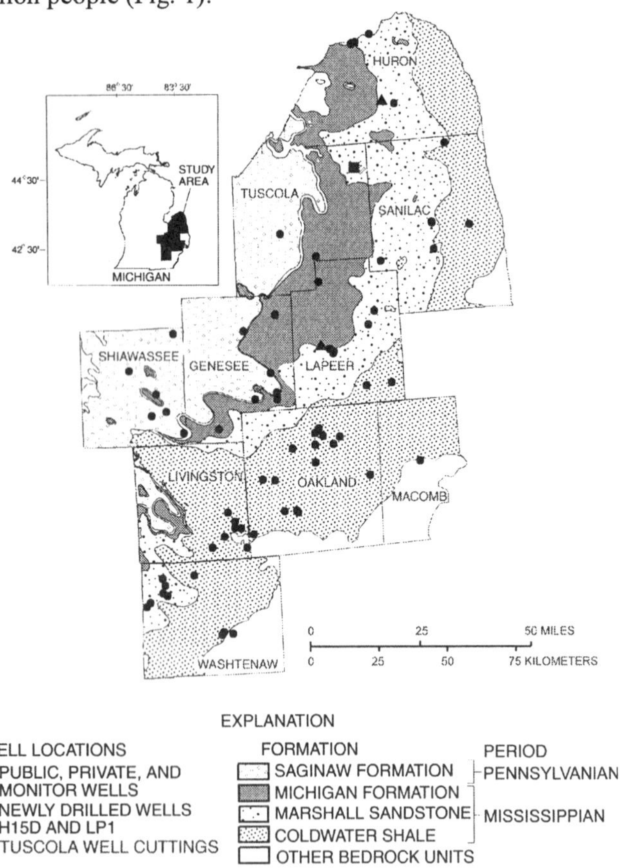

Figure 1. Location map showing study area and distribution of wells sampled, including test wells H-15D (Huron County) and LP-1 (Lapeer County), and Tuscola County well cuttings. Wells are completed in various formations (modified from Haack and Treccani, 2000).

Arsenic concentrations in about 70% of the water samples exceed 10 μg/L, including a few public wells. No public wells sampled exceed 50 μg/L (Haack and Rachol, 2000d; Haack and Trecanni, 2000; Kim, 1999). Most of the wells exceeding 50 μg/L are completed in the Mississippian Marshall Sandstone or its upper and lower confining units, the Michigan Formation and the Coldwater Shale, respectively. Some wells in glacial or Pennsylvanian bedrock aquifers also have elevated arsenic levels, but no glacial water sampled exceeds 50 μg/L. The distribution of wells with elevated arsenic contents shows considerable local variation, based on our results, and compilation of over 3,000 records by the State of Michigan and County health departments. Within the study area, median arsenic contents range from 2.9 μg/L in Washtenaw County (77.9% of wells ≤ 10 μg/L) to 16.6 μg/L in Genesee County (36.2% of wells ≤ 10 μg/L; Aichelle et al., 1999; Haack and Rachol, 2000a; Haack and Rachol, 2000b; Haack and Rachol, 2000c, d; Haack and Rachol, 2000e, f; Haack and Rachol, 2000g; Haack and Rachol, 2000h).

1.1 Hydrogeology of the study area

The Michigan Basin is a concentric structural depression containing sedimentary rocks ranging in age from Precambrian to Jurassic. Arcuate subcrop belts are present within the study area, at the eastern and southeastern margins of the basin (Fig. 1). Listed from southeast to northwest, these units are the Coldwater Shale, Marshall Sandstone, Michigan Formation (all Missisippian), and the Saginaw Formation (Pennsylvanian). The Marshall Sandstone is a significant bedrock aquifer, whereas the underlying Coldwater Shale and the overlying Michigan Formation are confining units for the Marshall. The Saginaw Formation consists of a lower confining unit and mid-to-upper units that are aquifer sandstones extending primarily over the central part of the Michigan Basin (Westjohn and Weaver, 1998).

Pleistocene glacial deposits in the study area consist primarily of coarse-textured tills, outwash deposits formed at the front of retreating ice lobes, and glacial lacustrine deposits (Westjohn and Weaver, 1998). Glacial deposits are thinnest or absent in the extreme northeastern and southeastern portions of the study area, and thicken progressively to the northwest. Tills and lacustrine deposits are generally viewed as confining units for glaciofluvial sands and gravels, but the confining units are highly discontinuous. Tills contain large amounts of Mississippian and Pennsylvanian rock fragments, in zones that are oxidized or unoxidized.

The distribution of wells having elevated arsenic levels generally corresponds to the eastern and southeastern parts of the Marshall Sandstone subcrop belt (Fig. 1). In this area, the Marshall Sandstone is in direct hydrologic contact with permeable Pleistocene glacial deposits that overlie much of the study region. Water quality in the Marshall Sandstone deteriorates downdip towards the center of the basin, where saline waters, and ultimately, brines are present at depth (Westjohn and Weaver, 1995, 1998). The majority of wells sampled in the present study are completed in the Marshall Sandstone, the Saginaw Formation, and Pleistocene glaciofluvial deposits. Together, these units constitute the major aquifers in the region.

1.2 Methods

To investigate the possible relation between the composition of host rocks and the arsenic content of ground water, well sampling was conducted in conjunction with sampling of aquifer materials in test cores and available well cuttings (Haack and Trecanni, 2000). Wells were selected to provide samples from a range of depths, in locations with both near-normal and elevated arsenic contents.

Water sample collection procedures used in this study are outlined by Koterba et al. (1995). Field measurements included temperature, dissolved oxygen, pH, specific conductance, redox potential, alkalinity, and field tests for total and ferrous iron, nitrate, ammonia, sulfate, sulfide, turbidity, and phosphate (Haack and Trecanni, 2000). Samples collected for trace-element analysis were filtered to pass 0.45 μm. Samples for arsenic analysis included sub-samples for total and filtered arsenic, arsenate, and arsenite (Kim, 1999).

Two test wells, H-15D and LP-1, were drilled in areas known to contain other wells having high arsenic levels (Fig. 1, 2). To minimize possible contamination from drilling mud, city water (H-15D) or ground water from a nearby well (LP-1), was used to circulate rock cuttings. Water samples were collected at 15.7 m intervals using a pump and packer system, with selected zones of interest sampled in packed 1.37 m or 5.33 m intervals, using a similar protocol.

Arsenic contents in filtered and whole-water sub-samples were determined by atomic absorption hydride generation (AAH), having a detection limit of 1 μg/L (Haack and Trecanni, 2000). Concentrations of most other metals were determined by ICP-MS, with 1 μg/L detection limits, except for iron, having a detection limit of 3 μg/L. Sulfate concentrations were determined by ion chromatography, with a detection limit of 0.1 mg/L. Arsenic speciation was determined at the University of Michigan using ion exchange methods modified and adapted by Kim (1999). Arsenic concentrations of speciated samples were determined by graphite-furnace atomic absorption spectrophotometry (Kim, 1999).

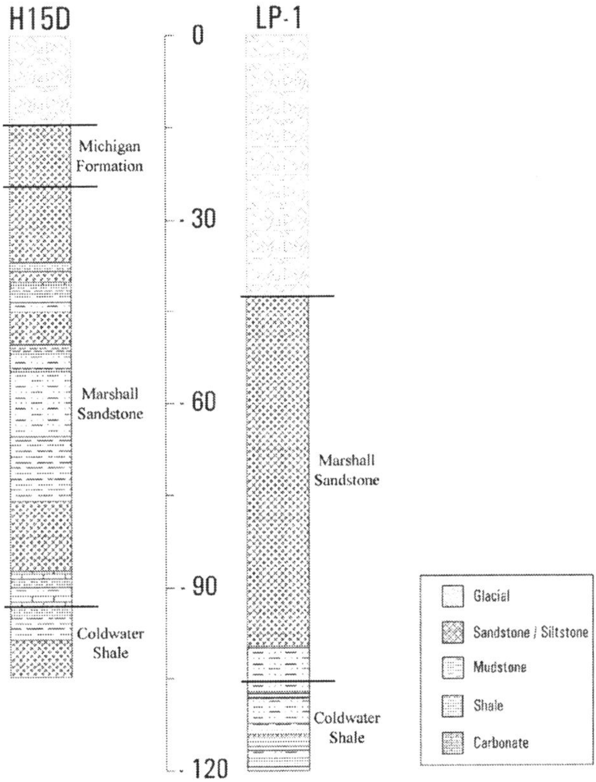

Figure 2. Generalized stratigraphic columns for test cores extracted from H-15D (left) and LP-1 (right) wells. Vertical scale is in meters below land surface.

For test wells H-15D and LP-1, retrieved core sections were inserted into plastic bags, flushed with nitrogen gas, and refrigerated until the samples were processed. Core samples were split, logged, and prepared for bulk analysis in intervals of approximately 1.2 m. Cuttings from 3 recently drilled wells penetrating the Marshall Sandstone in northeastern Tuscola County (Fig. 1), were also analyzed. For the H-15D core, arsenic concentrations were determined by ICP-MS, with detection limit of 4 ppm. For the LP-1 core and the Tuscola County well cuttings, multi-element analyses were obtained by ICP-AES, with a detection limit for arsenic of 10 ppm. Additional results for arsenic were determined for intervals of shale and sand separated from the LP-1 core at levels below 100.5 m. These results were obtained from a commercial laboratory, using semi-quantitative atomic absorption spectrophotometry (detection limit 5 ppm).

A fully-automated, 5 spectrometer electron microprobe (JEOL JXA 8900R Superprobe) was used to determine major- and minor-element concentrations in pyrite and iron oxyhydroxides, by the wavelength-dispersive method. Instrument conditions include a beam current of 3.0×10^{-8} amps, an accelerating voltage of 20 KeV, and a fully focussed beam, giving an actual working diameter of about 2-3 micrometers. Arsenic concentrations were determined for arsenic $L\alpha$ using a TAP analyzing crystal, and a ZAF correction procedure. A minimum detection limit of about 100 ppm was attained using a peak counting time of 60 seconds and 30 seconds for upper and lower backgrounds. Wavelength-dispersive arsenic $L\alpha$ elemental maps utilize a fully focussed beam, a beam current of 5.0×10^{-8} amps, and an accelerating voltage of 20 KeV. Other map parameters, such as dwell time, pixel size, and number of pixels, were chosen to best accommodate each of the grains studied.

2. RESULTS

2.1 Water Chemistry

A total of 114 private, public, and monitoring wells were sampled, including test wells H-15D and LP-1 (Haack and Trecanni, 2000; Kim, 1999). For nearly all of these samples, pH is circum-neutral (7.0 to 7.5) and dissolved oxygen is low (0.0 to 0.5 mg/L). We found several high-arsenic ($\geq$ 50 µg/L) wells completed in the Marshall Sandstone, but wells with similarly high levels of arsenic were identified in each of the major bedrock aquifers in the region. The highest arsenic concentration we determined is 220 µg/L, for a private well in Huron County (Haack and Trecanni, 2000).

Two test wells intersecting the Marshall Sandstone show a gradual decrease in arsenic contents with depth (Fig. 3). Arsenite (As III) is the dominant species, ranging from 70 to 95 % of total arsenic in wells throughout the study area having at least 10 µg/L arsenic. In the H-15D well, arsenite makes up 88–100% of total arsenic, and there are no distinct changes in oxidation state with depth. Redox conditions measured for H-15D waters are essentially uniform with depth. On an Eh-pH diagram, waters from H-15D and 12 other wells sampled plot slightly above the stability field for pyrite, but much below the Eh conditions at which sulfide oxidation has been found to occur in natural samples (Fig. 4; Sato, 1960). Sampling for aqueous arsenic at 15.7 m intervals agrees well with results taken at smaller intervals, with the exception of three samples near the bottom of H-15D,

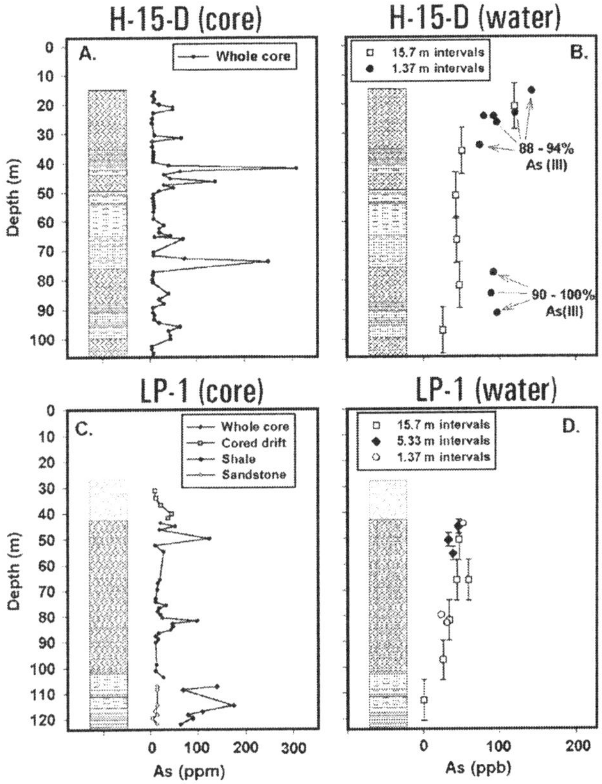

Figure 3. Arsenic in cores H-15D and LP-1 (left; A, C) and water in corresponding packed intervals (right, B, D). Brackets on water data indicate size of packed intervals represented. Results for arsenic speciation in H-15D waters are from Kim (1999).

showing higher arsenic contents in the 1.37 m sample intervals than in the 15.7 m intervals (Fig. 3B).

In the test wells and throughout the study area, aqueous arsenic shows no correlation with iron or sulfate (see Haack and Trecanni, 2000). This lack of correlation is unlike the case of the Fox River Valley in east central Wisconsin, where oxidation of a sulfide cement horizon is demonstrated locally by extreme arsenic contents and correlation of arsenic with iron, sulfate, and pH (Schreiber et al., this volume; Schreiber et al., 2000). More commonly, where the sulfide cement horizon does not intersect static well water levels, or is less prominent or less arsenic-rich, wells in the Fox River Valley show a scattered distribution of more moderate arsenic levels and evidence for pyrite oxidation is lacking (Gotkowitz et al., 2001). These results are similar to our findings in eastern and southeastern Michigan.

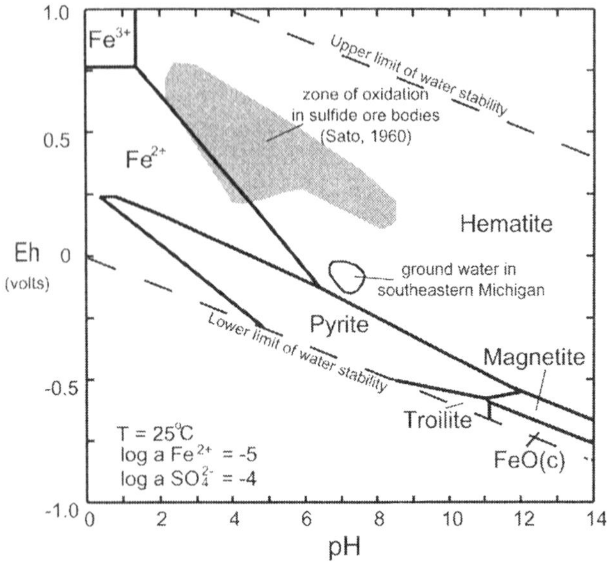

Figure 4. Eh – pH plot showing suboxic nature of southeastern Michigan ground water. Ground water field includes samples from H-15D and 12 other wells. Eh values are calculated from arsenic speciation analyses given in Haack and Treccani (2000), using geochemical reaction modeling software (Bethke, 2000) and assuming that the concentrations of the As(III) and As(IV) species approach redox equilibrium. Platinum electrode measurements taken in the field (not shown) range from 29 to 68 mV for H15D, and about 25-150 mV overall, slightly above the Eh calculated from arsenic species. The field measurements are considered less reliable than the calculated Eh values, due to the high reduced iron content of the water samples. Field platinum electrode measurements and calculated Eh values both show that redox conditions in southeastern Michigan wells are much more reducing than those required for in-situ pyrite oxidation (Sato, 1960).

2.2 Aquifer Materials

2.2.1 Bedrock Well Cuttings

Initial characterization of aquifer materials from the study area consisted of a suite of cuttings from 3 recently drilled wells in northeastern Tuscola County, in an area with high levels of arsenic in ground water (Fig. 1). Petrography of this material shows that pyrite is present locally as a cement for framework sand grains in the Marshall Sandstone (Fig. 5). Electron microprobe analyses and wavelength-dispersive elemental mapping show highly arsenic-enriched domains in this pyrite (Kolker et al., 1998). These domains occur as overgrowths on framboids having much lower arsenic contents. The framboids and their arsenic-rich overgrowths are incorporated into pyrite having low arsenic contents (Fig. 5). The overall compositional range for pyrite in electron microprobe spot analyses of well cuttings is <0.01 to 7.8 weight percent arsenic, more than enough to account for high levels of arsenic in that area, should this pyrite undergo decomposition (Table 1).

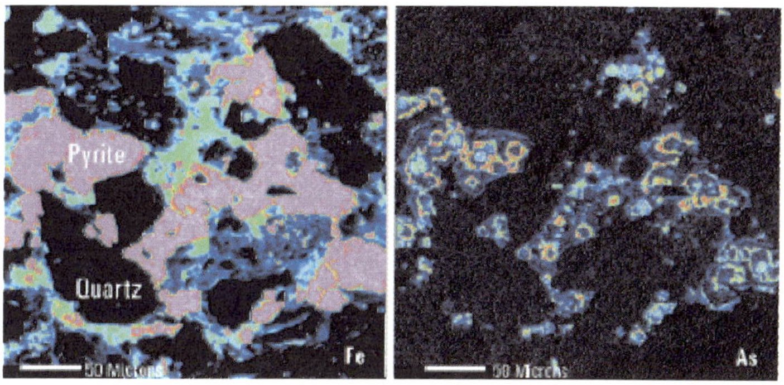

Figure 5. False-color wavelength-dispersive electron microprobe elemental maps of pyrite-cemented sandstone in Tuscola County Michigan well cuttings. Iron map (left) shows pyrite distribution. Arsenic map (right) shows irregular pattern of arsenic enrichment in pyrite (up to 7.8 wt. %), occurring as sausage-shaped overgrowths on arsenic-poor framboids. Sample is from well "A" at 54.9-59.4 meters, having a bulk arsenic content of 1020 ppm (Table 2). Maps consist of 325 x 325 1.0 μm pixels.

Bulk analyses of the Tuscola County well cuttings also show high arsenic contents (up to 1020 ppm) and good correlations between iron contents and concentrations of other chalcophile elements (Co, Ni, Cu, Zn, Cd, and Pb; Table 2). The high contents of these elements are clearly attributable to pyrite. For arsenic, the maximum value in the cuttings is roughly 3 times that of the most arsenic-enriched black (pyritic) shale determined in the test cores (see following section).

2.2.2 Bedrock Core Samples

The H-15D core penetrates the Michigan Formation, Marshall Sandstone, and Coldwater Shale (Fig. 2). The Marshall portion of the core includes a section of medium-coarse sand near its top (23.8 to 35.7 m depth), and a larger section below of interbedded sandstone, siltstone, and shale, including minor sections of black shale. The LP-1 core contains a thicker sand-dominated sequence of Marshall Sandstone (43.3 to 103.6 m depth), and below that, in the lower portion of the core, consists of interbedded sand, silt and clays, that includes the contact with Coldwater Shale (Fig. 2). The cores are largely unoxidized and contain fresh pyrite. A few oxidized (reddish) horizons are present locally in each core.

Arsenic-rich pyrite was found in each of the two test cores. Together with well cuttings from northeastern Tuscola County, arsenic-rich pyrite is present in at least 3 adjoining counties that make up a large portion of the study area (Fig. 1). In the H-15D core, arsenic-rich pyrite is present in both the Marshall Sandstone and in the Michigan Formation, where arsenic contents in pyrite up to about 2 weight percent were determined (Table 1). As in the well cuttings, the most arsenic-rich pyrite occurs as framboid overgrowths. In the LP-1 core, framboid overgrowths with spot arsenic concentrations as high as 8.5 weight percent were found to replace fossil fragments (Fig. 6; Kolker et al., 1999). For comparison, reconnaissance study of Marshall Sandstone in cores from Calhoun County, a control area in south-central Michigan where arsenic in wells is not widely problematic, shows that pyrite there is less abundant, and has lower arsenic contents (≤ 1.2 weight percent) (Kolker et al., 1998).

Table 1. Representative electron microprobe analyses of pyrite in eastern and southeastern Michigan samples of Marshall Sandstone (1-4, 6-8) and Michigan Formation (5) found in well cuttings (6-8) and USGS test cores (1-5) (in weight percent).

Element	1	2	3	4	5	6	7	8
S	47.26	50.94	50.67	52.04	51.48	49.20	51.66	50.07
Fe	44.52	45.99	34.06	47.53	46.63	45.56	46.27	46.28
Co	0.04	0.25	2.99	d.l.	d.l.	0.03	0.11	0.01
Ni	0.06	0.12	9.57	d.l.	d.l.	d.l.	0.36	d.l.
Cu	0.01	0.01	0.33	d.l.	0.01	0.03	0.09	d.l.
Zn	d.l.	0.01	d.l.	d.l.	d.l.	0.02	0.01	d.l.
As	7.08	2.00	1.68	0.02	2.12	5.96	1.31	4.43
Se	d.l.	d.l.	d.l.	d.l.	d.l.	d.l.	d.l.	d.l.
Cd	0.02	0.01	d.l.	0.02	d.l.	d.l.	d.l.	d.l.
Pb	0.02	d.l.	0.57	d.l.	d.l.	n.d.	n.d.	n.d.
Total	99.02	99.33	99.87	99.61	100.24	100.80	99.81	100.80

1 - 2 = Core samples, LP-1 test well, 81.2 m depth.

3 - 5 = Core samples, H15-D test well, 72.4m, 46.6 m, 21.0 m depth, respectively.

6 - 8 = Well cuttings, Tuscola County "C" well, 70.1-71.6 m, 71.6-73.2 m, 70.1-71.6 m, respectively. d.l. = below detection limit (0.04 wt. % for Co, 0.01 wt. % for other elements; results for Co are background-subtracted); n.d. = not determined

Table 2. Bulk geochemical data for well cuttings, Tuscola County, Michigan.

Well	Depth (m)	Fe*	As	Co	Ni	Cu	Zn	Cd	Pb
A	30.5 - 32.0	1.17	13.7	7.5	9.2	2.7	18	<2	4.1
A	27.4 - 29.0	1.6	<10	9	17.2	8.8	30.7	<2	7.1
C	76.2 - 77.7	5.34	100	19.3	44	10.2	61.5	4.8	11.9
B	65.5 - 67.1	5.99	54.7	21	52.2	11.8	109	5.3	16.2
C	74.7 - 76.2	6.07	118	18.1	43.1	13.7	169	5.6	11.5
C	68.6 - 70.1	6.1	188	25.2	62.1	15.6	346	5.6	20
C	70.1 - 71.6	6.18	166	25.1	58.4	9.2	204	5.8	15
C	71.6 - 73.2	6.21	212	23.8	56.5	16.9	200	5.3	20.9
C	61.0 - 62.4	7.16	146	20.2	47.5	7.1	332	6.1	15.8
A	54.9 - 59.4	9.51	1020	36.1	85.8	27.2	775	9	25.8
Correlation with Fe (r^2)		1.000	0.512	0.877	0.898	0.554	0.650	0.967	0.798

**Results for Fe are in weight percent, other results in ppm. Analysis by ICP-AES.*

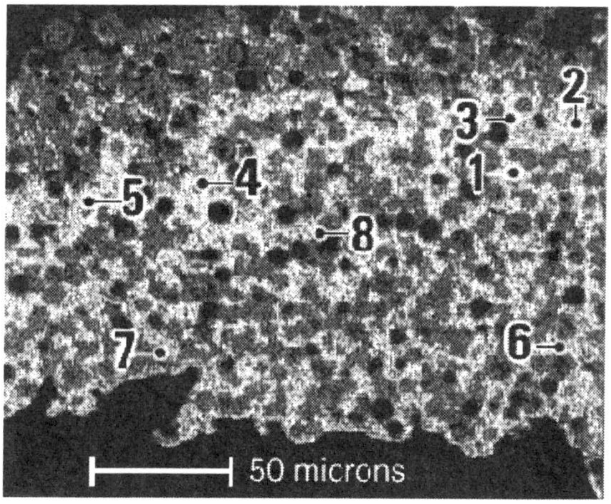

Figure 6. Grayscale wavelength-dispersive electron microprobe map of arsenic distribution in a segment of a pyritized fossil fragment, LP-1 core, 81.2 meters. Arsenic-rich pyrite occurs as overgrowths (bright bands) on framboids (darker spheres). Arsenic contents of the points shown are as follows: 1: 6.6 wt. %; 2: 7.2 wt. %; 3: 6.9 wt.%; 4: 6.7 wt. %; 5: 6.8 wt. %; 6: 8.5 wt. %; 7: 2.0 wt. %; 8: 1.0 wt. %. Area shown is approximately 400 x 450 0.50 μm pixels.

Bulk rock arsenic contents of the cores are plotted in Figure 3, based on sampling at intervals of approximately 1.2 m. For the H-15D core, arsenic ranges from <4 to 310 ppm, with the highest arsenic contents in black shales that constitute a small portion of the section. A similar range in arsenic contents was obtained for the LP-1 core (<10 to 283 ppm). The arsenic content of aquifer sands is generally less than 10 ppm, but pyrite-bearing intervals have arsenic contents in the 20-50 ppm range. There is no apparent correlation between levels of arsenic in the cores and aqueous arsenic (Fig. 3).

Analysis of sand and shale separates from the lower part of the LP-1 core (Fig. 3C) shows that shale intervals have consistently higher As contents than interbedded sandstones, but aqueous arsenic contents are no different in shale-hosted intervals than in aquifer sands. The lack of correlation between arsenic contents in water and cores, the unaltered character of pyrite within the cores, and the suboxic nature of the water, all suggest that pyrite oxidation is not occurring at depth in the test wells.

2.2.3 Till Samples

In order to determine the role of pyrite oxidation in glacial aquifers, we investigated several samples of till containing fragments of oxidized Marshall Sandstone. Electron microprobe analysis of iron oxyhydroxides in till samples shows arsenic contents up to about 0.7 weight percent (Fig. 7). If arsenic-rich iron oxyhydroxides are derived from weathering of arsenic-rich pyrite, the analysis indicates that a portion of the arsenic is retained during the weathering/oxidation process. Alternatively, arsenic may be concentrated on oxide surfaces by adsorption of aqueous arsenic. Therefore, depending on redox conditions, iron oxyhydroxides provide a potential source of arsenic in ground water, or a sink for aqueous arsenic.

3. DISCUSSION

If pyrite is the source of arsenic in the Marshall Sandstone aquifer, its oxidation must have occurred at surface or near-surface conditions. Weathering of detrital Marshall Sandstone and Michigan Formation shale in tills could explain the presence of arsenic-rich wells in glacial aquifers. However, high levels of arsenic are more common in the bedrock aquifer than in glacial aquifers, and in both aquifers, the distribution of such wells is scattered. Bedrock samples show the distribution of authigenic pyrite to be very patchy, and not all of the pyrite is arsenic-rich. Therefore, even in the case of pyrite oxidation at depth in a bedrock aquifer, we would expect local

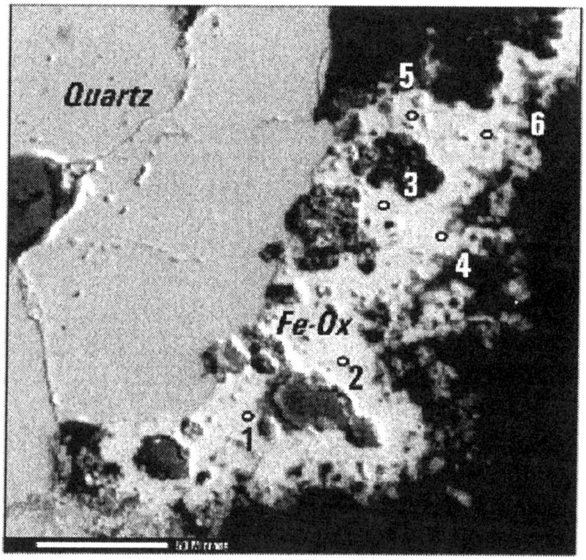

Figure 7. Grayscale wavelength-dispersive electron microprobe map of oxygen distribution showing iron oxyhydroxide rim (bright band) on quartz, in till sample LHP 8 from oxidized zone, Huron County, MI. Arsenic contents of points shown are as follows: 1: 1,200 ppm; 2: 1,300 ppm; 3: 3,300 ppm; 4: 1,400 ppm; 5: 2,800 ppm; 6: 1,000 ppm. The maximum As concentration determined for this sample is 7,300 ppm. Map area is 500 x 500 0.5 μm pixels.

variations in aqueous arsenic contents. In the absence of pyrite oxidation at depth, arsenic mobility could be the result of oxidation of Marshall Sandstone in tills, and/or reduction of arsenic-bearing iron oxyhydroxides in tills or weathered bedrock, by anoxic or suboxic ground water.

Although evidence for present-day pyrite oxidation is lacking, ground water levels following the last glaciation are inferred to be tens of meters lower than today, perhaps allowing paleo-oxidation of pyrite to have occurred in the Marshall Sandstone aquifer. This scenario is based primarily on the large difference in the levels of modern lakes such as Lake Huron, vs. their glacial precursors, which were as much as 120 meters lower (Flint, 1957).

Present-day drawdown in the vicinity of heavily used wells could also result in pyrite oxidation, on a scale corresponding to the extent of the drawdown. Because of the extreme arsenic contents of some Marshall Sandstone pyrite, even small amounts of oxidation could contribute arsenic to bedrock aquifers.

Insight into the oxidation process is given by Kim (1999) and Kim et al. (2000), who conducted leaching studies of H-15D core samples and found the amount of arsenic leached increases with the molar $NaHCO_3$

concentration of aqueous solutions. Kim et al. (2000) suggest that oxidation of arsenic-rich pyrite at shallow levels of the bedrock aquifer is enhanced by production of an arsenic-carbonate complex, or an iron bicarbonate complex as an intermediate. The $NaHCO_3$ concentrations used in these experiments range from about 7 to 120 times the average molar bicarbonate concentration in the Michigan ground water sampled. Arsenic dissolution by bicarbonate therefore seems possible in some natural water, but the extent to which this process might occur in the study area is unclear. There is not an apparent correlation between bicarbonate and arsenic contents in the wells sampled.

4. CONCLUSIONS

Two widespread natural sources of arsenic, arsenic-rich pyrite and arsenic-rich iron oxyhydroxides, have been identified in the study area, based on examination of bedrock samples and till derived from bedrock in the region. The irregular distribution of high arsenic values may be explained by the uneven distribution of these natural sources. Redox conditions of water in test wells, and comparison of arsenic contents in water and aquifer materials at the same depths in these wells, suggest that pyrite oxidation is very limited in the bedrock. However, we cannot rule out paleo-oxidation of pyrite in the bedrock aquifer, resulting from lower water table levels that existed following the last glaciation. Likewise, present-day oxidation may occur in some wells as a result of drawdown. Because there are multiple sources of arsenic, in glacial and bedrock aquifers that are in hydrologic continuity, no single process may explain the overall distribution of arsenic in eastern and southeastern Michigan wells.

ACKNOWLEDGEMENTS

The authors benefited from discussions with Alan Welch, Kirk Nordstom, Madeline Schreiber, Madeline Gotkowitz, Dave Johnson, Harvey Belkin, and Robert Finkelman. Reviews by Alan Welch, Robert Finkelman and Daniel Hayba helped improve the manuscript. We also acknowledge the respective contributions of Sandra Trecanni, Michael Sweat, Steve Aichelle, and Jason C. Willett to portions of this study. This study was conducted as part of the USGS Drinking Water Initiative, in cooperation with the Michigan Department of Environmental Quality, the Michigan Department of Community Health, and the Michigan counties in the study area.

Chapter 11

Occurrence of arsenic in ground water of the Middle Rio Grande Basin, central New Mexico

Laura M. Bexfield and L. Niel Plummer

U.S. Geological Survey

Chemical data from more than 400 ground-water sites in the Middle Rio Grande Basin of central New Mexico indicate that arsenic concentrations exceed the U.S. Environmental Protection Agency drinking-water standard of 10 micrograms per liter across broad areas of the Santa Fe Group aquifer system, which is currently the almost exclusive source of drinking-water supply for residents of the basin. Identification of sources of arsenic to ground water of the basin is complicated by multiple sources of ground-water recharge that differ substantially in chemical composition. Establishment of a clear hydrologic framework for the basin was useful in interpreting the significance of patterns in arsenic concentration. This investigation indicates that there are two main sources of high-arsenic water to the Middle Rio Grande Basin. One primary source is related to silicic volcanism in the Jemez Mountains to the north, where dilute recharge water likely flows through rocks that have been altered by contact with geothermal fluids. The other primary source is mineralized water of deep origin that mixes with shallower ground water in several locations around the basin, particularly along major structural features. Ground water that has not been affected by either of these two high-arsenic sources generally has low arsenic concentrations. In some areas of the basin, values of pH exceeding about 8.5 appear to contribute to elevated arsenic concentrations through desorption of arsenic from metal oxides.

1. INTRODUCTION

The more than 700,000 residents of the Middle Rio Grande Basin (MRGB), otherwise known as the Albuquerque Basin, of central New Mexico rely almost exclusively on ground water from the Santa Fe Group aquifer system for drinking-water supplies. Arsenic concentrations in ground water underlying the basin have been detected in excess of 600 μg/L, and concentrations exceeding 20 μg/L are present across large areas. The City of Albuquerque, which served more than 450,000 basin residents with its drinking-water supply wells in 1999 (City of Albuquerque, 2000), will be affected severely by the U.S. Environmental Protection Agency (USEPA) standard of 10 μg/L for As in drinking water (Federal Register, 2001). Just over half of the 92 City of Albuquerque wells currently in service meet that standard (City of Albuquerque, 2000); capital expenses for compliance with the new standard for the City of Albuquerque alone have been estimated at 150 million dollars (Soussan, 2001).

This investigation of the occurrence and behavior of As dissolved in ground water of the MRGB is based largely on ground-water sampling of 288 wells and springs between 1996 and 1998 as a part of the U.S. Geological Survey (USGS) MRGB study, the boundaries of which correspond to the boundaries of the basin (Fig. 1). The chemical and isotopic data collected as a part of the MRGB study are being used to better characterize the ground-water flow system of the basin (Plummer et al., 2001; Plummer et al., in press; Plummer et al., 1997). Samples were collected for analysis of a wide variety of constituents, including major- and minor-element chemistry, isotopic composition, and dissolved-gas content. Supplementary chemical data were obtained for the same study from monthly sampling at 14 surface-water sites. In order to fully characterize As distribution and possible factors affecting its occurrence in the basin, supplemental historical data for As and selected other constituents in ground and surface water were obtained from the USGS National Water Information System (NWIS) data base, and from a data base maintained by the City of Albuquerque on ground-water quality from its water-supply wells.

1.1 Previous Investigations

The presence of As in water and sediments of the MRGB has been long recognized as a problem relative to drinking-water standards (CH2M Hill, 1990, 1991; Chapin and Dunbar, 1994; Stanton et al., 2001a). Based on the spatial distribution of As and correlations between As and other water-quality parameters, As appears to originate from deep sources in most of the basin, and the volcanic center in the Jemez Mountains north of the basin also is a source of As-rich water (CH2M Hill, 1990, 1991). Proposed sources of As to ground water and surface water also include volcanic and potassium metasomatised rocks, and regional ground-water inflow to surface water (Chapin and Dunbar, 1994). Sorption of As onto sediments (particularly Fe, Mn, and Al oxides) is a likely method of removal of As from surface water (Chapin and Dunbar, 1994; Stanton et al., 2001a; Stanton et al., 2001b)

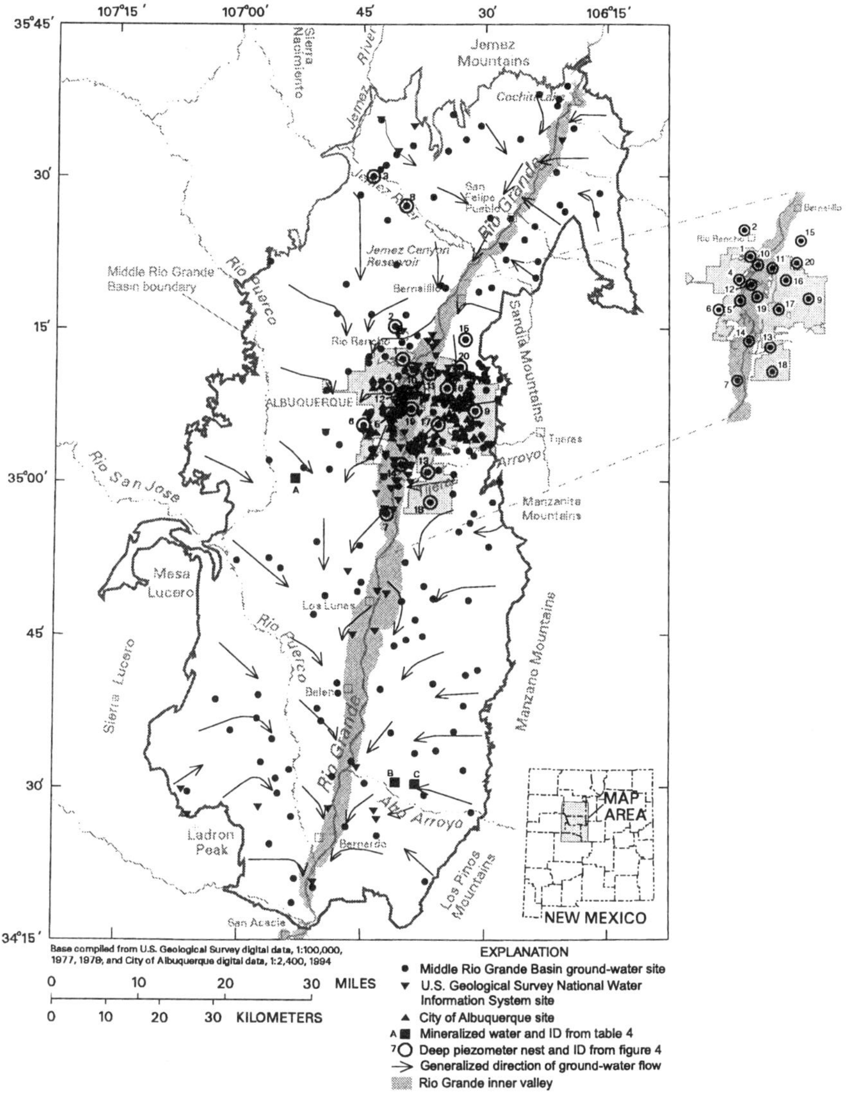

Figure 1. Selected features of the Middle Rio Grande Basin and vicinity.

examined sediment and rock samples from drill cores at three different locations in the MRGB for content and potential mobility of As. They concluded that most As was associated with acid-extractable amorphous and crystalline Fe oxides. For a study through the University of Houston, As speciation was performed on ground water from 87 City of Albuquerque wells; the study showed that samples from most wells contained only As (V), although some wells contained appreciable As (III) (Bill Lindberg, City of Albuquerque, written commun., 2000).

2. METHODS

2.1 MRGB Study of Ground-Water Chemistry

At wells sampled for this study, field parameters (pH, specific conductance, temperature, and dissolved-oxygen concentration) were determined and samples were passed through a 0.45-micron capsule filter for laboratory analysis of selected major and minor elements. At least three casing volumes of water were purged and field parameters were allowed to stabilize before sample collection. Minor-element samples were preserved with Ultrex nitric acid in the field. At nine ground-water sites, samples for minor elements were collected using a range of filter sizes. Differences among analytical results for samples filtered using 0.45-micron capsule filters, tangential filtration at 0.1 microns, and tangential filtration at 30,000 Daltons were negligible; As concentrations using the three methods differed by no more than 1 μg/L, and typically were identical within the analytical precision of +/- 5 percent.

In addition to ground-water samples, multiple surface-water samples also were collected for the study. Samples were collected as frequently as monthly at up to 14 surface-water sites, including sites on the Rio Grande and associated drains and irrigation canals, the Jemez River, the Rio Puerco, and Tijeras Arroyo (Fig. 1), between January 1997 and April 1999. Samples were dipped from streams rather than integrated across their widths and depths. Surface-water samples were filtered and preserved in the same manner as ground-water samples.

Analysis of major and minor elements for the MRGB study of ground-water chemistry was performed in the USGS Water Chemistry Laboratory in Reston, Virginia. Analysis of major cations and silica was performed using a multi-element direct-current plasma spectrometer; analysis of major anions was performed using ion chromatography. The analysis of minor elements was performed according to USEPA Method 200.8 (USEPA, 1994) using an inductively coupled plasma-mass spectrometer. The As detection limit was 0.1 μg/L. Further details of the analytical methods are given in Busenberg et al. (2000) and (Plummer et al., in press).

2.2 Additional Data Sources

Data available for ground-water samples in the NWIS and City of Albuquerque data bases were used for sites that had not been sampled as part of the MRGB study (Fig. 1). Surface-water samples having dissolved As data also were retrieved from the NWIS data base for selected sites to evaluate the presence of As in potential sources of recharge to the ground-water system. Chemical analysis of most samples from NWIS was performed at a USGS laboratory; methods of analysis vary because the dates of sample collection encompass the years 1972-97. All minor-element samples are believed to have been passed through 0.45-micron filters and acidified in the field. Bexfield et al. (1999) describe the data base compiled by the City of Albuquerque for its drinking-water supply wells (generally deep wells with long screened intervals) since 1988, including methods of sampling and analysis. Specifically, minor-element samples were unfiltered and were acidified on the day of collection. The median constituent concentrations presented by Bexfield et al. (1999) for each of 93 drinking-water supply wells were included in the data set used for this investigation, along with data for one additional well. City of Albuquerque data were retained instead of historical NWIS data, when present for the same site, because they were believed to better represent "typical" chemical compositions.

2.3 Acknowledgments

The successful completion of this study required the contributions of many people. The authors thank the numerous landowners, government agencies, and companies who provided access to their wells, in addition to those people who helped obtain sampling permission from well owners/operators. The authors also thank Bill Lindberg with the City of Albuquerque Water Utility Division, who provided access to chemical data for City of Albuquerque water-supply wells. Particular thanks go to the many individuals in the Reston, Virginia office of the USGS who performed most of the sampling and analysis that made this study possible. These people include Eurybiades Busenberg, Jerry Casile, Mike Doughten, Ward Sanford, Andrew Stack, Julian Wayland, and Peggy Widman. Finally, the study benefited greatly from the advice of Scott Anderholm and Blair Jones of the USGS, Fred Phillips and Rob Bowman of New Mexico Tech, and Nelia Dunbar of the New Mexico Bureau of Mines and Mineral Resources. The manuscript was improved by technical reviews from Scott Anderholm and Alan Welch of the USGS.

3. HYDROGEOLOGY OF THE MRGB

The MRGB extends between Cochiti Lake and San Acacia in the Basin and Range physiographic province of central New Mexico (Fig. 1). The basin, which contains basin-fill deposits up to about 14,000 feet thick (Thorn et al., 1993), is among a series of several alluvial basins located in the Rio Grande rift typified by high heat flow, young faulting, and recent volcanism (Lozinsky, 1988). The boundaries of the basin have been defined by the extent of Cenozoic deposits. The basin is surrounded partly by mountain ranges (Fig. 1); lower topographic relief occurs along the west side of the basin, which is bounded by the Lucero and Nacimiento uplifts and the Rio Puerco fault zone. Within the basin, the piedmont slopes from the eastern mountain fronts toward the main drainage, the Rio Grande, which is inset in a terraced valley with a flood plain up to about 4.5 miles wide. Most land in the MRGB is rangeland (Thorn et al., 1993), although agriculture is practiced in the Rio Grande flood plain, where depth to water generally is less than about 25 feet (Bexfield and Anderholm, 1997). Urban areas include the City of Albuquerque, which is the most populous city in New Mexico. The climate of the MRGB generally is categorized as semiarid, although the climate in surrounding mountainous areas ranges to humid continental (Thorn et al., 1993).

3.1 Geology

The primary aquifer underlying the MRGB consists of the generally unconsolidated to moderately consolidated basin-fill sediments of the Santa Fe Group. The overall aquifer system includes both the Santa Fe Group deposits, which are of Oligocene to middle Pleistocene age, and the more recent (post-Santa Fe Group) flood-plain, channel, and basin-fill deposits of Pleistocene to Holocene age that are up to about 130 feet thick and are in hydraulic connection with the Santa Fe Group deposits (Thorn et al., 1993). Hawley and Haase (1992) divided the deposits of the Santa Fe Group into three main hydrostratigraphic units representing various depositional environments, lithologic characteristics, and periods of deposition. Most ground-water samples for this study were collected from wells completed in the relatively permeable sediments of the upper Santa Fe Group (up to 1,500 feet thick), which consists primarily of piedmont-slope and fluvial basin-floor deposits, or in post-Santa Fe Group sediments. Several of the sampled wells were completed in the less permeable Middle Santa Fe Group.

Hawley and Haase (1992) found that the bulk composition of basin fill in the Albuquerque area was approximately 60 percent granitic-metamorphic detritus of Precambrian derivation, 30 percent volcanic detritus of middle Tertiary derivation, and less than 10 percent sedimentary detritus of Paleozoic or Mesozoic derivation. Below the northeast part of Albuquerque, sediments at depths from about 200 to 3,200 feet were described as volcanic-rich; the volcanic material was thought to be derived from the Jemez Mountains (a major center of young, silicic volcanism) and other sources farther to the north. Clay minerals present in fine-grained deposits of the basin were smectite, illite, kaolinite, and interlayered illite/smectite.

As defined, the MRGB consists of the Santo Domingo, Calabacillas, and Belen Subbasins (Grauch et al., 1999). Recent studies indicate the presence of structural highs between these subbasins, where the thickness of Santa Fe Group basin fill is small (less than 3,000 feet) compared to the thickness within the subbasins (up to 12,000 feet) (Grauch et al., 1999; Grauch et al., 2001). Numerous faults of widely varying offset bound and extend through parts of the MRGB with a general north-south strike. The potential of some of these faults to facilitate upward flow of deep water into relatively shallow parts of the aquifer has not been well characterized.

3.2 Hydrology

The Rio Grande extends the entire length of the MRGB (Fig. 1). A system of canals and drains in the flood plain carry irrigation water from the river to agricultural fields and intercept seepage from the river and irrigated fields. The mean annual discharge of the Rio Grande at Albuquerque was about 1,450 cubic feet per second for water years[1] 1974-98 (Ortiz et al., 1999). Although the Rio Grande is the only perennial stream in the MRGB, at least ten ephemeral streams can contribute substantial flow to the Rio Grande, and possibly contribute substantial quantities of recharge to the underlying aquifer.

The ground-water-flow system of the MRGB is quite complex, largely because of multiple sources of recharge to the basin, which many investigators have attempted to identify and quantify (Kernodle et al., 1995). These sources include mountain-front recharge, subsurface ground-water inflow from adjacent basins, and seepage from the Rio Grande and ephemeral streams. Ground water discharges from the MRGB to the Socorro Basin near San Acacia (Fig. 1). Ground-water discharge also occurs within the MRGB through evapotranspiration (particularly in the Rio Grande flood

[1] The water year is the 12-month period from October 1 through September 30. The water year is designated by the calendar year in which it ends and which includes 9 of the 12 months.

plain), ground-water pumpage, and discharge of ground water into drains and some reaches of the Rio Grande.

Predevelopment water-level maps (Bexfield and Anderholm, 2000; Titus, 1961) show that ground-water flow through the central part of the basin historically has been oriented primarily north to south, whereas flow near the basin margins historically has been oriented toward the central part of the basin (Fig. 1). In the vicinity of Albuquerque, a steady increase in ground-water pumping since about the mid-1940's has resulted in substantial declines in water levels and alterations in patterns of ground-water flow.

4. HYDROCHEMICAL FRAMEWORK OF THE MRGB

Chemical and isotopic characterization of ground water in the MRGB permits delineation of 13 separate water-quality zones (Fig. 2) (Plummer et al., in press). Generally, each zone has uniform characteristics of major-element chemistry that change little as water moves through the basin. These zones, which have been grouped by general location in the basin in Table 1, represent 12 source areas of ground-water recharge to the basin and one area of ground-water discharge. Two water-quality zones receive recharge from the northwest (Fig. 2 and Table 1). The compositions of most ground-water samples from the Northwestern zone indicate that the primary source of water is recharge from relatively low elevations along the western base of the Jemez Mountains (which consist largely of Tertiary volcanic rocks), although a few samples indicate small, localized areas of mixing with infiltration from the Jemez River. Ground water of the West-Central zone appears to extend most of the length of the MRGB and is present at depth under the Northwestern zone and probably parts of adjacent zones to the east. The compositions of most ground-water samples from the West-Central zone indicate that the water probably recharged in the Jemez Mountains. The light stable isotopes and old ages of ground water from the West-Central zone indicate that recharge to the zone occurs farther north (at higher elevations that likely exceed 8,000 feet) than recharge to the overlying Northwestern zone.

Three distinct water-quality zones are present along the western margin of the MRGB (Fig. 2 and Table 1). Ground water of the Western Boundary zone probably is a mixture of Na-Cl brine leaking into the basin from Paleozoic rocks (typically limestone, sandstone, and shale) along the western margin and infiltrating precipitation/arroyo flow within the basin. Ground water of the Rio Puerco zone likely is a mixture of water from the Western Boundary zone with surface water that infiltrates through the Rio Puerco

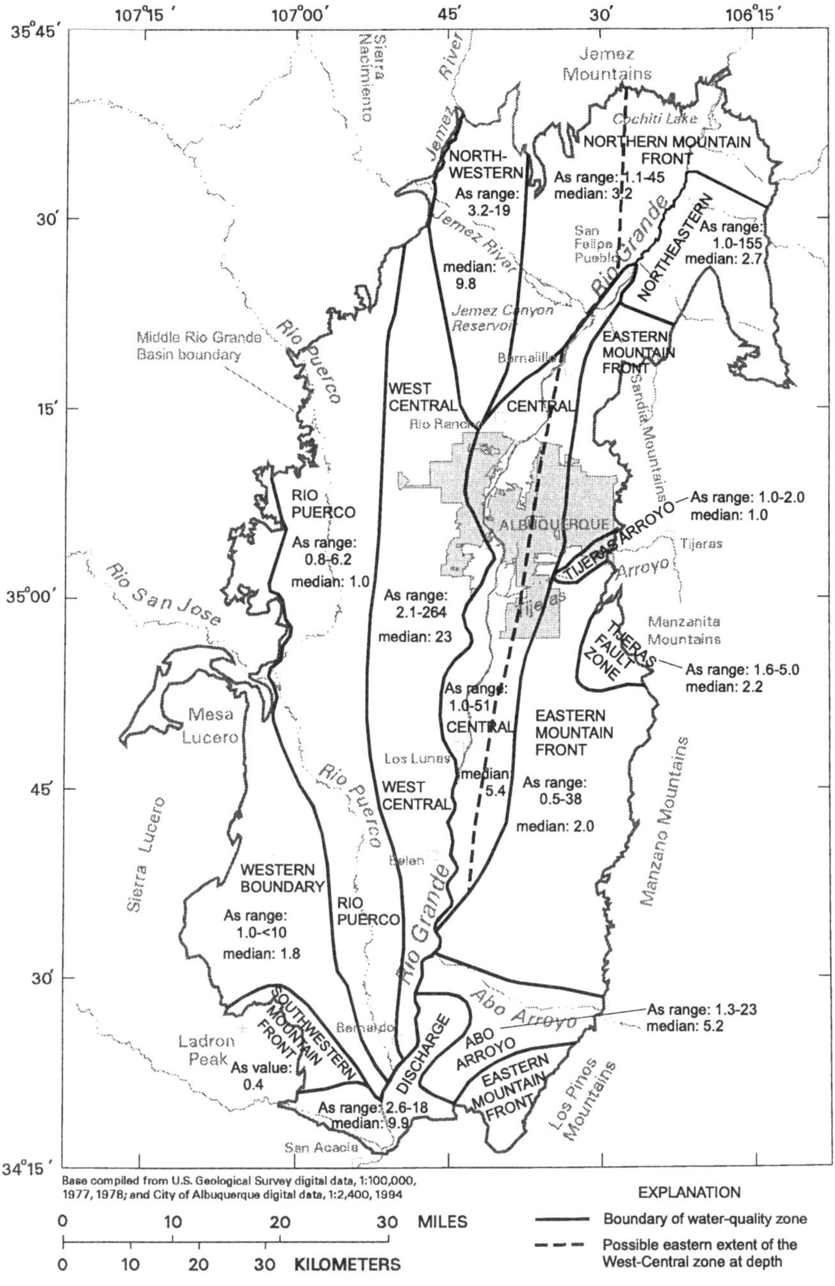

Figure 2. Water-quality zones defined for the Middle Rio Grande Basin (Plummer et al., 2001). Arsenic values are given in micrograms per liter.

Table 1. Selected characteristics and median constituent concentrations of water-quality zones in the Middle Rio Grande Basin [--, no data; µS/cm, microsiemens per centimeter at 25 degrees Celsius; mg/L, milligrams per liter; µg/L, micrograms per liter]

General basin area	Water-quality zone	Primary sources of recharge	Qualitative arsenic concentrations	Primary arsenic sources/ controls
Northwest	North-western	Low-elevation Jemez mountain front	Uniformly high	Geothermally-altered rocks in recharge area, desorption
Northwest	West Central	High-elevation Jemez Mountains	Uniformly high	Geothermally-altered rocks in recharge area, desorption
West	Western Boundary	Ground-water inflow from Paleozoic rocks	Uniformly low	Not relevant
West	Rio Puerco	Arroyo infiltration, ground-water inflow	Uniformly low	Not relevant
West	South-western Mountain Front	Ladron mountain front	Uniformly low	Not relevant
East and Central	Northern Mountain Front	Jemez mountain front	Variable	Upwelling of mineralized water, local desorption
East and Central	Abo Arroyo	Arroyo infiltration, mountain front	Variable	Upwelling of mineralized water, local desorption
East and Central	Eastern Mountain Front	Eastern mountain front	Variable	Upwelling of mineralized water, local desorption
East and Central	North-eastern	Mountain front, ground-water inflow, arroyo infiltration	Variable	Upwelling of mineralized water, local desorption
East and Central	Central	Rio Grande infiltration	Variable	Upwelling of mineralized water, local desorption
Tijeras	Tijeras Fault Zone	Mountain front, ground-water inflow	Uniformly low	Not relevant
Tijeras	Tijeras Arroyo	Arroyo infiltration, mountain front	Uniformly low	Not relevant
Discharge	Discharge	Ground water from upgradient zones	Uniformly high	Old, mineralized water

Table 1 (cont'd).

General basin area	Water-quality zone	Specific conductance (µS/cm)	pH, field (standard units)	Water temperature (deg. C)	Dissolved oxygen (mg/L)	Ca (mg/L)	Mg (mg/L)	Na (mg/L)	K (mg/L)	Alkalinity (mg/L as HCO_3)
Northwest	North-western	400	7.8	20.6	6.7	33.9	4.2	49.9	5.7	160
	West Central	535	8.2	23.8	3.0	12.0	2.5	103.	4.2	174
West	Western Boundary	4,572	7.7	22.0	4.1	135.	56.4	589.	15.2	300
	Rio Puerco	2,731	7.5	20.0	3.7	135.	42.7	290.	10.4	190
	South-western Mountain Front	462	8.1	19.1	4.4	52.6	13.5	27.8	2.5	202
East and Central	Northern Mountain Front	340	7.5	18.9	5.1	38.5	6.1	20.0	4.9	137
	Abo Arroyo	1,055	7.5	20.7	6.2	92.5	34.4	49.2	3.1	148
	Eastern Mountain Front	387	7.7	22.0	5.2	45.0	5.1	29.2	2.2	158
	North-eastern	1,221	7.5	19.4	6.4	141.	29.5	81.8	4.8	208
	Central	436	7.7	18.1	0.1	42.9	8.0	31.0	6.4	158
Tijeras	Tijeras Fault Zone	1,406	7.4	18.5	4.7	171.	36.0	95.0	6.1	599
	Tijeras Arroyo	677	7.4	16.1	7.0	89.4	24.5	29.3	3.8	240
Discharge	Discharge	1,771	7.7	20.6	0.1	93.0	31.0	190.	10.5	157

Table 1 (cont'd).

General basin area	Water-quality zone	SO_4 (mg/L)	Cl (mg/L)	F (mg/L)	SiO_2 (mg/L as SiO_2)	NO_3 (mg/L as N)	As (µg/L)	B (mg/L)	Fe (mg/L)	Li (mg/L)	Mn (mg/L)
Northwest	North-western	44.8	8.5	0.61	30.1	2.4	9.8	0.12	0.03	0.07	0.002
	West Central	92.0	13.4	0.99	34.5	1.2	23.2	0.24	0.03	0.05	0.002
West	Western Boundary	793.	820.	1.64	22.5	0.86	1.8	0.90	0.21	0.25	0.041
	Rio Puerco	1,080.	186.	0.63	21.8	0.88	1.0	0.29	0.13	0.25	0.015
	South-western Mountain Front	53.0	15.0	1.02	17.6	1.1	0.4	0.09	0.03	0.04	0.007
East and Central	Northern Mountain Front	19.5	5.6	0.35	53.3	0.56	3.2	0.04	0.06	0.06	0.005
	Abo Arroyo	346.	25.9	0.90	24.0	1.4	5.2	0.13	0.11	0.03	0.004
	Eastern Mountain Front	31.0	10.5	0.60	28.4	0.31	2.0	0.05	0.03	0.02	0.003
	North-eastern	390.	22.7	0.51	38.5	0.64	2.7	0.22	0.17	0.04	0.004
	Central	66.0	16.6	0.44	47.0	0.08	5.4	0.09	0.04	0.04	0.015
Tijeras	Tijeras Fault Zone	100.	139.	1.27	18.9	1.1	2.2	0.35	0.11	0.23	0.023
	Tijeras Arroyo	115.	56.6	0.60	19.5	3.8	1.0	0.06	0.05	0.02	0.005
Discharge	Discharge	290.	280.	1.40	39.0	0.42	9.9	0.63	0.08	0.33	0.010

Table 1 (cont'd).

General basin area	Water-quality zone	Mo (µg/L)	Sr (µg/L)	U (µg/L)	V (µg/L)	Zn (µg/L)	δD (per mil)	^{14}C (percent modern carbon)
Northwest	North-western	3.4	0.57	2.7	15.6	9	-65	29.6
	West Central	8.2	0.20	3.7	27.9	5	-97	8.8
West	Western Boundary	9.9	2.09	4.4	5.7	118	-64	6.2
	Rio Puerco	7.0	3.92	6.0	3.4	117	-63	36.4
	South-western Mountain Front	3.0	0.86	0.9	1.0	252	--	--
East and Central	Northern Mountain Front	1.7	0.31	1.0	6.4	258	-78	33.4
	Abo Arroyo	3.4	1.48	5.4	9.5	8	-65	24.1
	Eastern Mountain Front	2.0	0.32	3.6	7.5	7	-81	51.4
	North-eastern	6.7	1.72	8.5	3.8	100	-69	28.5
	Central	5.0	0.40	3.6	9.3	5	-95	61.0
Tijeras	Tijeras Fault Zone	3.7	1.11	7.3	6.3	62	-74	9.7
	Tijeras Arroyo	1.9	0.47	3.7	3.0	4	-75	72.8
Discharge	Discharge	10.3	3.02	3.9	7.1	16	-91	10.8

and/or ground water that seeps into the basin from Mesozoic rocks (typically Cretaceous sandstone and shale with local Jurassic gypsum and clastic units) along the northwestern boundary. Ground water of the Southwestern Mountain Front zone, delineated on the basis of only two samples, has much lower constituent concentrations than water from the Western Boundary and Rio Puerco zones and probably represents mountain-front recharge along the Ladron Mountains, which consist mainly of Precambrian granitic and metamorphic rocks.

Of the water-quality zones in the east and central areas of the basin, five are relatively large and receive recharge from either basin margins or surface water (Fig. 2 and Table 1). The major-element compositions of most samples from the Northern Mountain Front zone indicate that the primary source of recharge is mountain-front recharge along the eastern part of the Jemez Mountains. The major-constituent compositions of most ground-water samples from the Eastern Mountain Front zone are consistent with a mountain-front recharge source along the Sandia, Manzanita, and Manzano Mountains, which consist primarily of Precambrian metamorphic and igneous rocks that are overlain by Paleozoic limestone and sandstone (Anderholm, 1988; Hawley and Haase, 1992). Water compositions from much of the Abo Arroyo zone appear to be consistent with a primary recharge source being infiltration along the arroyo, which drains mostly Paleozoic sedimentary rocks and some crystalline Precambrian rocks (Anderholm, 2001). Ground water of the Northeastern zone appears to be derived from more than one primary source, including mountain-front recharge water, arroyo infiltration, and/or ground-water inflow from gypsum-containing rocks outside the basin. Finally, ground water of the Central zone, which extends parallel to the Rio Grande along much of the length of the basin, has a composition (including isotopic composition) generally consistent with a primary source from the Rio Grande.

Two additional water-quality zones (Tijeras Fault Zone and Tijeras Arroyo) that receive ground-water recharge along the eastern boundary of the basin are limited in areal extent (Fig. 2 and Table 1). Both the Tijeras Fault Zone and Tijeras Arroyo zones receive dilute, mountain-front recharge. However, ground water of the Tijeras Fault Zone zone appears also to contain a fraction of high-Cl water, such as from deep fracture systems in the Precambrian granitic rocks and Pennsylvanian limestone of the area, whereas ground water of the Tijeras Arroyo zone appears to contain a fraction of infiltration through Tijeras Arroyo, which drains primarily Paleozoic and Precambrian rocks.

The Discharge zone (Fig. 2 and Table 1) is located at the southern end of the MRGB. Ground water in this zone typically is relatively old and probably contains fractions of both ground water observed in adjacent water-quality zones and deep, high-Cl ground water that is moving upward as a result of the decrease in thickness and width of basin-fill deposits at the southern end of the basin.

5. ARSENIC IN SURFACE WATER

The Rio Grande and several ephemeral streams contribute recharge to the Santa Fe Group aquifer system. Therefore, data on the As concentrations in these streams (Table 2) can be used to characterize their potential to contribute As to the aquifer system. Although recharge from the Jemez River (Fig. 1) is very localized, this river is of particular interest with respect to As concentrations because it is fed partially by ground-water discharge in its upper reaches, including discharge from geothermal springs known to have As concentrations exceeding 1 mg/L (Shevenell et al., 1987; Trainer, 1974) and it flows into the Rio Grande. For the Jemez River below Jemez Canyon Dam (Fig. 1, at the downstream end of the Jemez Canyon Reservoir), the discharge-weighted average As concentration determined from NWIS data is 12 μg/L; the median As concentration for 12 samples collected as part of the MRGB study is 24 μg/L (Plummer et al., in press). Arsenic concentrations in the Rio Grande, which is believed to be the primary source of recharge to the Central zone, typically are lower at sites located above the inflow of the Jemez River than at sites located south of the inflow from the river. For the Rio Grande at San Felipe (above the Jemez River inflow) (Fig. 1), the discharge-weighted average As concentration from NWIS is 1.7 μg/L; for the Rio Grande at Albuquerque, the discharge-weighted average As concentration is 2.9 μg/L. Samples collected for the MRGB study at a site on the Rio Grande at the north end of Albuquerque have a median As concentration of 3.2 μg/L.

Some As data are available for other streams in the MRGB, although the data sets are relatively small (Table 2). NWIS and MRGB samples for sites on the Rio Puerco, Abo Arroyo, and Tijeras Arroyo indicate that As concentrations typically are about 2 μg/L or less. However, samples collected for the MRGB study show that As concentrations in the Rio Puerco can range up to 12 μg/L under certain conditions.

Table 2. Typical dissolved-arsenic concentrations in surface water at selected sites in the Middle Rio Grande Basin [μg/L, micrograms per liter; <, less than]

Selected statistics from USGS NWIS data (average concentrations are discharge-weighted)					
Surface-water site	Period of record	Number of samples	Minimum arsenic concentration (μg/L)	Average arsenic concentration (μg/L)	Maximum arsenic concentration (μg/L)
Jemez River below Jemez Canyon Dam	1974-96	23	8.0	12	59
Rio Grande at Albuquerque	1981-95	13	2.0	2.9	4.0
Rio Grande at San Felipe	1975-98	44	<1	1.7	4.0
Rio Puerco at Bernardo	1994-98	12	<1	1.1	2.0
Tijeras Arroyo above Four Hills Road	1989-91	12	<1	1.0	1.0

Selected statistics from the MRGB study (not discharge-weighted)					
Surface-water site	Period of record	Number of samples	Minimum arsenic concentration (μg/L)	Median arsenic concentration (μg/L)	Maximum arsenic concentration (μg/L)
Abo Arroyo at Hwy 60	1997	2	1.5	1.8	2.1
Jemez River below Jemez Canyon Dam	1998-99	12	21	24	28
Rio Grande at Alameda Blvd.	1997-99	22	1.9	3.2	5.3
Rio Puerco at Hwy 6	1998-99	9	1.1	5.5	12
Tijeras Arroyo at Four Hills Road	1997-99	22	0.5	0.9	1.8

6. ARSENIC CONCENTRATIONS IN GROUND WATER

6.1 Areal variations

Concentrations of As in ground water sampled in the MRGB range from less than 1 μg/L to more than 600 μg/L. The median As concentration for all ground-water samples compiled for this investigation within the MRGB is 5.3 μg/L. This median value probably is not representative of ground water throughout all locations and depths of the entire MRGB because the density of sample sites was larger in the Albuquerque area than across the rest of the basin, and most samples were obtained from wells completed in about the upper 600 feet of the aquifer. However, because most of the wells sampled were completed in parts of the aquifer used for drinking-water supplies, the median value of 5.3 μg/L probably is representative of the median As concentration in that part of the ground-water resource that is currently used.

Arsenic concentrations tend to be higher in the northwestern and central parts of the MRGB than along most basin margins (Fig. 3). Arsenic concentrations are consistently low (90th percentile of less than 10 μg/L) in zones grouped into the west and Tijeras areas of Table 1, which all have median As concentrations of 2.2 μg/L or less. Zones of the east and central areas include both high and low As concentrations; As concentrations typically range from about 1 μg/L near basin margins to more than 20 μg/L. Arsenic concentrations consistently are high (10th percentile of greater than 3 μg/L) in the northwest-area zones and the Discharge zone. These three zones all have median As concentrations greater than 9.8 μg/L; the West-Central zone has the highest median As concentration (23 μg/L) of any zone. No obvious or consistent trends in the areal patterns in As concentration (either increasing or decreasing) are seen with ground-water flow direction in the basin.

6.2 Variations with depth

For illustrative purposes, graphs of variability in As concentrations with depth (Fig. 4) were divided into two groups based on piezometer location. The amount of variability observed in As concentrations with depth ranges widely, from about 3 μg/L across 1,200 feet of aquifer to about 120 μg/L across 550 feet of aquifer (Fig. 4). Arsenic concentrations consistently increase by at least 5 micrograms per liter with depth in 11 piezometer nests (numbers 2, 4, 5, 7, 13, and 15-20). Arsenic concentrations show little

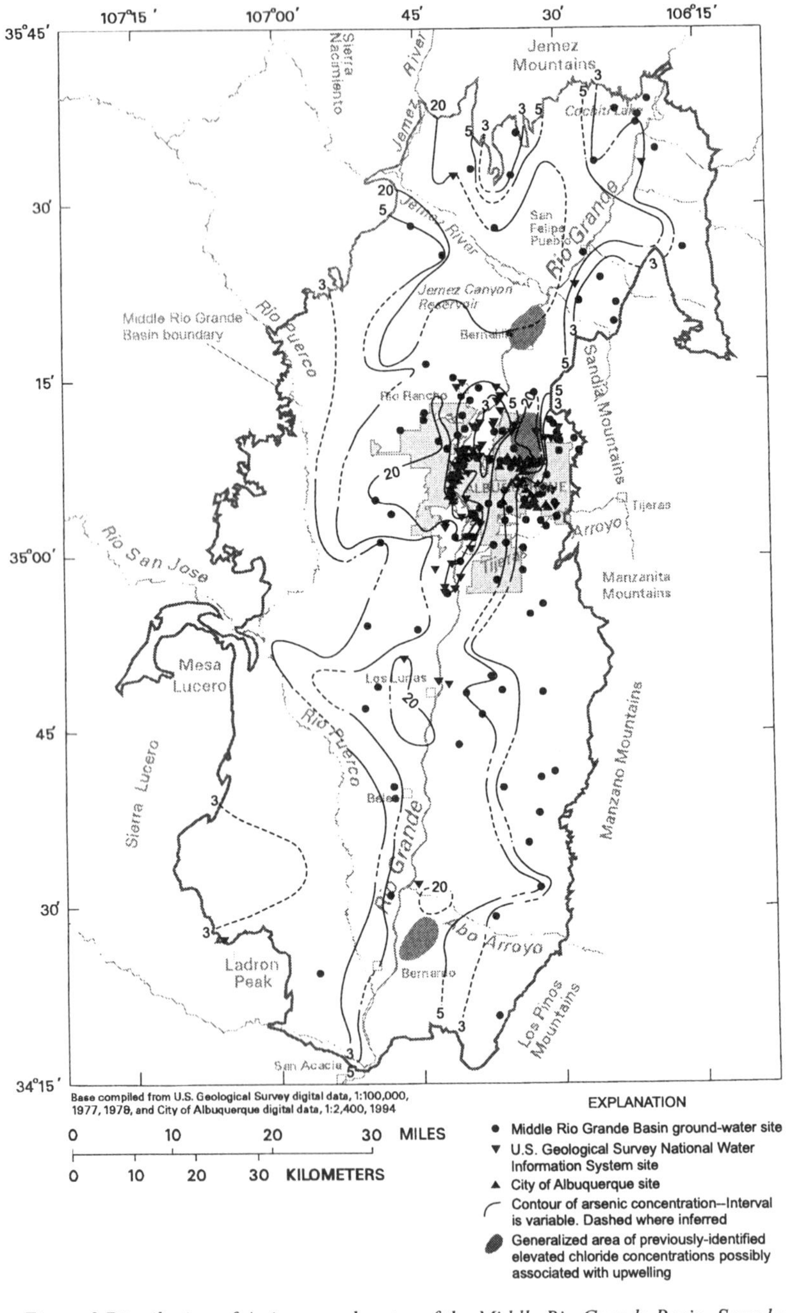

Figure 3.Distribution of As in ground water of the Middle Rio Grande Basin. Samples shown have As concentration less than or equal to 10 micrograms per liter and dissolved-solids concentration estimated as less than or equal to about 500 milligrams per liter.

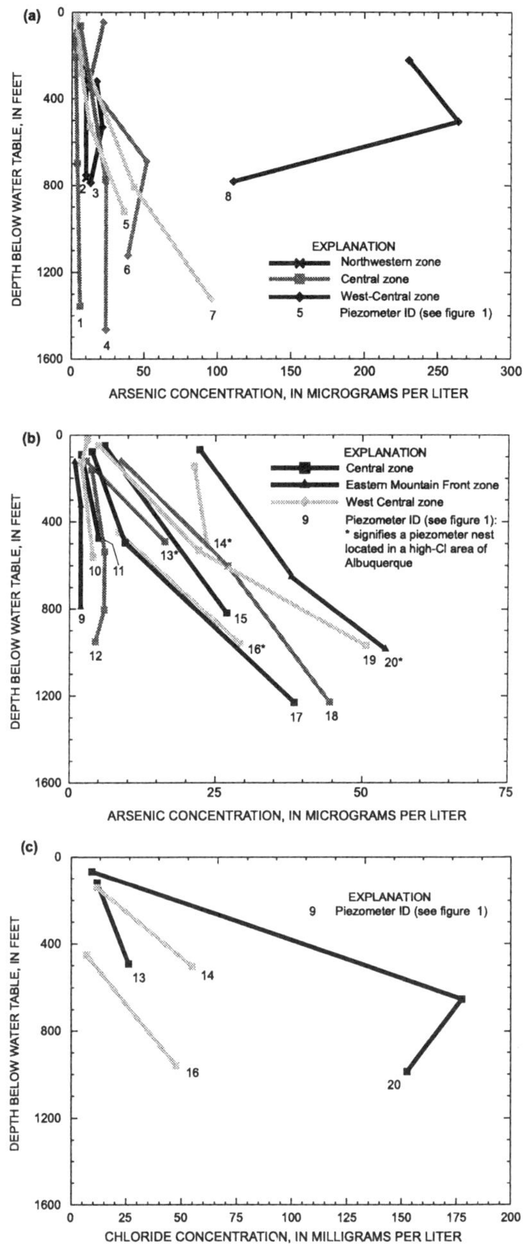

Figure 4. Variation in arsenic concentrations with depth in piezometer nests located (a) west of and (b) east of the Rio Grande, and (c) variation in chloride concentrations in piezometer nests located in high-Cl areas of Albuquerque.

change with depth in six piezometer nests (numbers 1, 9, 10, 11, 12, and 14), which produce mostly water of the Central or Eastern Mountain Front zone. In piezometer nests numbered 3, 6, and 8, completed entirely in water of the West-Central zone, the As concentration of the deepest piezometer is substantially smaller than the As concentration in at least one shallower piezometer. Only six piezometer nests produced ground water with an As concentration of 10 µg/L or less from all completions.

7. POSSIBLE FACTORS AFFECTING ARSENIC CONCENTRATIONS IN GROUND WATER

Because As concentrations greater than 10 µg/L are present across broad areas of the MRGB, which consists primarily of rangeland, natural processes rather than anthropogenic activities are most likely to affect As occurrence. Large-scale processes that can affect As concentrations in ground water include mineral dissolution and precipitation, adsorption and desorption processes, and the presence of water associated with thermal sources. Characteristics that are useful in determining the most important processes in selected water-quality zones of the MRGB are summarized in Tables 1 and 3.

7.1 Mineral dissolution

Various previous studies have suggested that the dissolution of As-bearing sulfide minerals or Fe oxides can be the primary source of As in ground water. For example, sulfide minerals have been proposed as the likely source of As to ground water in parts of the northeastern United States (Welch et al., 2000). Oxidation of As-containing pyrite as a result of increasingly oxidizing conditions over time has been proposed as a source of As to ground water in Wisconsin (Burkel and Stoll, 1999). Reductive dissolution of Fe oxides has been proposed as an important source of As to ground water in Bangladesh and West Bengal (Chowdhury et al., 1999; Nickson et al., 2000), and has been shown to release As from metal contaminated soils (Masscheleyn et al., 1991).

In the MRGB, the dissolution of As-bearing iron oxides appears unlikely to be a major source of dissolved As to ground water. The presence of dissolved oxygen and nitrate indicates oxidizing conditions throughout the aquifer at the depths sampled, except at relatively shallow depths in the Rio Grande inner valley (Fig. 1). The predominance of oxidizing conditions across large areas of the aquifer is supported by data for City of Albuquerque drinking-water supply wells (Bill Lindberg, City of Albuquerque, written

Table 3. Selected statistics for water-quality zones with elevated arsenic concentrations in the Middle Rio Grande Basin [fbls, feet below land surface; ns, correlation not statistically significant at the 0.05 level]

Water-quality zone:		Northern Mountain Front[1]	North-western	West Central[1]	Eastern Mountain Front[1]	Central[1]	
Number of samples with As data:		14	13	66	73	206	
Range in depth of wells, in fbls:		56-1,487	240-1,618	60-1,920	42-1,800	12.5-2,020	
Pearson correlation coefficients for selected constituents with As: [Coefficients are given for only those constituents where the p-value for correlation was less than 0.05]						Central east:	Central west:
	SC	0.69	ns	ns	0.24	ns	-0.26
	pH	ns	ns	0.46	ns	ns	0.29
	Temp	0.62	ns	ns	0.57	0.53	0.39
	DO	ns	ns	ns	-0.47	ns	ns
	Ca	ns	ns	-0.48	ns	-0.65	-0.61
	Mg	ns	ns	-0.55	ns	-0.28	-0.55
	Na	0.71	ns	ns	0.55	0.63	0.40
	K	0.62	ns	-0.42	0.50	0.31	ns
	SO_4	0.73	ns	ns	ns	ns	-0.29
	Cl	0.69	ns	ns	0.53	0.25	ns
	F	ns	ns	ns	ns	0.68	0.59
	SiO_2	ns	ns	ns	0.37	0.56	0.47
	NO_3	0.62	ns	ns	-0.35	ns	ns
	B	0.70	ns	0.39	0.58	0.65	0.39
	Fe	ns	ns	ns	ns	ns	-0.22
	Li	0.67	ns	ns	0.46	0.59	ns
	Mn	ns	ns	ns	ns	ns	-0.28
	Mo	ns	ns	-0.39	0.58	0.59	ns
	Sr	0.69	ns	-0.48	0.31	-0.29	-0.54
	V	ns	0.86	0.43	0.55	ns	0.58
	^{14}C	-0.60	ns	ns	-0.66	-0.72	-0.56

[1]*Pearson correlation coefficents calculated using logarithms of constituent concentrations*

commun., 2000), which show that more than 90 percent of the As present in ground water from 76 of 87 wells was in the form of As (V). Therefore, although Fe oxides are known to be present in sediments of the Santa Fe Group aquifer system (Stanton et al., 2001a; Stanton et al., 2001b), the reducing conditions that would favor Fe (III) reduction and dissolution of ferric oxide minerals generally are not present. Dissolution of sulfide minerals also is not a likely source of As because sulfide minerals are not common in sediments of the Santa Fe Group aquifer system (Hawley and Haase, 1992; Stanton et al., 1998; Stanton et al., 2001a; Stanton et al., 2001b). Furthermore, sulfur isotope data (Plummer et al., in press), indicate that most SO_4 present in ground water of the MRGB is derived from dissolution of evaporite deposits present along some basin margins rather than from dissolution of sulfide minerals. Also, Stanton et al. (2001a; Stanton et al., 2001b) found that 10 percent or less of As present in a core obtained from the western part of Albuquerque was associated with the sulfide/organic fraction of the core. Pearson correlation coefficients (also commonly known as r values) calculated for water-quality zones with substantial concentrations of As do not show the strong positive correlations between As and SO_4 and negative correlations between As and pH that would be expected if As concentrations were increasing as a result of sulfide dissolution (Table 3).

7.2 Adsorption processes

Arsenate ions ($HAsO_4^{2-}$ in the pH range of about 6.8 to 11.6 and $H_2AsO_4^-$ in the pH range of about 2.2 to 6.8), which likely are the predominant form of As in ground water of the MRGB, can be strongly sorbed onto metal oxides, particularly oxides of Fe and Al. Previous investigations have shown that arsenate ions are less strongly associated with metal oxides at pH values higher than about 7.0 to 8.0 than at lower pH values, probably because of competition from hydroxide ions for sorption sites (Boyle et al., 1998; Smith et al., 1998; Welch et al., 1988; Welch et al., 2000). A study by Robertson (1989) indicated that adsorption of arsenate ions on smectite or ferric oxyhydroxide was the major control on As in ground water of at least three alluvial basins in Arizona. The variation in pH values across the MRGB, which range from 6.4 to 9.8, could contribute to variation in the concentration of As in ground water of the basin. Stanton et al. (2001a) and Stanton et al. (2001b) showed that the most likely source of soluble As from a core obtained from the western part of Albuquerque is the "anion-exchangeable" fraction associated with clay and secondary Fe oxide surfaces that could be mobilized by ground water with high pH and/or low Eh values.

The West-Central zone (Table 3), where the core studied by Stanton et al. (2001a) and Stanton et al. (2001b) was obtained is the only water-quality zone with a statistically significant correlation between As and pH at the 0.05 level and a Pearson correlation coefficient greater than 0.3. The correlation coefficient for the West-Central zone is 0.46, and a graph of As versus pH (Fig. 5) for the zone shows that nearly all samples with As concentrations less than 20 μg/L have pH values of less than 8.5, whereas most samples with As concentrations exceeding 20 μg/L have pH values of 8.5 or greater. The observation that the largest As concentrations in the West-Central zone generally appear to be associated with pH values greater than about 8.5 may explain the lack of a strong overall relation between As and pH in other water-quality zones, where pH values generally do not exceed 8.5. It is nevertheless possible that desorption increases As concentrations in localized areas of elevated pH in other zones where "anion-exchangeable" As is available. Although adsorption/desorption of As associated with metal oxides appears likely to regulate dissolved-As concentrations in the West-Central zone, this process does not provide a full explanation for the source of As to the ground water and sediments of the zone.

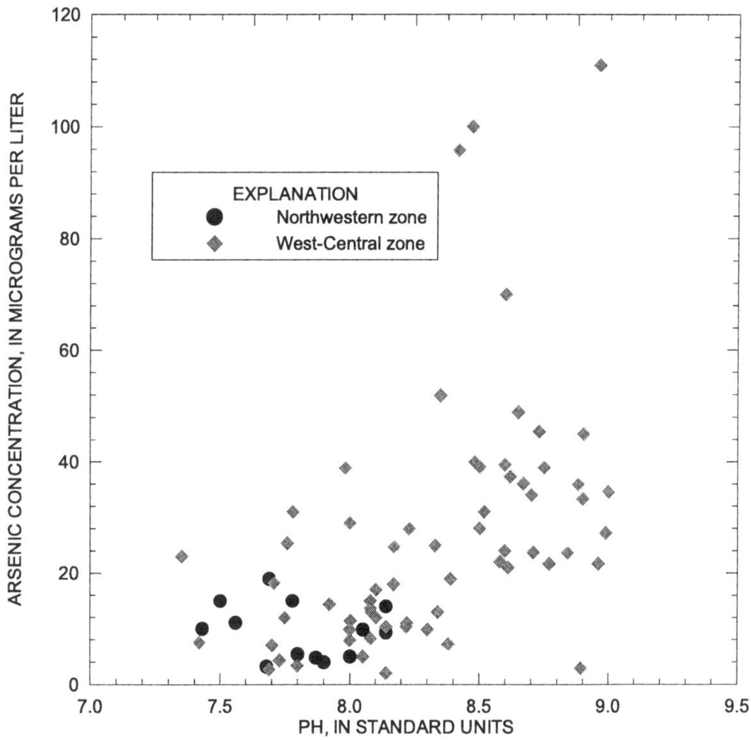

Figure 5. Relation of arsenic and pH in ground-water samples of the Northwestern and West-Central zones.

7.3 Water source

Median values and ranges of As concentration vary substantially among water-quality zones (Figs. 2 and 3), which indicates that water sources to the basin could be a primary factor affecting As concentrations. An association of As with source water would be consistent with observations of previous investigators indicating that thermal water and water from areas of intense evaporation commonly have high As concentrations (Welch et al., 2000). Ground water in zones grouped in the west and Tijeras areas of Table 1 has consistently small As concentrations. Therefore, in these zones, the water source is apparently low in As concentration; in particular, the generally low As concentrations in surface water of the Rio Puerco and Tijeras Arroyo are consistent with the generally low As concentrations in zones that receive recharge from these sources.

Ground water of zones in the east and central areas (Table 1) includes both high and low As concentrations. Within these zones, concentrations of As near basin margins, where much of the recharge occurs, are low (generally less than 2 μg/L), indicating that mountain-front recharge probably does not contain high As concentrations. Surface water in the Rio Grande and Abo Arroyo, which are important local sources of ground-water recharge, have low As concentrations (typically less than 4 μg/L). Therefore, surface-water recharge also does not appear to contribute to elevated As concentrations in ground water of the east and central areas. Previous investigators (Anderholm, 1988; Bexfield and Anderholm, 2002; Trainer et al., 2000) have proposed that deep (thousands of feet) and presumably old mineralized water with high Cl concentrations upwells in particular areas to mix with shallower ground water (Fig. 3). In some parts of the east and central areas, high As concentrations in ground water have been shown to coincide closely with these areas of elevated Cl concentration (Bexfield and Anderholm, 2002; Trainer et al., 2000). Upwelling may occur as leakage along faults or as the result of the movement of ground water over structural highs between subbasins, as indicated by the clustering of high Cl and As concentrations along these features (Fig. 6). Mixing with deep water having high Cl and As concentrations could explain the occurrence of elevated As concentrations in zones where the primary source water (surface water and/or mountain-front recharge) appears to have low As concentrations. Unfortunately, the exact origin and composition of deep ground water with elevated Cl and As (Table 4) is not clear. The observed decrease of both As and Cl concentrations with distance downgradient of these affected areas would be consistent with declining As concentrations as a result of dilution rather than adsorption.

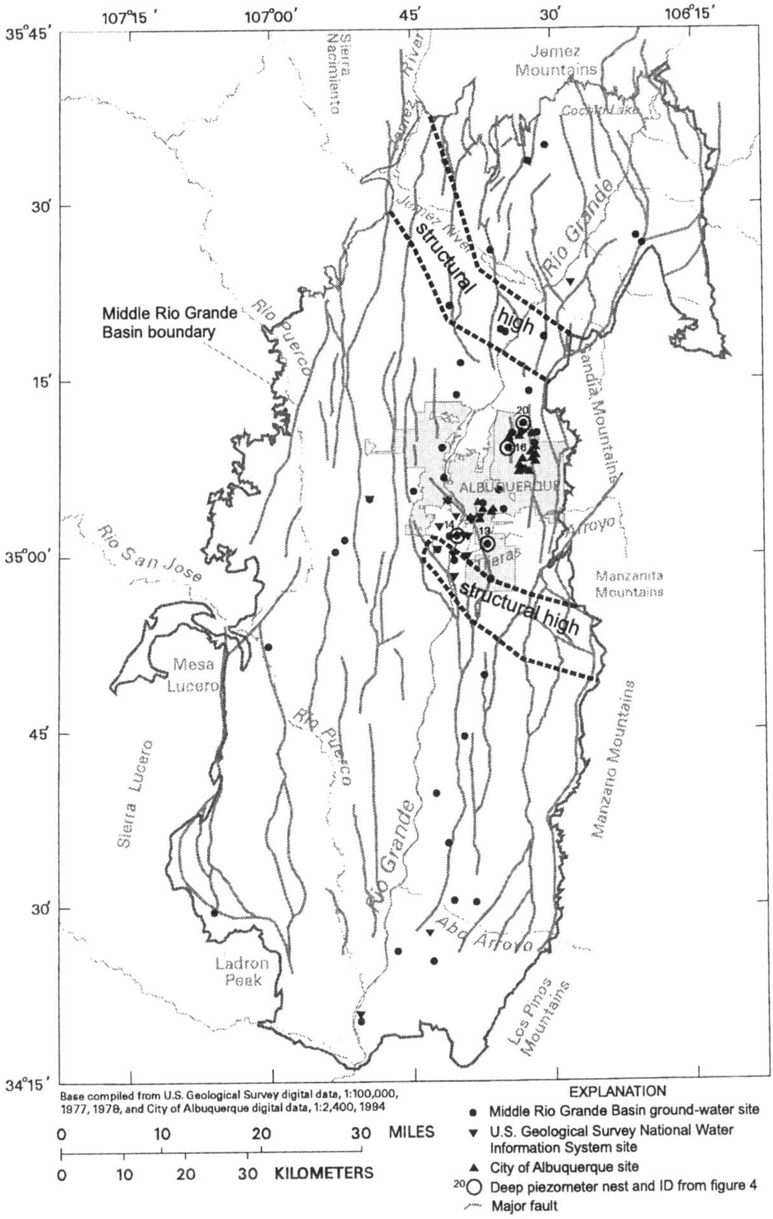

Figure 6. Locations of samples having water-quality characteristics indicative of mixing with mineralized water (generally chloride concentration greater than 10 milligrams per liter and arsenic concentration greater than 5 micrograms per liter), in relation to major structural features. (Major faults from Mark Hudson and Scott Minor, U.S. Geological Survey, written commun., 1999; structural highs generalized from Grauch et al. (1999)).

Table 4. Chemical characteristics of selected mineralized waters in the Middle Rio Grande Basin [ID, identifier; --, no data; µS/cm, microsiemens per centimeter at 25 degrees Celsius; deg. C, degrees Celsius; mg/L, milligrams per liter; µg/L, micrograms per liter; <, less than]

Sample ID on fig. 1	Sample type	Specific conductance (µS/ cm)	pH, field (standard units)	Water temperature (deg. C)	Dissolved oxygen (mg/L)	Ca (mg/L)	Mg (mg/L)	Na (mg/L)	K (mg/L)	Alkalinity (mg/L as HCO_3)	SO_4 (mg/L)	Cl (mg/L)	F (mg/L)	SiO_2 (mg/L as SiO_2)
A	mineralized	11,700	6.2	25.4	0.82	561.	108.	2,190.	135.	862	2,190.	2,680.	0.73	22.2
B	mineralized	8,200	7.0	27.0	0.08	185.	25.7	1,530.	107.	182	296.	2,520.	6.40	43.4
C	thermal/ mineralized	1,075	7.0	53.8	--	97.8	33.4	63.1	8.6	175	247.	95.6	0.43	43.0
none[1]	thermal/ mineralized	6,540	6.6	40.6	2.04	353.	32.5	834.	181.	1,538	42.1	1,500.	3.14	40.0

Sample ID on fig. 1	Sample type	NO_3 (mg/L as N)	As (µg/L)	B (mg/L)	Fe (mg/L)	Li (mg/L)	Mn (mg/L)	Mo (µg/L)	Sr (mg/L)	U (µg/L)	V (µg/L)	Zn (µg/L)	Deuterium content (per mil)	^{14}C (percent modern carbon)
A	mineralized	0.01	610.	1.65	2.74	1.70	1.01	20.5	13.0	5.5	36.4	76.1	-79.8	1.6
B	mineralized	<0.01	33.0	5.00	0.24	5.70	0.39	<1.0	5.77	0.1	17.4	<5.0	-82.7	29.3
C	thermal/ mineralized	0.04	51.9	0.14	0.31	0.15	0.06	60.9	1.26	0.1	0.6	586.	-59.7	13.6
none[1]	thermal/ mineralized	<0.02	1,500.	14.2	0.35	14.2	0.60	1.0	1.90	1.1	9.7	50.0	-85.9	--

[1] *Geothermal sample from Jemez Mountains used for mixing line in figure 9*

Areas of high As concentration appear to coincide closely with areas of high Cl concentration in the Northern Mountain Front, Eastern Mountain Front, and Central zones (Fig. 6). Statistical tests for correlation (Table 3) appear to support the conclusion that most of the higher As concentrations in these areas are associated with older, more mineralized water. Concentrations of As in the Northern Mountain Front zone are strongly associated with concentrations of SO_4, Na, B, Cl (Fig. 7), and Li, and with specific conductance and temperature. Anderholm (1988) and Trainer et al. (2000) describe high-Cl water near the southern end of this zone as conduit flow that may originate as geothermal fluid in the Valles Caldera of the Jemez Mountains. Chloride, Na, As, B, and Li are all found at high concentrations (greater than 1500 milligrams per kilogram for Cl and Na and 2.7 milligrams per kilogram for As, B, and Li) in geothermal water of the caldera (Goff et al., 1988). Similarly, in the Eastern Mountain Front zone, As is strongly correlated with carbon-14 (negative coefficient) and Mo, B, temperature, Na, V, Cl, and K (all positive coefficients). The inverse relation of As with carbon-14 in percent modern carbon (Fig. 8) indicates that As is positively correlated with ground-water age. In the eastern part of the Central zone (approximately east of the dashed line of Fig. 2), where nearly all ground-water samples with As concentrations greater than 10 μg/L are coincident with areas of high Cl concentrations, As is not strongly correlated with Cl concentrations or specific conductance values, but is strongly negatively correlated (-0.72) with carbon-14 content (Fig. 8). Other constituents with correlation coefficients greater than or equal to 0.60 with As concentration in the area are Ca (negative coefficient) and F, B, Na, Li, and Mo (all positive coefficients), which would be consistent with a source of deep, old water with elevated concentrations of Na and minor constituents.

Elevated As concentrations in western parts of the Central zone also may be associated with high-Cl upwelling (Figs. 3 and 6), but investigation of this possibility is complicated by the likely presence of As-rich ground water (median concentration of 23 μg/L) of the West-Central zone at depth below the western part of the Central zone. Analysis of samples from piezometer nests in the western part of the Central zone show a shift toward a more typical West-Central zone composition (higher pH, lower concentrations of Ca, Mg, and Sr, and higher concentrations of Na, V, F, and As; Table 1) with depth. Therefore, in western parts of the Central zone where low Cl concentrations indicate little or no effect from deep, mineralized water, elevated As concentrations probably are the result of mixing between the Central-zone water originating at the Rio Grande and the deeper West-Central zone water. Arsenic in this area shows strong negative correlations

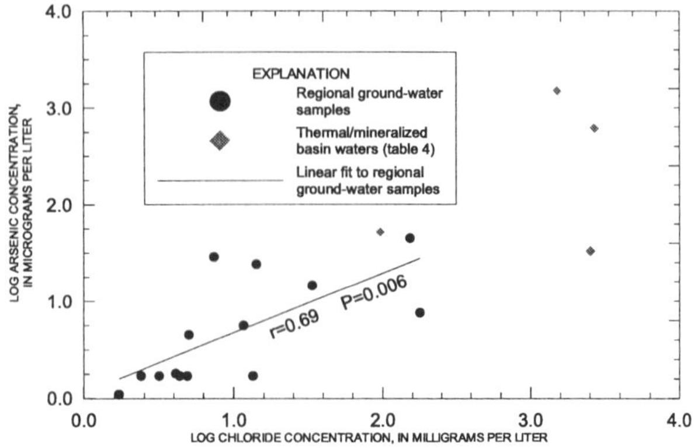

Figure 7. Relation of arsenic and chloride in ground-water samples of the Northern Mountain Front zone.

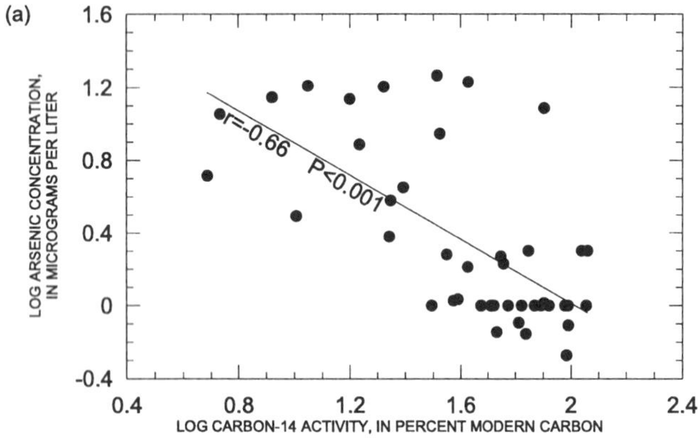

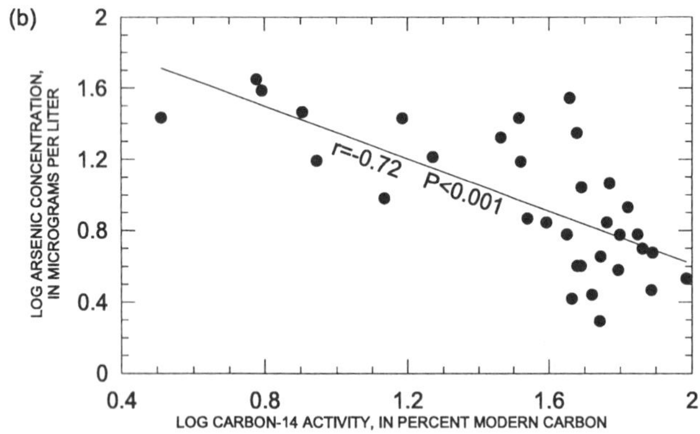

Figure 8. Relation of arsenic and carbon-14 activity in ground-water samples of the (a) Eastern Mountain Front zone and (b) eastern part of the Central zone.

with carbon-14 content and concentrations of Ca, Mg, and Sr and positive correlations with concentrations of V and F (Table 3), which appears consistent with this conclusion.

Additional observations appear to lend support to the conclusion that elevated As concentrations in parts of the east and central areas are associated with the upwelling of deep, mineralized water. Chloride, As, Na, and B concentrations tend to increase with depth in piezometer nests located in high-Cl areas of Albuquerque (Figs. 4 and 6). Also, As(III) constitutes about 10 to 50 percent of As in oxic ground water in the high-Cl northeast area of Albuquerque (Bill Lindberg, City of Albuquerque, written commun., 2000), suggesting a deep As source. Water from a deeper, more reducing zone, where As (III) predominates, apparently mixes with more shallow, oxic water. At the far southern end of the Eastern Mountain Front zone (Fig. 2), one ground-water sample with a temperature of 53.8 °C and a sample slightly downgradient with a specific conductance of 8,200 μS/cm (with As concentrations of 52 and 33 μg/L, respectively) suggest that deep, thermal water is upwelling in this area. Finally, the well that produces water with the highest As concentration (155 μg/L) in the Northeastern zone is near an intersection of major faults and has a relatively high Cl concentration (66.1 mg/L), indicating that upward movement of mineralized water along faults of the area could be the primary As source.

In the Abo Arroyo zone, the two ground-water samples with As concentrations greater than 3 μg/L have relatively high pH values and SiO_2 and K concentrations relative to the other two samples; Cl concentrations for these two samples are the lowest. It appears possible, although not certain, that a silicate reaction may have increased pH enough to allow some As to desorb from aquifer materials.

In contrast to zones with extensive areas of small As concentrations, nearly all ground-water samples from the northwest area (Table 1) have As concentrations greater than or equal to 5 μg/L. As discussed earlier, adsorption/desorption processes likely affect the As concentrations in ground water of the West-Central zone and could particularly increase As concentrations in areas where pH exceeds 8.5. However, the process does not provide a full explanation for the ultimate source of As to ground water and basin-fill deposits of the northwest area, particularly near the basin margins. Elevated As concentrations near basin margins indicate that the source water to these zones has relatively high As concentrations. The presence of silicic volcanic rocks and high-As geothermal waters in the Jemez Mountains, believed to be the primary recharge area to these zones, is consistent with this observation. However, the occurrence of elevated As in ground water near basin margins in these zones contrasts with the apparent lack of As in ground water near the recharge area of the Northern Mountain Front zone, in the eastern Jemez Mountains (Fig. 3). The near-surface geology of the eastern and western parts of the Jemez Mountains generally is quite similar, consisting mainly of Tertiary and Quaternary volcanic rocks. Therefore, some other feature must differ between these areas in order for As concentrations to be higher in recharge water from one area compared to the other.

The distribution of geothermal water from the Valles Caldera potentially could account for the difference in As concentrations observed near recharge areas for the Northern Mountain Front zone as opposed to the northwest-area zones. Previous investigations (Goff et al., 1988; Trainer, 1975, 1984; Trainer et al., 2000) have shown that mineralized water from the caldera probably flows primarily south and west, where it contributes to thermal springs, and probably does not leak from the caldera to the east. Mixing of mountain-front recharge water with small quantities of this geothermal water, typically with As concentrations higher than 1 mg/L, could increase As concentrations in the non-thermal recharge water.

A plot of As relative to Cl was used to examine the possibility that the mixing of geothermal water with mountain-front recharge water results in the As concentrations observed in ground water of the northwest-area zones (Fig. 9 and Table 4). Ground-water samples from these zones (Northwestern and West-Central) generally do not fall along the line representing conservative mixing between dilute mountain-front recharge water of modern age and water derived from Soda Dam Spring, a thermal spring located southwest of the Valles Caldera. Water from Soda Dam Spring has been shown to have an As-Cl ratio very similar to that found in geothermal water inside the Valles Caldera (Goff et al., 1988). This plot supports the conclusion that As concentrations in the northwest zones are controlled primarily by a process (or processes) other than mixing with geothermal water from the caldera.

Flow of mountain-front recharge water through rocks altered by contact with geothermal water is a possible source of As and other trace elements to ground water near the recharge area that appears consistent with the observed chemistry. Dunbar and Chapin (1994) state that local hydrothermal systems often lead to intense alteration of rocks and sediments, particularly those of volcanic origin; contact with geothermal water typically enriches the rocks in trace elements but not Cl (Nelia Dunbar, New Mexico Bureau of Mines and Mineral Resources, oral commun., 2000). Hydrothermally altered rocks have been observed both inside and south of the Valles Caldera (Charles et al., 1986; Goff and Gallaher, 2001). Therefore, local recharge water in the western part of the Jemez Mountains may acquire As by desorption or dissolution from volcanic rocks and sediments that have had

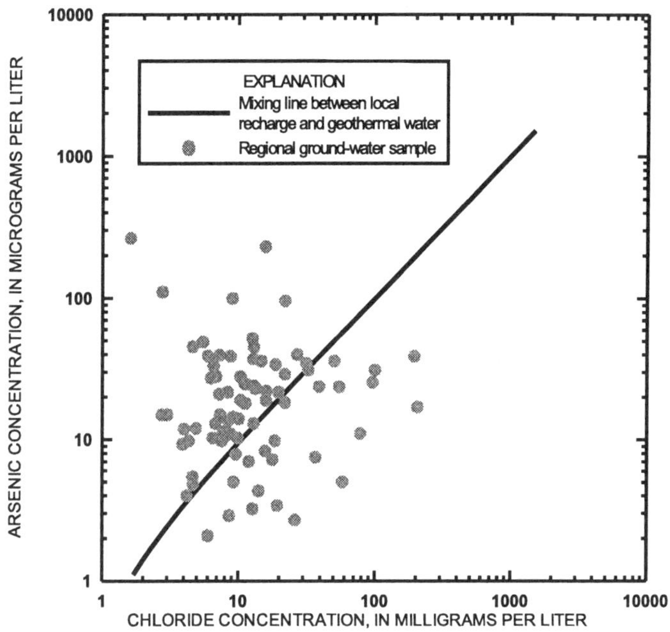

Figure 9. Relation of arsenic and chloride in ground-water samples of the Northwestern and West-Central zones and for representative local recharge and geothermal water.

contact with geothermal water, and then transport the As into the MRGB. Subsequently, adsorption/desorption processes involving Fe oxides in the basin-fill deposits affect the concentration at any particular location along the flow path.

Similar to zones of the northwest area, ground water of the Discharge zone has As concentrations, in general, greater than or equal to 5 µg/L. The relatively high As concentrations in the Discharge zone are not surprising given that this area represents the drain for the ground-water system for the entire basin. Ground water in this zone typically is quite old with high specific conductance values, indicating long flow paths and travel times. Interestingly, the lowest As concentrations are found in ground-water samples with the highest specific conductance values and Cl concentrations. These samples probably represent ground water with a substantial fraction of water from the Western Boundary zone, where the source of Cl and SO_4 is dissolution of Permian sedimentary rocks (Plummer et al., 2001; Plummer et al., in press).

8. IMPLICATIONS FOR DRINKING-WATER SUPPLIES

This investigation indicates that the quantity of potable ground water (total dissolved solids less than about 500 mg/L) in the Middle Rio Grande Basin that will meet the USEPA drinking-water standard of 10 μg/L for As is limited. Most wells known to produce water meeting these criteria are located in the vicinity of Albuquerque between the Rio Grande inner valley and the eastern mountain front and south of Albuquerque near the eastern mountain front (Fig. 3). Other clusters of sampled wells that meet the criteria are in the far northern part of the basin and at the northern end of the Sandia Mountains. However, even within these broad areas of generally low As concentrations, samples from some wells have As concentrations higher than 10 μg/L. Within Albuquerque in particular, many wells east of the Rio Grande that are located in areas of apparent mixing with deep, mineralized water or with water from the West-Central water-quality zone have As concentrations higher than 10 μg/L. In addition to Albuquerque, the municipalities of Rio Rancho, which is located largely over water of the West-Central and Northwestern zones, and Bernalillo, which is located largely over an area affected by the presence of mineralized water that may originate in the Valles Caldera, anticipate difficulty meeting the new USEPA standard for As. Arsenic concentrations exceed 10 μg/L across large areas near both of these municipalities (Fig. 3).

Although most of the wells sampled for this investigation were completed in the upper 1,750 feet of the Santa Fe Group aquifer, indications from deep piezometer nests are that As concentrations typically increase with depth. Also, unpublished data from two oil wells near the center of the basin at depths of 2,600 to 6,600 feet below land surface indicate As concentrations exceeding 50 to 100 μg/L (Scott Anderholm, USGS, written commun., 2001). Therefore, it is doubtful that wells completed at greater depths would yield water with lower As concentrations for drinking-water supply in those municipalities that anticipate difficulty meeting the new USEPA standard.

9. SUMMARY

Drinking-water supplies for residents of the Middle Rio Grande Basin, and particularly the City of Albuquerque, will be affected severely by the new U.S. Environmental Protection Agency (USEPA) of 10 μg/L for As in drinking water (Federal Register, 2001). Previous investigations and recent data collected at 288 sites as part of the USGS MRGB study indicate that As concentrations exceed 10 μg/L in ground water in the Santa Fe Group aquifer system (the almost exclusive source of drinking-water supply) across broad areas of the basin. Because the multiple sources of recharge to the aquifer system differ substantially in chemical composition, the basin was divided into 13 separate water-quality zones for analysis of As occurrence and sources.

This investigation of the distribution of As in ground water demonstrates that there are two main categories of high-As source water for the MRGB. One source is mineralized water of deep origin. Geologic structure appears to control the geographic distribution of this mineralized water, which upwells along faults or as the result of ground-water movement over structural highs. Mixing with such water elevates the As concentrations of ground water at shallower depths of the aquifer in parts of the central and east areas of the basin. The exact composition of this deep, mineralized water is not necessarily consistent throughout the basin, and its exact origin is unknown, although the water may be of thermal origin in some areas. The other high-As source affects ground water in the northwest part of the basin and is associated with silicic volcanism in the Jemez Mountains to the north. Upon entering the basin, the generally oxic, high-pH recharge water from the western portion of the Jemez Mountains contains substantial dissolved As that does not appear to have resulted primarily from mixing with geothermal fluids. Recharge of water through rocks that have been altered by contact with geothermal fluids could account for the occurrence of high As in ground water that originates in the western part of the Jemez Mountains, in contrast to low As in ground water that originates in the eastern part; more research is necessary to support this conclusion. Ground water that has not been affected by either of these two high-As sources, including water throughout most of the west and east areas of the basin, generally has low As concentrations.

Adsorption/desorption processes also appear to affect As concentrations along flow paths in the West-Central zone of the northwest part of the basin, where the highest As concentrations typically occur for samples with pH values of 8.5 or higher. Increasing As concentrations at higher pH values is consistent with data from a core studied by Stanton et al. (2001a and b), which indicate that "anion-exchangeable" As is present on Fe oxides in the area.

Chapter 12

Arsenic Contamination in the Water Supply of Milltown, Montana

Johnnie N. Moore and William W. Woessner
Department of Geology, University of Montana, Missoula, MT 59812

In 1981, four community supply wells in Milltown, Montana were found to contain arsenic levels ranging from 220 to 550 μg/l. The wells produced water from an unconfined aquifer composed of coarse sand, gravel and boulders, underlain by bedrock, adjacent to a small hydroelectric reservoir. Historical assessment showed that the contamination was long term. The reservoir sediments, that originated partially from large scale mining and milling operations up stream, were identified as the source of aquifer contamination. Reservoir water flowing through contaminated sediment produces arsenic-rich pore water that migrates vertically into the underlying highly conductive pre-dam alluvial floodplain sediments. This sulphur-limited, anoxic environment results in ground water arsenic concentrations up to ca. 1200 μg/l at the base of the fine-grained reservoir sediments. Arsenic is released at a rate of 1.6 MT/y from a study section covering 255,350 m^2 of the reservoir. The groundwater flowing north from beneath this section, and into the adjacent aquifer, contains arsenic at a median concentration of 740 μg/l and discharges arsenic at approximately 24.6 MT/y. Pre-dam floodplain groundwater originating from beneath other portions of the reservoir is suggested as one source of the additional arsenic flux. Ground water transport into the adjacent Milltown area results in a reduction of arsenic concentrations with distance. Based on relationships with chloride and sodium, mechanical dispersion does not account for the observed reduction in arsenic concentrations. Analyses of Fe(II)/Fe(Total), As (III)/As(Total), As/Fe and As/sulphate suggest co-precipitation/precipitation of arsenic with metal sulfides and/or co-precipitation/adsorption with iron oxyhydroxides dominate arsenic reduction during transport. Estimates of available arsenic within the reservoir sediments suggest only 15 to 25 y of discharge would remain if no additional arsenic was added. However, arsenic concentrations are not declining in wells with long term records and arsenic bound in contaminated sediment from upstream is delivered at an average of 590 to

740 MT annually. Current USEPA remediation alternatives range from leaving the existing system in place to total dam and reservoir sediment removal. However, it is not clear how well the consequences of such actions are understood.

1. DISCOVERY OF ARSENIC CONTAMINATION AND POTENTIAL SOURCES

The small community of Milltown, Montana, is located at the confluence of the Blackfoot and Clark Fork Rivers, about 10 km upstream and east of Missoula, Montana (population of 57,059). The Clark Fork River Valley trends E-W and cuts into Precambrian sedimentary (Belt Series) rocks of the Sapphire Mountains to the south and the Garnet Range to the north (Fig. 1). The valley is filled with 8 to 55 m of Quaternary–Recent alluvium (Woessner, 1995), an aquifer that supplies drinking water to the residents of Milltown. During newly-required sampling of community wells in November, 1981, the Montana Department of Health and Environmental Sciences (MT-DHES) determined that four community water wells serving 33 residences in Milltown contained levels of arsenic ranging from 220 to 510 µg/l. These values were above the USEPA MCL for arsenic of 50 µg/l. MT-DHES advised the residents not to drink or cook with water from wells in Milltown and bottled water was supplied to residents by Champion International, the landowner.

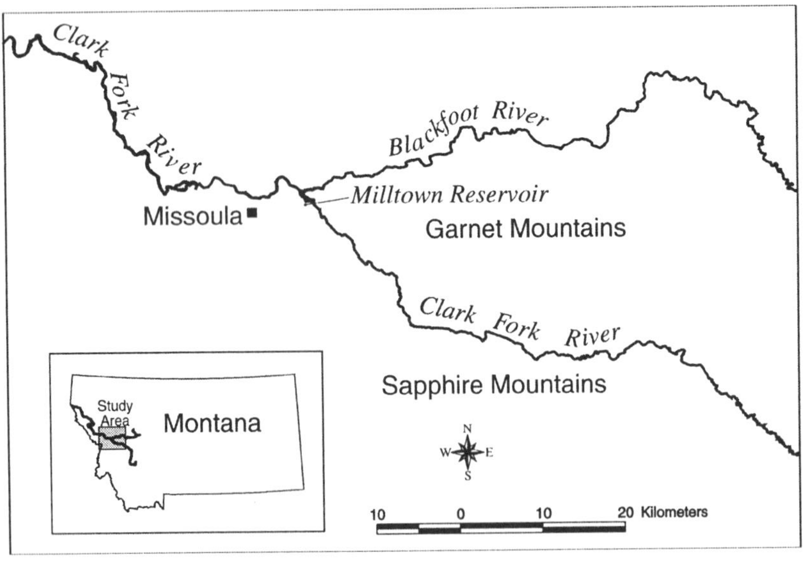

Figure 1: Location map.

In early 1982, we began a reconnaissance–level hydrogeologic study to identify potential sources of arsenic (Woessner and Popoff, 1982). Prior to the commencement of the study, local residents were interviewed to assess the timing of the contamination and possible sources. The occurrence of poor–quality water having a metallic taste, and red and black staining of porcelain fixtures appeared to be isolated to houses only in the southern part of Milltown (between the railroad tracks and I-90, Fig. 2). All residents identified the problem as long term. They were also eager to suggest a number of possible sources for contaminants. The Champion International plywood mill (now Stimpson Lumber) north of the effected area was immediately accused by some residents. In 1975, ground water adjacent to and just south of the plant site was contaminated with tannins and lignins when log treatment ponds overflowed. This resulted in abandoning a number of wells and the replacement of water supplies with well C-2 (Fig. 2).

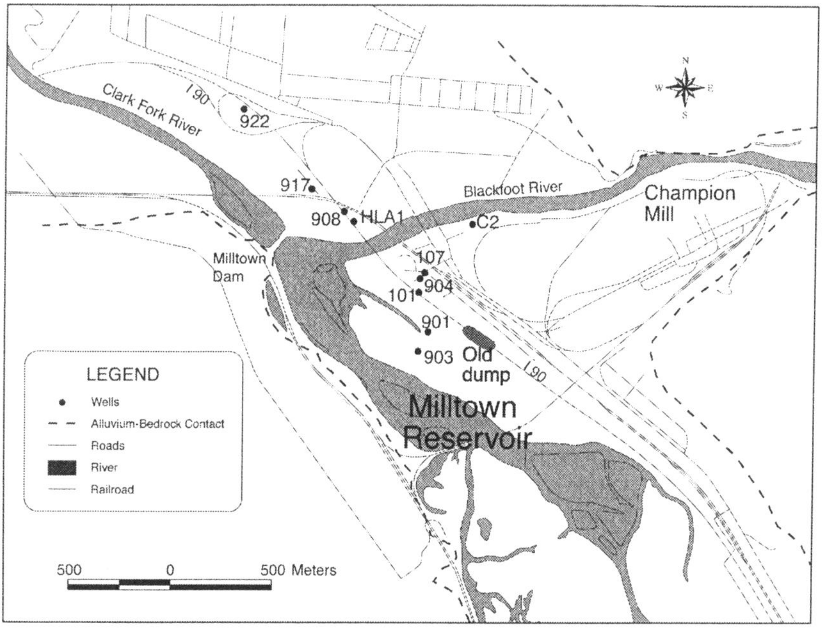

Figure 2: Detailed map of Milltown showing some of the features and wells mentioned in the text. Interstate 90 is the represented by the line labeled I90 passing near wells 101 and 908. The railroad runsbetween wells 917 and 908 and then SE paralleling I90.

The residents also noted that prior to Champion's purchase of the property, in the early 1970's, Anaconda Mining Company operated a sawmill on the property, producing timbers for the mines at Butte (a large mining operation upstream—see below). It was suspected that arsenic may have been used at the mill as a wood preservative. Additionally, residents were concerned about an old dump, which was used for mill wastes and household garbage, located southeast of the effected area on the bank of Milltown Reservoir, which is adjacent to the town.

Historical assessment showed that the problem was long term. No products with high-arsenic concentrations were currently being used at the plywood mill and log soak pond liquids did not contain arsenic. Anaconda Mining Company did use the site prior to Champion International, however, no records were available concerning use of arsenic. What was left of the dump wastes had been pushed out onto the reservoir sediments in the 1960's when I-90 was being constructed. Sediments beneath the dump had been used as borrow to construct a road grade to the bridge crossing the Blackfoot River arm of the reservoir. Woessner and Popoff (Woessner and Popoff, 1982) also investigated the occurrence of spills of arsenic–containing compounds resulting from a highway or rail accidents in Milltown and found none.

A few of the residents also wondered what was in the small reservoir to the south that dammed the Clark Fork River. The river had received mining and milling wastes for over 100 years from the Butte and Anaconda areas 190 to 240 km upstream. In 1975, Bailey, as part of her Master of Science at the University of Montana, sampled reservoir sediments (Bailey, 1976) and found elevated concentrations of copper, lead and zinc (Bailey and Weisel, 1976). However, she did not analyze the sediment for arsenic.

The preliminary 1982 hydrogeologic study found that ground water in the Milltown area originated from three likely areas: 1) the Blackfoot alluvial valley flowing beneath the mill site; 2) the Clark Fork River alluvium; 3) the reservoir (Woessner and Popoff, 1982). Although no wells existed between the old dump site/reservoir and Milltown, the potential for flow from the reservoir area was evident because the reservoir stage was 5 m higher than water levels in contaminated wells.

Because we thought it likely that arsenic accompanied the metal contamination found by Bailey in the reservoir sediment (Bailey and Weisel, 1976), we also took four grab samples of sediment through the ice on Milltown reservoir in the winter of 1982. Analyses of these samples, completed in the summer of 1982, revealed that reservoir sediment contained somewhat elevated concentrations of arsenic and lead: arsenic, 54-135 ppm; lead, 5-95 ppm. The sediment also contained very high levels of zinc, up to 2,880 ppm. Using these limited values we calculated the potential volume of contaminated sediment residing in the reservoir, about 4.4 million tons. This was sufficient to have Milltown Reservoir ranked as a CERCLA "Superfund" Site—#198 out of the initial 400 sites designated by the USEPA. Subsequently, MT-DHES and USEPA asked us to identify the source of arsenic and locate a new, arsenic–free water supply well for the community of Milltown.

We began that work in July of 1983 and successfully completed it by June 1984, identifying the reservoir sediment as the source of alluvial aquifer ground water contamination. Water quality sampling in 1983 revealed the highest arsenic concentrations yet recorded at the four community wells, in the 100 to 1000 µg/l range. However, background arsenic concentrations of less that 5 µg/l (the detection limit) were found immediately north of the contaminated wells. Sand points located in the reservoir sediment showed ground water arsenic concentrations between 500 and 10000 µg/l. In the portion of the ground water system in which arsenic exceeded background, manganese and iron usually exceeded drinking water standards. At the completion of this initial work, a new community water well was proposed and later constructed north of the ground water impacted by arsenic, iron and manganese, near community well, C2 (Woessner et al., 1984). Since completion of our work in 1984, numerous additional site evaluations have occurred as part of the CERCLA process (ENSR, 1991, 1992; HLA, 1986, 1987; Hydrometrics, 1985; Titan, 1994).

In this chapter, we review the processes and pathways releasing and transporting arsenic to the Milltown wells from the reservoir sediments. This interpretation is based on our and our graduate student's research and a variety of other sources on Milltown written over the last nineteen years.

2. RESERVOIR SEDIMENT SOURCE AND CHARACTER

Milltown Dam was built in 1907 and impounds the Blackfoot and Clark Rivers at their confluence at Milltown, Montana (Fig. 2). The rock and crib dam is underlain by bedrock (under its south half) and alluvium (north half). Milltown Dam is run as a hydroelectric facility by Montana Power Company (now PPL), who has owned the dam since 1929. The low head and minimal storage capacity (residence time of only 10 hours, Reynolds, 2001) of the reservoir severely limits its capability as a generation facility. It presently generates only about 1.4 KW of electricity, although upgrades to 4 KW have been proposed. The reservoir behind the dam is small, covering only 245 ha, and is nearly filled with fine–grained sediment transported mostly by the Clark Fork River and to a lesser extent by the Blackfoot River. It is these sediments that contain elevated concentrations of arsenic and metals.

About 200 km upstream from Milltown Reservoir, large–scale mining and processing of copper ores over the last ca. 100 years at Butte and Anaconda released over 100 million metric tons (MT) of wastes into the headwaters (Silver Bow and Warm Springs creeks) of the Clark Fork River (Moore and Luoma, 1990). Other tributaries of the Clark Fork River above Milltown also contain a large number of abandoned mines and some processing facilities that have added smaller quantities of wastes to those originating from the Butte/Anaconda region (Helgen and Moore, 1996). Milltown Dam is the first dam on the Clark Fork River downstream from the main inputs of waste from Anaconda and Butte (excluding tailings ponds). The sediment filling the dam is a mixture of "naturally-derived" sediment from uncontaminated tributaries and mining waste from upstream sources. The sediment package wedges out from approximately 8 meters thick near the dam face to 0.5 meters about 3 km upstream from the dam.

The stratigraphy of the reservoir sediment is complex. Inter-layered channel, floodplain and slough deposits create a complex mosaic of sediment types and grain sizes with a vertical intermixing of sandy and muddy deposits (ENSR, 1992; Woessner et al., 1984). All cores from the reservoir show discontinuous, interlayers of sand, mud and clay with isolated concentrations of organic material. The coarsest sediment is near the river channels and in the uppermost part of the reservoir. Finer-grained sediment is found in areas near the dam and farther from the active channel. The reservoir stage is over 5 m higher than the ground water found in the adjacent Milltown alluvial aquifer. This results in movement of ground water through the fine-grained sediments and into the underlying alluvium.

The fine-grained sediments filling the reservoir overlay the original fluvial plain sediments of the Clark Fork River. These now form an alluvial aquifer beneath the reservoir sediments. The fluvial sediments are composed of sand, gravel, cobbles and boulders. There is little known about their stratigraphic framework other than its origin as high-energy fluvial and catastrophic glacial-flood deposits (Alden, 1953). Based on borings to bedrock between lanes of the interstate and areas north of the reservoir, and interpretations of bedrock depth from geophysical studies by Titan (1994) and Schombel (1994), floodplain alluvium was estimated to be from 6 to 10 m thick, increasing from the dam to the southeast (Woessner, 1995). Ground water in these sediments appears to flow to the north towards Milltown, though only limited head data are available in these sediments.

The alluvial sediments adjacent to and surrounding the north side of the reservoir are dominated by sand, gravel and boulders. This material is connected with the pre-dam floodplain beneath the reservoir sediments and is of the same high-energy fluvial origin. The sediment package thickens from the reservoir to the north extending to depths of over 55 m. It is underlain by fractured bedrock that yields little ground water. Previous studies suggest that zones of hydraulic conductivity exceeding 18,000 m/d are most likely present in the alluvial aquifer (Woessner et al., 1984). Ground water occurs from 9 to

18 m below land surface and flow is generally from the reservoir area to the north-northwest.

The fine-grained sediments in the reservoir are highly elevated over background concentrations for a variety of elements. Arsenic values range from lows of ca. <10 to 1600 ppm, with mean values ca. 200-300 ppm (Titan, 1994; Woessner et al., 1984). Copper and zinc are much higher, with high values up to ca. 11,000 ppm and 14,000 ppm, respectively; sulfide sulfur ranges from ca. 750 to 6000 ppm (Titan, 1994; Woessner et al., 1984). Different workers have calculated the total amount of contaminated sediment in the reservoir. Estimates range from ca. 3.4 X 10^6 m^3 (Woessner et al., 1984), 5.0 X 10^6 m^3 (HLA, 1987), and 6.6 X 10^6 m^3 (EMC2, 2000) resulting in from ca. 1500 to 2650 MT of arsenic residing in the reservoir sediment. Pore water chemistry shows that arsenic and metals are definitely moving from the solid to the solute phase—arsenic pore water values range up to 10,000 µg/L and are commonly above 500 µg/l (Woessner, 1995).

The presence of elevated metals and arsenic in the sediments corresponds with the high concentration of dissolved constituents in the wells installed in the reservoir sediment (HLA, 1987; Titan, 1996; Woessner et al., 1984). In these wells, TDS is commonly over 1,400 mg/l, dissolved arsenic ranges from ca. 100 to 10,000 µg/l, Mn from ca. 0.1 to 25 mg/l and iron from ca. 1 to 60 mg/l. Arsenic speciation results show that samples from within reservoir sediment and the adjacent contaminated aquifer are dominated by As(III) (Moore et al., 1988).

3. CONDITIONS CONTROLLING THE RELEASE OF ARSENIC

Arsenic and metals are transported into the reservoir from the Clark Fork River as original sulfide minerals and as iron oxide/hydroxide crusts/coatings on silcates and other grains (Brook and Moore, 1988; Kuzel, 1993). However, elemental and mineralogic relationships of metals transported into and residing in the reservoir are complex and not well established. Once in the reservoir, river sediments reside dominantly in anoxic environments. Three conceptual geochemical models of the processes causing arsenic migration from the reservoir sediment have been proposed.

The geochemical environments within the upper 1.5 meters of reservoir sediments were first described by Moore et al. (1988). They found an upper oxidized zone in only the upper few centimeters of subaerially exposed sediment, beneath which saturated sediment was dominated by anoxic conditions. Udaloy (1988) found that the anoxic zone extended to the base of the reservoir sediment and that the boundary with the oxic zone fluctuated with reservoir stage. She developed a conceptual geochemical model based on microbially-mediated redox reactions and concluded that fluctuation in stage controlled redox conditions which mobilized arsenic from the reservoir sediments. The strong downward hydraulic gradients then transported arsenic, iron and manganese released by this redox pump (Moore, 1993) into the underlying alluvial aquifer. Moore et al. (Moore et al., 1988) originally suggested that sulfate reduction to sulfide and the subsequent precipitation of metal sulfides controlled arsenic transport to the ground water. Basically, they concluded that the system was sulfur limited, allowing arsenic to escape to the underlying aquifer because the amount or rate of sulfide formation was not sufficient to fix arsenic.

Subsequent work by Titan (1995) identified a transition zone between the upper oxic zone and lower anoxic zone. They discounted the redox pump as the main driver of arsenic and metal release because of the typically small annual fluctuations in reservoir stage (less than ca. 1.5 m) and instead used an advancing "geochemical front" to mobilize arsenic from iron oxides/oxyhydroxides in the sediment. Their model requires either a continual source of these phases (see below) or very slow kinetics for the dissolution of minerals deposited 130 to 90 years ago (the proposed period of major deposition). This mechanism is supported partially by the fact that arsenic concentration in the pore water locally increases with depth in the sediment (Moore, 1993; Titan, 1995). Titan also suggested that the system was sulfur limited allowing arsenic and iron to escape the anoxic-sulfidic zone at the base of the sediments. This mechanism is somewhat similar to those proposed to release arsenic from the deltaic sediments in the Bengal region of Bangladesh and India (Nickson et al., 2000), although those sediments contain much lower concentrations of arsenic than the Milltown Reservoir sediments.

CH2M Hill (1994) proposed similar mechanisms but suggested that sulfide was not limiting arsenic mobility. Instead they suggested that arsenic transport was "...more related to the repeated adsorption of dissolved AsV by remanant or precipitated oxyhydroxides within the oxidized part of the transition zone..." (CH2M Hill, 1994). They did, however, conclude that insufficient sulfide in their "reduced zone" (anoxic) allowed iron to remain mobile. Based on the brown color of some sediments from the lower portion of the stratigraphic package, they proposed that "...the lower part of the reservoir sediments appear to be more oxidized than reduced." They suggest that an oxidized zone of ground water exists beneath the reservoir sediment. However, none of the water geochemistry supports this conclusion. All waters collected from the aquifer beneath the reservoir sediment and immediately adjacent to the reservoir have signatures indicating an anoxic environment. Contaminated ground water in reservoir sediments and the alluvial aquifer contains from 70 to 100% As(III) and high concentrations of dissolved iron (commonly 1000-20,000 μg/l). Many of the deeper wells with high iron also have relatively low sulfate suggesting that the ground water affected by Milltown Reservoir sediments is anoxic sulfidic (Moore, 1993).

All the conceptual geochemical models depend on reduction of pre-deposited iron oxides to release arsenic in the anoxic ground water zone of the reservoir and then into the underlying alluvial aquifer. The relationships between iron, sulfate and arsenic in ground water give some insight into these processes (Fig. 3). In general, there is a strong correlation between iron and arsenic (Fig. 3A). These data support the concept that iron oxyhydroxides holding arsenic are dissolving under the anoxic conditions in the reservoir sediment and releasing arsenic. This results in arsenic-contaminated ground water also high in iron. Both arsenic and iron are reduced during this process resulting in the strong correlation between As(III) and Fe(II) (Fig. 3C). Although the sediment is anoxic there appears to be limited sulfur to form sulfides (Moore, 1993). When sulfate is high in ground water, arsenic and iron are both low (Fig. 3B & D), suggesting that sulfides are forming but leaving excess sulfate. Alternatively, sulfide in the ground water is oxidized and iron and arsenic are adsorbed/co-precipitated with oxyhydroxides. High concentrations of iron and arsenic are found only in ground water with low sulfate concentrations. Most likely the availability of sulfur limits the formation of metal sulfides (and hence the incorporation of iron and arsenic) (Moore et al., 1988) in the most contaminated areas of the reservoir sediment. This allows arsenic to escape from the reservoir sediment into the ground water flow system. Whatever the details of these release mechanisms, it is likely that these processes began almost immediately after dam completion, in 1907 and 1908, as the pool elevation rose above the pre-dam floodplain water table. As a result, reservoir water flowed through pre-dam contaminated floodplain sediments and newly deposited wastes. The long-term presence of poor water quality in Milltown, as reported by the residents, supports the concept of a continuous release of contaminants from a long–time source.

4. ARSENIC TRANSPORT AND FATE IN THE ALLUVIAL AQUIFER

Once arsenic is released from the contaminated reservoir sediments, it moves vertically, exiting the sediments and entering the pre-dam alluvium

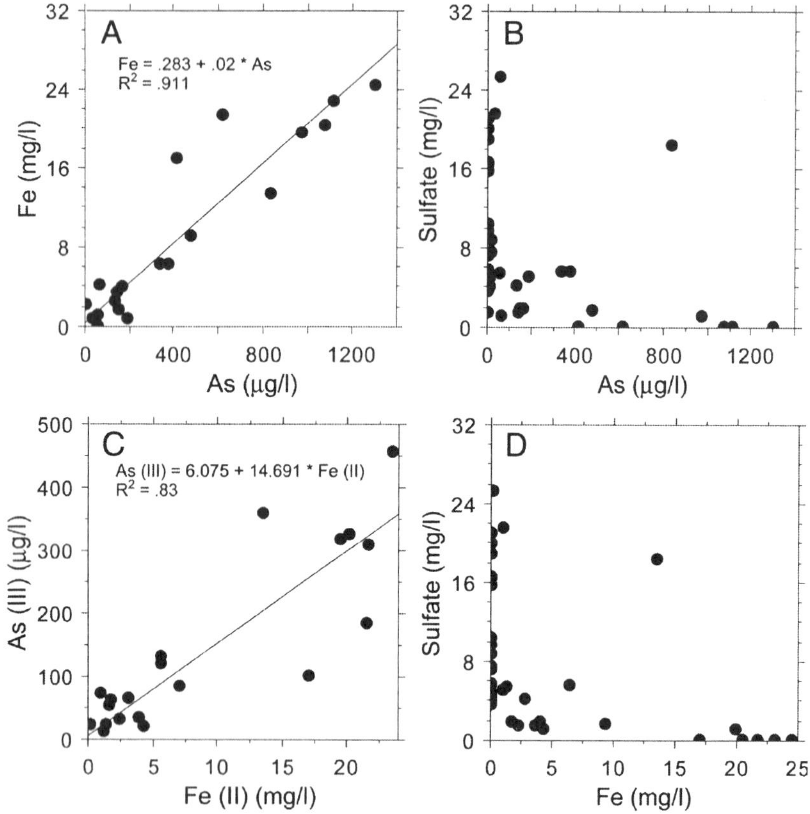

Figure 3: Geochemical data from wells sampled in Spring 1992. This data includes wells sampled within the reservoir and in the adjacent aquifer. Data from Titan (1995).

underlying the reservoir. Woessner et al. (1984) and Udaloy (1988) found that hydraulic gradients in the fine reservoir sediment were strongly downward, averaging between 0.15 to 0.40. Titan (1995) used four wells and computed an average vertical gradient of 0.25. Titan's (1995) analyses of Udaloy (1988) and Hydrometrics (1985) slug test data and their grain-size data results in mean hydraulic conductivity values that range from 0.061 to 0.64 m/d. Assuming these values represent horizontal hydraulic conductivity and that vertical conductivities are typically 0.1 of horizontal values, then sediment vertical hydraulic conductivity would range from 0.0061 to 0.064 m/d. Assuming a porosity of the reservoir sediments of 0.4, an average horizontal hydraulic gradient of 0.25, and the range of vertical hydraulic conductivity values, vertical ground water velocities would range from 0.004 to 0.04 m/d. Movement of river water through the channel sediments and in the pool area near the dam most likely occurs at a higher velocity because the sediment thickness is about one third the thickness of the exposed sediments, so has a larger hydraulic gradient.

Once ground water leaves the fine-grained reservoir sediments, it enters the underlying, conductive, pre-dam floodplain alluvium. These gravel-dominated deposits are seamlessly connected to the alluvial material just north of the reservoir forming the prolific aquifer beneath Milltown. Ground water flow in this system is principally horizontal, moving to the north and northwest (Fig. 4). The ground water flow directions reveal that reservoir-derived ground water flows north towards Milltown, then turns westward into a region of converging flow just west of the Blackfoot arm of the reservoir. A strong southwesterly component of ground water flow down the Blackfoot valley to the north converges with ground water coming from the alluvium beneath the reservoir, forcing ground water to the west. Head differences of >4.8 m between the stage of the Blackfoot reservoir arm and the underlying water table suggest that this water body is mostly perched and has little impact of the flow of ground water beneath it.

Valley alluvium has been influenced by the high-energy catastrophic Lake Missoula emptying events during the last ice age. As a result, zones of large diameter, well-sorted gravel and boulders make up parts of the sedimentary package. Woessner et al. (1984) and Titan (1995) report a range in hydraulic conductivity values from 1200 to 18,000 m/d in the alluvium. With hydraulic gradients ranging from 0.25 to 0.004, an assumed porosity of 0.2, velocities would likely range from 360 to 1,500 m/d. It is this aquifer that supplies water to Milltown, the surrounding area and 10 km down gradient, to the >50,000 residents of the city of Missoula.

We assessed the transport of arsenic in these two ground water systems, the fine-grained reservoir sediments-river channel, and the alluvial aquifer, using flux calculations for a portion of the reservoir and the corresponding flow tube in the alluvial aquifer (Fig. 5). The selected area of fine-grained reservoir sediments represents an area of 197,750 m^2 and the associated river channel an area of 51,600 m^2. The saturated cross sectional area of the underlying pre-dam floodplain alluvium parallel to the interstate is 4,640 m^2. Based on a maximum vertical hydraulic conductivity value of 0.064 m/d, a vertical gradient of 0.25 for the reservoir sediments and 0.75 for the river channel sediments, the discharge from the fine-grained reservoir sediments would be 3,160 m^3/d and 2,480 m^3/d, respectively. Discharge in the pre-dam floodplain alluvium was estimated at 91,590m^3/d based on a hydraulic conductivity value of 1645 m/d and a gradient of 0.012 (Woessner, 1995).

The mass of arsenic moving from the fine-grained reservoir sediments to the underlying alluvium was computed by using an average arsenic

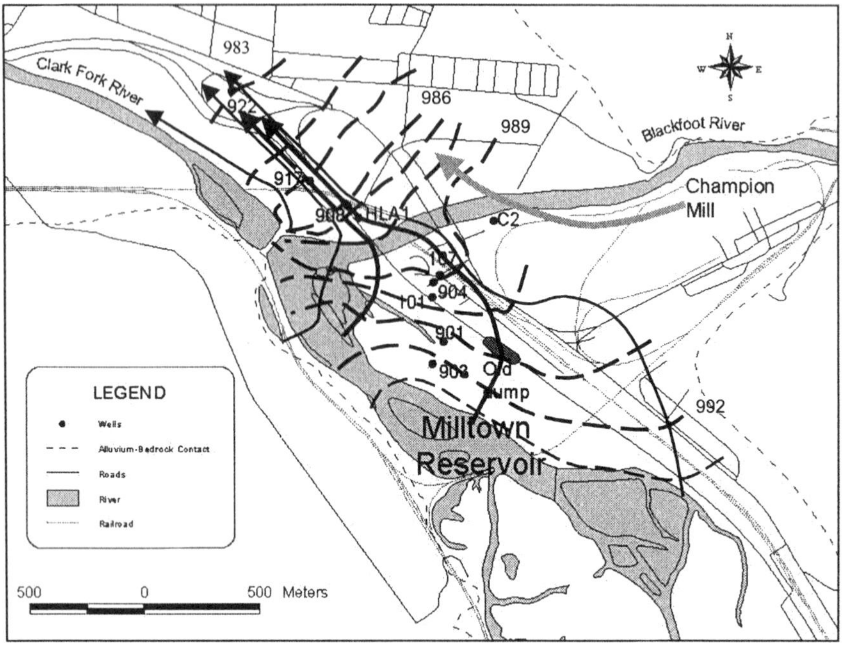

Figure 4: Groundwater elevation and flow within the Milltown aquifer during Spring 1992. Flow lines are represented by solid arrows and equipotential contours as dashed lines. The flow tube discussed in the text is designated by the heavier two center arrows containing numbered wells used for the discharge and mass loading calculations. The gray arrow represents groundwater flow originating from the Blackfoot River Valley. Contour interval is 1 m. Data from Titan (1995).

concentration of 1240 μg/l derived from monitoring well sampling (Titan, 1995), however, arsenic concentrations ranged widely with some values>5000 μg/l. A value of 200 μg/l was used for the stream channel sediments source (Titan, 1995). A median arsenic value for the cross section of alluvium of 740 μg/l was assigned (averages of Titan data are about 0.90 mg/). These values result in a flux of arsenic from the fine-grained sediments of 3.9 kg/d and from the river channel of 0.5 kg/d, for a total of 4.4 kg/d arsenic (Fig. 5). However, the arsenic flux through the saturated alluvial cross section adjacent to the reservoir is calculated to be 67.8 kg/d. Based on these computations, 24.6 tons/y of arsenic is flowing from beneath the reservoir, out of the flow tube, into the adjacent aquifer. This is considerably more than the 1.6 tons/y calculated to leave the fine-grained, reservoir sediment and flow downwards into the flow tube.

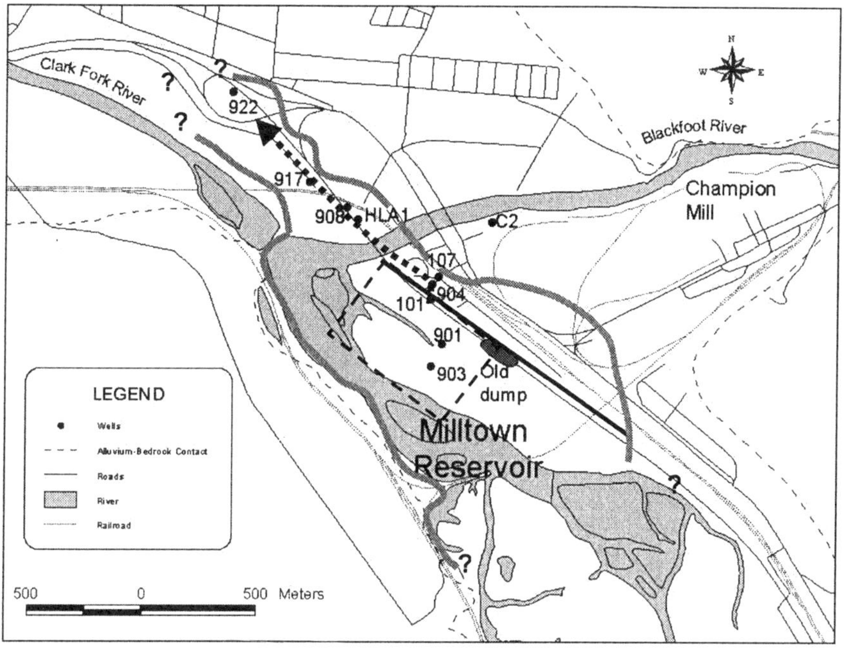

Figure 5: Map showing the area of the reservoir sediment used for flux calculations (dashed rectangle). The northern boundary length of the rectangle was used to compute the cross sectional area for the pre-dam floodplain alluvium. The dotted arrow represents a flow path from well 101 to well 922 along which geochemical data were plotted (Figs. 6, 7 and 8). The black diagonal line paralleling the interstate represents the length of the cross section area used for the general reservoir discharge and flux calculations. The thick, dark-gray lines represent the region where arsenic concentrations in ground water exceed background concentrations, 3 μg/l.

The water and mass flux analyses raise issues as to the source of arsenic in the flow tube (Figs. 4 and 5) within the reservoir area. The water flux from the reservoir sediments and river channel sediments is only 6 % of the water flowing in the underlying 8.2 m thick buried floodplain alluvium. Assuming that the hydraulic parameters used are correct and flux calculations adequately reflect the system, there must be another source of water and arsenic that is not accounted for in the analysis. The floodplain alluvial flow tube is defined only two-dimensionally and constrained by only 2 wells finished in the underlying gravel (within the reservoir). Possibly, a more complex, buried floodplain three-dimensional flow system adds arsenic-rich ground water from up-gradient (upstream within the reservoir) to the flow tube shown. Without additional nested wells in the coarse sediments beneath the reservoir sediment, this hypothesis can not be confirmed.

As the water flows from beneath the reservoir sediments along a flow path, (Fig. 5), arsenic concentrations change from 930 μg/l at the edge of the reservoir (well 101B) to 220 μg/l over a flow distance of about 250 meters (Fig. 6A). By 760 m, the concentration has decreased to 147 μg/l. Further changes during transport are unclear as limited well data suggest the plume centroid my not intersect the remaining wells (Fig. 4). However, a well about 1550 m down gradient (well 922), but slightly off the trend of the plume, contains ground water with 12 μg/l of arsenic (background values in the area are about 3 μg/l). If this well represents the arsenic concentration in the enter of the plume at this distance, then concentrations decreased approximately exponentially along the flow tube after exiting the reservoir. Iron follows a similar trend (Fig. 6B).

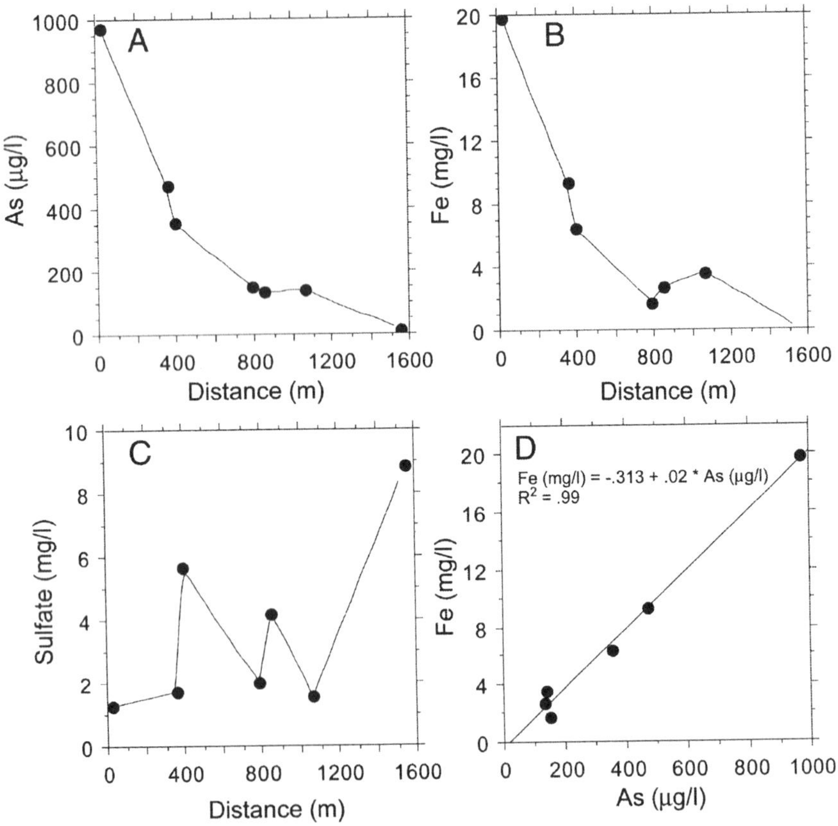

Figure 6: Groundwater geochemical data for wells along the flow path designated in Fig. 5. Data collected in Spring 1992 by Titan (1995). A, arsenic concentration along the flow path. B, iron concentration along the flow path. C, Sulfate concentration along the flow path. D, iron versus arsenic in the wells along the flow path.

Theoretical behavior of a conservative tracer originating from a constant and continuous source in a two-dimensional flow field would be expected to maintain source concentration (930 μg/l in this case) throughout the entire observed flow tube (1550 m) after only a short time period. However, potential conservative tracers in the plume (sodium and chloride, Fig. 7A) show the classical binary-mixing, dilution relationship (Faure, 1998) suggesting that three-dimensional dispersion (dilution) affects plume behavior. Mixing of un-impacted ground water from the Blackfoot River valley with contaminated water from the reservoir would explain the sodium-chloride relationships. However, plots of arsenic, iron and sulfate with chloride show no simple binary-mixing relationships, so dilution alone cannot explain the decrease in these constituents (Fig. 7B, C & D). Concentration changes along the flow tube must be controlled, at least somewhat, by geochemical processes. Durinng transport, the most likely control on arsenic is precipitation on and/or adsorption to aquifer materials. The water leaving the reservoir is anoxic (high dissolved iron and manganese and dominated by As(III) and Fe (II) (high values in Fig. 6 C). A plot of iron vs. arsenic shows that as water moves down the flow path, arsenic and iron decrease in concentration (Fig. 6A & B) and are strongly correlated (Fig. 6D). This suggests that a geochemical mechanism is decreasing arsenic and iron together because these relationships cannot be explained by dilution alone (Fig. 7).

One of two possible mechanisms are likely: 1) co-precipitation/precipitation of arsenic with metal sulfides; 2) co-precipitation/adsorption of arsenic with iron oxyhydroxides. Both mechanisms would produce the relationships seen in Fig. 6D. Sulfate is low at the reservoir margin and increases slightly and irregularly along the flow path (Fig. 6C) with the highest concentrations found farthest from the reservoir. This suggests that sulfur is not limiting precipitation in the plume. However, a simple trend is not evident. Plots of Fe(II)/Fe(Total) and As(III)/As(Total) were constructed to examine the redox state of these constituents along the flow path; they show somewhat similar trends (Fig. 8). Iron is dominated by Fe(II) throughout the plume and decreases slightly along the flow path. The last well in the plot (which also may not lie directly in the plume center) has very low iron concentrations (60 μg/l). The determination of Fe(II) in this sample is suspect because the Fe(II) value is four times the Fe(total) value. These data would suggest that anoxic conditions persist throughout most of the plume and oxic conditions exist only at the far end of the plume (low total Fe). The As(III)/As(Total) rations are somewhat higher in the plume at about 800 m from the reservoir. The sample farthest from the reservoir has the lowest ratio, but this sample is

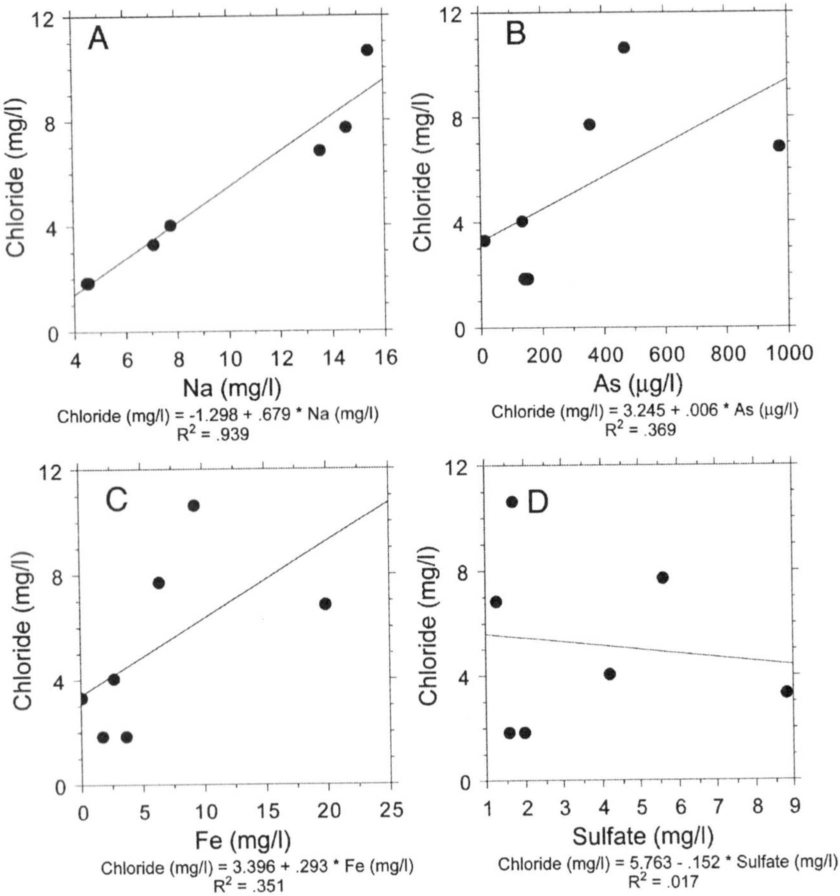

Figure 7: Groundwater geochemical data for wells along the flow designated in Fig. 5. Data collected in Spring 1992 by Titan (1995). A, chloride versus sodium. B, chloride versus arsenic. C, chloride versus iron. D, chloride versus sulfate.

suspect because of low As(Total) concentrations. Overall, the arsenic redox data are quite variable, but also indicate that at the end of the plume ground water may be more oxic than farther up gradient, although the last sample may not be directly in the center of the plume. Unfortunately, these inconclusive relationships in the geochemical data along the flow path, make distinguishing between sulfide and iron-oxyhydroxide precipitation very difficult. However, the trends and relationships in the geochemical data do show that the decrease in arsenic away from the reservoir is a combination of plume diffusion and precipitation/adsoprtion of arsenic along with iron within the aquifer as it moves along the flow path.

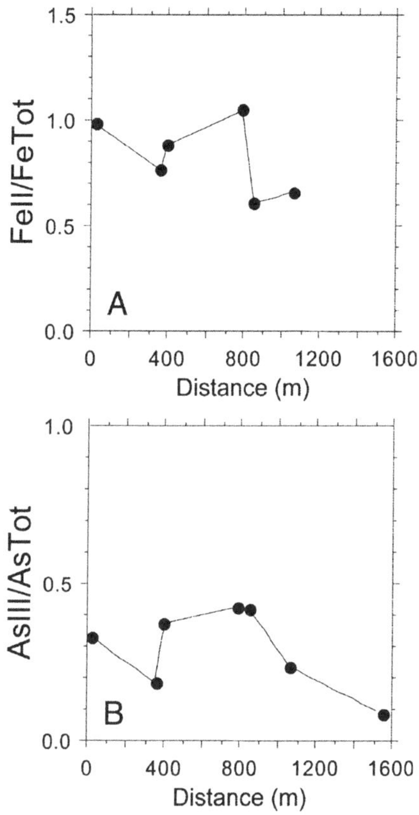

Figure 8: Arsenic and iron speciation data for groundwater collected from the wells along the flow path designated in Fig. 5. Data collected in Spring 1992 by Titan (1995). A, As(III) over total arsenic plotted against distance along the flow path. B, Fe(II) over total iron plotted against distance along the flow path.

5. ARSENIC FLUX AND TRENDS

Although the geochemical and hydrogeologic mechanisms of arsenic fixation and transport within and from the reservoir are not finally resolved, arsenic is certainly being released from the reservoir sediment at relatively high rates (tons per year). We estimate that 25.7 MT/yr of arsenic leave the reservoir sediment and enter the adjacent aquifer (Fig. 5). 24.6 MT/yr come from the lower reservoir area (essentially our flow tube in the analysis above). This is ca. 10-20% of the total amount of arsenic estimated to reside in the reservoir (1500-2650 MT, see second section above). At these rates all the arsenic should be removed from the reservoir sediment in 60 to 100 years. These estimates assume that all the arsenic would be removed from the sediment and the concentration would fall to zero. In actuality, concentrations would not fall below background concentrations, and may not reach this level. Extractions show that only 20-30% of the arsenic in the sediment is likely available (Titan, 1995; Woessner, 1995). If this estimate is correct, then all the currently-remaining, available arsenic should be removed from the reservoir sediment in from 15-25 years. If these rates are correct, then we might expect to see decreases in porewater/ground-water arsenic data collected over that last 17 years as sediment concentrations decreased. However, the arsenic concentrations in the aquifer do not appear to be decreasing (see below). We expect that there must be another source of arsenic: The continual addition of arsenic-contaminated sediment from upstream.

The USGS has collected sediment load data for the Clark Fork River (Hornberger et al., 1997; Lambing, 1991) above and below the reservoir (Gages 12334550 & 12340500) and the Blackfoot River above the reservoir (Gage 12340000) from 1985 to 1995. These eleven years had roughly similar peak flows to those for the seventy years of record: 315.5±94.9 m^3/s vs. 441.9±192 m^3/s (± 1 standard deviation), respectively. From 1985 to 1990, sediment accumulated in the reservoir for five of the six years sampled, ranging from 4,600 to 48,700 MT/year. Only during one year, 1987, was sediment eroded from the reservoir, when the reservoir stage was lowered for repairs for several months (Udaloy, 1988). During this year, 15,800 MT of sediment eroded from the reservoir. Even with this removal an average of 14,800 MT of sediment per year was deposited in the reservoir from 1985 to 1990. Measurements from 1991 to 1995 found that an average of 11,700 MT/y accumulated in the reservoir. This is likely a typical loading rate over the last seventy years, however, there have been high flow events that may have removed arsenic-rich sediment from reservoir. During 1997, peak flows were relatively high. There were only six other years with peak flows higher (including the estimated flood of record in 1908) in seventy years of record. We monitored sediment and arsenic flux during this flow event and found that only during approximately two weeks of the highest flow did the reservoir release more sediment than it received. Over the entire event (March to late June), we estimate that about 6.8 MT of arsenic accumulated in the reservoir (Shifflett, in preparation). Although sediment flux has not been determined for the highest flow years, modeling using HEC-6 for fifty-one years of flow records predicted an average yearly accumulation of 17,200 MT (EMC2, 2000; Titan, 1995). They found significant erosion only at flows above 850 m^3/s, which have occurred only three times during the seventy years of continuous record. The 100-year recurrence interval flow showed large amounts of erosion in the HEC-6 model runs, eroding 119,000 MT/y of sediment from the reservoir. However, this amount of sediment would be made up by deposition during only 7.4 years of average flows. All these data show that Milltown Reservoir is a major sink for sediment, and therefore, arsenic.

Because the USGS data are similar to the HEC-6 modeling and are actual data rather than simulated data, we used it for calculating the flux of arsenic through the reservoir. Using an average concentration of 50 mg/Kg of arsenic in the sediment coming into the reservoir from the Clark Fork River (data from Breuninger, 2000) and average 1985-90 data as one end member and the average 1990-95 data for the other, we calculate that from 590 to 740 MT of arsenic are added to the reservoir each year. Using the same percentage of available arsenic as used above, about 25% of this would possibly be available, or 148 to 185 MT/yr. This flux would be highly variable based on the USGS sediment flux data, but over the long term the reservoir is a sink for sediment-bound arsenic and prevents it from moving downstream. This of course provides a continually replenished source for ground water contamination originating from the reservoir sediments. Using our estimates of flux out of the reservoir, we estimate that an additional "5 to 7 years" of available arsenic are added to the reservoir each year. However, the disposition of this sediment within the reservoir is of primary importance in determining the long-term potential for arsenic contamination. Our flux estimates found that 38% of the ground water and 96% of the arsenic load to the adjacent aquifer comes from the lower section of the reservoir. However, the upper section of the reservoir likely accumulates more sediment than the lower section because of channel morphology and flow regime. Channels in the upper reservoir are sinuous, shallow and braided with sloughs and backwater areas common (Fig. 9) and therefore make excellent sediment traps. We would expect that concentrations in the lower reservoir should decrease over time as arsenic is rapidly removed by ground water transport through the sediment. However, we would only need ca. 10% of the total arsenic added to the reservoir each year to accumulate in the lower reservoir to maintain the yearly flux. Through time, we would expect the upper reservoir to play a more important role in supplying arsenic because it would continue to accumulate arsenic-contaminated sediment from upstream.

Although there have been hundreds of samples taken for chemical analyses at Milltown, there are very few long-term records comparing data from individual wells (Woessner, 1995). Woessner compared data from ten wells within the plume area and found that most concentrations increased between 1983-84 and 1990-92. Data from three of those wells with the largest number of samples (additional data from Titan, 1995) are plotted in Fig. 10. Well 101B shows the strongest difference in arsenic concentrations

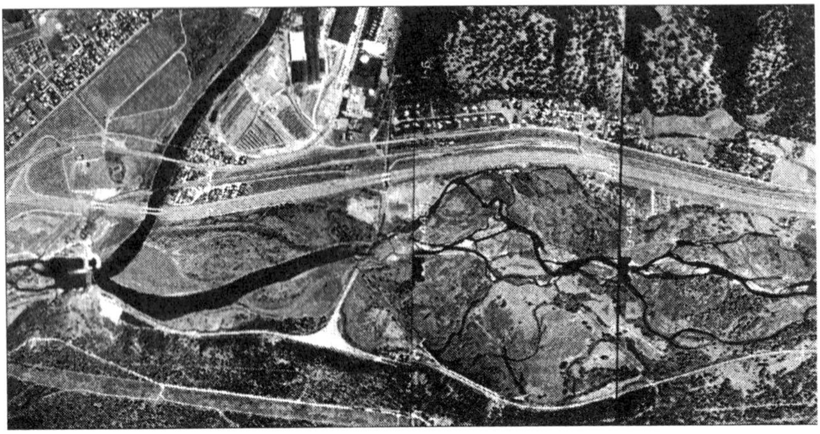

Figure 9: Aerial photograph mosaic showing Milltown reservoir at low flow in 1989. Main features are labeled in Fig. 2. The photograph covers approximately 3600 m in the horizontal (E-W) direction.

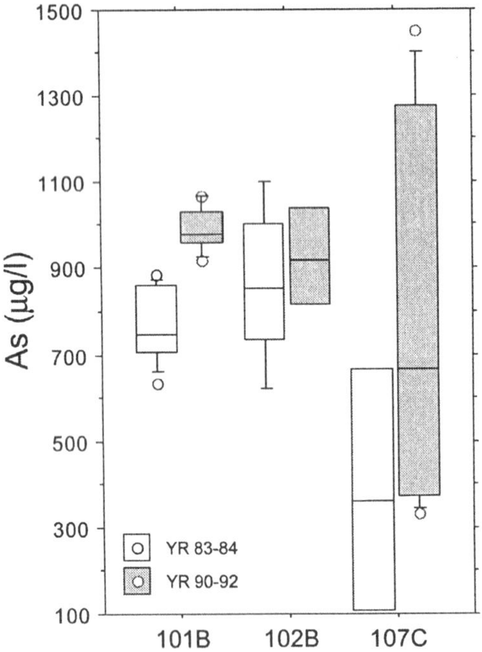

Figure 10: Box plot showing the difference between arsenic concentrations in three wells within the contaminant plume. Data collected by Woessner (1984) and Titan (1995) . The bottom and top of the box represents the 25th and 75th percentiles; the horizontal line dividing the box is the median concentration; the whiskers are the 5th and 95th percentiles and the circles data lying outside those percentiles.

between 1983-84 and 1990-92, with nearly complete separation of all data collected. The median concentration changed from 746 to 979 μg/l. The median arsenic values in wells 107C also increase through time, from 361 to 664 μg/l, but are less clearly separated when the variability of the data are considered (Fig. 10). Well 102B median values changed from 853 to 920 μg/l, but these differences are likely not significant (Fig. 10). Unfortunately, there is very little long-term data from wells located farther down the plume where changes in plume size and concentration would be easiest to see. But, the long term data within the plume does show that arsenic concentrations are increasing through time. This suggests that the geochemical and hydrogeologic mechanisms removing arsenic from the sediment and transporting it to the adjacent ground water are dynamic.

6. WHAT WILL THE FUTURE BRING?

The concern by government agencies and the public over Milltown Reservoir have moved well beyond ground water contamination. Potential integrity/safety of the dam, release of contaminated sediments downstream, effects on threatened fish species (bull trout), restricted fish passage, expansion of introduced fish species (northern pike), power generation and recreational use, are all now driving decisions for remediation. The discussion now focuses more on the fate of metal-contaminated sediments than on arsenic levels in ground water. The objectives of any remediation at the site are now to eliminate the release of contaminated sediment downstream, protect aquatic life below the dam, and protect wetland resources while considering power generation needs and recreational use (EMC2, 2000). The USEPA is considering a range of alternative remediation scenarios from total removal of the contaminated sediment and the dam to no action. It is assumed that under all these alternatives the effects on ground water contamination will be positive or negligible and that the system is defined well enough to predict outcomes (both in the surface water and the ground water). To date, the work at Milltown has failed to resolve the key geochemical and hydrologic processes responsible for arsenic and metal transport and fate. We believe that there are major hydrogeologic and geochemical questions that need answering before predictions of final outcomes of remedial actions can be made. Questions, such as: 1) What are the geochemical processes controlling fixation of arsenic in the reservoir and will they remain stable? (It is especially crucial to establish whether oxyhydroxides or sulfides fix arsenic in the aquifer.) 2) What is the long-term fate of arsenic in the plume/aquifer? (The present records show increases in arsenic concentrations and we need to know what controls these concentration changes and what we can expect after remediation.) 3) How will dam/sediment removal alter the ground water flow paths and the transport of arsenic from the upper reservoir? (That area now supplies a small proportion of the arsenic to the aquifer, but changed hydrologic and geochemical regimes may alter arsenic release/flux from this area; it is important to know if this will become an substantial source for surface water and ground water contamination.)

Beyond these and other technical questions, we also need to consider the possibility that the geochemical and hydrogeologic system at Milltown (and similar sites) is so complex that we will be unable to make reliable predictions, but only measure the response to our actions. In that case, we should consider Milltown as a large field laboratory. Appropriately-scaled experiments need to be conducted as remediation progresses to provide feedback into the remediation process. This "adaptive remediation" would better assure success as scientists and regulators continue to deal with complex systems like Milltown reservoir. The goal should not be just the remediation of one site to meet specific criteria, but a deep understanding of process and function that can be applied to the restoration of contaminated aquatic systems anywhere in the world.

Chapter 13

Natural Remediation Potential of Arsenic-Contaminated Ground Water

Kenneth G. Stollenwerk and John A. Colman
U.S. Geological Survey

Migration of leachate from a municipal landfill in Saco, Maine has resulted in arsenic concentrations in ground water as high as 647 µg/L. Laboratory experimental data indicate the primary source of arsenic to be reductive dissolution of arsenic-enriched iron oxyhydroxides in the aquifer by organic carbon in landfill leachate. A core from an uncontaminated part of the aquifer yielded no dissolved iron or arsenic when leached with oxic ground water. Eluent ground water spiked with organic carbon in order to create reducing conditions mobilized both ferrous iron and arsenite from this core. The landfill was capped in early 1998 to eliminate the source of leachate. Cores from the contaminated portion of the aquifer were collected and leached with uncontaminated ground water in the laboratory to simulate natural remediation conditions. Data from these experiments show that significant concentrations of labile organic carbon have accumulated on aquifer solids, causing significant biological oxygen demand. In laboratory leaching experiments of the most contaminated core, the organic carbon caused complete consumption of the influent dissolved oxygen (6 mg/L) for 220 pore volumes. Arsenic leaching from contaminated cores rapidly decreased in concentration initially in response to flushing with uncontaminated ground water. Subsequent leaching produced more gradual decreases in dissolved arsenic concentrations, controlled by a combination of reductive dissolution of arsenic-enriched iron oxyhydroxides and adsorption/desorption. In leachate from the most contaminated core, arsenic concentrations exceeded the new United States Environmental Protection Agency drinking-water standard of 10 µg/L for more than 200 pore volumes. A geochemical model simulated the concentration of selected constituents as uncontaminated ground water eluted through contaminated aquifer solids. Concentrations of dissolved oxygen, arsenic, and iron, in leachate from one core were used to calibrate the model. This model was validated by successfully simulating constituent concentrations in leachate from cores collected from other contaminated areas of this aquifer.

1. INTRODUCTION

Concentrations of arsenic (As) that exceed the new Federal drinking-water standard of 10 μg/L (U.S. Environmental Protection Agency, 2001) are a common problem in New England water supply wells (Zuena and Keane, 1985). Peters et al. (1999) reported As concentrations as high as 180 μg/L in water from domestic wells in New Hampshire that were completed in bedrock. They found an association between high As concentrations in ground water and pegamatite dikes in granitic bedrock. Ayotte et al. (1999) observed that there was a correlation between concentrations of As greater than 5 μg/L in water from domestic bedrock wells in eastern New England and calcareous metasedimentary rocks. In both of these studies the source of As appears to be from weathering of bedrock and not from anthropogenic contamination. Anthropogenic induced changes in ground water chemistry are also suspected to have caused high As concentrations in some areas of New England. Arsenic concentrations as high as 800 μg/L have been measured in ground water associated with plumes of leachate from many landfills in contact with bedrock in New England (Edward Hathaway, Remedial Project Manager, U.S. Environmental Protection Agency, oral commun., 1997). The As appears to have been mobilized from aquifer solids by leachate from these landfills.

The aqueous chemistry of As can be complex. Arsenic is stable in four oxidation states (+5, +3, 0, -3) under Eh conditions prevalent in aqueous systems (Ferguson and Gavis, 1972). Arsenic metal rarely occurs and the -3 oxidation state is found only in very reducing environments. Arsenate [As(V)] is stable in oxidizing environments. For pH values common in gound water, the predominate As(V) species in solution are $H_2AsO_4^-$ between pH 2.2 and 6.9, and $HAsO_4^{-2}$ between pH 6.9 and 11.5. Arsenite [As(III)] is stable in moderately reducing environments; $H_3AsO_3^o$ predominates up to pH 9.2, and $H_2AsO_3^-$ from pH 9.2-12 (Ferguson and Gavis, 1972). If conditions are favorable, biomethylation can form a variety of methylated species that are stable under oxidizing or reducing conditions. The more common methylated species include: methylarsonic acid [$CH_3AsO(OH)_2^o$], dimethylarsinic acid [$(CH_3)_2AsO(OH)^o$], methylarsonous acid [$CH_3As(OH)_2^o$], and dimethylarsinous acid [$(CH_3)_2AsOH^o$] (Braman and Foreback, 1973; Cullen and Reimer, 1989; National Research Council, 1999). In sulfidic solutions, thioarsenic species such as $H_2AsOS_2^-$ and $H_2AsS_3^-$ can form (Clarke and Helz, 2000; Helz et al., 1995).

The oxidation state of As has a significant effect on its rate of transport in ground water, with As(V) adsorbing to a greater extent than As(III) at lower pH values and As(III) adsorbing to a greater extent than As(V) at higher pH values (Bowell, 1994; Manning et al., 1998; Pierce and Moore, 1982; Raven et al., 1998). Relatively low concentrations of As in ground water that result from weathering of geologic materials can become concentrated via adsorption onto various mineral phases including iron (Fe), aluminum (Al), and manganese (Mn) oxyhydroxides, and clay minerals (Anderson et al., 1976; Lin and Puls, 2000; Oscarson et al., 1983; Pierce and Moore, 1982). Changes in pH or the concentration of competing ions can result in desorption of As (Darland and Inskeep, 1997; Fuller and Davis, 1989; Schlottmann and Breit, 1992). A change from an oxidizing to a reducing environment can cause reduction of Fe oxyhydroxides and release of adsorbed As to ground water (Nimick, 1998; Panno et al., 1994).

Recent experimental observations suggest that bicarbonate (HCO_3^-) can facilitate mobilization of As from As-containing sulfides such as orpiment (As_2S_3) in both oxic and anoxic environments (Kim et al., 2000). Mobilization of As was facilitated by increasing HCO_3^- concentrations and pH. The authors found that As was not significantly leached from aquifer solids when HCO_3^- was 305 mg/L and pH was 7.4. Increasing HCO_3^- to 1200 mg/L at pH 8.5 resulted in leaching of 1.5% of the As from the aquifer solids; for 36,000 mg/L HCO_3^- at pH 8.5, 14.8% of the As was leached from the aquifer solids.

This chapter presents results of an investigation of As contamination of ground water at the Saco Municipal Landfill Superfund Site, Saco, Maine. The source of As was investigated as well as the potential for natural remediation of selected constituents in contaminated ground water following closure of the landfill.

2. SITE DESCRIPTION

The Saco Municipal Landfill (Fig. 1) was operated from the early 1960's to 1989 and consists of three separate landfill areas (Nielsen et al., 1995). Landfill Areas 1 and 2 (not shown on Fig. 1) are located on the east side of Sandy Brook and have been capped with clay since 1976 and 1989. Areas 1 and 2 are underlain by a fine sand facies of distal marine fan deposits 1.5 to 6 m thick (Colman and Lyford, submitted). Beneath the sand are glaciomarine silt and clay deposits 7.5 to 45 m thick. Arsenic concentrations in ground water that has been contaminated by leachate from both of these landfills are low (average 21 μg/L, n = 6, range <3.0-50 μg/L) (Woodard and Curran, 1998). Area 4 (formerly 3 and 4) is on the west side of Sandy Brook. This landfill has been inactive since 1989 and was covered with an impermeable membrane in 1998. The geology of Area 4 is distinctly different from Areas 1 and 2. Area 4 is underlain by gravel and till-like

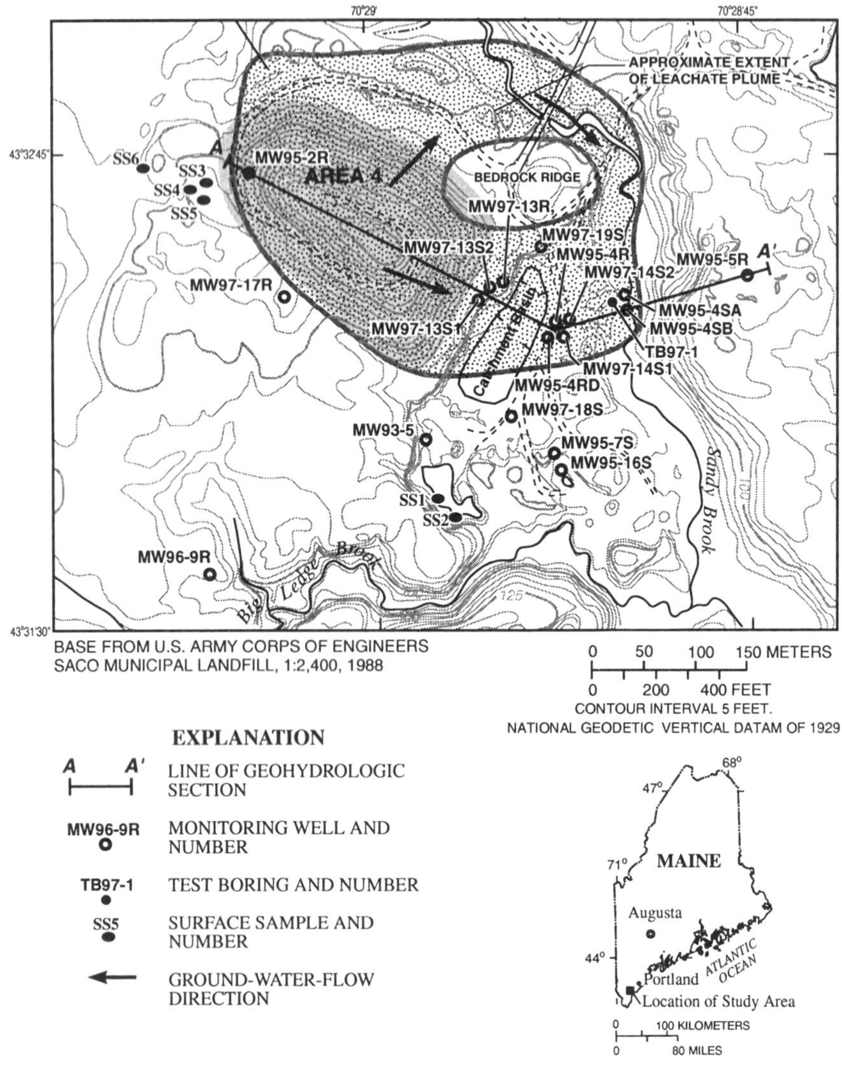

Figure 1. Site map for the Saco landfill Area 4, Saco, ME, with 1997 core and ground water sampling locations.

facies of glaciomarine fan deposits 0 to 11 m thick. Bedrock underlies the surficial deposits and in places directly underlies the landfill. The bedrock is a fine- grained hornfels consisting of thinly-layered, calcareous, quartz-muscovite phyllite locally interlayered with thin beds of micaceous quartzite. Ground water beneath and downgradient from Area 4, which is the focus of this investigation, contains As concentrations as high as 647 μg/L (average =

326 μg/L, n=12, range 52-647 μg/L). The plume of contaminated ground water from Area 4 discharges to Sandy Brook (Fig. 1).

Leachate from the Area 4 landfill has contaminated surficial alluvium and the upper part of the bedrock aquifer at the Saco site (Fig. 2). Contaminated ground water contains greater concentrations of dissolved solids than uncontaminated ground water from either the bedrock or alluvium (Table 1). Organic carbon from the landfill has caused reducing conditions in the aquifer as indicated by low concentrations of dissolved oxygen (O_2) and high concentrations of dissolved organic carbon (DOC), Fe, Mn, and ammonia (NH_3). Sulfate (SO_4^{2-}) reducing conditions may be present in the immediate vicinity of the landfill; however, no sulfide species were measured in any of the ground water samples (detection limit = 0.1 mg/L). It is possible that any sulfides that formed were scavenged through precipitation of iron sulfides; however, iron sulfides were not analyzed for in core material. High concentrations of dissolved Fe and Mn indicate that reduction of the oxyhydroxide phases of these elements by organic carbon dominate the redox potential in contaminated areas of the aquifer sampled during this study.

High concentrations of As are generally associated with ground water contaminated by leachate from the landfill (Table 1). Arsenic concentrations in uncontaminated ground water from bedrock and alluvial wells near the landfill were <0.5 μg/L. Ground-water samples collected for As speciation were filtered in-line through a 0.1-μm filter and quick frozen in liquid nitrogen (Colman and Lyford, submitted). About 80% (range = 64-94%, n=9) of the total dissolved As in contaminated ground water was As(III), the remainder was As(V). However, the 20% As(V) may have been a result of analytical error associated with warming of the samples in an oxygen environment prior to analysis (Colman and Lyford, submitted). Seven of the nine contaminated ground water samples had O_2 concentrations below the field detection level of 0.05 mg/L. The other two samples had O_2 concentrations of 0.10 and 0.15 mg/L. Dissolved organic carbon concentrations were much greater in contaminated ground water than in uncontaminated ground water. Given these ground water chemistries, geochemical modeling indicates that all As should have been As(III). Methylated species of As were not detected.

In general, As in landfill leachate is found in very low concentrations (Christensen et al., 2001). The source of As at the Area 4 site could be waste deposited in the landfill or the geologic materials on which waste was deposited. No wells were drilled within the Area 4 landfill; therefore, its potential as a source of As could not be determined. Samples representative of geologic formations from both sides of Sandy Brook were collected from

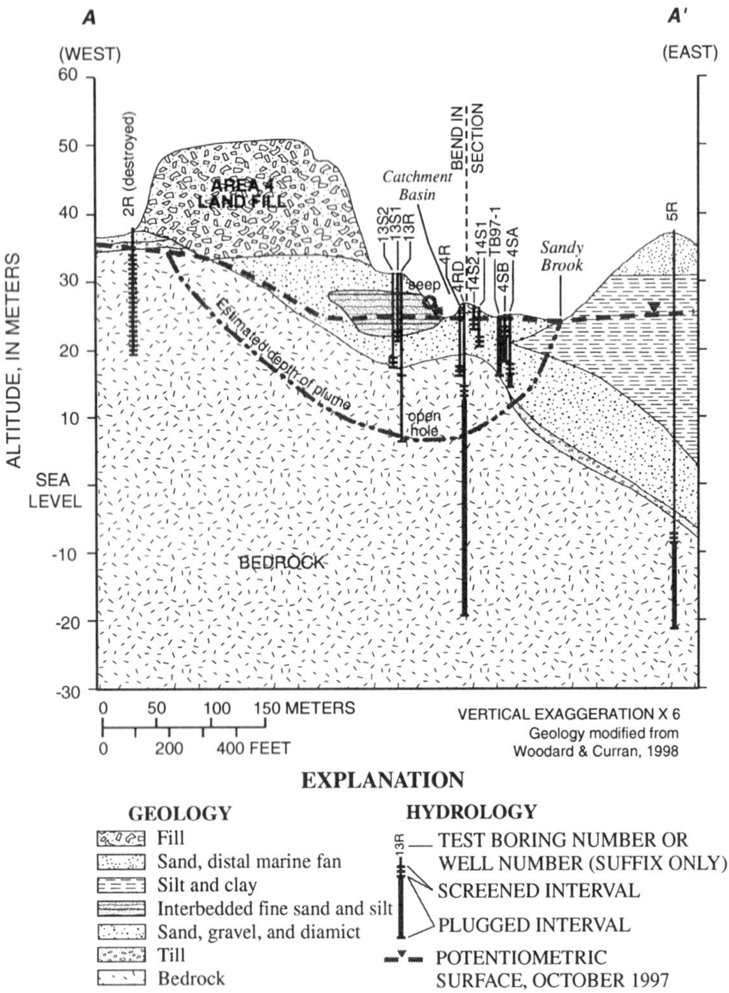

Figure 2. Geohydrologic section line A-A' for Area 4.

areas outside the contaminated ground water plumes and anlyzed for total As and Fe. Hydroxylamine hydrochloride (NH_2OH-HCl) extractions were also used to assess the leachable As and Fe associated with non-crystalline Fe oxyhydroxides (Colman and Lyford, submitted). Their results show much greater concentrations of As in the geologic formations associated with Area 4 than Areas 1 and 2. Whole rock As concentrations in the unconsolidated deposits east of Sandy Brook (representative of material underlying Areas 1 and 2) contained 3.0 to 6.8 μg/g As. The NH_2OH-HCl leachable fraction contained 0.2 to 0.4 μg/g As. Unconsolidated sediments west of Sandy

Table 1. Composition of ground water (August-October, 1997) contaminated by leachate from the Area 4 landfill, and uncontaminated ground water in bedrock and alluvial deposits adjacent to the Area 4 landfill. (Concentrations in milligrams per liter unless otherwise noted)

Constituent	Nine bedrock and alluvial wells contaminated by Area 4 leachate		Uncontaminated bedrock well MW97-17R	Uncontaminated alluvial well MW95-7S
	Range	Average		
pH, unitless	6.2-7.0	6.7	7.0	6.6
Specific conductance, μS/cm	419-6430	2220	282	126
Ca	35-160	91	47	15
Mg	7.9-103	40	3.5	3.2
Na	29-612	206	4.8	6.0
K	4.7-140	33	0.67	2.2
Fe (total dissolved)[a]	13.0-59.6	31.9	<0.003	0.004
Mn (total dissolved)[a]	1.13-12.2	4.70	0.009	0.133
As,μg/L (total dissolved)[a]	52.1-647	276	[b]	0.43
As(III), μg/L	39.4-536	213	[b]	[b]
NH_3 (as N)	2.4-417	94	[c]	0.03
Cl	24-870	253	4.6	2.9
SO_4	0.5-12	4.4	9.3	7.8
HCO_3	102-1590	535	53	25
NO_3 +NO_2 (as N)	<0.05-0.22	0.06	[c]	0.09
SiO_2	17-47	32	7.3	7.9
Dissolved oxygen as O_2	<0.05-0.15	[d]	4.0	3.6
Dissolved organic carbon	4.6-140	41	0.8	0.5

[a]*Concentrations in contaminated ground water based on eight wells. One of the nine samples had dissolved Fe and Mn concentrations that were only 2% of the total Fe an Mn in an unfiltered sample, possibly indicating O_2 contamination and precipitation of solid phases during sample collection. Therefore, Fe, Mn, and As concentrations for this sample were excluded from the reported data.*
[b]*Below detection limit.*
[c]*Not analyzed.*
[d]*Only one well contained O_2 greater than the field detection limit of 0.05 mg/L.*

Brook (representative of material underlying Area 4) had whole rock As concentrations ranging from 6.8 to 51.0 μg/g and the NH_2OH-HCl leachable fraction had As concentrations from 0.7 to 5.5 μg/g. Bedrock west of Sandy Brook had whole rock As concentrations ranging from 8.5 to 122 μg/g, and NH_2OH-HCl leachable concentrations from 0.2 to 15.5 μg/g (Colman and Lyford, submitted). We hypothesize that prior to landfill construction, naturally occurring As in ground water was adsorbed by Fe and possibly Al oxyhydroxides, coating the bedrock fractures and surfaces of alluvial grains.

Migration of organic carbon from the landfill is thought to have caused reductive dissolution of Fe oxyhydroxides, releasing As to ground water.

After Area 4 was covered in 1998, natural dewatering of the landfill was estimated to take about two years. Pre-landfill ground-water-flow conditions were expected to become reestablished and contaminants removed as uncontaminated ground water from upgradient flowed through the aquifer. The primary objective of this study was to assess the rate of natural remediation of the aquifer with respect to As. A second objective was to determine if aquifer solids could be the primary source of As in the landfill plume beneath and downgradient from Area 4.

3. METHODS

3.1 Experimental

3.1.1 Remediation experiments

Cores were collected from contaminated areas in the aquifer downgradient of Area 4 (Figs. 1 and 2) and leached with uncontaminated ground water to estimate the volume of water needed to decrease As concentrations to acceptable levels. The cores were collected in aluminum core barrels and immediately frozen with dry ice for transport to the laboratory where they were stored in a freezer maintained at minus 20°C until needed for experiments. Cores were 4.6-cm inside diameter and ranged in length from 28 to 61 cm. Breakthrough curves for selected constituents from two of the contaminated cores are shown in this paper. Both cores were collected from the alluvial glaciomarine fan deposits and consist of fine sand and silt. The first core is from site MW97-13S1 at a depth of about 7 m below land surface, just below the water table; the second core is from near the glaciomarine fan/bedrock contact at site MW97-13S2 at a depth of 10 m below land surface (Fig. 2).

For each leaching experiment, a core was removed from the freezer and the plastic caps on each end of the core were replaced with thick-walled rubber caps containing inlet and outlet ports and a 2-mm-thick porous disk to ensure an even flow of eluent across the core surface. Nylon-mesh filter disks were placed inside each cap to contain the aquifer solids. The core was allowed to thaw for three days and reach room temperature (~22°C) prior to leaching.

A peristaltic pump was used to move uncontaminated ground water through the cores at an average velocity of 0.2 m/d, the average velocity of ground-water flow in the aquifer. Uncontaminated ground water (Table 2) was synthesized in the laboratory and is an average of the ground-water composition from four wells in an area that was uncontaminated by the landfill. The experimental apparatus was too large to fit in a glove box; therefore, it was necessary to collect samples manually in order to maintain an oxygen-free environment. Leachate from cores was transferred to a sample collection bottle through Teflon tubing to minimize oxygen contamination. (Use of brand names in this chapter is for identification purposes only and does not constitute endorsement by the U.S. Geological Survey.) Collection bottles were 60-mL high density polyethylene bottles for most constituents and 125 mL amber glass bottles for DOC analysis. Leachate samples that were to be analyzed for redox-sensitive constituents were collected manually under an oxygen free, nitrogen atmosphere. During the early phase of the experiments, all leachate was collected and analyzed to capture rapid changes in chemistry. Therefore, samples were collected only during the day; flow through the cores was stopped overnight. To maintain an average velocity of 0.2 m/d, cores were leached at 0.8 m/d during the day. After the rate of change in concentrations of leachate constituents became small, cores were leached at a constant flow rate of 0.2 m/d, 24 hours a day. During the day, samples were collected as described above. At night the effluent was routed to waste. The total volume of leachate and the mass balance of constituents were tracked throughout the experiments. Concentrations in the leachate routed to waste were interpolated from concentrations determined during manual collection.

Samples were analyzed immediately after collection for O_2, pH, [Fe(II)], [Fe(III)], (NH_3), nitrite (NO_2^-), nitrate (NO_3^-), HCO_3^-, and specific conductance. Dissolved organic carbon samples were stored at 4°C and generally analyzed within one week. Aliquots for As(III), As(V) and phosphate [(P(V)] were acidified to pH 2 with ultrapure hydrochloric acid (HCl) and stored in 30-mL polyethylene bottles until analyzed. Periodic analyses of unacidified samples immediately after collection verified that acidification with ultrapure HCl had no effect on the oxidation state of As. Samples were also collected for cation analysis and acidified with HCl. Samples for anion analysis were acidified with nitric acid (HNO_3).

3.1.2 Iron reduction core experiment

An experiment was conducted to evaluate the potential for mobilization of As from uncontaminated aquifer solids as a result of reduction of Fe oxyhydroxides by organic carbon. A core from site MW97-16S in the till near Area 4 (Fig. 1) that had not been affected by the leachate plume was collected and prepared as described in the previous section. Uncontaminated ground water containing sucrose (120 mg/L DOC) was eluted through the core at 0.2 m/d. The intent was to establish anoxic conditions in the core in order to initiate reductive dissolution of Fe oxyhydroxide, releasing any associated As. Leachate samples were collected as described in the previous section.

Table 2. Composition of pore water from two contaminated cores and uncontaminated ground water eluent. (Concentrations in milligrams per liter unless otherwise noted)

Constituent	Synthetic uncontaminated ground water	Core 1	Core 2
pH, unitless	6.60	6.52	6.73
Temperature °C	22.0	22.0	22.0
Specific conductance, µS/cm	125	1410	1380
Ca	8.8	35	140
Mg	3.2	15	19
Na	12	110	140
K	0	18	28
Fe(II)	0	50	32
Mn	0	2.7	2.4
As(III), µg/L	0	390	480
NH_3 (as N)	0	50	26
Cl	20	100	98
SO_4	11	49	35
HCO_3	30	630	1240
NO_3	0	<0.1	<0.1
PO_4	0	0.3	0.1
SiO_2	0	20	30
Dissolved oxygen as O_2	6	<0.005	<0.005
Dissolved organic carbon[a]	0/120	45	31

[a]*The dissolved organic carbon concentration in uncontaminated ground water eluent for the Fe reduction core experiment was 120 mg/L C, and 0 mg/L C for the remediation experiments.*

3.2 Analytical

3.2.1 As(III), As(V), P(V)

A modified phosphomolybdate method was used to determine As(III), As(V) and P(V) in the laboratory experiments (Johnson, 1971; Murphy and

Riley, 1962; Oscarson et al., 1980; Strickland and Parson, 1968). This method is based on the spectrophotometric measurement of the blue As(V) and P(V) molybdate complexes. Phosphate is measured in one sample after reducing As(V) to As(III), which does not form the color complex. In a second sample, both P(V) and As(V) are measured and the As(V) concentration determined by subtracting the P(V) concentration. A third sample is oxidized to convert As(III) to As(V), and the As(III) concentration obtained by subtracting As(V) and P(V).

The limit of detection for As(III) and As(V) with this technique is 3 µg/L with an error of $\pm$ 70% at the 95% confidence interval. Concentrations between 3 and 10 µg/L have an error of $\pm$ 40% at the 95% confidence interval. Concentrations between 10 and 40 µg/L have an error of $\pm$ 15% at the 95% confidence interval, and concentrations greater than 40 µg/L have an error of $\pm$ 10% at the 95% confidence interval. Error estimates were determined for 15 different concentrations ranging from 3 µg/L to 500 µg/L. There were 15 replicates at each concentration. Periodic analyses of replicate samples for total As by ICP were within the confidence intervals stated above.

Ground water samples collected from the aquifer were analyzed for total inorganic As and As(III), and methylated As species by a combination of hydride generation, gas chromatography, and atomic absorption spectroscopy (Crecelius et al., 1986). Inorganic As(V) was determined by difference between total inorganic As and As(III).

3.2.2 Other solutes

Dissolved oxygen, Fe(II), Fe(III), NH_3, NO_2^-, and NO_3^- were measured using the colorimetric CHEMets^{TM} method. This method uses glass ampoules sealed under a vacuum containing an appropriate reagent. The tip of the ampoule is snapped off while immersed in the sample. A fixed amount of sample is drawn into the ampoule where the solute reacts with the reagent, forming a color complex. The intensity of color is measured with a photometer for O_2 and by visual comparison with standards of a known concentration for the other solutes. The specific methods used for each analyte are described in the product catalog (CHEMetrics, 1998).

Solution pH was measured with a Ross combination electrode and Orion model 601A pH meter. Specific conductance was measured with a Yellow Springs Instruments model 32 conductance meter and cell. Cations were determined by inductively coupled atomic-emission spectroscopy. Sulfate and chloride were determined by ion chromatography. Bicarbonate was determined by titration with H_2SO_4 to pH 4.5.

3.3 Computer Model

The one-dimensional reaction-transport model PHREEQC version 2 (Parkhurst and Appelo, 1999) was used to simulate chemical changes resulting from elution of ground water through the cores. PHREEQC has the capability to model advective transport of water in combination with a variety of chemical reactions including homogeneous aqueous reactions, mineral equilibria, and surface-complexation reactions. Version 2 of PHREEQC has added capabilities to simulate kinetic reactions.

3.4 Geochemical Model

The composition of pore waters from contaminated cores 1 and 2 were used to initialize the model (Table 2). Concentrations represent leachate collected from the initial half pore volume of each core. Eluent specified in the transport simulations had the composition of uncontaminated ground water in Table 2. Reactions proposed to describe concentration changes for selected constituents within the cores are based on comparisons between eluent and leachate chemistry and analysis of selected constituents in the core samples. Equilibrium constants and kinetic rates for the reactions were adjusted to give the best fit to leachate concentrations from core 1. The same reactions, equilibrium constants, and kinetic rates were then tested by modeling the concentrations of constituents in leachate from core 2. This geochemical model will be used in the future to simulate evolution of contaminated ground water associated with the Area 4 landfill at the aquifer scale.

Reactions included in the geochemical model were those that had the most significant impact on As concentrations in the core experiments. The primary biogeochemical reaction in the cores is oxidation of organic carbon by a succession of electron acceptors: O_2, MnO_2, and $Fe(OH)_3$. Sulfide concentrations in ground water were below the detection level of 0.1 mg/L, indicating the absence of any significant sulfate reduction. Theoretically, after reduction of O_2, Mn oxyhydroxides should be reduced prior to Fe oxyhydroxides. In reality, both Mn and Fe were measured in core leachate indicating simultaneous reduction of both solid phases. Because of the difficulty in fitting rate constants for both reactions to the experimental data, reduction of Mn oxyhydroxides was excluded from the geochemical model. The resulting impact on As concentrations was considered to be negligible because Fe oxyhydroxides generally have a greater capacity for As adsorption than do Mn oxyhydroxides (Stollenwerk, this volume). Also, Fe oxyhydroxide concentrations in these aquifer solids were about 16 times greater than Mn oxyhydroxides (Colman and Lyford, submitted). Reduction of Mn oxyhydroxides was considered in calculating the concentration of reactive organic carbon for the geochemical model (discussed later).

Reduction of O_2 by organic carbon was modeled according to:

$$O_2 + CH_2O = H_2CO_3 \quad (1)$$

In reality, the organic compounds are more complex; however, information on the composition of the organic phase was not collected. The rate of O_2 reduction represented by reaction 1 was modeled using Monod kinetics:

$$\text{Rate} = (r_{max})(\text{mol/LO}_2)/(2.5\text{e-}05 + \text{mol/LO}_2) \quad (2)$$

The maximum rate constant (r_{max}) for decay of organic carbon by O_2 was 2.0e-07 1/s. When O_2 concentrations decreased below 0.02 μmol/L, Fe oxyhydroxides became the dominant electron acceptor according to:

$$4Fe(OH)_3 + CH_2O + 7H^+ = 4Fe^{+2} + 10\,H_2O + HCO_3^- \quad (3)$$

Decay of organic carbon by $Fe(OH)_3$ was modeled at a constant rate of 6.7e-09 1/s.

Arsenic was modeled using a combination of adsorption on $Fe(OH)_3$ and reductive dissolution of a separate, modified $Fe(OH)_3$ solid phase with a fixed concentration of As. Although Fe oxyhydroxide is represented as $Fe(OH)_3$ in reaction 3, the actual formula for this solid phase in the geochemical model is $Fe_{0.995}As_{0.005}(OH)_3$. This As/Fe ratio is based on results from the Fe reduction core experiment which is discussed later.

A diffuse layer surface-complexation model was used to model adsorption of As. Two different types of sites were used; $Surf_T\_OH$ represents the total concentration of surface complexation sites, which were used to model adsorption/desorption of H^+:

$$Surf_T\_OH + H^+ = Surf_T\_OH_2^+ \quad (4)$$

$$Surf_T\_OH = Surf_T\_O^- + H^+ \quad (5)$$

These sites act as a buffer for pH control. Equilibrium constants for both reactions were obtained from Dzombak and Morel (1990). The Log $K_{H+(1)}$ for reaction 4 is 7.29; the Log $K_{H+(2)}$ for reaction 5 is -8.93.

A subset of sites ($Surf_S\_OH$) were used to model adsorption of As species:

$$\text{Surf}_S\text{\_OH} + H_2AsO_4^- = \text{Surf}_S\text{\_HAsO}_4^- + H_2O \quad (6)$$

$$\text{Surf}_S\text{\_OH} + HAsO_4^{-2} = \text{Surf}_S\text{\_AsO}_4^{-2} + H_2O \quad (7)$$

For the pH values of leachate in these experiments (6.5 to 7.5), the As(V) species in reactions 6 and 7 predominate. Arsenate concentrations were below detection limits throughout the experiments; however, after reactive organic carbon concentrations in the core decreased to the level where reduction of O_2 was incomplete, geochemical modeling predicted oxidation of As(III) to As(V). Because As(V) was not detected in core effluent, it was not possible to derive equilibrium constants for these adsorption reactions on this core material, and the equilibrium constants listed in Dzombak and Morel (1990) were used in the model. For reaction 6, the Log $K_{As(V)1}$ was 23.51 and for reaction 7, the Log $K_{As(V)2}$ was 10.58. Arsenite adsorption was modeled by:

$$\text{Surf}_S\text{\_OH} + H_3AsO_3^o = \text{Surf}_S\text{\_H}_2\text{AsO}_3^o + H_2O \quad (8)$$

The species $H_3AsO_3^o$ predominates for the pH range of these experiments. The equilibrium constant, Log $K_{As(III)} = 7.00$, was obtained by fitting the As(III) data in leachate from core 1. This constant is greater than the Log K of 5.41 reported by Dzombak and Morel (1990) for adsorption of As(III) by x-ray amorphous Fe oxyhydroxide (ferrihydrite).

Phosphate was present in leachate from the cores and likely competes with As for adsorption sites. The concentration of P(V) in the pore water and first 25 pore volumes of effluent from core 1 was 0.3 mg/L, gradually decreasing to about 0.05 mg/L by the end of the experiment. Phosphate in the pore water and first 30 pore volumes of core 2 was 0.1 mg/L, gradually decreasing to about 0.02 mg/L by the end of the experiment. The effects of P(V) on adsorption of As, as well as other potential competing species such as HCO_3^- and $H_4SiO_4^o$, were not included in the geochemical model to reduce its complexity. Any effect that these species might have on As adsorption was included in the Log K for As(III) that was fit to the experimental data. The same holds true for potential effects of cations.

A summary of the parameters used in the model is provided in Table 3. Dispersivity was determined from breakthrough of nonreactive bromide. Input for PHREEQC requires units of mass per liter of solution. Therefore, concentrations that were directly measured in the solid phase including the mass of solids, and concentrations of $Fe_{0.995}As_{0.005}(OH)_3$ and reactive organic

Table 3. Parameters used in reactive transport modeling

Parameter	Core 1	Core 2
Core length, m	0.38	0.46
Number of cells	19	23
Dispersivity, m	0.0096	0.0096
Average velocity, m/d	0.2	0.2
Solids, g/L	3750	3750
Porosity	0.4	0.4
$Fe_{.995}As_{.005}(OH)_3$, mol/L	0.05	0.05
Surface sites ($Surf_T\_OH$), μmol/L	5000	5000
Surfact sites ($Surf_S\_OH$), μmol/L	100	100
Surface complexation		
Log $K_{H+(1)}$	7.29	7.29
Log $K_{H+(2)}$	-8.93	-8.93
Log $K_{As(V)1}$	23.51	23.51
Log $K_{As(V)2}$	10.58	10.58
Log $K_{As(III)}$	7.00	7.00
Reactive organic carbon, mmol/L	60	38
Maximum rate constant for decay of organic carbon by O_2, 1/s	2.0e-07	2.0e-07
Rate constant for decay of organic carbon by Fe(III), 1/s	6.7e-09	6.7e-09

carbon, were converted using a measured porosity of 0.4 and an assumed solid density of 2.65 g/cm^3.

The concentration of $Fe_{0.995}As_{0.005}(OH)_3$ was estimated to be 0.05 mol/L, about 20% of the average Fe concentration measured in the NH_2OH-HCl leaches. The As/Fe ratio was calculated from the Fe reduction core experimental data. Adsorption reactions were assumed to occur on a separate Fe oxyhydroxide phase. Surface site concentrations were based on data from Dzombak and Morel (1990). They estimated the total concentration of surface sites for ferrihydrite as 0.205 mol/mol Fe. However, aquifer solids generally contain Fe oxyhydroxides such as goethite that have a smaller adsorption capacity than ferrihydrite. For example, Sun and Doner (1998) measured about 40% less adsorption of As(III) and As(V) by goethite than ferrihydrite. The concentration of sites used to model As adsorption in our experiments was used as a fitting parameter. We chose a total surface site concentration for cores 1 and 2 that was half of the concentration suggested by Dzombak and Morel (1990).

The concentrations of reactive organic carbon for the model were 20% of the total organic carbon measured in the cores. This percentage was calculated from the measured amount of O_2, Fe(III) and Mn(IV) reduced during the core 1 experiment.

4. RESULTS

4.1 Iron Reduction Core Experiment

The purpose of this experiment was to test the hypothesis that As in ground water that has been contaminated by leachate from the landfill could be a result of reductive dissolution of an As-enriched Fe oxyhydroxide phase. An alluvial core consisting of fine sand and silt (MW95-16S, Fig. 1) uncontaminated by landfill leachate was used. Uncontaminated ground water (Table 2) was spiked with 120 mg/L DOC in the form of sucrose ($C_{12}H_{22}O_{11}$) and eluted through the core. In this experiment, about 8 pore volumes were required to establish a sufficient bacterial population to completely reduce O_2 in the eluent and initiate reductive dissolution of Fe oxyhydroxide. Ferrous iron was first detected in pore volume 8, and As(III) in pore volume 10 (Figs. 3a-3b). Concentrations of both As(III) and Fe(II) rapidly increased until about pore volume 50. After pore volume 50, there was considerable variability in the concentrations of both As(III) (100-300 μg/L) and Fe(II) (30-60 mg/L); however, the average concentrations of As(III) (about 200 μg/L) and Fe(II) (about 45 mg/L) were relatively constant between pore volume 50 and 160. The large range in concentrations of As(III) and Fe(II) after pore volume 50 was a result of the effect of different velocities on the reductive dissolution of Fe oxyhydroxide. During periods of no flow, the eluent had more time to equilibrate with the solid phases; the initial pore volume that eluted upon resumption of flow contained high concentrations of As(III) and Fe(II). During periods of continuous flow, there was less time for reaction and concentrations decreased, usually approaching steady state for a particular flow rate. By pore volume 160, flow through the core became impeded to the extent that water could no longer be eluted through the core, presumably because accumulation of bacterial biomass and reaction by-products occluded porosity. Examination of the core material after the experiment was stopped revealed significant amounts of black material coating the grains. Geochemical modeling indicated supersaturation with respect to siderite; therefore, precipitation of an iron carbonate phase could also have contributed to blockage of flow through the core.

The concentrations of As(III) and Fe(II) in leachate from this experiment were simulated with the same geochemical model calibrated to the data from core 1 (discussed in a later section). The rate constant used to model

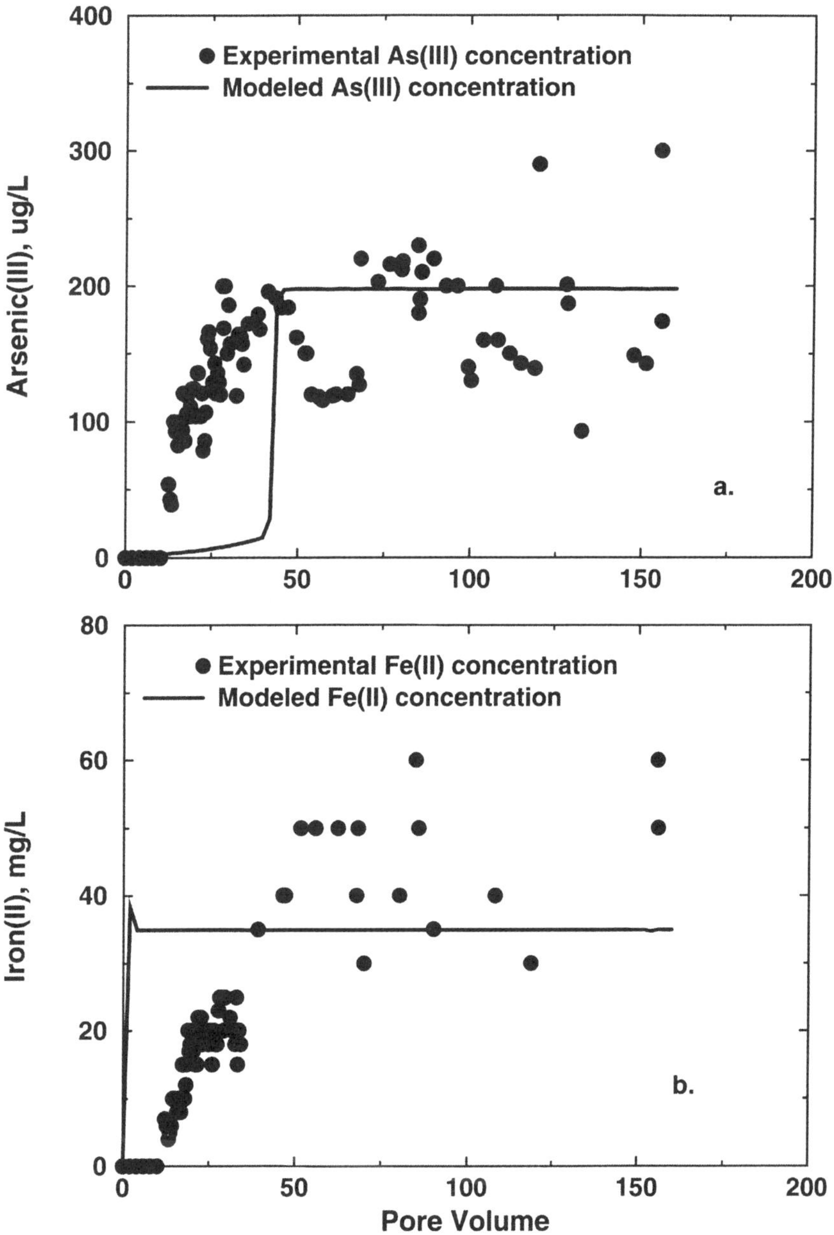

Figure 3. Experimental and modeled As(III) concentrations (a) and Fe(II) concentrations (b) in leachate from the Fe reduction core experiment.

reduction of $Fe_{0.995}As_{0.005}(OH)_3$ in core 1 did not account for the time required to establish a bacterial colony; therefore, the model predicted an earlier increase in Fe(II) concentrations than was observed (Fig. 3b). After steady-state concentrations were reached at about pore volume 50, the model underpredicted average Fe(II) concentrations by about 8 mg/L. Transport of As(III) through the core was mediated by adsorption/desorption reactions. The model predicted greater adsorption of As(III) during the first 50 pore volumes than was observed (Fig. 3a). However, the model did simulate the average As(III) concentrations after pore volume 50.

There was a correlation between As(III) and Fe(II) concentrations in this experiment (correlation coefficient of 0.79; Fig. 4), indicating an association between As and Fe oxyhydroxides. The average As/Fe ratio was 0.67% when converted to a weight basis (0.5% based on molar concentrations). This ratio is about 7 times greater than the average measured by Colman and Lyford (submitted) in their NH_2OH-HCl extractions of uncontaminated aquifer solids. Their results measured As/Fe ratios, on a weight basis, ranging from 0.008-0.29% (average 0.095%, n=18). The differences in As/Fe ratios between these two studies could be a result of differences in the initial ratios of As/Fe, and/or the NH_2OH-HCl dissolution procedure. In addition to dissolving the As-enriched fraction, NH_2OH-HCl extractions also might have dissolved Fe oxyhydroxides that had much lower concentrations of As. Readsorption of As during the NH_2OH-HCl extractions was evaluated by adding a known concentration of As to replicates for about 20% of the core samples. Complete recovery of all added As indicated no readsorption during the extraction procedure.

The higher As/Fe ratio in the Fe reduction core experiment also could have been caused by precipitation of siderite ($FeCO_3$) within the core. Bicarbonate concentrations were not measured in core leachate; however, geochemical modeling indicated that pore water should have been supersaturated with respect to $FeCO_3$ based on the amount of organic carbon that had to be oxidized to produce the observed Fe(II) concentrations in leachate. Precipitation of siderite was not modeled. The As/Fe ratio from the Fe reduction experiment was used in the geochemical model rather than the lower ratios from the NH_2OH-HCl extractions because geochemical conditions in the core experiment were more likely to be representative of those occurring in the aquifer.

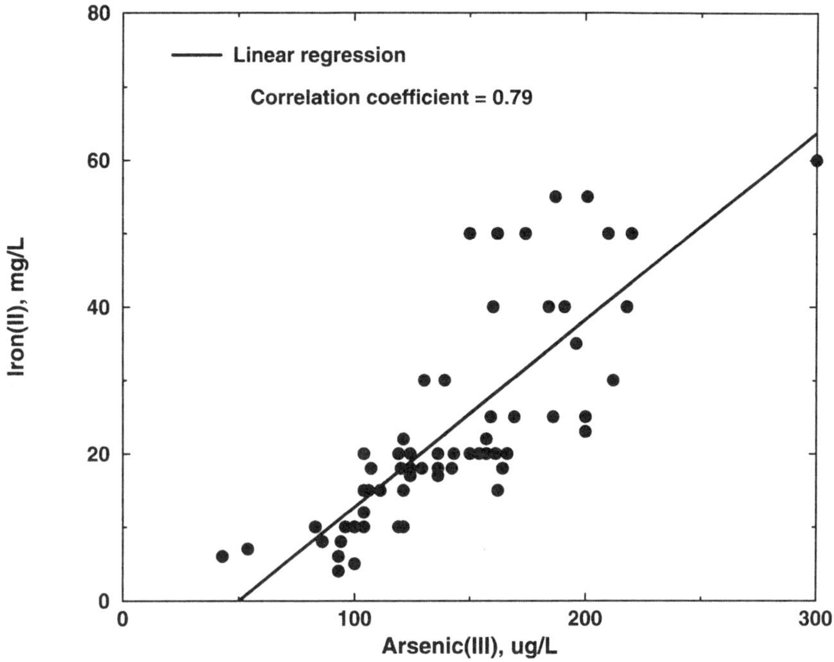

Figure 4. Correlation between Fe(II) and As(III) concentrations in leachate from the Fe reduction core experiment

4.2 Remediation Experiments

4.2.1 Dissolved oxygen

Dissolved oxygen concentrations in leachate from cores 1 and 2 are shown in Figs. 5a and 5b. Dissolved oxygen in the uncontaminated ground water eluent was 6 mg/L. Almost all of the O_2 in the first 220 pore volumes of ground water eluted through core 1 and the first 150 pore volumes eluted through core 2 were consumed within the core. Dissolved oxygen (<0.2 mg/L) measured in leachate during the first 25 pore volumes from both cores may be a result of reaction kinetics related to the variable flow rate during this early phase of both experiments. During periods of no flow, reactions within the cores had from 16 h (overnight) to 60 h (overweekend) to equilibrate. Dissolved oxygen concentrations in the initial pore volume of leachate following these periods of no flow were below the laboratory detection limit of 0.005 mg/ L. Subsequent pore volumes leached at 0.8 m/d

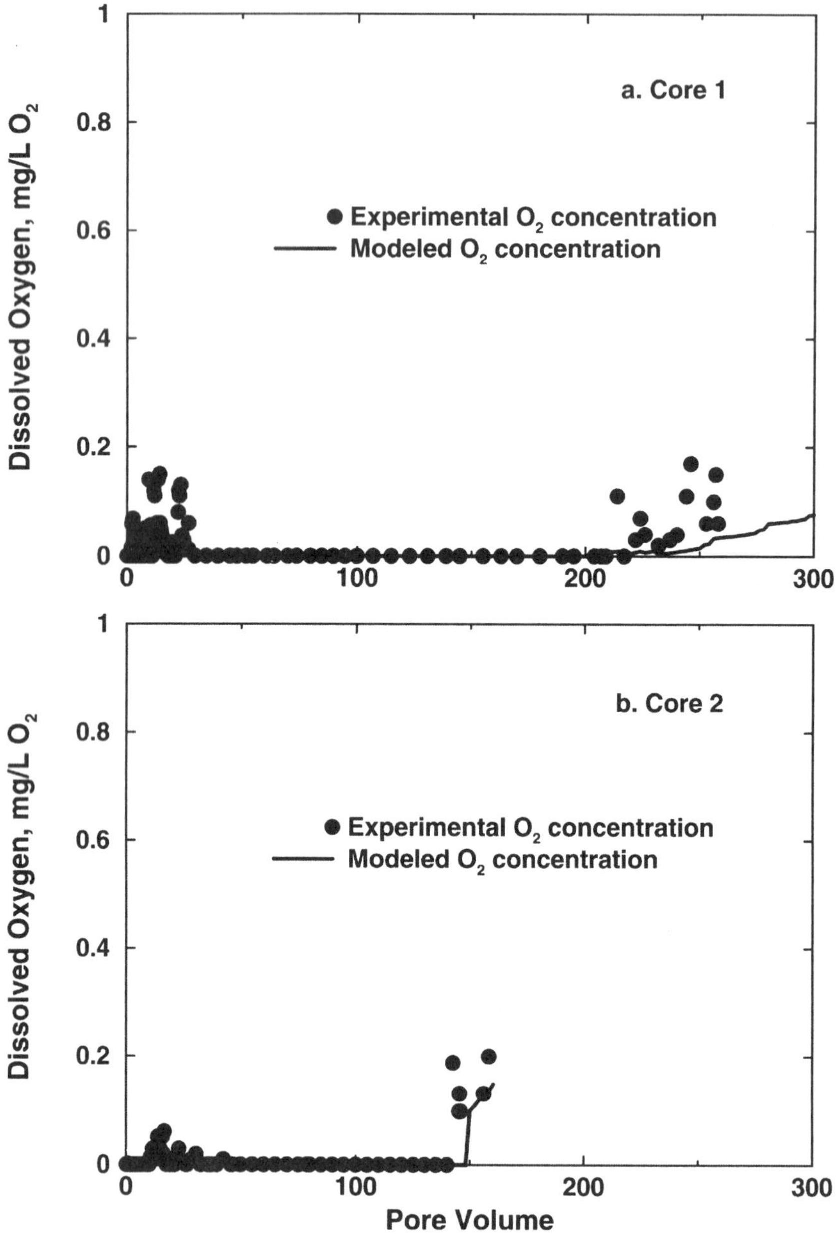

Figure 5. Experimental and modeled O_2 concentrations in leachate from core 1 (a) and core 2 (b). Dissolved oxygen in uncontaminated ground water eluent was 6 mg/L.

resulted in insufficient time for complete reaction and O_2 was detected in leachate. After a constant steady-state velocity of 0.2 m/d was established, O_2 concentrations in leachate remained below the detection limit until O_2 began to break-through at about pore volume 220 for core 1 and pore volume 150 for core 2. The earlier breakthrough of O_2 from core 2 than from core 1 is indicative of the lower concentration of reactive organic carbon in core 2.

The most plausible explanation for loss of oxygen in the eluent ground water is reduction by organic carbon. Core 1 contained 0.08 mol/kg total organic carbon and core 2 contained 0.05 mol/kg total organic carbon. Whether the reaction between O_2 and organic carbon occurred in the solid phase or in solution could not be determined; however, DOC was detected in leachate throughout the experiment (Fig. 6). Initial pore water DOC concentrations were 45 mg/L in core 1 and 31 mg/L in core 2 (not plotted in Fig. 6). Concentrations in leachate from both cores rapidly decreased to less than 1 mg/L within 20 pore volumes; DOC concentrations then gradually decreased to about 0.1 mg/L by the end of the experiments.

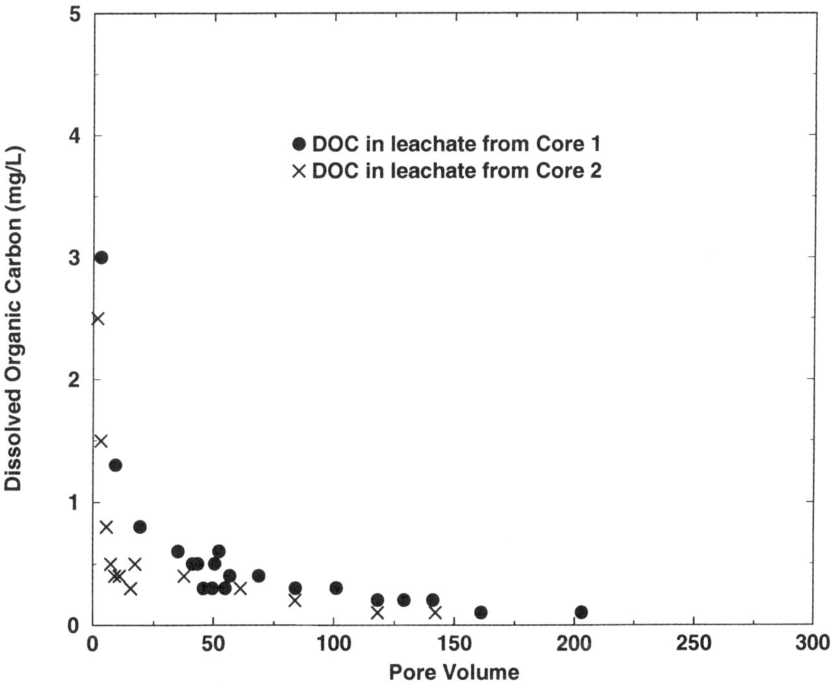

Figure 6. Dissolved organic carbon concentrations in leachate from cores 1 and 2. Initial DOC concentration in pore water of core 1 was 45 mg/L C, and in pore water from core 2 was 31 mg/L C. Dissolved organic carbon in uncontaminated ground water eluent was below detection.

Reduction of O_2 by organic carbon according to reaction 1 produces H_2CO_3, most of which dissociates to H^+ and HCO_3^- at the pH values of these experiments (between 6.5 and 7.5). Therefore, evidence for reaction 1 can be inferred from leachate HCO_3^- concentrations. The stoichiometry of reaction 1 indicates that one mole of HCO_3^- should be produced for every mole of O_2 that is reduced. Thus, reduction of 6 mg/L O_2 in the eluent should produce about 11 mg/L HCO_3^-. Reduction of Fe and Mn oxyhydroxides (discussed later) would have produced an additional 3 mg/L of HCO_3^-. In addition, dissolution of carbonate minerals in the cores could also have contributed HCO_3^- to solution. Geochemical modeling indicates that siderite was supersaturated in the pore water of both cores (saturation index = 1.4 for core 1 and 2.3 for core 2), an indication that $FeCO_3$ could have been present in both cores prior to the start of the experiments. Calcite was undersaturated in core 1 (saturation index = -0.91), but was supersaturated in core 2 (saturation index = 0.93).

Bicarbonate in the initial pore water of core 1 was 630 mg/L, and the pore water of core 2 contained 1240 mg/L HCO_3^-. Eluent HCO_3^- was 30 mg/L. Concentrations rapidly decreased to less than 100 mg/L within the first 10 pore volumes (Fig. 7). Between pore volumes 10 and 90, HCO_3^- decreased from 100 mg/L to about 60 mg/L. These concentrations are greater than can be accounted for by oxidation of organic carbon by O_2, Fe(III), and Mn(III,IV) and indicate an additional contribution, likely from dissolution of carbonate minerals and/or desorption of HCO_3^-. After pore volume 90, HCO_3^- concentrations remained relatively constant between 43 mg/L and 50 mg/L. These concentrations are close to what would be expected when adding the eluent HCO_3^- (30 mg/L) to the HCO_3^- produced from reduction of O_2 (11 mg/L), and Fe and Mn oxyhydroxides (3 mg/L). Although this explanation simplifies the combination of reactions that most likely were occurring in the cores, it supports O_2 reduction by organic carbon.

An estimate of the concentration of reactive organic carbon in each core was determined for the geochemical model. This concentration (20% of the total organic carbon in the cores) was based on the amount required to reduce O_2 in core 1 (about 15%), the amount required to reduce Fe and Mn oxyhydroxides in core 1 (about 3%), plus an additional 2% that was included to account for the fact that O_2 reduction was continuing when the experiments were stopped. The concentration of reactive organic carbon used in the model was 0.016 mol/kg (0.06 mol/L) for core 1 and 0.01 mol/kg (0.038 mol/L) for core 2. These results indicate that measurements of total organic carbon alone do not provide sufficient information for predicting the reducing capacity of aquifer solids contaminated by organic matter. It is the reactive fraction of the organic carbon that is an important indicator of the persistence of reducing conditions.

The maximum rate that organic carbon reacted with oxygen (2.0e-07 1/s) was determined by fitting the rate constant to the O_2 concentrations in leachate from core 1 (Fig. 5a). The model used a constant velocity of 0.2 m/d which did not account for the velocity fluctuations during the first 25 pore volumes. Therefore, the O_2 concentrations measured in leachate during the first 25 pore volumes were not simulated. The same rate constant was then used to simulate O_2 in leachate from core 2 (Fig. 5b). The model successfully simulated the observed breakthrough of O_2 at pore volume 150.

These results are derived from only two cores in the contaminated aquifer; the nature of breakthrough curves for O_2 from other parts of the aquifer might differ depending on the reactivity and amount of organic carbon. However, the results do indicate that, given a reasonable estimate of reactive organic carbon (with respect to redox reactions), it should be possible to model the persistence of reducing conditions after the source of contamination has stopped.

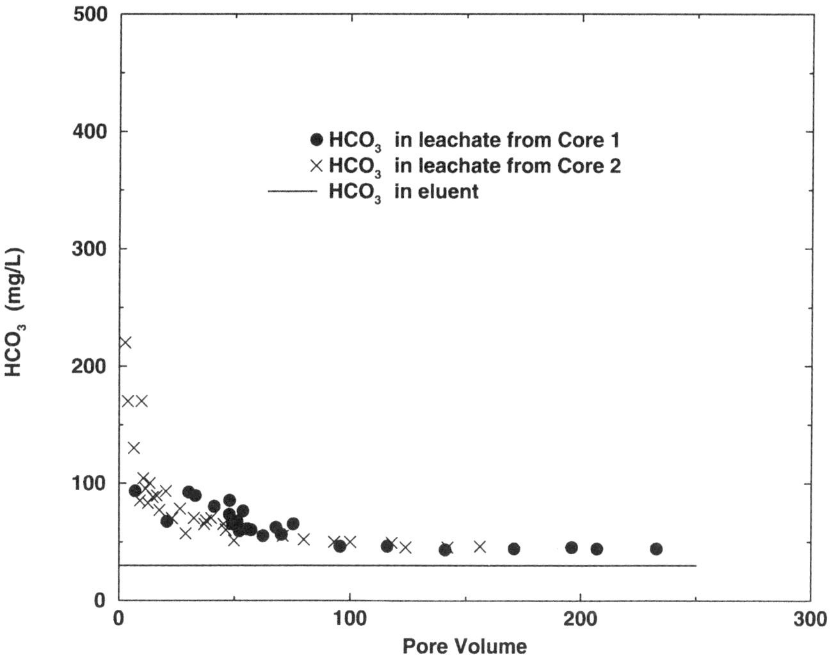

Figure 7. Bicarbonate concentrations in leachate from cores 1 and 2. Initial HCO_3^- concentration in pore water of core 1 was 630 mg/L, and in pore water from core 2 was 1240 mg/L. Bicarbonate concenetration in the uncontaminated ground water eluent was 30 mg/L.

4.2.2 Iron

Total dissolved Fe concentrations in ground water contaminated by Area 4 leachate were much greater than dissolved Fe in uncontaminated groundwater (Table 1). Speciation of Fe in these ground water samples was not determined; however, geochemical modeling indicated that the majority (>99%) of Fe was in the ferrous oxidation state. Although some of this Fe might have originated in the landfill, most appears to result from the reduction of Fe oxyhydroxides in the aquifer by organic carbon. In water from seven of the eight wells with high concentrations of dissolved Fe, O_2 concentrations were below the field detection limit of 0.05 mg/L. Dissolved O_2 in water from the other well was 0.15 mg/L. Dissolved Mn concentrations in water from these wells were also significantly greater than in uncontaminated ground water (Table 1), an indication of reductive dissolution of Mn oxyhydroxides.

The concentration of Fe in pore water from core 1 was 50 mg/L, and the concentration in pore water from core 2 was 32 mg/L (Table 2). Chemical analyses measured only Fe(II); Fe(III) was below the detection limit. The concentration of Fe(II) in leachate from the contaminated cores rapidly decreased as Fe-free uncontaminated ground water displaced the contaminated pore water (Figs. 8a-8b). Within a few pore volumes, Fe(II) concentrations were less than 5 mg/L. For the remainder of the experiments, Fe(II) concentrations decreased at a much slower rate. The low concentrations (<5 mg/L) measured in the first 25 pore volumes of leachate coincided with higher flow velocities and measurable O_2 concentrations. Based on the mass of Fe(II) in leachate from core 1, the concentration of reactive organic carbon necessary to reduce Fe(III) to Fe(II) was about 2% of the total organic carbon.

Adsorption/desorption could have affected Fe(II) concentrations in leachate from both columns; however, adsorption reactions involving Fe(II) were not considered in the geochemical model. Separation of the effects of adsorption on aqueous Fe(II) concentrations from reductive dissolution of Fe oxyhydroxides was not possible given the data collected during these experiments. Therefore, iron concentrations were only modeled with reaction 3, reduction of $Fe_{0.995}As_{0.005}(OH)_3$ by organic carbon.

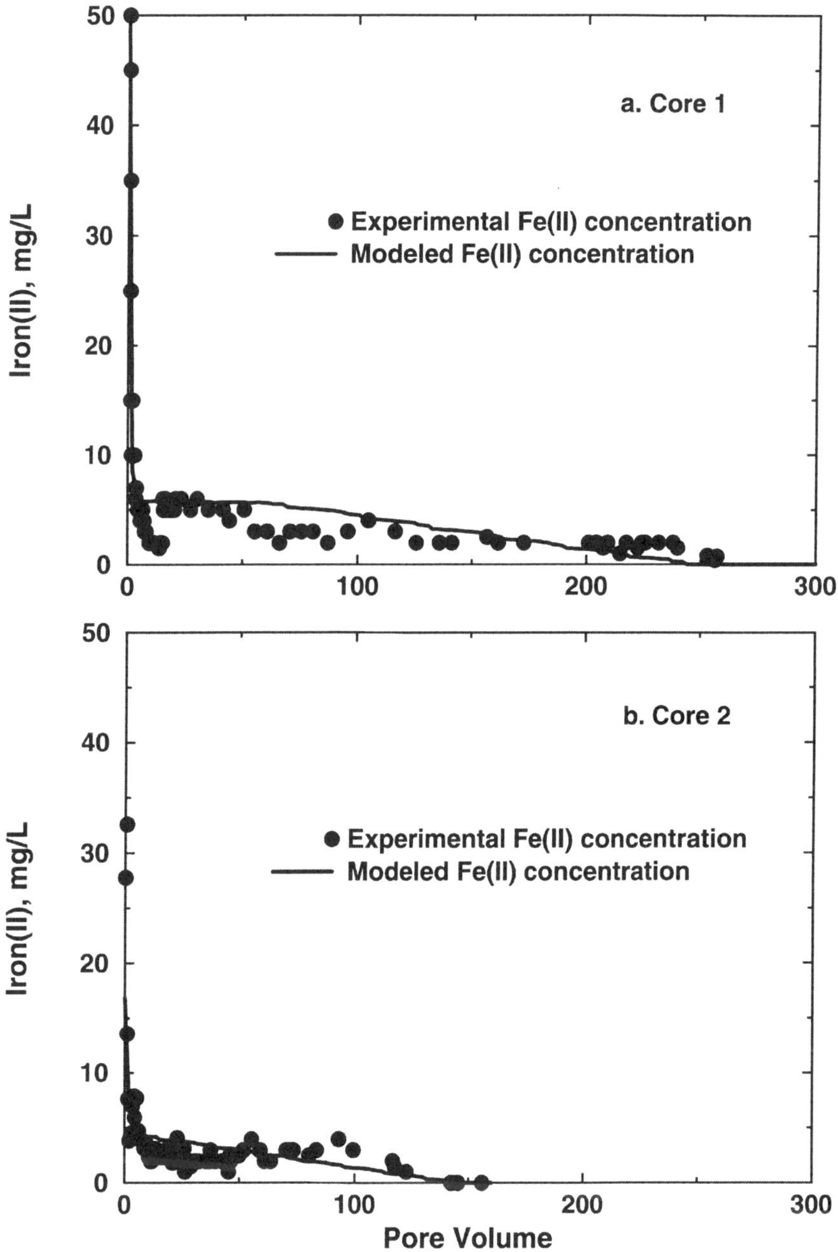

Figure 8. Experimental and modeled Fe(II) concentrations in leachate from core 1 (a) and core 2 (b). Iron in uncontaminated ground water eluent was below detection.

Data from the Fe reduction experiment were used to define the As concentration of $Fe(OH)_3$. The PHASES data block of the input file for PHREEQC redefined $Fe(OH)_3$ according to the following reaction:

$$(Fe)_{0.995}(As)_{0.005}(OH)_3 = 0.995Fe^{+3} + 0.005H_3AsO_3 + 2.985OH^- \quad (9)$$

Log K = -41.3

Ferrous iron concentrations decreased to below detection in leachate from both cores at the approximate pore volume that O_2 began to be detected in leachate. The modeled Fe(II) concentrations in leachate from core 1 (Fig. 8a) were obtained by fitting the rate constant for reaction 3 [using $Fe_{0.995}As_{0.005}(OH)_3$] and the Log K for reaction 8 to the experimental data. When O_2 concentrations in the cores had decreased to less than 0.02 μmol/L, reduction of As-enriched Fe oxyhydroxide was initiated at a rate of 6.7e-09 1/s. This rate constant was then used to model the data from core 2, and produced comparable results for Fe(II) concentrations in leachate and for the pore volume at which Fe(II) concentrations decreased to below detection (Fig. 8b).

4.2.3 Manganese

Reduction of Mn oxyhydroxides concurrent with Fe oxyhydroxides was apparent in both cores (data not shown). Manganese concentrations followed a pattern similar to Fe(II); however, initial Mn concentrations were much lower than Fe(II) (Table 2). Manganese concentrations in leachate were about one-half of the Fe concentrations. Although reduction of Mn oxyhydroxides was not simulated in the geochemical model, the total mass of Mn in leachate from core 1 was used to estimate the amount of reactive organic carbon that had been oxidized. Assuming that all of the Mn was reduced from Mn(IV) to Mn(II), the concentration of organic carbon oxidized by Mn (IV) was about 1% of the total organic carbon concentration in the core 1 solids.

4.2.4 Arsenic

Arsenite concentrations in leachate from cores 1 and 2 are shown in Figs. 9a and 9b. Although 20% of the total As concentration measured in contaminated ground water samples collected in 1997 (Table 1) was determined to be As(V), all of the As in the core experiments was As(III). Arsenite concentrations in leachate from core 1 (initial pore water

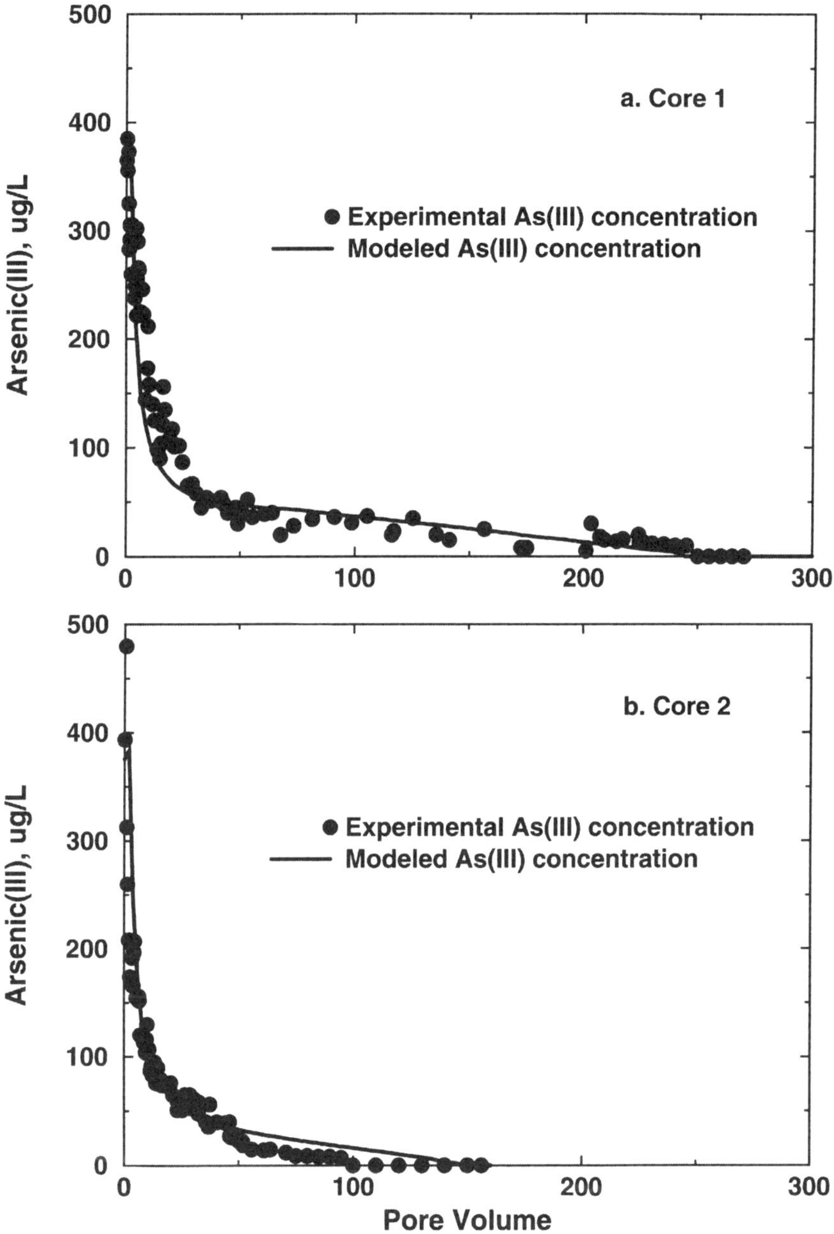

Figure 9. Experimental and modeled As(III) concentrations in leachate from core 1 (a) and core 2 (b). Arsenic in uncontaminated ground water eluent was below detection.

concentration = 390 μg/L) and core 2 (initial pore water concentration = 480 μg/L) were controlled by three mechanisms. The rapid decrease in As(III) during the first 15 pore volumes of both experiments can be attributed to displacement of As(III) in the pore water by As-free uncontaminated ground water and desorption of high concentrations of As(III) from the solid phase. Between pore volumes 15 and 50, As(III) decreased at a slower rate; concentrations were likely dominated by desorption. The rate of decrease in As(III) was much more gradual after pore volume 50; concentrations were controlled by a combination of release of As(III) to solution from the reductive dissolution of $Fe_{0.995}As_{0.005}(OH)_3$ and adsorption/desorption equilibria. Arsenite concentrations decreased to below detection at about the same pore volume that Fe(II) concentrations became undetectable (coincident with breakthrough of O_2).

The equilibrium constant for adsorption of As(III) was adjusted to give the best match to As(III) concentrations in leachate from core 1 (Fig. 9a). The geochemical model predicted that as O_2 moved through the cores and conditions became oxidizing, As(III) should be oxidized to As(V) and strongly adsorbed, thus resulting in As concentrations below the detection limit at the pore volume when O_2 broke through.

The geochemical model simulated As(III) concentrations in the first 50 pore volumes of leachate from core 2. However the model simulated measurable As(III) to about pore volume 150, whereas the data in Fig. 9b show that As(III) concentrations decreased to the detection limit by pore volume 100.

The As(III) data from these core experiments have significant implications for remediation. More than 200 pore volumes were required to lower concentrations in leachate from core 1 to the drinking-water standard of 10 μg/L for As. Persistence of anoxic conditions in these cores appears to have had the most significant impact on the mobility of As. When conditions in the core were reducing, As(III) dominated and reductive dissolution of As-containing Fe oxyhydroxides provided a continual source of As. When the redox environment became oxidizing, as evidenced by breakthrough of O_2, As concentrations decreased to below detection. Geochemical modeling predicted that any remaining As(III) in the core oxidized to As(V) and was strongly adsorbed.

5. CONCLUSIONS

High concentrations of As in ground water beneath and downgradient of a municipal landfill appear to have been caused by organic carbon in leachate migrating from the landfill. Bedrock and alluvium on which the landfill was located contain Fe oxyhydroxide coatings that are enriched in As. Data from laboratory experiments support the hypothesis that reductive dissolution of these Fe oxyhydroxides releases As to solution. Arsenic concentrations in ground water are moderated however by adsorption on the solid phase.

Capping the landfill, thereby blocking the source of more reactive organic carbon, followed by allowing natural recovery of the aquifer, is expected to eventually restore the water quality of the aquifer to pre-landfill conditions. However, analyses of core material from contaminated areas of the aquifer show that significant concentrations of organic carbon have accumulated on the aquifer solids during the years of landfill operation. Leaching of contaminated cores with oxygenated, uncontaminated ground water indicated that some of this organic carbon is reactive to oxidants including O_2, Fe(III), and Mn(III, IV). Data from these experiments have shown that more than 200 pore volumes of uncontaminated ground water may be needed to reestablish oxidizing ground water conditions.

When O_2 concentrations in ground water began to increase in the laboratory core leaching experiments, reductive dissolution of Fe oxyhydroxides stopped, eliminating mobilization of As. Arsenite remaining in solution or adsorbed as an As(III) species is predicted to oxidize to As(V) which is more strongly adsorbed. In these experiments, a decrease in As concentrations to below the drinking-water standard of 10 µg/L was associated with increasing concentrations of O_2, and with Fe concentrations that were below detection.

Data from one of the laboratory experiments were used to develop and calibrate a geochemical model. This model was successful in simulating concentrations of O_2, Fe, and As in leachate from other cores and may be suitable for modeling natural remediation processes in the contaminated aquifer.

Results from this study could be applicable to other areas in New England or elsewhere where high concentrations of As are found in ground water associated with landfills. The data show that, even after the source of landfill leachate is eliminated, lowering As to pre-landfill concentrations via natural remediation could require several years.

ACKNOWLEDGMENTS

The authors thank James Tindall, Roger Lee, and Robert Frod for providing constructive reviews of this manuscript.

Chapter 14

Modeling in situ iron removal from groundwater with trace elements such as As

C.A.J. Appelo[1] and W.W.J.M. de Vet[2]
[1]Hydrochemical Consultant, Valeriusstraat 11
1071 MB Amsterdam, The Netherlands
[2]Hydron-ZH, Postbus 122, 2800 AC Gouda, The Netherlands

The cyclic injection of oxygenated water in an aquifer may induce in situ iron removal from groundwater. During injection of aerated water, sorbed ferrous iron is displaced by cations, oxidized in the pore space, and precipitated as ferric iron oxyhydroxide. During pumping, ferrous iron is sorbed from groundwater on the exchange and sorption sites, and the breakthrough of dissolved iron is retarded. Other trace elements such as arsenic may be eliminated jointly with iron by sorption or co-precipitation.
The volume of iron-free groundwater that can be pumped per volume of injected, aerated water defines the efficiency of the process. The efficiency is determined by the ratio of the retardations of oxygen during injection and of iron during pumping. This chapter shows how these retardations can be calculated for given water qualities and aquifer compositions.
The first seven cycles of an in situ iron removal project in The Netherlands were simulated with the hydrogeochemical transport model PHREEQC (version 2). The concentration changes of CH_4, NH_4^+, Mn^{2+}, Fe^{2+}, PO_4^{3-} and As are discussed in detail. Arsenic shows concentration jumps in pumped groundwater which are related to oxidation/reduction and sorption/desorption reactions resulting from the water quality variations.

1. INTRODUCTION

In situ iron removal is a useful technique for reducing the iron concentration in groundwater pumped for consumption or industrial purposes (Hallberg and Martinell, 1976; Meyerhoff, 1996; Rott and Lamberth, 1993; Van Beek, 1980). The technique entails the periodic injection of a volume of aerated or oxygenated water in an aquifer, followed by pumping of the injected water and subsequently of groundwater in which the iron concentration is lower than in native groundwater. Iron in the aquifer is oxidized during injection of the oxygenated water and precipitates as iron-oxyhydroxide. The loss of iron also liberates cation exchangers which are filled again when pumping is resumed and groundwater with dissolved iron contacts the cation exchange sites in the aquifer (Appelo et al., 1999). The precipitated iron forms a sorber which augments the exchange capacity of the aquifer. Moreover, iron-oxyhydroxide is a well known scavenger for heavy metals, and by analogy with above ground operations (Benjamin et al., 1996) it can be expected that together with iron, the concentrations of heavy metals in groundwater will decrease as well. A loss of arsenic with in situ iron removal has been reported already (Rott et al., 1996).

However, it is still fairly uncertain which chemical or biochemical process takes the lead in changing the concentrations of heavy metals in aquifers. Sorption on ferrihydrite can now be modeled in great detail using data from laboratory experiments, but to discern between sorption and desorption and uptake and release in a growing and reordering precipitate is difficult even in the laboratory (Eick et al., 1999; Fuller et al., 1993; Gerth et al., 1993; McKenzie, 1980; Nesbitt et al., 1995; Waychunas et al., 1993; Zachara et al., 1987). It is also clear that other iron-oxyhydroxides than ferrihydrite exhibit other sorption behavior, and have other constants in the surface complexation model (Mathur, 1995). Another drawback is that complexation constants for Fe^{2+} and HCO_3^- are lacking, although these are important species in groundwater that will influence sorption of other elements. Thus, it is important to perform field studies which show the applicability of our models for sorption in natural systems (Davis et al., 1998; Kent et al., 2000; Runkel et al., 1999; Stollenwerk, 1998) and this chapter follows that line.

In particular, data were interpreted from the groundwater pumping station Schuwacht of Hydron-ZH (Gouda, The Netherlands) where the water quality was monitored for the first 7 cycles of in situ iron removal. Emphasis was placed on the effects for arsenic concentrations, prompted by fear that the subsoil might become contaminated by trace elements which are incorporated in the iron precipitates. The general principles of in situ iron removal are not difficult and the efficiency of the operation can be estimated with simple formulas. However, for arsenic also redox effects and complicated displacements from sorption sites are significant which cannot be calculated without numerical models. This chapter describes the general principles of in situ iron removal and discusses details of the Schuwacht plant starting with simple hand calculations for basic insight and continuing with more comprehensive modeling for As.

2. PRINCIPLES OF IN SITU IRON REMOVAL

Concentration profiles of oxygen and iron at three successive steps of in situ iron removal are shown in Figure 1. During injection of 1000 m^3 aerated water, groundwater with dissolved iron is displaced in the aquifer. If sorption sites would be absent, ferrous iron would simply move along with groundwater, and the reaction between oxygen and Fe^{2+} would be limited to small amounts due to mixing at the front. However, cations from injected water exchange for sorbed Fe^{2+} and the oxidation of this iron consumes oxygen. Thus, the oxygen front lags behind the injection water front. When the operation is switched to pumping, first the injected volume is withdrawn, with some front spreading as result of dispersion. Then, for some time, groundwater can be pumped with a reduced iron concentration, because ferrous iron is lost to the exchange sites as groundwater flows through the oxidized zone. After a fixed time, or when the iron concentration passes some limit, another volume of aerated water is injected and the next cycle (or run) of in-situ iron removal begins.

The efficiency of the process can be calculated if we neglect dispersion and limit the sorption reactions to the zone where oxygen has penetrated, and then consider how much iron can be sorbed in that part of the aquifer. Thus, first the position of the oxygen front at the end of the injection stage must be located. This position can be found from the reaction of oxygen with ferrous iron:

$$O_2 + 4Fe^{2+} + 4H^+ = 2H_2O + 4Fe^{3+} \quad (1)$$

All dissolved Fe^{2+} for this reaction comes from sorbed iron and the retardation of oxygen amounts to:

$$R_{O_2} = 1 + q_{Fe}/4m_{O_2} \quad (2)$$

where q_{Fe} is sorbed iron (mol/l pore water) and m_{O_2} is dissolved oxygen in injection water (mol/l). Sorbed iron in eqn. 2 is in exchange equilibrium with dissolved iron in groundwater and can be calculated with standard geochemical programs as will be demonstrated in section 3.

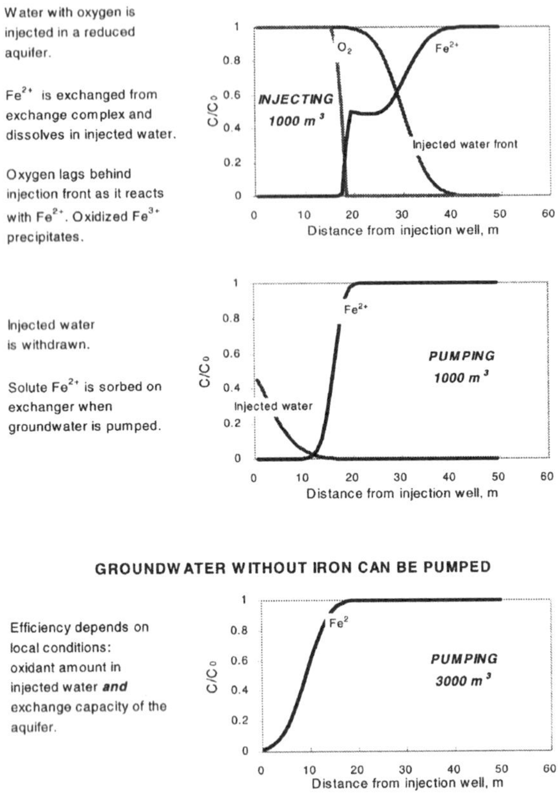

Figure 1. Oxygen- and iron-concentration profiles at three stages of an in-situ iron removal cycle.

For the case of linear flow, the distance between the oxygen front and the injection well is:

$$x_{O_2} = x_{inj} / R_{O_2} \tag{3}$$

where x_{inj} is the distance traveled by the injected water. The fraction of injected water from which oxygen is activily utilized in the process is:

$$f_{inj} = (x_{inj} - x_{O_2}) / x_{inj} = 1 - 1/R_{O_2} \tag{4}$$

Note that eqn. 4 implies that oxygen may not be reacting at all when the retardation equals 1, *i.e.* when sorbed iron is zero, or $q_{Fe} = 0$ in eqn. 2. On the other hand, all oxygen is used when sorbed iron is infinite. Thus, the efficiency of in situ iron removal depends on the sorption capacity of the aquifer for iron, and it will be low in a coarse, gravelly aquifer.

Now during pumping, if native groundwater with dissolved iron returns and flows along the emptied sorption sites, Fe^{2+} will be sorbing again, and iron is retarded with respect to groundwater flow. The retardation equals:

$$R_{Fe} = 1 + q_{Fe}/m_{Fe} \tag{5}$$

Equation (5) allows to calculate the volume of groundwater that can be pumped until the sorption sites are all filled again and the iron concentration of native groundwater arrives at the well. The volume of water in between the injection well and x_{O_2} is:

$$V_{inj}\,(1 - f_{inj}) = V_{inj} / R_{O_2} \tag{6}$$

And the volume of groundwater which can be pumped is:

$$V_{gw} = V_{inj} \times R_{Fe} / R_{O_2} \tag{7}$$

Note again, that eqn. 7 implies that iron arrives immediately at the well when the ratio of sorbed and dissolved iron is very small, and, of course, that it will never arrive when the concentration of dissolved iron is zero. Thus, the efficiency of in situ iron removal also depends on the concentration of iron in groundwater.

We can define the efficiency of in situ iron removal as:

$$E = V_{gw} / V_{inj} \tag{8}$$

which, according to eqn. 7, equals:

$$E = R_{Fe} / R_{O_2} \tag{9}$$

For a hand calculation, we may assume that q_{Fe} in eqn. 5 is the same as was used for calculating the reaction of oxygen in eqn. 2, and initially neglect the additional sorption of iron on the precipitate. Otherwise, if the sorption on the precipitated iron-oxyhydroxide is known, it can be used to increase q_{Fe} in eqn. 5 with respect to eqn. 2 by the appropriate factor.

3. SORPTION REACTIONS OF Fe^{2+}

Sorption in aquifers takes place mainly on clay minerals, organic matter and oxides, and is differently modeled depending on the properties of the solid. Clay minerals such as smectite and illite carry a charge due to structural substitutions which is constant at the pH's of groundwater. Sorption of cations is, in this case, a cation exchange reaction which does not affect the charge of the clay mineral. For example when Na^+ from injection water displaces Fe^{2+} from the exchange sites of the clay mineral, the reaction is:

$$Na^+ + \frac{1}{2}Fe\text{-}X_2 = Na\text{-}X + \frac{1}{2}Fe^{2+}; \quad \log K = -0.22 \tag{10}$$

Here, X indicates the cation exchanger, with a charge of X^-. The cation exchange capacity (*CEC*, $mmol_c/kg$) can be estimated with empirical formulas such as:

$$CEC\ (mmol_c/kg) = 7 \cdot (\%\ clay) + 35 \cdot (\%\ C) \tag{11}$$

where (% clay) and (% C) are the weight percentages of clay < 2µm and organic carbon, respectively (Appelo and Postma, 1993). The *CEC* can be recalculated to a capacity per liter groundwater by multiplying with the bulk density ρ_b (kg/dm^3) and dividing by the water filled porosity ε_w (-):

$$X^-\ (mmol_c/l) = CEC\ (mmol_c/kg) \cdot \rho_b / \varepsilon_w \tag{12}$$

Cation exchange in groundwater is a multicomponent process in which all the solute cations participate. It can be calculated easily with geochemical models such as PHREEQC-2 (Parkhurst and Appelo, 1999) which have databases with representative values of the exchange constants. An example PHREEQC-2 input file for calculating exchangeable iron is given in Table 1.

Sorption to iron-oxyhydroxide can be computed with the surface complexation model of Dzombak and Morel (1990). This model assembles the results of numerous laboratory experiments on sorption of trace elements to ferrihydrite (Hfo, hydrous ferric oxide, FeOOH). Ferrihydrite is a more or less amorphous substance which is found in nature in seepage zones of reduced, iron containing groundwater. Probably, it will be representative for the iron-oxyhydroxide which forms during in situ iron removal in aquifers, but this has not been verified yet.

The main difference between surface complexation and exchange is that the surface complexer will acquire a charge depending on the ions which are sorbed to the surface. The proton is an important charge-determining ion, and in distilled water ferrihydrite will carry a charge depending on the pH and be chargeless at a pH of about 8.1. Thus, at pH = 7, ferrihydrite has a positive charge and cations are bound only when their chemical affinity to the surface oxygens is sufficient to overcome the electrostatic repulsion. The

Table 1. PHREEQC-2 input file for calculating exchangeable and sorbed iron, and the undimensional distribution coefficient for iron.

```
# Download PHREEQC-2 via links in www.xs4all.nl/~appt
# or directly from www.geo.vu.nl/users/posv/phreeqc.html
# or wwwbrr.cr.usgs.gov/projects/GWC_coupled/phreeqc/index.html

# Part A: define iron oxide and groundwater composition
PHASES
 Ferrihydrite; FeOO2H3 + 3H+ = Fe+3 + 3H2O; log_k 2.0
EQUILIBRIUM_PHASES 1
 Ferrihydrite 0.0 32e-3     # 300 ppm Fe in sediment / 56 * 6
SOLUTION 1
 -units mg/l; -temp 10
 pH     7.1; pe    0.0 Ferrihydrite
 Na    77.3; K     5.3; Mg 13.1; Ca 93
 Amm    3.8; Fe    6.0; Mn 1.1
 Cl   134.0; S(6) 43.4; Alkalinity 308 as HCO3
 P     3.16   as PO4; As  10 ug/l
SAVE solution 1
END

# Part B: define exchangeable and sorbed iron
USE solution 1
EXCHANGE 1
 -equil 1; X 0.03     # fine sand, CEC = 5 mmolc/kg
SURFACE 1
 -equil 1; Hfo_w Ferrihydrite 0.2 5.3e4; Hfo_s Ferrihydrite 0.005
USER_PRINT
 10 Cplx_Fe = mol("Hfo_sOFe+") + mol("Hfo_wOFe+") + mol("Hfo_wOFeOH")
 20 print "FeX2 =", mol("FeX2"), " Fe on Hfo =", Cplx_Fe
 30 print " Kd  =", (mol("FeX2") + Cplx_Fe) / tot("Fe")
END
---------------------------------User print--------------------------
FeX2 =   1.8662e-04  Fe on Hfo =   5.2851e-04
 Kd  =   6.6517e+00
```

electrostatic effect is calculated with the Boltzmann factor, exp($-zF\psi/RT$), where the potential ψ is a function of the charge and the surface area of the ferrihydrite, and, in Dzombak and Morel's model, of the ionic strength of the solution. Chemical binding is distributed over weak and strong sites which exist in a proportion of 0.2 and 0.005 mol sites / mol ferrihydrite. The increasing complexation capacity with increasing amounts of ferrihydrite that precipitate during in situ iron removal can be modeled with PHREEQC-2 by coupling Hfo, the moles of the surface complex, to the mass of ferrihydrite in the system, cf. Table 1.

Surface complexation is a typical multi-component reaction, similar to cation exchange. The database for surface complexation includes complexation constants for major elements in groundwater such as Ca^{2+} and SO_4^{2-}, but not for Fe^{2+} and HCO_3^-. In the first instance, constants for these ions can be estimated with linear free energy relations (*LFER*'s) in which the properties of similar chemical systems are compared and interpolated (Dzombak and Morel, 1990). Thus, the surface complexation constant for Fe^{2+} is expected to lie in between the ones for Cd^{2+} and for Zn^{2+}, in line with the known differences of the association constants of these heavy metals with OH^- in water. For the weak sites, the *LFER* gives:

$$\text{Hfo\_wOH} + Fe^{2+} = \text{Hfo\_wOFe}^{+} + H^{+}; \qquad \log K = -2.5 \qquad (13)$$

The *LFER*-derived complexation constant *K* enabled to fit the recently published experimental sorption isotherm of Liger et al. (1999) very well, except at pH > 8 (Fig. 2). The deviation at higher pH is due to sorption of a hydroxy complex and the data of Liger et al. were used for optimizing new constants for the database which are given in Table 2. The complexation constant for the strong sites could not be derived from Liger's data because the concentration was low for ferrihydrite and high for Fe^{2+}. The value in Table 2 was obtained from new measurements by C. Tournassat with more appropriate concentrations in the experiment. It may be noted that the experimental value is much different from the one which can be estimated by interpolation with Cd^{2+} and Zn^{2+}, thus indicating that the *LFER* value needs to be checked whenever possible.

Another important ion in groundwater, HCO_3^-, is also lacking from the database in Dzombak and Morel's model. The bicarbonate ion is usually avoided in sorption experiments with oxides because of pH buffering and slow, kinetic exchange at the air-solution interface. The experimental data of Van Geen et al. (1994) for goethite were used to derive complexation constants for ferrihydrite which are listed in Table 2. Although goethite has a much higher crystallinity than ferrihydrite, it was found that the sorption envelope of HCO_3^- on ferrihydrite measured by Zachara et al. (1987) could be modeled with nearly the same constants (Appelo et al., 2002). The major effect of sorption of HCO_3^- on ferrihydrite is identical to what was noted already by Van Geen et al. for goethite, *i.e.* a large proportion of the surface sites will be occupied by HCO_3^- at the usual bicarbonate concentrations in groundwater.

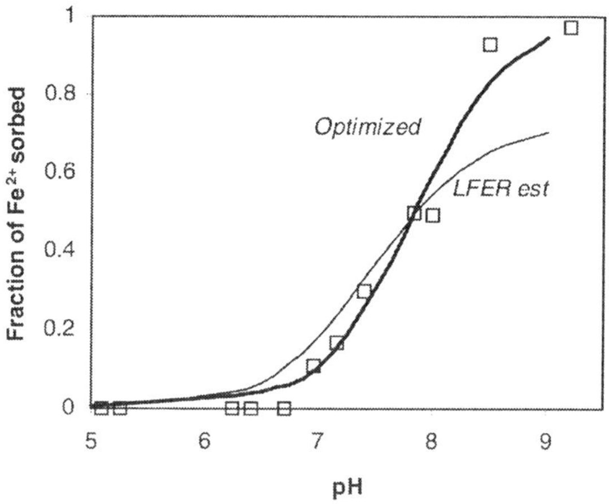

*Figure 2. Sorption edge of Fe^{2+} on ferrihydrite, data from Liger et al. (1999), model with surface complexation constants estimated from linear free energy relation (*LFER*) and model optimized on the data of Liger et al.*

Table 2. Surface complexation constants of Fe^{2+} and HCO_3^- on ferrihydrite (Appelo et al., 2002).

Hfo_wOH + CO_3^{2-} + H^+ = Hfo_wOCO_2^- + H_2O	log K 12.56[1]
Hfo_wOH + CO_3^{2-} + 2H^+ = Hfo_wOCO_2H + H_2O	log K 20.62[1]
Hfo_wOH + Fe^{2+} = Hfo_wOFe^+ + H^+	log K = -2.98[2]
Hfo_wOH + Fe^{2+} + H_2O = Hfo_wOFeOH + 2H^+	log K = -11.55[2]
Hfo_sOH + Fe^{2+} = Hfo_sOFe^+ + H	log K = -0.95[3]

[1]Optimized using data from Van Geen et al. (1994), slightly different in Appelo et al. (2002)
[2]Optimized using data from Liger et al. (1999)
[3]Optimized using data from C. Tournassat

4. A SPECIFIC EXAMPLE: PUMPING STATION SCHUWACHT, HYDRON-ZH

Pumping station Schuwacht of the drinking water company Hydron-ZH is a bank filtration unit with the pumping wells located at a distance of 70 - 200 m from the river Rhine near Gouda, The Netherlands. In 1998 and 1999, pumping from filter PP8 was periodically interrupted to inject a volume of treated and aerated groundwater. Water was injected at 30 m^3/h for 2 days and immediately afterwards pumping was resumed at 23 m^3/h for 40 days. The purpose of the injection was to ameliorate the conditioning of the raw water for the water treatment plant where iron, manganese, ammonia and methane are removed.

Pumping well PP8 is located in coarse sandy sediments of the Sterksel formation at a depth of 20-30 m bs. Chemical analysis of the sand provided an average of 0.5% calcite, 0.03% organic carbon, 300 ppm Fe as Fe-oxide, no pyrite, and less than 1 ppm As. The lower 7 m sediment contained 1% silt, the upper 3 m 5-7% silt and the porosity was 0.34. The sediment's exchange capacity was estimated to be 5 $mmol_c/l$ in the coarse sandy lower part and 30 $mmol_c/l$ in the upper 3 m.

Water in the well originates for almost 100% from the Rhine, but the water quality has changed during infiltration in the riverbed and passage through the aquifer. The concentrations of Ca^{2+}, Fe^{2+}, Mn^{2+}, NH_4^+ and the alkalinity have increased, and the groundwater contains methane. Methane is important for judging the suitability of an aquifer for in situ iron removal and for modeling the chemical processes, since it often indicates a redox state in which sulfides have formed and are stable. When sulfides are present, a rapid reaction with injected oxygen will have various adverse effects for the water quality, varying from decrease of pH to increase of heavy metals. However, methane formation is not expected in the Sterksel aquifer for the given, low concentration of organic carbon, but it may originate in the organic rich layers of the riverbed. Water sampled at various times showed that the methane concentration was inversely coupled to the SO_4^{2-} concentration and increased linearly as the ratio of SO_4^{2-} to Cl^- decreased. This indicates that methane is probably formed locally during infiltration in the river bed in part of the water after SO_4^{2-} has been reduced, and that this water is mixed with water without methane but with SO_4^{2-} in the well. In this case, sulfides may be present in the riverbed, but they are not expected in the aquifer and need not be considered for the in situ iron removal operation. Water quality analyses are given in Table 3.

4.1 Efficiency calculations

The oxygen concentration in the injected water was 0.28 mmol O_2 / l. According to reaction 1, this amount can oxidize 1.12 mmol Fe^{2+} / l. The iron concentration in native groundwater is 0.1 mmol Fe^{2+} / l, and the simplest calculation would predict an efficiency of 11.2, *i.e.* for every liter of injected water, 11.2 liters groundwater without iron may be pumped. However, it was noted above (reaction 2) that the efficiency is limited by the intermediate reaction with sorbed iron.

The concentration of sorbed Fe^{2+} in the fine sandy part of the aquifer can be calculated to be $q_{Fe} = 0.19$ mmol/l for the initial groundwater quality (first analysis in Table 3). Using eqn. 2, the retardation of the oxygen front is $R_{O_2} = 1 + 0.19/(4\times0.28) = 1.17$. At the end of the first cycle, 1440 m^3 water were injected and the front had radiated to 11.8 m from the well, with 9.7 m filter length and porosity 0.34. The oxygen front was then located at the position which conforms to the injection of 1440/1.17 = 1230 m^3, or 10.9 m. The fronts were also calculated with PHREEQC-2 (Parkhurst and Appelo, 1999) and are given for cycles 1 and 7 in Figure 3. The modeled oxygen front for the first run shows slightly more retardation than was calculated by hand, because injected water was oxidizing dissolved Fe^{2+} at the front by mixing with groundwater. Mixing by dispersion in the aquifer is included in the computer model, but the reactions due to this process are difficult to account for in hand calculations.

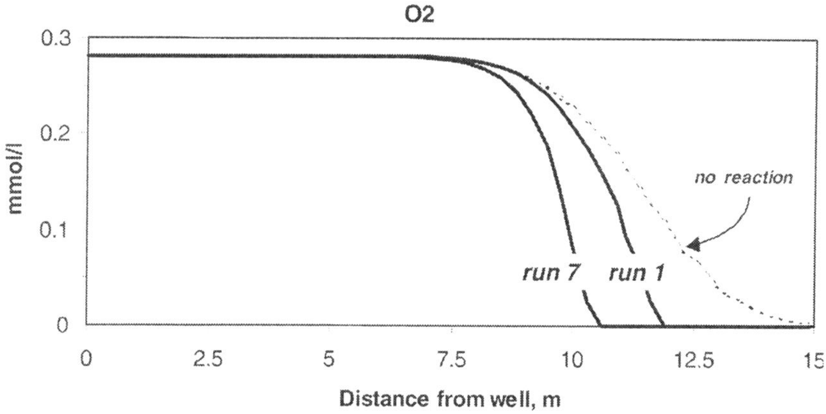

Figure 3. Calculated oxygen profiles at the end of the injection of aerated water in cycles 1 and 7 of in situ iron removal in Schuwacht.

During pumping, ferrous iron is sorbed to 0.19 mmol/l exchange sites, and furthermore to 0.19 mmol/l precipitated iron-oxyhydroxide. The amount sorbed to iron-oxyhydroxide was calculated to be 0.2 mol Fe^{2+} per mol iron-oxyhydroxide for the water composition of PP8 (discussed later). The total iron sorbed during the first cycle is thus $0.19 + (0.2 \times 0.19) = 0.23$ mmol/l. The expected efficiency is therefore, from eqn. 9, $E = R_{Fe}/R_{O_2} = (1 + 0.23/0.1) / 1.17 = 2.8$. Note that it is the dynamics of the system, expressed by the retardations of the ions, that makes the efficiency decrease to a much smaller value than was estimated earlier when assuming simply that all O_2 was available for oxidation of iron.

In the second run, sorbed iron had increased to $q_{Fe} = 0.23$ mmol/l, and therefore the oxygen front was more retarded than in the first cycle. Also,

0.23 mmol/l iron-oxyhydroxide precipitated and the sorption capacity amounted to 0.19 + (0.2 × (0.19 + 0.23) = 0.27 mmol/l in the third run. The increased sorption capacity resulted in a more retarded O_2 front which is illustrated for the 7th run in Figure 3. The retardation and hence the efficiency will continue to increase until in the final end all oxygen is reacting immediately at the injection well, and the theoretical limit of 11.2 is reached for the efficiency.

4.2 Amounts and distribution of precipitate

The total amount of iron-oxyhydroxide which precipitated during a cycle in Schuwacht, can be estimated from:

$$[Fe(OH)_3]_n = (0.19 + 1.2 \times [Fe(OH)_3]_{n-1}) \tag{14}$$

where $[Fe(OH)_3]_n$ indicates mol precipitate/l at run number n. For run 7 the amount is 2.5 mmol $Fe(OH)_3$/l, or 23 ppm Fe in the sediment, which is in good agreement with the model calculation shown in Figure 4. The total precipitate after run 7 would occupy 0.009% of the porosity, assuming a density of 3.0 kg/dm^3 for the precipitate. The amounts are small and, indeed, clogging has not been observed in systems which have been operating for over 20 years (Meyerhoff, 1996). Clogging is probably also prevented because closing pores will force the injected water to pass around, thus evening out the reaction in space.

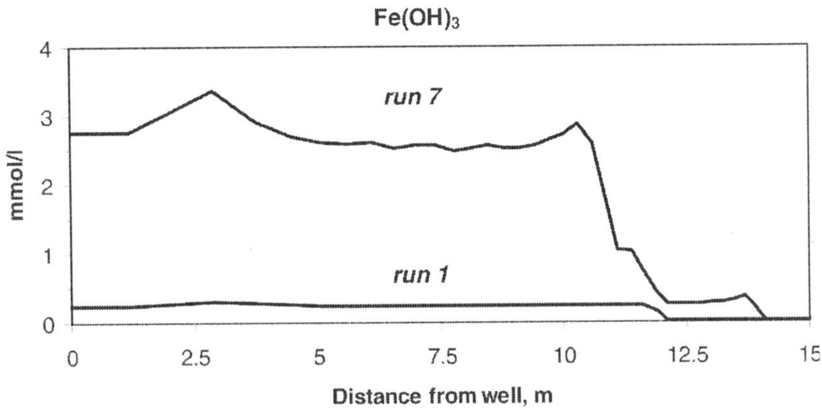

Figure 4. Calculated profiles of precipitated iron-oxyhydroxide in cycles 1 and 7 in Schuwacht.

4.3 Concentration fronts of selected elements during pumping

Cation exchange will modify the concentrations in groundwater during pumping of groundwater, and it will be different for cations which sorb on clay minerals (such as NH_4^+), and for ions which also sorb on iron-oxyhydroxide. Elements which do not sorb, will be a tracer for the groundwater front and it was found that methane behaved conservatively in Schuwacht. Equilibria with calcite, siderite and rhodochrosite were also considered in separate model runs, but the carbonates seemed to be unimportant in Schuwacht. Thus, the model results described here, are for a system with cation exchange and surface complexation only.

Methane

Figure 5 shows the methane concentration for cycles 1 and 7 as function of the ratio of pumped and injected volume, V/V_{inj}. The methane concentration was half of the groundwater concentration at $V/V_{inj} = 1$, and close to the final concentration at $V/V_{inj} = 2$. The end concentration was 2 mg CH_4/l in the first run, and 0.8 mg/l in the 7th run due to variations in the groundwater quality. However, the concentration patterns were identical in both runs, and conform to a conservative substance, with a dispersivity of 0.3 m of the aquifer. The concentrations of HCO_3^- and dissolved organic carbon also showed conservative behavior, which indicates that injected water did not react with sedimentary organic carbon.

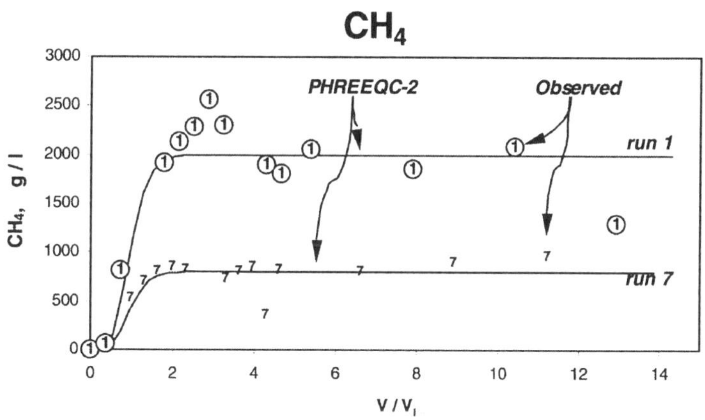

Figure 5. Observed and modeled methane concentrations in cyles 1 and 7 in Schuwacht.

Ammonium

Figure 6 shows the ammonium concentration for cycles 1 and 7. The groundwater concentration of ammonium changed from 3.8 in cycle 1 to 3.3 mg/l in cycle 7. Figure 6 shows a dotted line for conservative flow, and a full line for the case that cation exchange is included in the model. The insignificant difference is due to the small exchange capacity of 5 $mmol_c/l$ for the lower 7 m and 30 $mmol_c/l$ for the upper 3 m of the aquifer. The heterogeneity was modeled in separate computer runs for the upper and the lower part, adapting only the cation exchange capacity and combining the model results in the proportion of 3:7. The model results for runs 1 and 7 would coincide exactly when normalized to the final concentration, because the exchange capacity of the sediment is invariant and iron-oxyhydroxide does not adsorb ammonium.

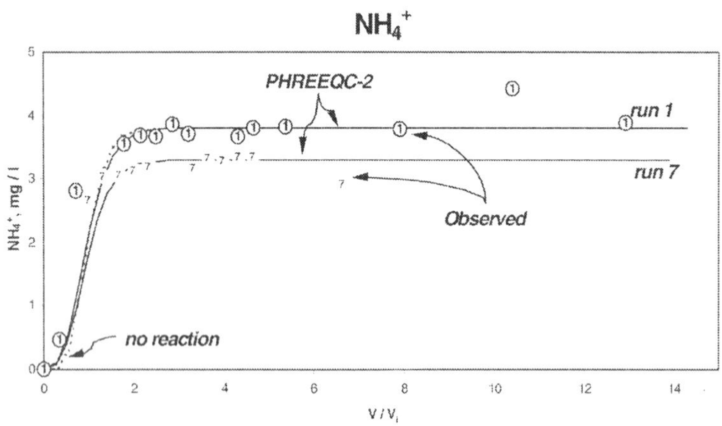

Figure 6. Observed and modeled ammonia concentrations in cyles 1 and 7 in Schuwacht.

Manganese

The Mn^{2+} ion sorbs both to clay minerals and to iron-oxyhydroxide, and the sediment's exchange capacity increases in proportion to the amount of iron-oxyhydroxide which precipitated in the previous runs. The increasing retardation associated with the increase of the sorption capacity is clearly visible in Figure 7 in the delayed increase of the Mn concentration in run 7 compared to run 1. The retardation is directly related to the number of sorption sites on the iron-oxyhydroxide precipitate which is discussed next. The model shows a reasonable match of the observed concentrations but tends to be less disperse.

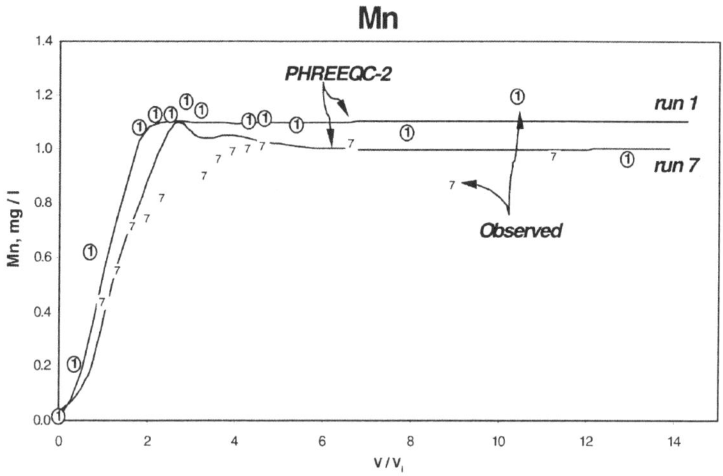

Figure 7. Observed and modeled manganese concentrations in cyles 1 and 7 in Schuwacht.

Iron

The iron concentrations during pumping in the first and seventh run are shown in Figure 8. Again, the end-concentration in groundwater fluctuated and was 6 and 5.2 mg Fe^{2+}/l for run 1 and 7, respectively. In the first run, a retardation of approximately 2.5 with respect to conservative behavior is indicated in Figure 8, which is close to what was calculated by hand before. The increased retardation in subsequent runs can be modeled if the sorption sites amount to 1.4 mol sites per mol Fe. This is quite high, Dzombak and Morel (1990) proposed 0.2 mol sites per mol ferrihydrite. Of course, the latter number is valid for laboratory conditions where the oxides are

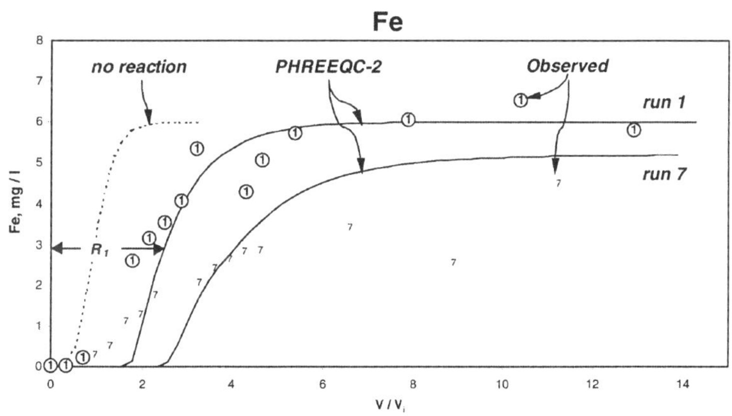

Figure 8. Observed and modeled iron concentrations in cyles 1 and 7 in Schuwacht.

prepared for more or less standardized experiments and aged and sometimes heated. It may be that during in situ iron removal, the precipitate forms a thin layer on existing surfaces of only a few atom layers thick, thus acquiring a specific surface area that is larger than in the laboratory preparates. Also, not all the sites are covered by ferrous iron, but phosphate and carbonate occupy more than 90% of the 1.4 mol sites per mol iron, which suggests that a complicated, mixed solid grows in the aquifer, and not a neat iron-oxyhydroxide with properties which remain invariant over months and for which sorption and desorption can be calculated with one and the same model. It is also probable that the efficiency increase will be less in later runs because the specific surface area diminishes when the precipitate grows in thickness and ripens out and incorporates and rejects various elements in a more crystalline solid. It can be noted here that in model runs with calcite equilibrium the pH is slightly higher and more iron is sorbed because the surface charge is more negative. The optimized number of sites then decreases to about 1.3 mol per mol iron. However, the modeled pH is somewhat higher than was measured.

The effect of sorption of HCO_3^- on ferrihydrite in the model also requires discussion. With the derived complexation constants, HCO_3^- was, together with phosphate, the dominant species on the weak sites of the ferrihydrite surface. Sorption had no effect on the groundwater concentrations of HCO_3^-, but the high HCO_3^- concentrations did affect the sorption of the other surface species, in particular of Fe^{2+} and PO_4^{3-}. Without sorption of HCO_3^-, the increased sorption of Fe^{2+} in run 7 compared to run 1 can be modeled with fewer sorption sites, 0.8 mol sites per mol Fe would be sufficient. However, even that number is high compared to the usual estimates for laboratory prepared ferrihydrite, and the formation of a mixed solid still seems the most plausible. The existing iron-oxyhydroxide in the aquifer was given an unreactivity factor of 3e-3, meaning that 300 ppm Fe initially amounted to (1.4 mol sites/mol Fe) × (300mg/kg / 56g/mol) × (ρ_b/ε = 6 kg/l) × 3e-3 = 0.135 mmol sites/l groundwater. The (un)reactivity factor was bracketed on the high end by retardation for iron in the first cycle, and on the low end by having sufficiently sorbed As available for desorption (discussed later).

The modeled concentration lines in Figure 8 are for a combination of a coarse sand (2/3 of aquifer thickness) and silty sand (1/3 of thickness). The higher initial exchange capacity of the silty sand increases the retardation, and smears out the concentration pattern. This does fit the observed slow increase to the final concentration in the first run, but the modeled pattern is too steep for the seventh run. It may be that the pattern is more spread out in the field because in the aquifer more layers exist with a different initial exchange capacity than in the simple two-units model. Such differences will lead to a faster initial breakthrough, and to a more retarded arrival of the final concentration. However, it is also likely that kinetics of the Fe^{2+} oxidation and of the exchange and sorption reactions will play a role.

Phosphate

The 1.4 sorption sites per mol precipitated iron was determined by comparing the iron as well as the phosphate concentrations (Fig. 9). The sorption behavior of the two ions on ferrihydrite is intimately coupled. Sorption of HPO_4^{2-} and $H_2PO_4^-$ increases the negative charge of ferrihydrite, and stimulates sorption of Fe^{2+} (the surface of ferrihydrite would be positively charged at the pH of groundwater if phosphate would not be present). On the other hand, sorption of Fe^{2+} compensates the negative charge increase and thus tends to stimulate sorption of phosphate as well. Just as for iron, HCO_3^- has important effects in limiting sorption and in even more markedly influencing the concentration increases of phosphate. Without sorption of HCO_3^-, the modeled concentration increase of phosphate is much steeper as is shown in Figure 10, and the assumed two-layer structure of the model aquifer becomes quite conspicuous in two steps in the model lines. The stepwise character is a result of the steep sorption isotherm of phosphate on ferrihydrite when sorption of HCO_3^- is not in the model.

Arsenic

The arsenic concentration in pumped water varied from less than 2 µg As/l to 14 µg/l. The concentration pattern of As was different from the other elements and showed a marked concentration increase at the time that iron and phosphate arrived at the well (Fig. 11). The highest concentrations of 13-14 µg As/l were found in the first two cycles only, in later runs the concentrations decreased to less than 9 µg/l, although still showing a concentration jump when the iron and phosphate concentrations started to increase in the well. Only total arsenic concentrations were measured, but the modeled pattern in Figure 11 is the result of redox reactions of As(III) and As(V), of the different affinity of these species for ferrihydrite, and further marked by the displacement by phosphate. For an insight in the competing reactions which affect the behavior of As, it is helpful to calculate distribution coefficients for the groundwater quality at hand (first cycle in Table 3). The distribution coefficients relate the concentration sorbed to solute, and are given in undimensionalized form in Table 4.

Inspecting Table 4, it can be noted that arsenite is much less sorbed than arsenate in groundwater without phosphate. However, in the presence of PO_4^{3-} the sorption of arsenate is diminished to a very small quantity due to negative charging of the ferrihydrite surface. The effect is greater for arsenate than for arsenite because arsenate is sorbed as a negative species, while arsenite forms a neutral surface complex. The HCO_3^- surface complex also is mainly neutral, and the ion floods the weak sites of ferrihydrite and

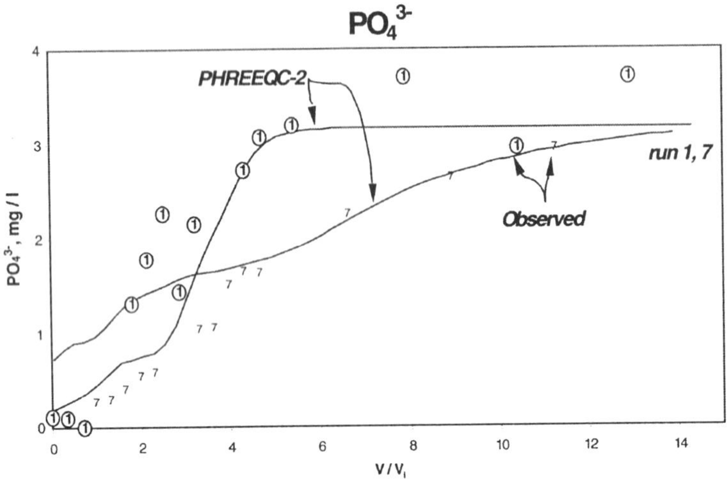

Figure 9. Observed and modeled phosphate concentrations in cyles 1 and 7 in Schuwacht.

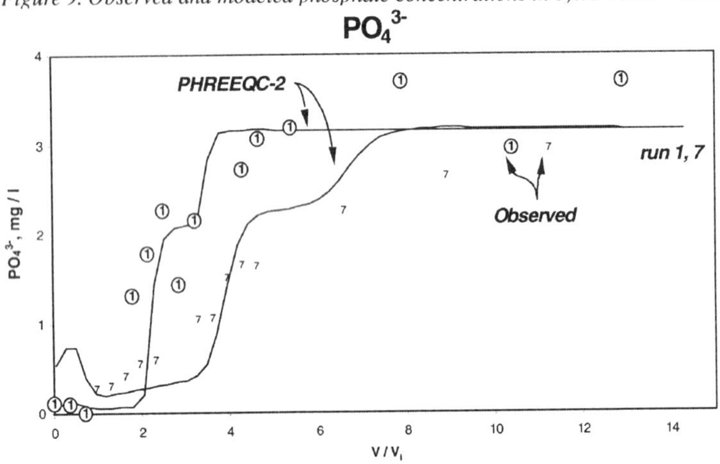

Figure 10. Observed and modeled phosphate concentrations in cyles 1 and 7 in Schuwacht, without sorption of HCO_3^- on ferrihydrite.

displaces arsenite while the effect on arsenate is somewhat less. On the other hand, sorption of Fe^{2+} charges the surface positively, and enhances sorption of arsenate, while it merely competes for sorption sites with arsenite. The smaller distribution coefficient for groundwater compared to injected water, is entirely due to the presence of more phosphate and HCO_3^- in groundwater.

Now, the effect of injection of oxidized water is that As(III) is oxidized, and since As(V) is sorbed stronger than As(III), dissolved concentrations would diminish. However, when PO_4^{3-} contacts the surface to which As(V)

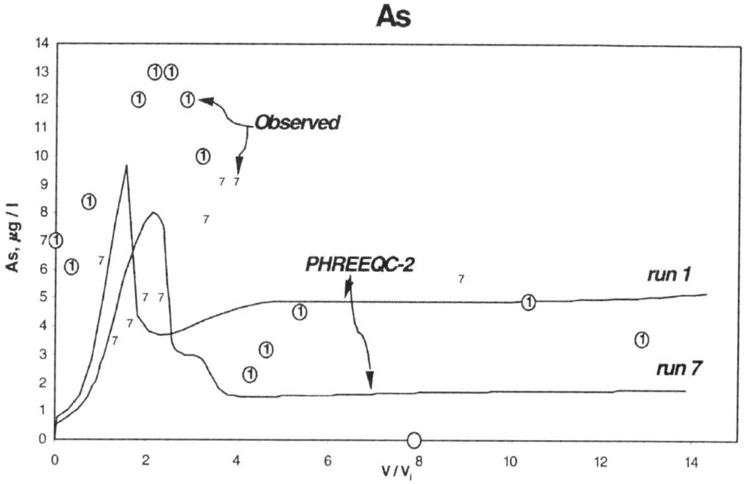

Figure 11. Observed and modeled arsenic concentrations in cyles 1 and 7 in Schuwacht.

Table 3. Composition of native groundwater and injected waters, in pH units and mg/l. Temperature is 10°C.

	Groundwater		*Injected water*	
	Cycle 1-3	Cycle 4-7	Cycle 1-2	Cycle 3-7
pH	7.1	7.1	7.4	7.65
Na	77.3	77.3	65.5	65.5
K	5.3	4.8	3.6	3.6
Mg	13.1	12.2	11.5	11.5
Ca	93	87	83	83
NH_4^+	3.8	3.3	0.01	0.01
Fe	6.0	5.2	0	0
Mn	1.1	1.0	0	0
Cl	134	137	124	124
HCO_3^-	308	260	214	201
SO_4	43.4	51.0	44.0	44.0
PO_4	3.16	3.16	0.1	0.1
As (µg/l)	10	5	0	0

is attached, it desorbs As, as indicated by the low distribution coefficient (Table 4). According to the model calculations, PO_4^{3-} displaces As in the form of a wave which arrives in the well just before the increase of the phosphate concentration. As soon as Fe^{2+} arrives, As(V) is reduced to As(III) which is less affected by PO_4^{3-} and sorbs again in agreement with the higher distribution coefficient. To model the concentration wave of arsenic, sufficient As(III) must be present in the system. Thus, in the first run the number of sorption sites on iron-oxyhydroxide had to be adequate and this bracketed the (un)reactivity factor for already present iron-oxyhydroxide in the aquifer. Also, the surface complexation constant for H_3AsO_3 was increased threefold to obtain sufficiently sorbed arsenic.

Table 4. Model distribution coefficients for As, (mol adsorbed/l) / (mol solute/l), on 1 mmol ferrihydrite for groundwater at Schuwacht.

	As(III)	As(V)
Fe, Alk, PO_4, all 0 mg/l	22	211
6 mg Fe/l[1]	18	234
3.16 mg PO_4/l[1]	6	2
308 mg HCO_3/l[1]	6	36
Fe, Alk, PO_4 all at groundw. conc.[1]	4	1
Injected water[1]	12	53

[1]Groundwater concentration from Table 3

The modeled pattern in Figure 11 shows the observed concentration trends, but, clearly, it does not match the details. Notably, the predicted As peaks arrive too early and they are too small. The column experiments of Isenbeck-Schröter (1995) and Darland and Inskeep (1997) required kinetic reactions for Freundlich or Langmuir sorption isotherms for As, but a kinetic model appears to spread out the As concentrations in Figure 11 only, and it does not shift the position of the peak to later arrival times. Therefore, other reactions might explain the discrepancy.

First, it is remarkable that the As concentration was quite high already when injected water was backpumped. It may be that As was sorbed to colloidal iron-oxyhydroxide particles formed during oxidation, but were too small to be removed by filtration over 0.45 μm before analysis. This mechanism was suggested by Rott et al. (1996) who observed similar As peaks during the first cycles of an in situ iron removal system. When samples were analyzed from a later cycle in Schuwacht, the As concentrations had decreased to about 2 μg As/l and they were similar in unfiltered and 0.1 μm filtered subsamples. Thus, sorption to colloidal iron cannot be ruled out as a mechanism and it should be investigated thoroughly in the incipient cycles of another system.

Second, the combined presence of Fe^{2+} and Mn^{2+} in groundwater, and precipitation of both Mn-oxides and Fe-oxides would lead to separation in space of the precipitates, with Fe^{2+} reducing the Mn-oxides (Postma and Appelo, 2000). This may have special, but as yet not quantified effects on the behavior of As.

Third, the surface complexation model of Dzombak and Morel (1990) is based on data published by Pierce and Moore (1982) and is excellent for calculating laboratory experiments with As by Wilkie and Hering (1996) and Manning and Goldberg (1996). The data of Manning and Goldberg only require a small adaptation of the ionic strength to use the double layer model and to arrive at a similar potential/charge relation for ferrihydrite as for the constant capacity model which they applied. These authors investigated competition of SO_4^{2-} and PO_4^{3-} for sorption of As. However, the multicomponent effects of a natural groundwater are still to be verified and a column experiment with well defined conditions and water qualities is highly desirable for elucidating the combined effects of the various processes on As during in situ iron removal.

5. SUMMARY

The principles of in situ iron removal from groundwater were explained in simple terms, based on the reaction of sorbed ferrous iron in the aquifer. The efficiency of an in situ iron removal system was defined as the ratio of the volume of groundwater pumped to the volume of oxygenated water injected. For the easy case of sharp fronts without dispersion or smearing by kinetics, the efficiency is given by the ratio of the retardation of Fe^{2+} over the retardation of O_2, $E = V_{gw}/V_{inj} = R_{Fe}/R_{O_2}$. A specific case of in situ iron removal in the Netherlands was discussed, showing many of the details of in situ iron removal common to these systems. The features are connected with exchange and sorption reactions in the aquifer. Sorption for ammonium was small and identical for all cycles. Retardation of Mn^{2+}, Fe^{2+} and PO_4^{3-} increased in successive cycles due to sorption to the iron-oxyhydroxide which was precipitated in the previous cycles. The concentration patterns of these elements could be modeled well with known reactions. Arsenic showed a complicated behavior with initial concentration jumps which appear to be related to redox transitions and displacement by PO_4^{3-}. A relatively high sorption capacity of the freshly precipitated ferrihydrite suggests that the precipitate in the aquifer is a mixed solid of Fe, P and C, rather than a neat iron-oxyhydroxide to which Fe^{2+} and other elements are sorbed.

ACKNOWLEDGEMENTS

The cooperation of ir D. van der Woerdt for initiating this study is highly appreciated. Helpful reviews were provided by Kees van Beek, Margot Isenbeck-Schröter, and Ken Stollenwerk.

Chapter 15

In situ Arsenic Remediation in a fractured, alkaline aquifer

Alan H. Welch, Kenneth G. Stollenwerk, Douglas K. Maurer and Lawrence S. Feinson

U.S. Geological Survey

In situ removal of arsenic from ground water used for water supply has been accomplished in circum-neutral ground water containing high dissolved iron concentrations. In contrast, the ground water at our study site is alkaline, contains measurable dissolved oxygen and little dissolved iron. Because the dissolved iron concentration is low in the basalt aquifer, the iron oxide content of the aquifer would not increase with successive pumping cycles unless iron is added to the injected water. Additionally, the high pH limits adsorption onto iron oxide present in the aquifer. Having the ability to lower arsenic concentrations in high-pH, oxic ground water could have wide application because similar high arsenic ground water is present in many parts of the world.

Laboratory and field results show that the basalt has limited capacity for adsorption of As(V), presumably by naturally occurring hydrous ferric oxide (HFO). However, addition of HFO can significantly increase As(V) adsorption. Lowering the pH combined with increasing the iron oxide content in the basalt aquifer reduces arsenic concentrations in produced ground water. Arsenic removal was very effective in laboratory experiments and during early part of the initial push-pull experiment. Moderate arsenic removal after increasing the iron oxide content of the basalt aquifer and lowering the pH during the cross-flow experiments was limited to about 50% at best. This relatively low removal may be due to a variety or combination of factors, including local hydraulics of the aquifer, chemical reaction kinetics, and changing aqueous chemistry caused by the lowered pH.

1. INTRODUCTION

In situ removal of arsenic from ground water used for water supply has been accomplished by using co-precipitation or adsorption involving iron oxide (Appelo et al., 1999; Rott and Friedle, 1999). The most common approach consists of withdrawing iron-rich ground water and introducing atmospheric oxygen or potassium permanganate (Matthess, 1981; Meyerhoff, 1996; Rott and Friedle, 1999) to promote the formation of iron oxide and conversion of any dissolved As(III) to As(V) followed by injection in an aquifer and subsequent withdrawal for use. Ground water in these systems is moderately to strongly reducing, as evidenced by the presence of dissolved Fe(II) (Rott and Friedle, 1999) and methane (Appelo and deVet, this volume). The process is usually cyclic, with a period of injection followed by a period of pumping of water with lowered arsenic and iron concentrations continuing long after the volume of injected water has been pumped. The efficiency of arsenic removal has been noted to increase after continued cycles without well or aquifer clogging even after operation for decades (Appelo et al., 1999; Meyerhoff, 1996). HFO (hydrous ferric oxide) appears to be the most important phase responsible for removing the arsenic from the ground water (Appelo and deVet, this volume). The arsenic removal process associated with iron removal may be described as a series of reactions involving dissolved oxygen, aqueous and exchangeable cations including Fe(II), and arsenic (Appelo and deVet, this volume). Injection of water containing dissolved oxygen can lead to rapid exchange of Fe(II) for cations in the injected water with subsequent Fe(II) oxidation to form HFO. When the flow direction is reversed, the injected water has lower arsenic and iron concentrations.

The initial hypothesis for this study was that arsenic concentrations in water pumped from a basalt aquifer could be lowered by adsorption onto iron oxide. The hydraulic and geochemical conditions at the study site have some important differences compared with the approach described above. Water at the test site is alkaline (pH = 9.1), contains measurable dissolved oxygen (about 1 mg/L; wells B-3, B-5, and B-6 in Table 1), contains little dissolved iron, and the arsenic is dominantly As(V) (site number 69 of Maurer, 2001 is well B-3). In contrast, reported successes with in-situ arsenic removal generally involve Fe(II)-rich water with a circum-neutral pH. Because the dissolved iron concentration is low in the basalt aquifer, the iron oxide content of the aquifer will not increase with successive pumping cycles unless iron is added to the injected water. Additionally, the high pH limits adsorption of anions onto iron oxide present in the aquifer. Nonetheless, remediation of arsenic in high pH, oxic ground water may have wide application because similar high-arsenic ground water is present in many parts of the United States (Welch et al., 2000) and elsewhere such as northern Argentina (Smedley et al., 2002).

2. HYDROLOGY AND GEOCHEMISTRY OF THE BASALT AQUIFER NEAR RATTLESNAKE HILL

A basalt aquifer underlying Fallon, Nevada (Fig. 1), is the sole source of municipal supply for the city of Fallon, the Fallon Naval Air Station, and the Fallon Paiute-Shoshone Tribe; serving a total population of about 11,000 people. Concentrations of dissolved arsenic in water pumped from the basalt range from 70 to 120 μg/L (Maurer, 2001, p. 52; Maurer and Welch, 1996). The basalt aquifer is a mushroom-shaped body of basalt, its top exposed at Rattlesnake Hill, an eroded volcanic cone about 1.6 km in diameter and 61 m high, and the remainder buried as deeply as 183 m below land surface near its periphery (Fig. 1). In plan view, the lateral extent of the basalt is roughly peanut-shaped, oriented southwest to northeast, and is about 6.4 km wide and 16 km long. The age of the basalt ranges from about 1 million years for a sample taken from Rattlesnake Hill, to 2.5 million years for a sample taken from a test hole about 2.4 km south of the cone. The aquifer is highly productive, with transmissivities of over 100,000 ft^2/day, and varies in lithology from dense, fractured flows and breccias to highly porous, scoriaceous zones (Glancy, 1986, p. 15 and 18).

3. EXPERIMENTAL METHODS

3.1 Well Construction

The study concentrated at a site on the eastern side of Rattlesnake Hill (Fig. 1). The site was chosen because the basalt is present at shallow depths and the location is down gradient and relatively distant from municipal supply wells. Initially three wells were installed for the study; two (B-3 and B-6) separated by 6.1 m and a third (B-5) 122 m south of B-6. Plans were to use the two nearby wells for injection and sampling, and to use the third well as a source of water from the basalt for injection.

The three initial wells were drilled using a 15-cm diameter air-hammer and odex casing that was removed as the wells were constructed. Basalt was encountered at a depth of about 9.1 m below land surface at all wells, static water levels lie at about 15.2 m below land surface, and the wells are from 21.3 to 22.9 m deep. The wells were constructed of 5-cm PVC casing, and PVC screen with 0.5-cm machine slots was emplaced in the bottom 3 m of the borehole. Sand pack was emplaced from 1.5 to 4.6 m above the screen and the remainder of the borehole was filled with bentonite grout. Preliminary injection and sampling using a submersible pump, capable of

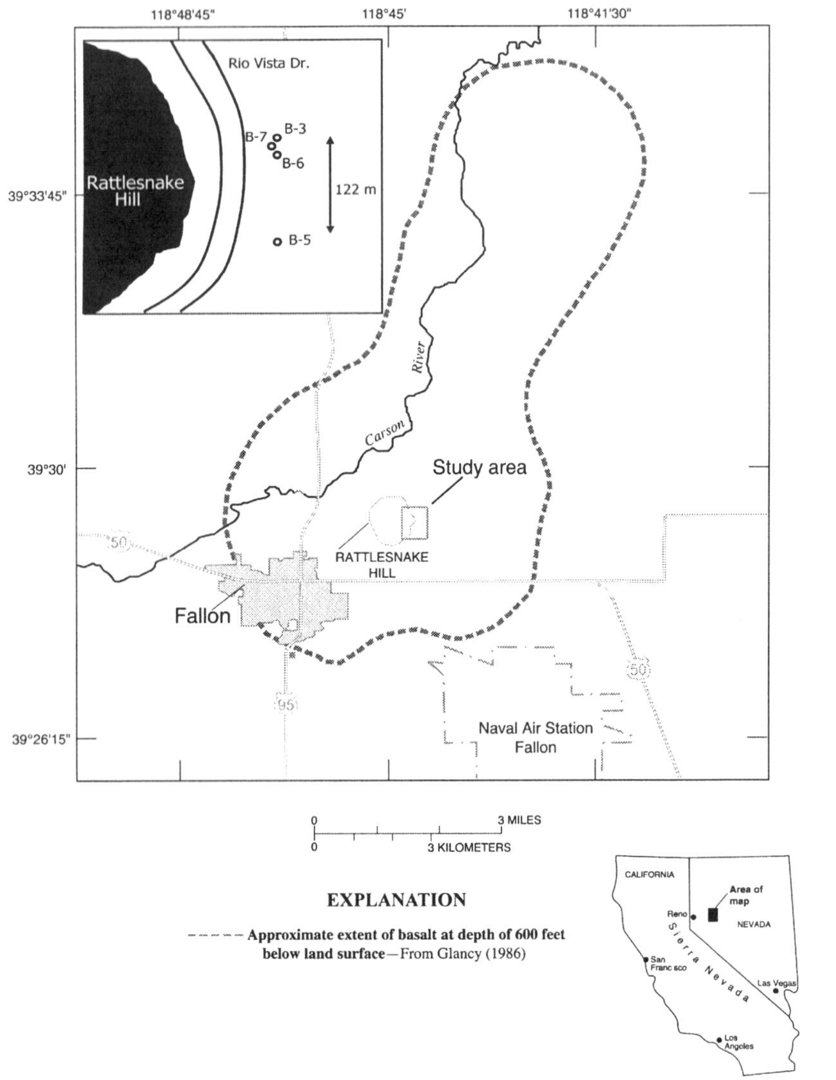

Figure 1. Location of remediation experiment.

pumping about 5 gal/min, suggested that a larger pumping capacity would be desirable for continued testing. A fourth well (B-7) was installed approximately equidistant from wells B-3 and B-6 to shorten the flow path between injection and pumping wells. The fourth well was constructed using the 15-cm diameter air hammer and by driving 15-cm diameter steel casing to a depth of 22.9 m below land surface. The lower 3 m of the steel casing has .32-cm wide machine slots.

Basalt cores were collected from a depth of 21 to 22 m during construction of well B-3. Recovered basalt ranged from dense, intact sections of core to less dense, vesicular pieces of basalt 1 to 5-cm diameter, which may represent a loosely consolidated scoriaceous zone. The intact cores contained vesicles, which were not interconnected. The cores were impermeable to water flow; therefore, column experiments were not possible. Batch reactor experiments were conducted on the smaller pieces of basalt. Some of these pieces had visible coatings of calcite and reddish-brown coatings that are assumed to be iron oxide on one or two surfaces. The presence of calcite, phillipsite, and clay minerals is consistent with X-ray diffraction analysis and thin-section petrography reported by Lico and Seiler (1994) for samples from other wells penetrating the basalt aquifer. The coatings were most likely deposited on fracture surfaces exposed to ground water. Although these basalt pieces contained numerous vesicles, interconnectivity was poor. Water-saturated porosity of individual pieces, approximately 3 to 5-cm in diameter, ranged from 1.5 to 5.6%, and averaged 3.7% (n=5).

3.2 Laboratory Simulations

3.2.1 Buffering capacity of the basalt

The buffering capacity of basalt surfaces exposed to ground water was estimated by acid titration. Larger pieces were broken into 0.1 to 3 cm-size pieces for experimental convenience. Although breaking the pieces created new surfaces, the greater part of the buffering capacity is most likely due to the presence of calcite. The newly broken samples were not crushed to minimize the area of fresh basalt surfaces which were geochemically different from in situ surfaces previously exposed to ground water. Approximately 50 g from five separate basalt samples were added to 100 mL deionized water in 250 mL polyethylene bottles. Initial pH values of the slurries ranged from 9.0 to 9.9 and were related to the abundance of visible calcite coatings. Hydrochloric acid (0.1 N) was rapidly added to each slurry to achieve a pH of 4.5. The bottles were placed on a reciprocating mixer and allowed to equilibrate for 3 days. This HCl addition was repeated seven times for each of the five samples.

3.2.2 Arsenate adsorption

The capacity of basalt to adsorb As(V) as a function of pH was measured for three basalt samples. The samples were initially equilibrated with a synthetic ground water with an arsenic concentration of 145 μg/L (pH = 9.3),

Table 1. Analysis of ground water. The temperature of the ground water ranged from 21.5 to 23.5 Celsius.

WELL	DATE	V[1], L x 1000	D.O.[2] mg/L	pH	Ca mg/L	Mg mg/L	K mg/L	Na mg/L	HCO_3 mg/L	CO_3 mg/L	Cl mg/L	SiO_2 mg/L	SO_4 mg/L	P mg/L	As ? g/L	Fe ? g/L	Mn ? g/L
Initial samples before injection of $FeCl_3$																	
B-3	10/20/97	--	0.9	9.1	1.7	0.9	7.8	236	245	30	130	24.6	102	0.27	--	--	<1
B-5	6/29/99	--	1.0	9.2	1.7	0.8	7.9	232	235	37	130	24.7	102	--	121	<10	<3
B-3	6/30/99	--	0.4	9.3	1.7	0.7	7.9	232	250	45	127	24.7	100	--	123	<10	<3
B-6	6/30/99	--	0.6	9.3	1.6	0.8	7.9	233	226	42	131	24.8	102	--	117	<10	<3
Recovery from push-pull test																	
B-3	9/30/99	0.4	--	5.5	8.5	2.2	9.0	251	54	--	265	19.7	163	0.02	<2.0	90	145
B-3	9/30/99	1	--	6.5	5.4	1.5	8.6	245	156	--	217	16.9	125	<.010	<2.0	40	72
B-3	9/30/99	6	--	8.3	2.7	0.9	8.4	246	256	12	163	22.1	105	0.21	51	20	6.5
B-3	9/30/99	12	--	8.8	2.2	0.8	8.3	245	249	24	145	23.6	104	0.32	99	<10	<2.2
Pumped water during pH adjustment following $FeCl_3$ injection																	
B-7	8/30/00	16	--	6.4	9.6	2.2	10.2	262	151	--	278	25.1	101	0.55	52	80	81
B-7	8/30/00	26	--	6.5	10.9	2.6	10.6	265	153	--	269	28.4	103	0.94	80	580	72
B-7	8/30/00	55	--	8.5	3.7	1.3	9.1	256	272	--	163	27.5	103	0.57	106	10	14
B-7	8/30/00	77	--	6.8	4.5	1.5	9.3	257	227	--	210	26.1	102	0.19	58	70	47
B-7	8/30/00	85	--	6.9	3.9	1.3	9.1	259	227	--	203	27.7	104	0.09	78	50	24
B-7	8/30/00	115	--	8.2	2.5	0.9	8.7	255	268	--	156	26.8	104	0.31	102	10	5.5
Pumped water during pH adjustment																	
B-7	5/30/01	40	--	8.8	1.6	0.6	8.2	231	264	18	140	25.1	103	0.21	120	20	0.4
B-7	5/30/01	60	--	7.9	1.8	0.7	8.3	236	263	--	166	25.5	103	0.22	115	<10	0.8
B-7	5/30/01	102	--	7.0	3.0	1.0	9.3	242	211	--	206	26.4	103	0.28	103	30	7.6
B-7	5/30/01	108	--	6.9	3.7	1.2	9.0	240	199	--	217	26.5	103	0.32	106	40	12
B-7	5/30/01	115	--	6.9	4.1	1.3	9.5	240	194	--	404	26.8	113	0.43	105	60	15
Pumped water after flushing pH-adjusted water																	
B-7	5/31/01	--	--	9.1	1.8	0.7	8.3	240	257	22	130	26	100	0.27	120	9	0.8

[1] *Volume of water pumped during withdrawal.*

[2] *Dissolved oxygen*

which is somewhat greater than in the ambient ground water (see analysis of B-3 collected prior to the addition of ferric chloride in Table 1). The arsenic concentration was then determined at pH values ranging from about 6 to 9.5. The iron oxide content of these samples was then increased through the addition of ferric chloride ($FeCl_3$). After each iron addition, arsenic concentrations were then determined over the same pH range. Finally, the arsenic concentration of the solution was increased to 1000 μg/L with subsequent determination of the pH-arsenic relation. Three basalt samples were examined using this experimental procedure.

Basalt pieces weighing about 100 g and ranging in size from 0.1 to 3 cm were added to 1000 mL polyethylene bottles and equilibrated with 500 mL of the synthetic ground water. Initial pH values were adjusted to about 6.5 with 0.1 N HCl; pH was then gradually increased with 0.1 N NaOH. Slurry samples were equilibrated for one day on a reciprocating mixer following each addition of NaOH. Aliquots (20 mL) were removed, filtered through a 0.4 μm membrane and analyzed for As(V) and P(V).

$FeCl_3$ was then added to the slurries to evaluate the effect of HFO precipitation on As(V) adsorption. Additional synthetic ground water was added to each slurry to make the final volume 500 mL. In one experiment, 10 mg/L Fe(III) was added. The slurries were then equilibrated for 3 days to allow complete precipitation of HFO. NaOH was then added to slowly increase pH. Slurries were equilibrated for 1 day following each addition of 0.1 N NaOH, followed by 0.4 μm filtering of 20 mL aliquots for As(V) and P(V) analysis. Additional $FeCl_3$ was then added to increase the Fe(III) concentration to 100 mg/L and the experiment was repeated as above. Arsenate adsorption for the 100 mg/L Fe(III) concentration was measured for initial As(V) concentrations of 145 and 1000 μg/L. The iron additions of 10 and 100 mg/L are equivalent to 0.1 and 1.0 mg/g of basalt.

3.3 Field Sampling and Analytical Methods

Ground-water samples were collected from four wells in the study area. Data for these sites, along with previously collected water-quality data, are given in Table 1 and locations are shown on Fig. 1. Measurements of temperature, pH, and dissolved oxygen were made in a flow-through chamber. Field meters were calibrated using appropriate pH standards (Wilde and Radtke, 1998). Alkalinity was determined on-site by incremental titration of filtered water with sulfuric acid (Wilde and Radtke, 1998).

Water samples collected for determination of inorganic constituents were processed in the field and shipped to the USGS National Water Quality Laboratory for analysis (Fishman, 1993; Fishman and Friedman, 1989; Garbarino, 1999). Filtered samples were collected after passing through a

0.45 μm pore-size capsule filter. Samples analyzed for cations were acidified to a pH of about two in the field with ultra pure HNO_3. Samples for analysis of nitrogen species and phosphate were collected in opaque plastic bottles and chilled to about 3 °C. Unfiltered samples also were collected, primarily for analysis of arsenic and iron.

A modified phosphomolybdate method was used to determine As(III), As(V) and P(V) (Johnson, 1971; Murphy and Riley, 1962; Oscarson et al., 1980; Strickland and Parson, 1968) in the field and in water associated with batch experiments. This method is based on the spectrophotometric measurement of the blue As(V) and P(V) molybdate complexes. Phosphate is measured in one sample after reducing As(V) to As(III), which does not form the color complex. In a second sample, both P(V) and As(V) are measured and the As(V) concentration is determined by subtracting the P(V) concentration. A third sample is oxidized to convert As(III) to As(V), and the As(III) concentration is obtained by subtracting As(V) and P(V).

The limit of detection for arsenic with this technique is 3 μg/L with an error of ±70% at the 95% confidence interval. Concentrations between 3 and 10 μg/L have an error of ±40% at the 95% confidence interval. Concentrations between 10 and 40 μg/L have an error of ±15% at the 95% confidence interval, and concentrations > 40 μg/L have an error of ±10% at the 95% confidence interval. Error estimates were determined for 15 different concentrations ranging from 3 to 500 μg/L. There were 15 replicates at each concentration.

3.4 Field Experiments

Field experiments designed to evaluate the removal of arsenic from ground water included lowering the pH of injected water and increasing the adsorption capacity of the basalt aquifer. The experiments are described below.

1. Two push-pull experiments consisted of pumping water from B-5 into B-3 followed by recovery of the injected water. The pH of the injected water was adjusted between 5.1 and 6.4 through the addition of HCl. Recovery and field-testing of water from B-3 indicated minimal or no removal of arsenic. The results are consistent with the laboratory experiments, which indicated that the basalt has limited capacity for arsenic adsorption. A similar experiment consisted of adding of $FeCl_3$ to increase the adsorption capacity of the aquifer and to decrease the pH. Arsenic concentrations were significantly lower in the recovered water.
2. A cross-flow experiment, consisting of continuous pumping and injection, included addition of $FeCl_3$ followed by pH adjustment.

Field and laboratory testing indicated significant arsenic removal, although the arsenic concentrations increased as the pH increased.

3. Another cross-flow experiment nine months after the addition of $FeCl_3$. Water was again injected into the aquifer following pH adjustment. Arsenic was removed from the pumped water, although to a lesser degree.

4. RESULTS OF LABORATORY AND FIELD EXPERIMENTS

4.1 Laboratory Experiments

The pieces of basalt recovered from well B-3 neutralized an average of 8.7 μmol H^+/g basalt (range 5.8 to 10.0 μmol/g, n = 5; final pH = 7.0). Placing this neutralization capacity in the context of the in situ conditions for the basalt aquifer is difficult because the large range in porosity and permeability for the aquifer. However, an approximation can be obtained if the assumption is made that the average effective porosity of 3.7% measured for the basalt used in this experiment is comparable to the effective porosity of basalt in the aquifer. Assuming a bulk density of 2.6 g/cm^3 for the basalt, the aquifer basalt would be able to neutralize about 610 mmol H^+/L. For comparison, the HCO_3^- and CO_3^{2-} in the basalt aquifer ground water can neutralize about 5 mmol H^+/L. This observation has important implications for pH adjustment alone as a technique for lowering As(V) concentrations. The natural buffering capacity of the aquifer may rapidly neutralize any low pH ground water injected into the aquifer, thereby maintaining high pH values with minimal As(V) adsorption.

Decreasing the pH of synthetic ground water did not have a significant effect on As(V) adsorption by unaltered basalt (Fig. 2). In one basalt sample, the As(V) concentration decreased from 145 μg/L at pH 9.3 to 100 μg/L at pH 6.6. There was essentially no As(V) removed by the second basalt sample, and As(V) concentrations in the third sample were actually greater than the initial concentration (145 μg/L) in the synthetic ground water. Apparently, the basalt initially contained adsorbed As(V), which was released to solution during the experiment.

Precipitation of HFO caused a decrease in aqueous As(V) for the three basalts at all pH values. For an initial Fe(III) concentration of 10 mg/L, there was a 30 to 50% increase in As(V) adsorption for pH values >7. At pH 6.2, almost all of the As(V) was removed from solution. Increasing the initial Fe(III) concentration to 100 mg/L resulted in sufficient HFO precipitation to

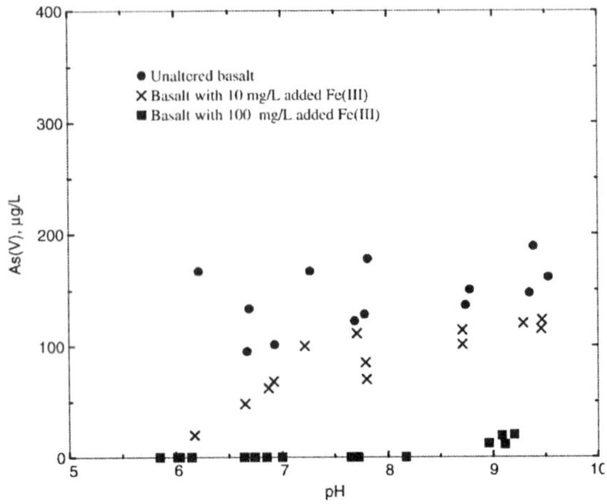

Figure 2. As(V) concentration in equilibrium with three basalt slurries. The initial As(V) concentration was 145 µg/L.

completely adsorb As(V) at all pH values below 9. Even at pH 9, almost 85% of the As(V) was removed from solution. The substantial adsorption capacity of the added HFO [100 mg/L Fe(III)] is evident in Fig. 3. In this experiment, the initial As(V) concentration was 1000 µg/L. All As(V) was adsorbed at pH values <7. Even at pH 9.4, 40% was removed.

The presence of other strongly adsorbing solutes would be expected to compete with As(V) for adsorption sites. Phosphate is one such solute. Phosphate concentration in the synthetic ground water used in the laboratory experiments was 0.38 mg/L. Phosphate actually increased in ground water equilibrated with unaltered basalt, and basalt from two of the 10 mg/L Fe(III) experiments (Fig. 4). At pH >9, P(V) concentrations ranged from 0.41 to 0.62 mg/L. Phosphate was adsorbed as pH decreased; however, even at pH 6, there was still substantial P(V) in solution. P(V) concentrations decreased to below the detection level after the addition of 100 mg/L Fe(III).

4.2 RESULTS OF FIELD EXPERIMENTS

The second push-pull experiment 'pushed' about 610 L from well B-5 at a rate of about 8.3 L/min into well B-3 with the addition of about 55 g of Fe in the form $FeCl_3$. After about one hour, well B-3 was pumped until more than twice the injected volume was recovered. The low pH in the water during the beginning of the recovery period was caused by precipitation of iron oxide, as described by the reaction: $Fe^{3+} + 2\ H_2O = FeOOH + 3H^+$. Field colorimetric analysis of arsenic indicated that arsenic was removed

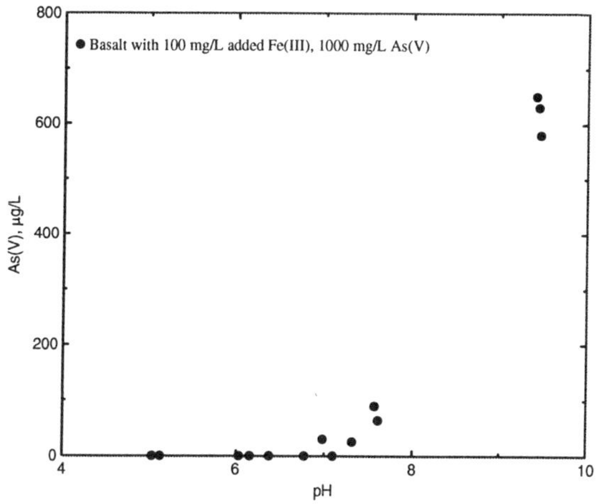

Figure 3. As(V) concentration in equilibrium with basalt.

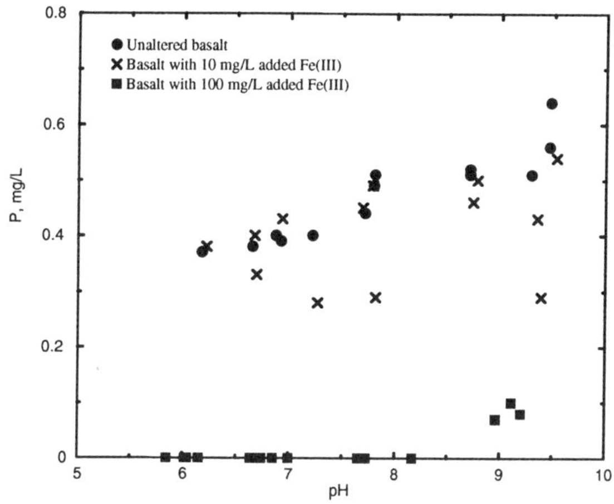

Figure 4. Phosphate concentration in equilibrium with three basalt slurries. The initial As(V) and P concentrations were 145 µg/L and 0.38 mg/L, respectively.

throughout the pumping (Fig. 5). Arsenic removal was confirmed by laboratory analysis on four samples collected during the recovery period (Table 1), although arsenic removal was less efficient as the pH increased. Arsenic removal continued after a volume equivalent to that injected (about 610 L) was pumped and the pH exceeded 8. The following removal estimates are based on the assumption that the ambient arsenic concentration was 120 µg/L, the mean concentration of wells B-3, B-5 and B-6 in June 1999 (Table 1). Nearly 70% of the arsenic was removed from the first 1,000 liters. About 44% of arsenic was removed from the water when the total volume pumped ranged from 610 to 1,000 L owing to less efficient removal as the pH increased to above 8.

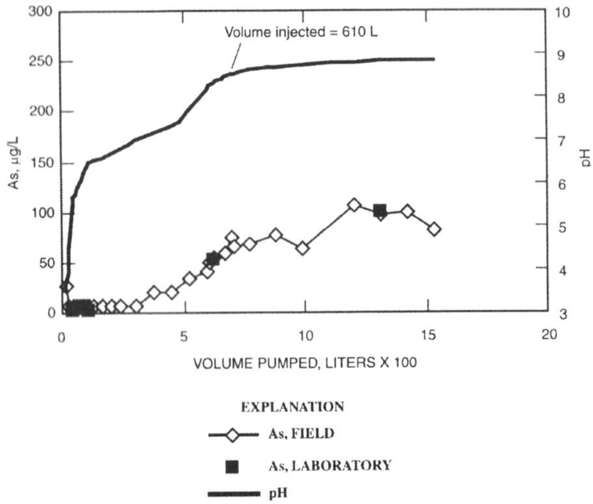

Figure 5. Arsenic removal during push-pull experiment. Pumping rate was 2.5 L/min for the first 130 L and 7.2 L/min thereafter.

These results provide empirical evidence that reaction rates are sufficiently rapid to produce substantial arsenic removal, presumably due to adsorption on iron oxide. Although the laboratory experiments indicated that this approach could reduce arsenic concentrations, the hydraulics and chemistry of the aquifer are not sufficiently well understood to accurately predict the results. Of particular concern was the fractured character of the aquifer. If two wells are connected by a wide fracture then the surface area in contact with the ground water could be insufficient to produce significant arsenic adsorption. At the other extreme, even closely spaced wells may not be hydraulically well connected. The latter situation is the case for wells B-7 and B-6. Well B-7, which has a larger diameter casing and therefore allows greater pumping rates, was drilled to allow injection into wells B-3 and B-6. However, injection of about 135,000 liters of water from B-7 into B-6 at a rate of about 150 L/min with the addition of HCl showed no measurable effect in water quality withdrawn from B-7. This apparent limited hydraulic connection between these wells resulted in changing the design of the subsequent remediation experiment to include only wells B-7 and B-3 (Fig. 6).

A cross-flow experiment increased the iron oxide content of the basalt aquifer followed by pH adjustment. Two cycles of iron addition, pH adjustment, and recovery (Fig. 7) consisted of: 1.) Addition of 1 kg of $FeCl_3$ (2 kg total) with a pumping rate of about 100 L/min, 2.) HCl addition to lower the pH of the recovered ground water to < 7.0, and 3.) Pumping to waste at a rate of about 200 to 230 L/min with no injection. About one-half

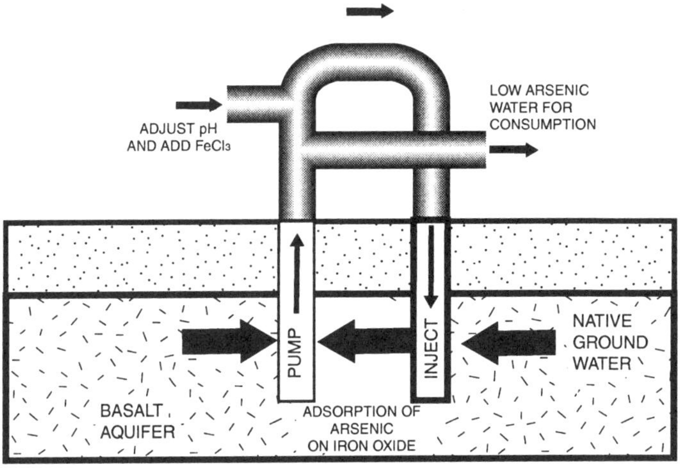

Figure 6. Remediation experiment pumping scheme.

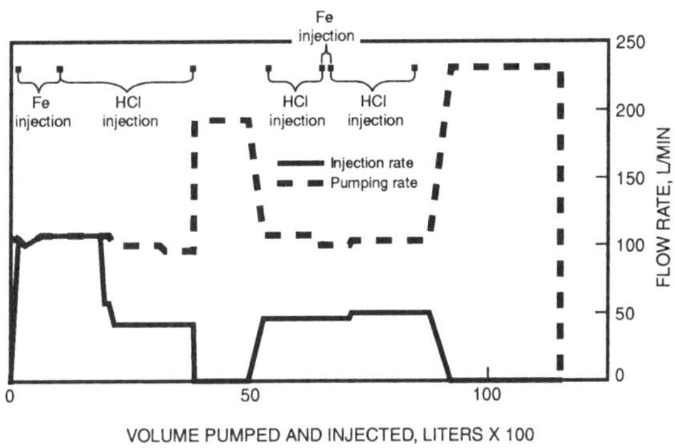

Figure 7. Rates of pumping and injection of water from well B-7 to B-3

of the water pumped from well B-7 was not re-injected during the HCl addition to simulate some withdrawal for water use.

Filtered arsenic concentrations in ground water pumped during the HCl addition generally ranged from about 50 to 90 µg/L (Fig. 8A), which corresponds to removal of about 25 to 50% (assuming an arsenic concentration of about 120 µg/L in the ambient ground water). The arsenic removal was not greater after the second addition of $FeCl_3$ compared to the removal after the first addition. Filtered arsenic concentrations were lower than the ambient concentration of 120 µg/L after the cessation of the HCl addition even when the pH exceeded 8.0.

Arsenic transport on particulate iron has been suggested as the cause of higher arsenic concentrations observed in the water pumped immediately after injection of iron- and arsenic-rich water (Appelo and deVet, this volume; Rott and Meyerhoff, 1996). During the first two periods of HCl addition, unfiltered samples contained more arsenic than filtered samples (Fig. 8A). Immediately after the second $FeCl_3$ treatment, the water had a distinct reddish iron oxide color and the amount of particulate iron was very high (>10 mg/L), as indicated by the high iron content in unfiltered samples (Fig. 8B). Shortly after the cessation of the $FeCl_3$ addition the filtered and unfiltered arsenic concentrations were about 90 µg/L and both the filtered and unfiltered iron concentrations rapidly decreased (Fig. 8B). Taken together, the arsenic and iron concentrations during the latter part of the final period of HCl addition suggest that the arsenic was primarily dissolved and that about 25% of the arsenic was removed.

Nine months after the first cross-flow test, the effect of pH adjustment on arsenic concentration was evaluated during another cross-flow experiment. The pumping rate from well B-7 was about 210 L/min and about one-third of the water was diverted to waste. Arsenic concentrations in filtered samples decreased to values in the range 101 to 105 µg/L near the end of pumping when the pH was about 7.0, which corresponds to a removal efficiency of about 10 to 15% (Fig. 9). Unfiltered arsenic and iron concentrations near the end of the pumping (pumping volume = 108,000 L) were 103 and 36 µg/L respectively, suggesting that arsenic transport on particulate iron oxide was not an important source of arsenic in the pumped ground water.

Calcium and magnesium concentrations were higher in lower pH ground water pumped during the last two field experiments (Fig. 10A). Additionally, the lower pH ground water is undersaturated with respect to calcite, which has been identified in the basalt aquifer (Lico and Seiler, 1994). Because the dissolution rates increase with increasing undersaturation, calcite dissolution is expected to be greater in the lower pH water (Fig. 10A). The somewhat higher sodium concentrations in the lower pH ground water (Table 1) are likely a result of cation exchange, which appears to be an important control on the cation composition of ground water in the Carson Desert (Lico and Seiler, 1994). If cation exchange is the primary cause of the higher sodium concentrations, then the amount of calcite dissolution is equal to the amount involved in the exchange plus the increase in the aqueous calcium. Because phosphate can substitute for carbonate in calcite, the increase in phosphate may be related to the amount of calcite dissolution. Because phosphate competes for adsorption sites on iron oxide, the increased phosphate concentrations may be partially responsible for limiting the arsenic removal. If calcite is the source of the phosphate, then greater arsenic removal efficiencies may be achieved by adjusting the pH to a value in the range of about 7.0 to 8.0, which could produce much less calcite dissolution than appears to occur in water with a pH less than about 7.0. Additionally, long-term injection of ground water with a lower pH may eventually dissolve

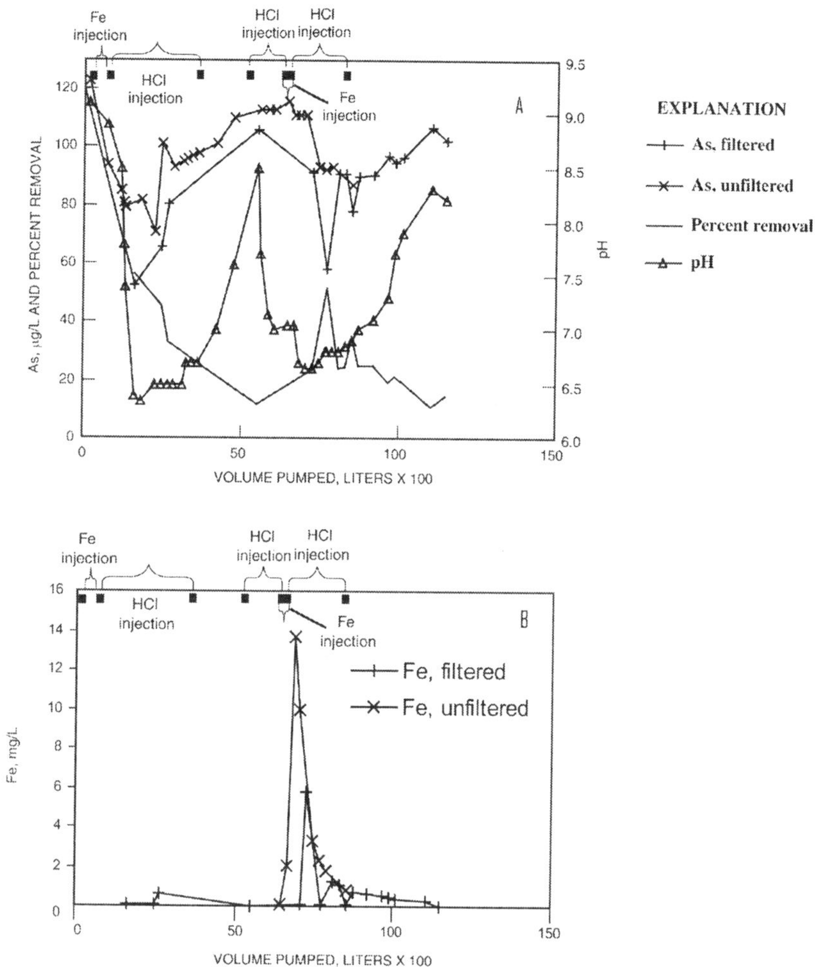

Figure 8. Arsenic and iron concentrations and pH during cross-flow experiment.

much of the calcite leading to decreasing amounts of dissolution, which could improve the arsenic removal.

5. DISCUSSION

The laboratory and field results show that the aquifer basalt has limited capacity for adsorption of As(V), presumably by naturally occurring HFO. However, addition of HFO can significantly increase As(V) adsorption. The effectiveness of HFO addition to this aquifer depends on the HFO concentration and the pH because the concentration of adsorption sites on HFO increases with decreasing pH. Initially, pH adjustment will only be

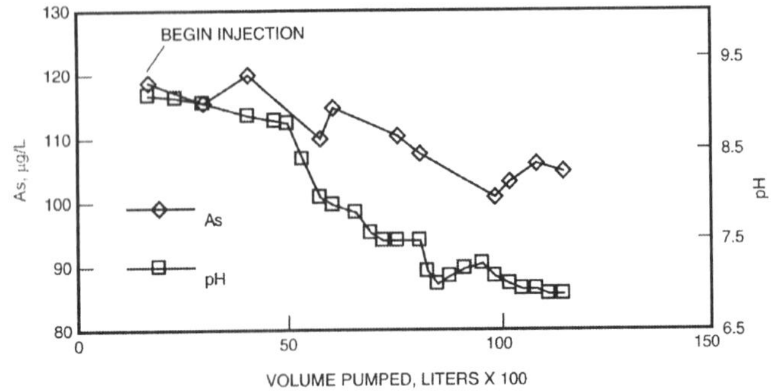

Figure 9. Arsenic concentrations in ground water pumped during pH adjustment.

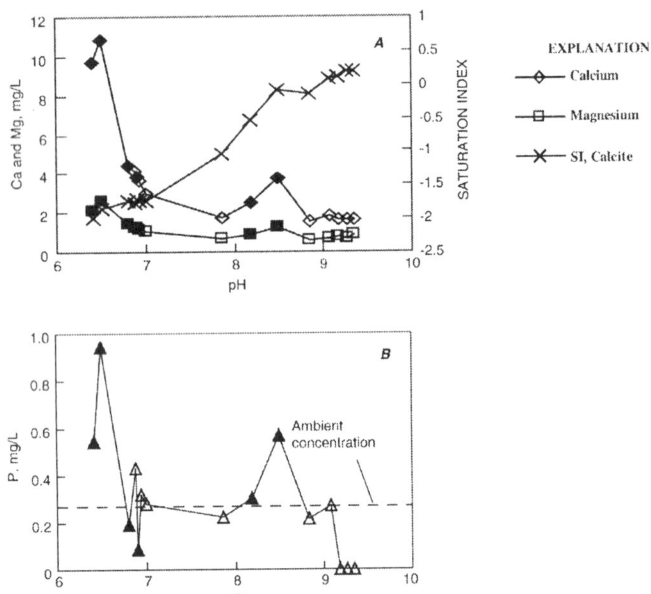

Figure 10. Relations between pH and (A) calcium and magnesium concentrations, and the saturation index for calcite, and (B) phosphate concentration. The solid and open symbols represent samples collected during the first (8/30/00) and second (5/30/01) cycling experiments, respectively.

effective for a short time because of the buffering capacity of the basalt. However, repeated treatments of the aquifer with Fe(III) and supplemental pH adjustments should reduce the carbonate concentrations in the basalt reducing the buffering capacity and allowing lower pH conditions to persist for longer periods.

The cross-flow and laboratory experiments demonstrate that decreasing the pH combined with increasing the iron oxide content in the basalt aquifer reduces arsenic concentrations in produced ground water. Arsenic removal was very effective in laboratory experiments and during the early part of the initial push-pull experiment. Moderate arsenic removal after increasing the iron oxide content of the basalt aquifer and lowering the pH during the cross-flow experiments was limited to about 50%. This relatively low removal may be due to a variety or combination of factors, including local hydraulics of the aquifer, chemical reaction kinetics, and changing aqueous chemistry caused by the lowered pH.

The fractured character and high hydraulic conductivity of the basalt aquifer has been emphasized by various investigators (Glancy, 1986; Maurer and Welch, 1996). Ground water in fractured media is in much less intimate contact with the surface of the aquifer material compared with most sedimentary aquifers. This low surface area could limit the amount of adsorption by the iron oxide that was introduced into the aquifer. As noted above, the arsenic removal efficiency did not appear to increase after a second addition of $FeCl_3$. Because of the relatively low surface area, the iron oxide may have coated much of the fracture surface near the pumped well during the earlier iron additions. Iron oxide formed by the last $FeCl_3$ addition may largely form a coating on the previously formed oxide, limiting any increase in effective oxide surface area available for adsorption.

ACKNOWLEDGEMENTS

The funding from the Fallon Paiute-Shoshone Tribe through the U.S. Bureau of Reclamation's Native American Technical Assistance Program for the arsenic remediation experiments is gratefully acknowledged. The Churchill County Road Department and the U.S. Bureau of Reclamation allowed access to the field site. The manuscript benefited from constructive and insightful reviews by Mike Lico and Steve Sando of the U.S. Geological Survey and Carol Boughton of the U.S. Bureau of Reclamation.

REFERENCES CITED

Acharyya, S. K., 1997, Arsenic in groundwater - geological overview, Consultation on arsenic in drinking water and resulting arsenic toxicity in India and Bangladesh., Proceedings of WHO conference, New Delhi, May 1997, p. 12.

Acharyya, S. K., Chakraborty, P., Lahiri, S., Raymahashay, B. C., Guha, S., Bhowmik, A., Chowdhury, T. R., Basu, G. K., Mandal, B. K., Biswas, B. K., Samanta, G., Chowdhury, K., Chanda, C. R., Lodh, D., Roy, S. L., Saha, K. C., Roy, S., Kabir, S., Quamruzzaman, Q., Chakraborti, D., and McArthur, J. M., 1999, Brief communications: Arsenic poisoning in the Ganges delta. The natural contamination of drinking water by arsenic needs to be urgently addressed: Nature, v. 401, p. 545.

Acharyya, S. K., Lahiri, S., Raymahashay, B. C., and Bhowmik, A., 2000, Arsenic toxicity of groundwater in parts of the Bengal basin in India and Bangladesh: the role of Quaternary stratigraphy and Holocene sea-level fluctuation: Environmental Geology, v. 39, no. 10, p. 1127-1137.

Adyalkar, P. G., Ghosh, P. C., and Mehta, B. C., 1981, On the salinity of groundwater in South 24 Parganas District, West Bengal, India: *in* van Duijvenbooden, W., Glasbergen, P., and Lelyveld, H., eds., Quality of Groundwater, p. 63-67.

Affolter, R. H., and Hatch, J. R., 2002, Characterization of the quality of coals from the Illinois Basin, Resource Assessment of the Springfield, Herrin, Danville, and Baken Coals in the Illinois Basin, U.S. Geological Survey Professional Paper 1625-O Chapter E.

Aggett, J., and Aspell, A. C., 1978, Release of arsenic from geothermal sources: New Zealand Energy Research & Development Report, 35, 12 p.

-, 1980, Arsenic from geothermal sources in the Waikato catchment: New Zealand Journal of Science, v. 23, p. 77-82.

Aggett, J., and Kreigman, M. R., 1988, The extent of formation of arsenic(III) in sediment interstitial waters and its release to hypolimnetic waters in Lake Ohakuri: Water Resources, v. 22, p. 407-411.

Aggett, J., and O'Brien, G. A., 1985, Detailed model for the mobility of arsenic in lacustrine sediments based on measurements in Lake Ohakuri: Environmental Science & Technology, v. 19, p. 231-238.

Aggett, J., and Roberts, L. S., 1986, Insight into the mechanism of accumulation of arsenate and phosphate in hydro lake sediments by measuring the rate of dissolution with ethylenediaminetetraacetic acid: Environmental Science & Technology, v. 20, p. 183-186.

Aichelle, S., Hill-Rowley, R., and Malone, M., 1999, Arsenic, nitrate, and chloride in ground water, Oakland County, Michigan: U. S. Geological Survey Fact Sheet 135-98, 6 p.

Ainsworth, C. C., Girvin, D. C., Zachara, J. M., and Smith, S. C., 1989, Chromate adsorption on goethite: Effects of aluminum substitution: Soil Science Society of America Journal, v. 53, p. 411-418.

Aiuppa, A., Dongarra, G., Capasso, G., and Allard, P., 2000, Trace elements in the thermal groundwaters of Vulcano Island (Sicily): Journal of Volcanology and Geothermal Research, v. 98, p. 189-207.

Alam, M. K., Hasan, A. K. M. S., Khan, M. R., and Whitney, J. W., 1990, Geological map of Bangladesh, scale 1:1,000,000., Geological Survey of Bangladesh, Dhaka.

Alberta Health and Wellness, 2000, Arsenic in groundwater from domestic wells in three areas of northern Alberta: Alberta Health and Wellness--Health Surveillance, 35 p.

Alden, W. C., 1953, Physiography and glacial geology of western Montana and adjacent areas: U.S. Geological Survey Profressional Paper 231, 200 p.

Allaby, A., and Allaby, M., 1990, The Consise Oxford Dictionary of Earth Sciences: Oxford Paperback Reference: New York, Oxford University Press, 410 p.

Allison, J. D., Brown, D. S., and Novo-Gradac, K. J., 1991, MINTEQA2/PRODEFA2; A geochemical assessment model for environmental systems, Version 3.0 user's manual: U.S. Environmental Protection Agency, 106 p. p.

Alpers, C. N., and Nordstrom, D. K., 1999, Geochemical modeling of water-rock interactions in mining environments: *in* Plumlee, G. S., and Logsdon, M. J., eds., The Environmental Geochemistry of Mineral Deposits, Part A. Processes, Techniques, and Health Issues, Society for Economic Geologists, Reviews in Economic Geology, p. 289-323.

Anderholm, S. K., 1988, Ground-water geochemistry of the Albuquerque-Belen Basin, Central New Mexico: U.S. Geological Survey Water-Resources Investigations Report 86-4094, 110 p.

-, 2001, Mountain-front recharge along the east side of the Albuquerque Basin, central New Mexico (revised): U.S. Geological Survey Water-Resources Investigations Report 00-4010, 36 p.

Anderson, C. T., 1930, The heat capacities of arsenic, arsenic trioxide and arsenic pentoxide at low temperatures: Journal of the American Chemical Society, v. 52, p. 2296-2301.

Anderson, G. M., 1983, Some geochemical aspects of sulfide precipitation in carbonate rocks: *in* International conference on Mississippi Valley Type lead-zinc deposits, Rolla, Missouri, p. 61-76.

Anderson, M. A., and Benjamin, M. M., 1990, Surface and bulk characteristics of binary oxide suspensions: Environmental Science & Technology, v. 24, p. 692-698.

Anderson, M. A., Ferguson, H. F., and Gavis, J., 1976, Arsenate adsorption on amorphous aluminum hydroxide: Journal of Colloid and Interface Science, v. 54, p. 391-399.

Anderson, M. A., and Malotky, D. T., 1979, The adsorption of protolyzable anions on hydrous oxides at the isoelectric pH: Journal of Colloid and Interface Science, v. 72, p. 413-427.

Anderson, M. A., Tejedor-Tejedor, M. I., and Stanforth, R. R., 1985, Influence of aggregation on the uptake kinetics of phosphate on goethite: Environmental Science & Technology, v. 9, p. 632-637.

Angeli, J., and Souchay, P., 1960, Sur les thioarsenites en solution: Compte Rendu, v. C 250, p. 713-715.

Anonymous, 1999, Arsenic in Drinking Water: Washington DC, National Academy Press, 263 p.

Anthony, J. W., Bideaux, R. A., Bladh, K. W., and Nichols, M. C., 1990, Elements, Sulfides, Sulfosalts, Handbook of Mineralogy: Tucson, Arizona, Mineral Data Publishing.

-, 1997, Halides, Hydroxides, Oxides, Handbook of Mineralogy, Mineral Data Publishing, p. 628.

Antikainen, P. J., and Tevanen, K., 1961, The effect of temperature on the ionization of arsenious acid: Suomen Kemistilehti, v. 34B, p. 3-4.

Aoki, M., and Yui, S., 1981, Mineralogical Properties and genesis of scorodite in the Osorezan geothermal area: Hirosaki University, p. 104-111 p.

Appelo, C. A. J., Drijver, B., Hekkenberg, R., and De Jonge, M., 1999, Modeling in situ iron removal from ground water: Ground Water, p. 811-817.

Appelo, C. A. J., and Postma, D., 1993, Geochemistry, groundwater and pollution, .A.A. Balkema, Rotterdam, p. 536.

Appelo, C. A. J., Van Der Weiden, M. J. J., Tournassat, C., and Charlet, L., 2002, Surface complexation of ferrous iron and carbonate on ferrihydrite and the mobilization of arsenic: Environmental Science & Technology, v. 36, p. 3096-3103.

Appold, M. S., and Garven, G., 1999, The hydrology of ore formation in the Southeast-Missouri District; numerical models of topography-driven fluid flow during the Ouachita Orogeny: Economic Geology, v. 94, no. 6, p. 913-935.

Arcand, G. M., 1957, Distribution of tripositive arsenic between hydrochloric acid solutions and 1,1'-dichlorodiethyl ether: Journal of the American Chemical Society, v. 79, p. 1865-1870.

Archer, D. G., 1992, Thermodynamic properties of the NaCl + H_2O system. I. Thermodynamic properties of NaCl(cr): Journal of Physical Chemistry, v. 21, p. 1-21.

-, 1998, Thermodynamic properties of import to environmental processes and remediation. I. Previous thermodynamic property values for cadmium and some of its compounds: Journal of Physical Chemistry, v. 27, p. 915-946.

-, 2000, Thermodynamic properties of the $NaNO_3$ + H_2O system: Journal of Physical Chemistry, v. 29, p. 1141-1156.

Archer, D. G., Filor, D., Oakley, E., and Cotts, E. J., 1996, Enthalpy increment measurements from 4.5 K to 350 K and the thermodynamic properties of the titanium silicide Ti5Si3(cr): Journal of Chemical and Engineering Data, v. 41, p. 571-575.

Archer, D. G., and Nordstrom, D. K., in press, Thermodynamic properties of selected arsenic species of import to ground-water applications: Journal of Chemical and Engineering Data.

Armienta, M. A., Rodriques, R., Aguayo, A., Ceniceros, N., Villasenor, G., and Cruz, O., 1997, Arsenic contamination of groundwater in Zimapan, Mexico: Hydrogeology Journal, v. 5, no. 2, p. 39-46.

Arnorsson, S., and Linvall, R., 2001, The distribution of arsenic, molybdenum and tungsten in natural waters in basaltic terrain, North Iceland: *in* Proceedings, 10th Water-Rock Interaction Symposium, Sardinia.

Astor, B., 1999, A Study Of The Effects Of Strip Mining On Stream Water Chemistry In The Black Warrior Basin: Masters: University Of Alabama, 92 p.

Aurelius, L., 1988, Investigation of arsenic contamination of groundwater occurring near Knott, Texas: Texas Department of Agriculture, 53 p.

Axness, K. A., Potokar, J., and Van Drasek, T., 2002, When the well runs dry: examining the water supply issues in Brown County, Wisconsin: Water Resources IMPACT, v. 4, no. 2, p. 4-8.

Ayotte, J. D., Montgomery, D. L., Flanagan, S. M., and Robinson, K. W., in review, Arsenic in ground water in eastern New England: Occurrence, controls, and human health implications: Pembroke, New Hampshire, U.S. Geological Survey Water-Resources Investigations Report.

Ayotte, J. D., Nielsen, M. G., and Robinson, G. R., Jr., 1998, Relation of arsenic concentrations in ground water to bedrock lithology in eastern New England: *in* Geological Society of America Annual Meeting, Toronto, Ontario, p. A-58.

Ayotte, J. D., Nielsen, M. G., Robinson, G. R. J., and Moore, R. B., 1999, Relation of arsenic, iron, and manganese in ground water to aquifer type, bedrock lithogeochemistry, and land use in the New England coastal basin: U.S. Geological Survey Water-Resources Investigations Report 99-4162, 60 p.

Ayotte, J. D., Ryker, S. J., and Proctor, A., 2001, Variation in bedrock ground-water arsenic concentrations and implications for human-health studies in New England: *in* Geological Society of America, 2001 annual meeting, Boston, MA.

Azcue, J. M., and Nriagu, J. O., 1994, Arsenic: Historical perspectives: *in* Nriagu, J. O., ed., Arsenic in the Environment Part 1: Cycling and Characterization: Advances in Environmental Science and Technology: New York, John Wiley and Sons, p. 430.

Bagla, P., and Kaiser, J., 1996, Epidemiology - India's spreading health crisis draws global arsenic experts: Science, v. 274, no. 5285, p. 174-175.

Bailey, A. K., 1976, Concentration of heavy metals in the sediments of a hydroelectric impoundment: M.S.: University of Montana, 62 p.

Bailey, A. K., and Weisel, G. F., 1976, Completion report: Concentration of heavy metals in sediments of a hydroelectric impoundment: Montana State University Joint Water Resources Research Center, 75 p.

Bajpai, S., and Chaudhuri, M., 1999, Removal of arsenic from ground water by manganese dioxide-coated sand: Journal of Environmental Engineering, v. 125, p. 782-784.

Baker, M. D., Wong, P. T. S., Chau, Y. K., Mayfield, C. I., and Inniss, W. E., 1983, Methylation of arsenic by freshwater green algae: Canadian Journal of Fisheries and Aquatic Sciences, v. 40, p. 1254-1257.

Ball, J. W., and Nordstrom, D. K., 1985, Major and trace-element analyses of acid mine waters in the Leviathan Mine drainage basin, California/Nevada: U.S. Geological Survey Water-Resources Investigations Report 85-4169, 46 p. p.

-, 1998, Critical evaluation and selection of standard state thermodynamic properties for chromium metal and its aqueous ions, hydrolysis species, oxides and hydroxides: Journal of Chemical and Engineering Data, v. 43, p. 895-918.

Ball, J. W., Nordstrom, D. K., Jenne, E. A., and Vivit, D. V., 1998, Chemical analyses of hot springs, pools, geysers and surface waters from Yellowstone National Park, Wyoming, and vicinity, 1974-75: U.S. Geological Survey Open-File Report, 98-182, 45 p.

Ball, J. W., Parks, G. A., Haas, J. L., Jr., and Nordstrom, D. K., 1988, A personal computer version of PHAS20, for the multiple simultaneous multiple regression of thermochemical data: U.S. Geological Survey Open-File Report, 88-489-A, 119 p.

Ballantyne, J. M., and Moore, J. N., 1988, Arsenic geochemistry in geothermal systems: Geochimica et Cosmochimica Acta, v. 52, p. 475-483.

Barnes, H. L., 1983, Ore-depositing reactions in Mississippi Valley-type deposits: *in* Kisvarsanyi, G., Grant, S. K., Pratt, W. P., and Koenig, J. W., eds., International conference on Mississippi Valley-type lead-zinc deposits; proceedings volume: Rolla, Missouri, United States, University of Missouri Press, p. 77-85.

Barringer, J. L., Szabo, Z., Barringer, T. H., and Holmes, C. W., 2001, Mobility of arsenic in agricultural and wetlands soils and sediments, northern Coastal Plain of New Jersey: *in* USGS Workshop on Arsenic in the Environment, Denver, Colorado.

Barton, P. B., 1969, Thermochemical study of the system Fe-As-S: Geochimica et Cosmochimica Acta, v. 33, p. 841-857.

Batten, W. G., and Bradbury, K. R., 1996, Regional groundwater flow system between the Wolf and Fox Rivers near Green Bay, Wisconsin: Wisconsin Geologic and Natural History Survey Information Circular 75, 28 p.

Bauer, H. H., and Vaccaro, J. J., 1990, Estimates of ground-water recharge to the Columbia Plateau regional aquifer system, Washington, Oregon, and Idaho, for predevelopment and current land-use conditions: U.S. Geological Survey Water-Resources Investigations Report 88-4108, 37 p.

Beezer, A. E., Mortimer, C. T., and Tyler, E. G., 1965, Heats of formation and bond energies. Part XIII. Arsenic tribromide, arsenious and arsenic oxides, and aqueous solutions of sodium arsenite and sodium arsenate: Journal of the Chemical Society (A), p. 4471-4478.

Belkin, H. E., and Anonymous, 1998, Mineralogy and speciation of arsenic in coals of the Upper Permian Longtan Formation, Guizhou Province, P.R. China: Abstracts with Programs - Geological Society of America, v. 30, no. 7, p. 59.

Belzile, N., and Tessier, A., 1990, Interactions between arsenic and iron oxyhydroxides in lacustrine sediments: Geochimica et Cosmochimica Acta, v. 54, p. 103-109.

Benjamin, M. M., Sletten, R. S., Bailey, R. P., and Bennett, T., 1996, Sorption and filtration of metals using iron-oxide coated sand: Water Resources Research, v. 30, p. 2609-2620.

Beran, A., Goetzinger, M. A., and Rieck, B., 1994, A fluid inclusion study of realgar from the Allchar deposit, FYR Macedonia: Neues Jahrbuch fur Mineralogie. Abhandlungen, v. 167, p. 345-348.

Berg, M., Tran, H. C., Nguyen, T. C., Pham, H. V., Schertenleib, R., and Giger, W., 2001, Arsenic contamination of groundwater and drinking water in Vietnam: A human health threat: Environmental Science & Technology, v. 35, no. 13, p. 2621-2626.

Berger, B. R., Goldhaber, M. B., Hildenbrand, T., Wanty, R. B., and Anonymous, 1998, Origin of carlin-style gold deposits; coupled regional fluid flow, core-complex related extension, strike-slip faults, and magmatism: Abstracts with Programs - Geological Society of America, v. 30, no. 7, p. 371.

Bethke, C. M., 1986, Hydrologic constraints on the genesis of the Upper Mississippi Valley mineral district from Illinois Basin brines: Economic Geology, v. 81, p. 233-249.

Bethke, C. M., and Marshak, S., 1990, Brine migrations across North America -- the plate tectonics of groundwater: Annual Review of Earth and Planetary Sciences, v. 18, p. 228-315.

Bevans, H. E., Lico, M. S., and Lawrence, S. J., 1998, Water quality in the Las Vegas Valley area and the Carson and Truckee River Basins, Nevada and California, 1992-96: U.S. Geological Survey Circular 1170, 1170, 47 p.

Bexfield, L. M., and Anderholm, S. K., 1997, Water-quality assessment of the Rio Grande Valley, Colorado, New Mexico, and Texas-Ground-water quality in the Rio Grande flood plain, Cochiti Lake, New Mexico, to El Paso, Texas, 1995: U.S. Geological Survey Water-Resources Investigations Report, 96-4249, 93 p.

-, 2000, Predevelopment water-level map of the Santa Fe Group aquifer system in the Middle Rio Grande Basin between Cochiti Lake and San Acacia, New Mexico, U.S. Geological Survey Water-Resources Investigations Report, p. 1 sheet.

-, 2002, Spatial patterns and temporal variability in water quality from City of Albuquerque drinking-water supply wells and piezometer nests, with implications for the ground-water flow system: U.S. Geological Survey Water-Resources Investigations Report 01-4244, 101 p.

Bexfield, L. M., Lindberg, W. E., and Anderholm, S. K., 1999, Summary of water-quality data for City of Albuquerque drinking-water supply wells, 1988-97, U.S. Geological Survey Open-File Report 99-195.

Beyer, H., 1989, Arsenolite as alteration product of realgar at low temperatures: Aufschluss, v. 40, p. 215-220.

BGS and DPHE, 2001, Arsenic contamination of groundwater in Bangladesh: *in* Kinniburgh, D. G., and Smedley, P. L., eds., BGS Technical Report WC/00/19: Keyworth, British Geological Survey.

Bhattacharya, P., Chatterjee, D., and Jacks, G., 1997a, Occurrence of arsenic-contaminated groundwater in alluvial aquifers from delta plains, Eastern India: options for safe drinking water supply: Water Resources Development, v. 13, p. 79-92.

-, 1997b, Occurrence of As-contaminated groundwater in alluvial aquifers from the Delta Plains, Eastern India: options for safe drinking water supply: Water Resources Development, v. 13, p. 79-82.

Bierens de Haan, S., 1991, A review of the rate of pyrite oxidation in aqueous systems at low temperature: Earth-Science Reviews, v. 31, no. 1, p. 1-10.

Birkle, P., and Merkel, B., 1998, Environmental effects of surface outflow of geothermal water at the geothermal field of Los Azufres, Mexico: Geothermal Resources Council Transactions, p. 287-291.

Blachnik, R. v., Hoppe, A., and Wickel, U., 1980, The systems arsenic-sulfurand arsenic-selenium and the thermodynamic data for their compounds: Zeitschrift fur Anorganische Chemie, v. 463, p. 78-90.

Blackburn, W. H., and Dennen, W. H., 1997, Encyclopedia of Mineral Names: Mineralogical Association of Canada, 360 p.

Bonazzi, P., Menchetti, S., and Pratesi, G., 1995, The crystal structure of pararealgar, As_4S_4: American Mineralogist, v. 80, p. 400-403.

Bostick, B. C., and Fendorf, S. E., 1999, Arsenite surface complexes on metal sulfides: *in* Proceedings of the Soil Science Society of America Annual Meeting, p. 178.

Bottomley, D. J., 1984, Origins of some arseniferous groundwaters in Nova Scotia and New Brunswick, Canada: Journal of Hydrology, v. 69, no. 1-4, p. 223-257.

Bowell, R. J., 1994, Sorption of arsenic by iron oxides and oxyhydroxides in soils: Applied Geochemistry, v. 9, p. 279-286.

Boyle, D. R., Turner, R. J. W., and Hall, G. E. M., 1998, Anomalous arsenic concentrations in groundwaters of an island community, Bowen Island, British Columbia: Environmental Geochemistry and Health, v. 20, no. 4, p. 199-212.

Bragg, L. J., Oman, J. K., Tewalt, S. J., Oman, C. L., Rega, N. H., Washington, P. M., and Finkelman, R. B., 1997, U. S. Geological Survey Coal Quality (COALQUAL) database; version 2.0, U.S. Geological Survey Open-File Report, 97-134.

Braman, R. S., and Foreback, C. C., 1973, Methylated forms of arsenic in the environment: Science, v. 182, p. 1247-1249.

Brammer, H., and Brinkman, R., 1977, Surface-water gley soils in Bangladesh: Environment, landforms and soil morphology: Geoderma, v. 17, p. 91-109.

Brannon, J. C., Podosek, F. A., and Cole, S. C., 1997, Radiometric dating of Mississippi Valley-type ore deposits: *in* Sangster, D. F., ed., Carbonate-hosted lead-zinc deposits: Special Publication - Society of Economic Geologists, p. 536-545.

Brannon, J. M., and Patrick, W. H., 1987, Fixation, transformation and mobilization of arsenic in sediments: Environmental Science & Technology, v. 21, p. 450-459.

Breit, G. N., Foster, A. L., Sanzalone, R. F., Yount, J. C., Whitney, J. W., Welch, A. H., Islam, M. K., and Islam, M. N., 2001, Arsenic cycling in eastern Bangladesh: the role of phyllosilicates: Geological Society of America Abstracts with Program, v. 32, no. 7, p. A192.

Breuninger, A. B., 2000, Effects of floodplain remediation on bed sediment contamination in the upper Clark Fork River basin, western, Montana.: M.S. Thesis: University of Montana, 182 p.

Breward, N., and Williams, M., 1994, Preliminary appraisal of the environmental geochemistry of the Bau mining area, Sarawak, Malaysia: British Geological Survey, WC/94/67R, 18 p.

Britton, H. T. S., and Jackson, P., 1934, Physicochemical studies of complex formation involving weak acids. Part X. complex formation between tartaric acid and (a) arsenic acid, (b) arsenious acid, (c) antimonous hydroxide, in acid and alkaline solutions. The dissociation constants of arsenious and arsenic acids: Journal of the Chemical Society, p. 1048-1062.

Britzke, E. V., Kapustinsky, A. F., and Tschenzowa, L. G., 1933, The affinity of metals for sulfur. Part III. Combustion- and formation-enthalpies of the sulfides of arsenic and the compounds $As_2O_3.As_2O_5$ and $As_2O_3.SO_3$: Zeitschrift fur Anorganische Chemie, v. 213, p. 58-64.

Brook, E. J., and Moore, J. N., 1988, particle-size and chemical control of As, Cd, Cu, Fe, Mn, Ni, Pb, and Zn in bed sediment from the Clark Fork River, Montana, U.S.A.: The Science of the Total Environment, v. 76, p. 247-266.

Brookins, D. G., 1986, Geochemical behavior of antimony, arsenic, cadmium, and thallium: Chemical Geology, v. 54, p. 271-278.

Brown, G. E., Jr., 1990, Spectroscopic studies of chemisorption reaction mechanisms at oxide-water interfaces: *in* Hochella, M. F., and White, A. F., eds., Mineral-Water Interface Geochemistry: Washington, D. C., Mineralogical Society of America, p. 309-363.

Brown, K. L., and Bacon, L. G., 2000, Cascade development and mineral extraction - alternatives to reinjection. In Environmental Safety and Health Issues in Geothermal Development: *in* World Geothermal Congress, Kazuno, Japan.

Bruce, B. W., and Oelsner, G. P., 2001, Contrasting water quality from paired domestic/public supply wells, central High Plains: Journal of the American Water Resources Association, v. 37, no. 5, p. 1389-1403.

Bryndzia, L. T., and Kleppa, O. J., 1988, Standard molar enthalpies of formation of realgar (a-AsS) and orpiment (As_2S_3) by high-temperature direct-synthesis calorimetry: Journal of Chemical Thermodynamics, v. 20, p. 755-764.

Burgess, W., Burren, M., Cuthbert, M., Mather, S., Perrin, J., Hasan, M., Ahmed, K., Ravenscroft, P., and Rahman, M., 2000, Field relationships and models of arsenic in aquifers of southern Bangladesh: *in* Proceedings of the XXX IAH Congress on Groundwater: Past Achievements and Future Challenges, Cape Town, South Africa, p. 707-712.

Burkel, R. S., and Stoll, R. C., 1999, Naturally occurring arsenic in sandstone aquifer water supply wells of northeastern Wisconsin: Ground Water Monititoring & Remediation, v. 19, no. 2, p. 114-121.

Burns, P. C., and Percival, J. B., 2001, Alacranite, As_4S_4: A new occurrence, new formula, and determination of the crystal structure: Canadian Mineralogist, v. 39, p. 809-818.

Busenberg, E., Plummer, L. N., Doughten, M. W., Widman, P. K., and Bartholomay, R. C., 2000, Chemical and isotopic composition and gas concentrations of ground water and surface water from selected sites at and near the Idaho National Engineering and Environmental Laboratory, Idaho, 1994-97: U.S. Geological Survey Open-File Report, 00-81, 55 p.

Cannon, W. F., Kolker, A., Westjohn, D. B., Miller, J. D., Jr. (chairperson), and Jirsa, M. A. c., 1998, The geological source of arsenic in ground water in southeastern Michigan: *in* Institute on Lake Superior Geology, 44th annual meeting, Minneapolis, MN, United States, p. 50-51.

Carlson, L., Lindstrom, E. B., Hallberg, K. B., and Tuovinen, O. H., 1992, Solid phase products of bacterial oxidation of arsenical pyrite: Applied and Environmental Microbiology, v. 58, p. 1046-1049.

Carlson, R. C., 1999, Trends in the United States marketplace for individually owned private water wells: Spectrum Economics Inc., for the National Ground Water Educational Foundation, 41 p.

Carrillo, A., and Drever, J. I., 1998, Adsorption of arsenic by natural aquifer material in the San Antonio-El Triunfo mining area, Baja California, Mexico: Environmental Geology, v. 35, p. 251-257.

Carroll, R. E., Pashin, J. C., and Kugler, R. L., 1995, Burial history and source-rock characteristics of Upper Devonian through Pennsylvanian strata, Black Warrior basin, Alabama: Geological Survey of Alabama Circular, v. 187, p. 29.

Cartwright, I., and Oliver, N. H. S., 2000, Metamorphic fluids and their relationship to the formation of metamorphosed and metamorphogenic ore deposits: *in* Spry, P. G., Marshall, B., and Vokes, F. M., eds., Metamorphosed and Metamorphogenic Ore Deposits: Reviews in Economic Geology, p. 81-96.

Cebrian, M. E., Albores, A., Garcia-Vargas, G., and Del Razo, L. M., 1994, Chronic arsenic poisioning in humans: The case of Mexico: *in* Nriagu, J. O., ed., Arsenic in the Environment, Part II: Human Health and Ecosystem Effects: Advances in Environmental Science and Technology: New York, John Wiley and Sons, p. 93-107.

Cederberg, G. A., Street, R. L., and Leckie, J. O., 1985, A groundwater mass transport and equilibrium chemistry model for multicomponent systems: Water Resources Research, v. 21, p. 1095-1104.

Celico, P., Dall'Aglio, M., Ghiara, M. R., Stanzione, D., Brondi, M., and Prosperi, M., 1992, Geochemical monitoring of the thermal fluids in the Phlegraean Fields from 1970 to 1990: Bollettino della Scoieta Geologica Italiana, v. 111, p. 409-422.

CGWB, 1999, High Incidence of arsenic in groundwater in West Bengal: Central Ground Water Board, Ministry of Water Resources, Goverment of India.

CH2M Hill, 1990, Spatial distribution and temporal variations of arsenic in the Albuquerque Basin (subtasks 3D and 3E report): Albuquerque, CH2M Hill report.

-, 1991, Executive Summary, Task 3-Arsenic characterization (subtask 3F): CH2M Hill report to the City of Albuquerque Public Works Department, April 5, 1991, 7 p.

-, 1994, Milltown Supefund Site supplemental data interpretation of geochemical release mechanisms and arsenic fate and transport: Helena, Montana, USEPA.

Chague-Goff, C., Rosen, M. R., and Eser, P., 1999, Sewage effluent discharge and geothermal input in a natural wetland, Tongariro Delta: New Zealand, Ecological Engineering, v. 12, p. 149-170.

Chakraborty, A. K., and Saha, K. C., 1987, Arsenical dermatitis from tubewell water in West Bengal: Indian Journal of Medical Research, v. 85, p. 326-334.

Chapin, C. E., and Dunbar, N. W., 1994, A regional perspective on arsenic in waters of the Middle Rio Grande Basin, New Mexico, *in* New Mexico Water Resources Research Institute, The water future of Albuquerque and the Middle Rio Grande Basin: New Mexico Water Resources Research Institute, p. 257-276.

Charles, R. W., Buden, R. J. V., and Goff, F., 1986, An interpretation of the alteration assemblages at Sulphur Springs, Valles Caldera, New Mexico: Journal of Geophysical Research, Solid Earth and Planets, v. 91, no. B2, p. 1887-1898.

Chatterjee, A., Das, D., Mandal, B. K., Chowdhury, T. R., Samanta, G., and Chakraborti, D., 1995, Arsenic in groundwater in 6 districts of West Bengal, India - the biggest arsenic calamity in the world. 1. Arsenic species in drinking water and urine of the affected people: Analyst, v. 120, no. 3, p. 643-650.

CHEMetrics, 1998, Product Catalog.

Chen, C. J., Chuang , Y. C., Lin, T. M., and Wu, H. Y., 1985, Malignant neoplasms among residents of a blackfoot disease-endemic area in Taiwan: high-arsenic artesian well water and cancers: Cancer Research, v. 45, p. 5895-5899.

Chen, C.-J., Hsu, L.-I., Tseng, C. H., Hsueh, Y.-M., and Chiou, H.-Y., 1999, Emerging epidemics of arseniasis in Asia. In Arsenic: Exposure and Health Effects: *in* Proceedings, 3rd International Conference, Oxford, p. 113-121.

Chen, S. L., Dzeng, S. R., Yang, M. H., Chlu, K. H., Shieh, G. M., and Wal, C. M., 1994, Arsenic species in groundwaters of the Blackfoot Disease Area, Taiwan: Environmental Science & Technology, v. 28, p. 877-881.

Chen, S. L., Yeh , S. J., Yang, M. H., and Lin, T. H., 1995, Trace element concentration and arsenic speciation in the well water of a Taiwan area with endemic Blackfoot disease: Biological Trace Element Research, v. 48, no. 3, p. 263-274.

Chen, X., 2001, Migration of induced-infiltrated stream water into nearby aquifers due to seasonal ground water withdrawal: Ground Water, v. 39, no. 5, p. 721-728.

Cheng, L., Fenter, P., Sturchio, N. C., Zhong, Z., and Bedzyk, M. J., 1999, X-ray standing wave study of arsenite incorporation at the calcite surface: Geochimica et Cosmochimica Acta, v. 63, p. 3153-3157.

Cheng, L., Lyman, P. F., Sturchio, N. C., and Bedzyk, M. J., 1997, X-ray standing wave investigation of the surface structure of selenite anions adsorbed on calcite: Surface Science, v. 382, p. L690-L695.

Cherry, J. A., Shaikh, A. U., Tallman, D. E., and Nicholson, R. V., 1979a, Arsenic species as an indicator of redox conditions in groundwater: Journal of Hydrology, v. 43, p. 373-392.
Cherry, J. A., Shaikh, D. E., Tallman, D. E., and Nicholson, R. V., 1979b, Arsenic species as an indicator of redox conditions in ground water: Journal of Hydrology, v. 43, p. 373-392.
Chesley, J. T., Halliday, A. N., Kyser, T. K., and Spry, P. G., 1994, Direct dating of Mississippi Valley-Type Mineralization; use of Sm-Nd in fluorite: Economic Geology and the Bulletin of the Society of Economic Geologists, v. 89, no. 5, p. 1192-1199.
Chitrakar, R. L., and Neku, A., 2001, The scenario of arsenic in drinking water in Nepal: Department of Water Supply & Sewerage, Nepal, 11 p.
Chiu, V. Q., and Hering, J. G., 2000, Arsenic adsorption and oxidation at manganite surfaces: 1. Method for simultaneous determination of adsorbed and dissolved arsenic species: Environmental Science & Technology, v. 34, p. 2029-2034.
Choi, Y. S., 1998, Sequence stratigraphy and sedimentology of the Middle to Upper Ordovician Ancell and Sinnipee Groups, Wisconsin: Ph.D. Dissertation: University of Wisconsin-Madison, 284 p.
Chormann, F. H., Jr.;, 1985, The occurrence of arsenic in soils and stream sediments, Town of Hudson, New Hampshire: M.S.: University of New Hampshire, 222 p.
Chowdhury, T. R., Basu, G. K., Mandal, B. K., Biswas, B. K., Samanta, G., Chowdhury, U. K., Chanda, C. R., Lodh, D., Roy, S. L., Saha, K. C., Roy, Sibtosh, Kabir, Saiful, Quamruzzaman, Qazi, and Chakraborti, D., 1999a, Arsenic poisoning in the Ganges delta, brief communication, Nature, p. 545-546.
Chowdhury, T. R., Basu, G. K., Mandal, B. K., Biswas, B. K., Samanta, G., Chowdhury, U. K., Chanda, C. R., Lodh, D., Roy, S. L., Saha, K. C., Roy, S., Kabir, S., Quamruzzaman, Q., and Chakraborti, D., 1999b, Arsenic poisoning in the Ganges delta: Nature, v. 401, p. 545-546.
Chowdhury, T. R., Manal, B. K., Samanta, G., Basu, G. K., Chowdhury, P. P., Chanda, C. R., Karan, N. K., Lodh, D., Dhar, R. K., Das, D., Saha, K. C., and Chakraborti, D., 1997, Arsenic in groundwater in six districts of West Bengal, India:the biggest arsenic calamity in the world: the status report up to August 1995: *in* Chappell, W. R., ed., Arsenic: exposure and health effects: London, Chapman and Hall, p. 93-111.
Christensen, B. T., Bertelsen, F., and Gissel-Nielsen, G., 1989, Selenite fixation by soil particle-size separates: Journal of Soil Science, v. 40, p. 641-647.
Christensen, T. H., Kjeldsen, P., Bjerg, P. L., Jensen, D. L., Christensen, J. B., Baun, A., Albrechtsen, H.-J., and Heron, G., 2001, Biogeochemistry of landfill leachate plumes: Applied Geochemistry, v. 16, p. 659-718.
City of Albuquerque, 2000, Water quality report: City of Albuquerque annual report to customers: City of Albuquerque, 12 p.
Clark, A. H., 1970, Alpha-arsenic sulfide, from Mina Alacrán, Pampa Larga, Chile: American Mineralogist, v. 55, p. 1338-1344.
-, 1972, Mineralogy of the Alacran deposit, Pampa Larga, Chile. V. Dimorphite of supergene origin: Neues Jahrbuch Fur Mineralogie. Monatshefte, p. 423-429.
Clarke, F. W., 1924, The data of geochemistry: U.S. Geological Survey Bulletin, 770, 841 p.
Clarke, M. B., and Helz, G. R., 2000, Metal-thiometalate transport of biologically active trace elements in sulfidic environments. 1. Experimental evidence for copper thioarsenite complexing: Environmental Science & Technology, v. 34, p. 1477-1482.
Cobb, J. C., 1979, Timing and development of mineralized veins during diagenesis in coal beds: *in* Neuvieme Congres International de Stratigraphie et de Geologie du Carbonifere, Washington and Champaign-Urbana, p. 371-376.
Colman, J. A., and Lyford, F. P., submitted, Geochemical associations and sources of arsenic in landfill leachate from a landfill in Saco, Maine.

Conlon, T. D., 1998, Hydrogeology and simulation of ground-water flow in the sandstone aquifer, northeastern Wisconsin: U.S. Geological Survey Water-Resources Investigations Report 97-4096, 60 p.

Cook , J., 1987, Left Bank Outfall Drain Stage I Project: scavenger well studies and pilot project: British Geological Survey, 14 p.

Copard, Y., Disnar, J. R., Becq-Giraudon, J. F., and Boussafir, M., 2000, Evidence and effects of fluid circulation on organic matter in intramontane coalfields (Massif Central, France): International Journal of Coal Geology, v. 44, no. 1, p. 49-68.

Cordy, G. E., Gellenbeck, D. J., Gebler, J. B., Anning, D. W., Coes, A. L., Edmonds, R. J., Rees, J. A. H., and Sanger, H. W., 1998, Water quality in the Central Arizona Basins, Arizona, 1995-98: U.S. Geological Survey Circular 1213, 1213, 38 p.

Coston, J. A., Fuller, C. C., and Davis, J. A., 1995, Pb2+ and Zn2+ adsorption by a natural aluminum- and iron-bearing surface coating on an aquifer sand: Geochimica et Cosmochimica Acta, v. 59, p. 3535-3547.

Coveney, R. M., Jr., Goebel, E. D., and Ragan, V. M., 1987, Pressures and temperatures from aqueous fluid inclusions in sphalerite from Midcontinent country rocks: Economic Geology, v. 82, no. 3, p. 740-751.

Cox, C. D., and Ghosh, M. M., 1994, Surface complexation of methylated arsenates by hydrous oxides: Water Research, v. 28, p. 1181-1188.

Cox, J. D., Wagman, D. D., and Medvedev, V. A., 1989, CODATA Key Values for Thermodynamics, Hemisphere, 271 p.

Craig, H., Boato, G., and White, D., 1956, Isotopic geochemistry of thermal waters: *in* Proceedings of 2nd Conference on Nuclear Processes in Geological Settings, p. 29-38.

Crecelius, E. A., Bloom, N. S., Cowan, C. E., and Jenne, E. A., 1986, Speciation of selenium and arsenic in natural water and sediments, volume 2: Arsenic speciation: Palo Alto, California, Battelle, Pacific Northwest Laboratories, p. variously paginated.

Criaud, A., and Fouillac, C., 1989, The distribution of arsenic(III) and arsenic(V) in geothermal waters: Examples from the Massif Central of France, the Island of Dominica in the Leeward islands of the Caribbean, the Valles Caldera of New Mexico, USA, and southwest Bulgaria: Chemical Geology, no. 76, p. 259-269.

CSME, 1997, Geology and geochemistry of arsenic occurrences in groundwater of six districts of West Bengal: Centre for Study of Man & Environment, Calcutta.

Cullen, W. R., and Reimer, K. J., 1989, Arsenic speciation in the environment: Chemical Reviews, v. 89, p. 713 - 764.

Culvert, H. V., 1967, Low-temperature specific heat of arsenic and antimony: Physical Review, v. 157, p. 560-563.

Cummings, D. E., Caccavo, F. J., Fendorf, S., and Rosenweig, R. F., 1999, Arsenic mobilization by the dissimilatory Fe(III)-reducing bacterium *Shewanella alga* BrY: Environmental Science & Technology, v. 33, p. 723-729.

Dana, E. S., and Ford, W. E. A., 1949, Textbook of Mineralogy: New York, John Wiley and Sons, 851 p.

D'Angelo, D., Norton, S. A., and Loiselle, M. C., 1996, Historical uses and fate of arsenic in Maine: M.S.: Water Research Institute, 33 p.

Darby, d. E. C., 1980, Arsenic and boron in the Tongonan environment: Geothermal Resource Council Transactions, v. 4, p. 671-674.

Darland, J. E., and Inskeep, W. P., 1997a, Effects of pH and phosphate competition on the transport of arsenate: Journal of Environmental Quality, v. 26, p. 1133-1139.

-, 1997b, Effects of pore water velocity on the transport of arsenate: Environmental Science & Technology, v. 31, p. 704-709.

Das, D., Chatterjee, A., Mandal, B. K., Samanta, G., Chakraborti, D., and Chanda, B., 1995, Arsenic in ground-water in 6 districts of West Bengal, India - the biggest arsenic calamity in the world. 2. Arsenic concentration in drinking-water, hair, nails, urine, skin-scale and liver-tissue (biopsy) of the affected people: Analyst, v. 120, no. 3, p. 917-924.

Das, D., Samanta, G., Mandal, B. K., Chowdhury, T. R., Chanda, C. R., Chowdhury, P. P., Basu, G. K., and Chakraborti, D., 1996, Arsenic in groundwater in six districts of West Bengal, India: Environmental Geochemistry and Health, v. 18, no. 1, p. 5-15.

Davies, J., 1994, The hydrogeochemistry of alluvial aquifers in Central Bangladesh: *in* McCall, G. J. H., ed., Groundwater Quality: AGID Special Publication, Chapman and Hall, London, p. 9–18.

-, 2001, Geology and sedimentology: *in* Kinniburgh, D. G., and Smedley, P. L., eds., Arsenic contamination of groundwater in Bangladesh: British Geological Survey Technical Report WC/00/19: Keyworth, p. 17-46.

Davis, C. C., Knocke, W. R., and Edwards, M., 2001, Implications of aqueous silica sorption to iron hydroxide: Mobilization of iron colloids and interference with sorption of arsenate and humic substances: Environmental Science & Technology, v. 35, p. 3158-3162.

Davis, J. A., Coston, J. A., Kent, D. B., and Fuller, C. C., 1998, Application of the surface complexation concept to complex mineral assemblages: Environmental Science & Technology, v. 32, p. 2820-2828.

Davis, J. A., James, R. O., and Leckie, J. O., 1978, Surface ionization and complexation at the oxide/water interface. I. Computation of electrical double layer properties in simple electrolytes: Journal of Colloid and Interface Science, v. 63, p. 480-499.

Davis, J. A., and Kent, D. B., 1990, Surface complexation modeling in aqueous geochemistry: *in* Hochella, M. F., and White, A. F., eds., Mineral-Water Interface Geochemistry: Washington, D. C., Mineralogical Society of America, p. 177-248.

Davis, M. K., Reich, K. D., and Tikkanen, M. W., 1994, Nationwide and California arsenic occurrence studies: *in* W.R. Chappell, C. O. A., and C.R. Cothern, ed., Arsenic exposure and health: Northwood, England, Science and Technology Letters, p. 31-40.

DCH, 1997, Arsenic pollution in groundwater of Bangladesh: Dhaka, Dhaka Community Hospital Trust.

De Carlo, E. H., and Thomas, D. M., 1985, Removal of arsenic from geothermal fluids by adsorptive bubble flotation with colloidal ferric hydroxide: Environmental Science & Technology, v. 19, p. 538-544.

de Senarmont, H., 1851, Sur la formation des mineraux par voie humide dans les gits metallifered concretionnes.: Annales de Chimie et de Physique, v. 32, p. 129-175.

De Vitre, R., Belzile, N., and Tessier, A., 1991, Speciation and adsorption of arsenic on diagenetic iron oxyhydroxides: Limnology and Oceanograpy, v. 36, p. 1480-1485.

Deely, J. M., and Sheppard, D. S., 1996, Whangaehu River, New Zealand: Geochemistry of a river discharging from an active crater lake: Applied Geochemistry, v. 11, p. 447-460.

DeKonnick, L. L., 1909, The precipitation of arsenic with hydrogen sulfide: Bulletin des Societes Chimiques Belges, v. 23, p. 88-94.

Del Razo, L. M., Arellano, M. A., and Cebrain, M. E., 1990, The oxidation states of arsenic in well-water from a chronic arsenicism area of northern Mexico: Environmental Pollution, v. 64, p. 143-153.

Demuth, J., and Avouris, P., 1983, Surface spectroscopy: Physics Today, p. 62-68.

Dennehy, K. F., Litke, D. W., and McMahon, P. B., 2002, The High Plains Aquifer, USA: groundwater development and sustainability: *in* Hiscock, K. M. R., M.O.; Davison, R.M., ed., Sustainable Groundwater Development: London, The Geological Society of London, p. 99-119.

Dipankar, D., Amit, C., Samanta, G., Mandal, B. K., Chowdhury, T. R., Samanta, G., Chowdhury, P. P., Chanda, C. R., Gautam, B., Dilip, L., Swarup, N., Chakroborty, T., Mandal, S., Bhattacharya, S. M., and Chakraborti, D., 1994, Arsenic contamination in groundwater in six districts of West Bengal, India: the biggest arsenic calamity in the world: Analyst, v. 119, no. 12, p. 168N-170N.

Douglass, D. L., Shing, C., and Wang, G., 1992, The light-induced alteration of realgar to pararealgar: American Mineralogist, v. 77, p. 1266-1274.

Dove, P. M., and Rimstidt, J. D., 1985, The solubility and stability of scorodite, $FeAsO_4$ 2 H_2O: American Mineralogist, v. 70, p. 838-844.

Dowdle, P. R., Laverman, A. M., and Oremland, R. S., 1996, Bacterial dissimilatory reduction of arsenic(V) to arsenic(III) in anoxic sediments: Applied and Environmental Microbiology, v. 62, p. 1664-1669.

Doyle, T. A., Davis, A., and Runnells, D. D., 1994, Predicting the environmental stability of treated copper smelter flue dust: Applied Geochemistry, v. 9, p. 337-350.

DPHE/BGS/MML, 1999, Groundwater studies for arsenic contamination in Bangladesh. Phase I: Rapid Investigation Phase.

Drever, J. I., 1997, The Geochemistry of Natural Waters: Surface and Groundwater Environments, Prentice-Hall, 436 p.

Drummond, S. E., and Ohmoto, H., 1985, Chemical evolution and mineral deposition in boiling hydrothermal systems: Economic Geology, v. 80, p. 126-147.

Dunning, G. E., 1988, Calcium arsenate minerals. New to the Getchell mine, Nevada: Mineralogical Record, v. 19, p. 253-257.

Dziegielewski, B., Sharma, S., Bik, T., Margono, H., and Yang, X., 2002, Analysis of water use trends in the United States: 1950-1995: Southern Illinois University, under 1999 U.S. Geological Survey National Competitive Grant No. 99HQGR0222, Subgrant No. 00-312.

Dzombak, D., and Morel, F., 1990a, Surface Complexation Modeling: Hydrous Ferric Oxide: New York, Wiley, 393 p.

Dzombak, D. A., and Morel, F. M. M., 1990b, Surface complexation modeling: Hydrous Ferric Oxide: Wiley, p. 393.

Eary, L. E., 1992, The solubility of amorphous As_2S_3 from 25 to 90°C: Geochimica et Cosmochimica Acta, v. 56, p. 2267-2280.

Eary, L. E., and Schramke, J. A., 1990, Rates of Inorganic Oxidation Reactions Involving Dissolved Oxygen: *in* Melchior, D. C., and Bassett, R. L., eds., Chemical Modeling of Aqueous Systems II, American Chemical Society Symposium 416, p. 379-396.

Eckel, E. B., 1997, Minerals of Colorado, Fulcrum, 665 p.

Egawa, H., Nonaka, T., and Maeda, H., 1985, Studies of selective absorption resins XXII: Removal and recovery of arsenic ion in geothermal power waste solution with chelating resin containing mercapto groups: Separation Science and Technology, v. 20, p. 653-664.

Eggleton, R. A., and Fitzpatrick, R. W., 1988, New data and a revised structural model for ferrihydrite: Clays and Clay Minerals, v. 36, p. 111-124.

Ehrlich, H. L., 1996, Geomicrobiology: New York, Marcel Dekker Inc.

Eick, M. J., Peak, J. D., Brady, P. V., and Pesek, J. D., 1999, Kinetics of lead adsorption/desorption on goethite: Residence time effect: Soil Science Society of America Journal, v. 42, p. 28-39.

Elkhatib, E. A., Bennett, O. L., and Wright, R. J., 1984a, Arsenite sorption and desorption in soils: Soil Science Society of America Journal, v. 48, p. 1025-1030.

-, 1984b, Kinetics of arsenite sorption in soils: Soil Science Society of America Journal, v. 48, p. 758-762.

Elliott, W. C., and Aronson, J. L., 1987, Alleghanian episode of K-bentonite illitization in the southern Appalachian Basin: Geology, v. 15, p. 735-739.

Ellis, A. J., 1967, The chemistry of some explored geothermal systems: *in* Barnes, H. L., ed., Geochemistry of Hydrothermal Ore Deposits: New York, Holt, Rinehart, and Winston Inc., p. 465-514.

Ellis, A. J., and Mahon, A., 1977, Chemistry and Geothermal Systems: New York, Academic Press, 392 p.

Ellis, A. J., and Mahon, W. A. J., 1964, Natural hydrothermal systems and experimental hot-water/rock interactions: Geochimica et Cosmochimica Acta, v. 28, p. 1323-1357.

-, 1967, Natural hydrothermal systems and experimental hot water/rock interactions (Part II): Geochimica et Cosmochimica Acta, v. 31, p. 519-538.

EMC2, 2000, Milltown Reservoir Draft Focused Feasibility Study Report: Anaconda, ARCO.

Emerson, S., Kalhorn, S., Jacobs, L., Tebo, B. M., Nealson, K. H., and Rosson, R. A., 1982, Environmental oxidation rate of manganese(II): bacterial catalysis: Geochimica et Cosmochimica Acta, v. 46, p. 1073-1079.

Engesgaard, P., and Kipp, K. L., 1992, A geochemical transport model for redox-controlled movement of mineral fronts in groundwater flow systems: A case of nitrate removal by oxidation of pyrite: Water Resources Research, v. 28, p. 2829-2843.

Engesgaard, P., and Traberg, R., 1996, Contaminant transport at a waste residue deposit, 2, Geochemical transport modeling: Water Resources Research, v. 32, p. 939-951.

ENSR, 1991, Preliminary phase I remedial investigation report- Milltown reservoir sediments superfund site: Executive summary for exhibits I, II and III.: Fort Collins, Colorado, ENSR Consulting and Engineering.

-, 1992, Milltown reservoir sediments superfund site, remedial investigation: Fort Collins, Colorado, ENSR Consulting and Engineering.

Erickson, R. L., Chazin, B., and Erickson, M. S., 1988, Summary geochemical maps of the Harrison 1 x 2 degree quadrangle, Arkansas and Missouri:, U. S. Geological Survey Miscellaneous Field Studies Map MF-1994-A, p. 1 sheet.

Erickson, R. L., Erickson, M. S., Chazin, B., Baxter, J. W., Masters, J. M., and Eidel, J. J., 1987, Subsurface geochemical investigation in western and southern Illinois; a pilot study: Illinois Mineral Notes, v. 98, p. 45.

Erickson, R. L., Erickson, M. S., Mosier, E. L., and Chazin, B., 1991, Geochemical studies of subsurface carbonate rocks: *in* Martin, J. A., and Pratt, W. P., eds., Geology and mineral-resource assessment of the Springfield 1 by 2 degrees quadrangle, Missouri, as appraised in September 1985: U. S. Geological Survey Bulletin 1942, p. 51-52.

Erickson, R. L., Mosier, E. L., Odland, S. K., and Erickson, M. S., 1981, A favorable belt for possible mineral discovery in the subsurface Cambrian rocks in southern Missouri: Economic Geology, v. 76, p. 921-933.

Erickson, R. L., Mosier, E. L., and Viets, J. G., 1978, Generalized geologic and summary geochemical maps of the Rolla 1 x 2 degree quadrangle, Missouri: U. S. Geological Survey Miscellaneous Field Studies Map MF-1004-A, scale 1:250,000.

Erickson, R. L., Mosier, E. L., Viets, J. G., Odland, S. K., and Erickson, M. S., 1983, Subsurface geochemical exploration in carbonate terraines; Midcontinent, U.S.A.: *in* Kisvarsanyi, G., Grant, S. K., Pratt, W. P., and Koenig, J. W., eds., International conference on Mississippi Valley-type lead-zinc deposits; proceedings volume: Rolla, Missouri, University of Missouri Press, p. 575-583.

Eskenazy, G. M., 1995, Geochemistry of arsenic and antimony in Bulgarian coals: Chemical Geology, v. 119, no. 1-4, p. 239-254.

Evans, M. A., and Battles, D. A., 1999, Fluid inclusion and stable isotope analyses of veins from the central Appalachian Valley and Ridge province: Implications for regional synorogenic hydrologic structure and fluid migration: Geological Society of America Bulletin, v. 111, no. 12, p. 1841-1860.

Ewers, G. R., 1977, Experimental hot water-rock interactions and their significance to natural hydrothermal systems in New Zealand: Geochimica et Cosmochimica Acta, v. 41, p. 143-150.

Ewers, G. R., and Keays, R. R., 1977, Volatile and precious metal zoning in the Broadlands geothermal field, New Zealand: Economic Geology, v. 72, p. 1337-1354.

Farmer, V. C., Krishnamurti, G. S. R., and Huang, P. M., 1991, Synthetic allophane and layer-silicate formation in SiO_2-Al_2O_3-FeO-Fe_2O_3-MgO-H_2O systems at 23°C and 89°C in a calcareous environment: Clays and Clay Minerals, v. 39, p. 561-570.

Faure, G., 1998, Principles and Applications of Geochemistry: Upper Saddle River, New Jersey, Prentice Hall, 600 p.

Federal Register, 2001, National Primary Drinking Water Regulations; Arsenic and Clarifications to Compliance and New Source Contaminants Monitoring; Final Rule: Office of the Federal Register, v. 66, no. 14, p. 6975-7066.

Feldman, P., and Rosenboom, J. W., 2000, Cambodia drinking water quality assessment: Government of Cambodia.

Fendorf, S., Eick, M. J., Grossl, P., and Sparks, P. D. L., 1997, Arsenate and chromate retention mechanisms on goethite. 1. Surface structure: Environmental Science & Technology, v. 31, no. 2, p. 315-326.

Ferguson, J. F., and Gavis, J., 1972, A review of the arsenic cycle in natural waters: Water Research, v. 6, p. 1259-1274.

Finlayson, J. B., and Webster, J. G., 1989, Arsenic, iron and silica in effluent and sinters of the Wairakei borefield: DSIR Chemistry Division, 89/17, 22 p.

Fishman, M. J., 1993, Methods of analysis by the U.S. Geological Survey National Water Quality Laboratory--Determination of inorganic and organic constituents in water and fluvial sediments: U.S. Geological Survey Open-File Report 93-125, 217 p.

Fishman, M. J., and Friedman, L. C., 1989, Methods for the determination of inorganic substances in water and fluvial sediments: U.S. Geological Survey Techniques of Water-Resources Investigations, Book 5, Chapter A1, 545 p. p.

Fishman, N. S., 1997, Basin-wide fluid movement in a Cambrian paleoaquifer: Evidence from the Mt. Simon Sandstone, Illinois and Indiana.

Flint, R. F., 1957, Glacial and Pleistocene geology: London, John Wiley and Sons, Inc., 553 p.

Focazio, M. J., Welch, A. H., Watkins, S. A., Helsel, D. R., and Horn, M. A., 2000, A retrospective analysis on the occurrence of arsenic in public ground-water resources of the United States and limitations in drinking-water-supply characterizations: U.S. Geological Survey Water-Resources Investigations Report 99-4179, 21 p.

Foerster, F., and Pressprich, H., 1927, On the electromotive force behavior of the arsenious-arsenic acid solutions: Z.eitschrift fur Elektrochemie, v. 33, p. 176-181.

Forbes, G. S., Estill, H. W., and Walker, O. J., 1922, Induction period in reactions between thiosulfate and arsenite or arsenate: a useful clock reaction: Journal of the American Chemical Society, v. 44, p. 97-102.

Fordham, A. W., and Norrish, K., 1979, Arsenate-73 uptake by components of several acidic soils and its implications for phosphate retention: Australian Journal of Soil Research, v. 17, p. 307-316.

Fordyce, F. M., Williams, T. M., Paijitprapapon, A., and Charoenchaisri, P., 1995, Hydrogeochemistry of arsenic in an area of chronic mining-related arsenism, Ron Phibun District, Nakhon Si Thammarat Province, Thailand: preliminary results: British Geological Survey, 73 p.

Foster, A. L., 1999, Partitioning and Transformation of Arsenic and Selenium in Natural and Laboratory Systems: Ph.D.: Stanford University, 225 p.

Foster, A. L., Breit, G. N., Welch, A. H., Whitney, J. W., Yount, J. C., Islam, M. S., Alam, M. M., Islam, M. K., and Islam, M. N., 2000a, In-situ identification of arsenic species in soil and aquifer sediment from Ramrail, Brahmanbaria, Bangladesh: American Geophysical Union Fall Meeting Supplement, p. H21D-01.

Foster, A. L., Breit, G. N., Whitney, J. W., Welch, A. H., Yount, J. C., Alam, M. K., Islam, M. N., Islam, M. S., Karim, M., and Manwar, A., 2000b, X-ray absorption fine structure spectroscopic investigation of arsenic species in soil and aquifer sediments from Brahmnabaria, Bangladesh: *in* 4th Annual Arsenic Conference, San Diego, Jun 18-22, 2000.

Foster, A. L., Brown, G. E., Jr., Parks, G. A., and Tingle, T. N., 1998a, Quantitative speciation of arsenic in mine tailings using x-ray absorption spectroscopy: American Mineralogist, v. 89, p. 553-568.

Foster, A. L., Brown, G. E. J., and Parks, G. A., 1998b, X-ray absorption fine-structure spectroscopy study of photocatalyzed, heterogeneous As(III) oxidation on kaolin and anatase: Environmental Science & Technology, v. 32, p. 1444-1452.

Fournier, R. O., and Potter, R. W., 1982, A revised and expanded silica (quartz) geothermometer: Geothermal Resource Council Bulletin, v. 11, no. 10, p. 3-12.

Fournier, R. O., Thompson, J. M., and Hutchison, R. A., 1986, Fluctuations in composition of Cistern Spring, Norris Geyser Basin, Yellowstone National Park, Wyoming - variable boiling and mixing 1962-1985: *in* Proceedings, 5th Water-Rock Interaction Symposium, Reykjavik, Iceland, p. 206-209.

-, 1992, The geochemistry of hot spring waters at Norris Geyser Basin, Yellowstone National Park, USA: *in* 7th Water-Rock Interaction Symposium, Park City, Utah, USA., p. 1289-1292.

Fournier, R. O., and Truesdell, A. H., 1973, An empirical Na-K-Ca geothermometer for natural waters: Geochimica et Cosmochimica Acta, v. 37, p. 515-525.

Fowler, A. D., and Anderson, M. T., 1991, Geopressure zones as proximal sources of hydrothermal fluids in sedimentary basins and the origin of mississippi valley-type deposits in shale-rich sequences: Institution of Mining and Metallurgy, Transactions, Section B: Applied Earth Science, v. 100, p. B14-B18.

Franke, O. L., DeHan, R. S., Cleaves, E. T., Job, C. A., Anzzolin, A. R., Wilber, W. G., and Lapham, W. W., 1997, Conceptual frameworks for ground-water-quality monitoring, Denver, p. 94.

Frazier, W. J., and Schwimmer, D. R., 1987, Regional stratigraphy of North America, Plenum Press, New York, 719 p.

Freeman, M. C., 1985, The reduction of arsenate to arsenite by an Anabaena-bacteria assemblage isolated from the Waikato River: New Zealand Journal of Marine and Freshwater Research, v. 19, p. 277-282.

Frey, M. M., and Edwards, M. A., 1997, Surveying arsenic occurrence: Journal of the American Water Works Association, v. 89, no. 3, p. 105-117.

Frost, F., Frank, D., Pierson, K., Woodruff, L., Raasina, B., Davis, R., and Davies, J., 1993, A seasonal study of arsenic in ground water, Snohomish County, Washington, USA: Environmental Geochemistry and Health, v. 15, p. 209-214.

Frost, R. R., and Griffin, R. A., 1977, Effect of pH on adsorption of arsenic and selenium from landfill leachate by clay minerals: Soil Science Society of America Journal, v. 41, p. 53-57.

Fuller, C. C., and Davis, J. A., 1989, Influence of coupling of sorption and photosynthetic processes on trace element cycles in natural waters: Nature, v. 340, p. 52-54.

Fuller, C. C., Davis, J. A., Coston, J. A., and Dixon, E., 1996, Characterization of metal adsorption variability in a sand and gravel aquifer, Cape Cod, Massachusetts, U.S.A.: Journal of Contaminant Hydrology, v. 22, p. 165-187.

Fuller, C. C., Davis, J. A., and Waychunas, G. A., 1993, Surface-chemistry of ferrihydrite. 2. Kinetics of arsenate adsorption and coprecipitation: Geochimica et Cosmochimica Acta, v. 57, p. 2271-2282.

Gao, Y., and Mucci, A., 2001, Acid base reactions, phosphate and arsenate complexation, and their competitive adsorption at the surface of goethite in 0.7 M NaCl solution: Geochimica et Cosmochimica Acta, v. 65, p. 2361- 2378.

Garbarino, J. R., 1999, Methods of Analysis by the U.S. Geological Survey National Water Quality Laboratory--Determination of Dissolved Arsenic, Boron, Lithium, Selenium, Strontium, Thallium, and Vanadium Using Inductively Coupled Plasma-Mass Spectrometry: U.S. Geological Survery Open-File Report 99-093, 31 p.

Garrett, A. B., Holmes, O., and Laube, A., 1940, The solubility of arsenious oxide in dilute solutions of hydrochloric acid and sodium hydroxide. The character of the ions of trivalent arsenic. Evidence for polymerization of arsenious acid: Journal of the American Chemical Society, v. 62, p. 2024-2028.

Garven, G., 1985, The role of regional fluid flow in the genesis of the Pine Point Deposit, Western Canada sedimentary basin: Economic Geology and the Bulletin of the Society of Economic Geologists, v. 80, no. 2, p. 307-324.

Garven, G., Ge, S., Person, M. A., and Sverjensky, D. A., 1993, Genesis of stratabound ore deposits in the Midcontinent basins of North America; 1, The role of regional groundwater flow: American Journal of Science, v. 293, no. 6, p. 497-568.

Gaus, I., Webster, R., and Kinniburgh, D. G., 2001, Scales of variation: *in* Kinniburgh, D. G., and Smedley, P. L., eds., British Geological Survey Technical Report WC/00/19: Keyworth, p. 161-173.

Gavrieli, I., Zheng, Y., van Geen, A., Stute, M., Dhar, R., Ahmed, M. A., Simpson, J., and Goldstein, S. L., 2000, Hydrogeochemical study of arsenic contamination in Bangladesh groundwater - the role of redox condition: *in* Goldschmidt 2000, Oxford, UK, p. 435.

Gayer, R., Rickard, D., 1994, Colloform Gold in Coal from Southern Wales: Geology, v. 22, p. 35-38.

Gayer, R. A., Cole, J. E., Frodsham, K., Hartley, A. J., Hillier, B., Miliorizos, M., and White, S. C., 1991, The role of fluids of the evolution of the South Wales Coalfield foreland basin: *in* Proceedings of the Annual conference of the Ussher Society, Bristol, p. 380-384.

Gayer, R. A., Rose, M., Dehmer, J., and Shao, L. Y., 1999, Impact of sulphur and trace element geochemistry on the utilization of a marine-influenced coal; case study from the South Wales Variscan foreland basin: *in* Proceedings, Geochemistry of coal and its impact on the environment and human health, Amsterdam, p. 151-174.

German, J. M., 1989, Geologic setting and genesis of gold deposits of the Dahlonega and Carroll County gold belts, Georgia: Economic Geology and the Bulletin of the Society of Economic Geologists, v. 84, no. 4, p. 903-923.

Gerth, J., Bruemmer, G. W., and Tiller, K. G., 1993, Retention of Ni, Zn and Cd by Si? associated goethite: Z. Pflanzenern. Bodenk, v. 156, p. 123-129.

Ghosh, M. M., and Teoh, R. S., 1985, Adsorption of arsenic on hydrous aluminum oxide: *in* 7th Mid-Atlantic Industrial Waste Conference, Lancaster, Pennsylvania, p. 139-155.

Ghosh, M. M., and Yuan, J. R., 1987, Adsorption of inorganic arsenic and organoarsenicals on hydrous oxides: Environmental Progress, v. 6, p. 150-157.

Ghosh, S., and De, S., 1995, Source of the arsenious sediments at Kachua and Itina, Habra Block, North 24 Paraganas, West Bengal - a case study: Indian Journal of Earth Sciences, v. 22, p. 183-189.

Gibbons, R. D., 1994, Statistical Methods for Groundwater Monitoring: New York, Wiley and Sons, 286 p.

Giggenbach, W. F., 1971, Isotopic composition of waters of the Broadlands Geothermal field: New Zealand Journal of Science, v. 14, p. 961-969.

1988, Geothermal solute equilibria. Derivation of Na-K-Mg-Ca geoindicators: Geochimica et Cosmochimica Acta, v. 52, p. 2749-2765.

Gihring, T. M., Druschel, G. K., McCleskey, R. B., Hamers, R. J., and Banfield, J. F., 2001, Rapid arsenite oxidation by Thermus aquaticus and Thermus thermophilus: Field and laboratory investigations: Environmental Science & Technology, v. 35, p. 3857-3862.

Glancy, P. A., 1986, Geohydrology of the basalt and unconsolidated sedimentary aquifers in the Fallon area, Churchill County, Nevada: U.S. Geological Survey Water-Supply Paper 2263, 62 p.

Gluskoter, H. J., Kuhn, J. K., and Miller, W. G., 1977, Organic and inorganic affinities of elements in coals: Abstracts with Programs - Geological Society of America, v. 9, no. 7, p. 990.

God, R., and Zemann, J., 2000, Native arsenic - realgar mineralization in marbles from Saualpe, Carinthia, Austria: Mineralogy and Petrology, v. 70, p. 37-53.

Goff, F., and Gallaher, B., 2001, Volcanic and hydrothermal evolution of Valles Caldera, New Mexico: Geological Society of America Field Trip Guide, Geological Society of America field trip #401.

Goff, F., Shevenell, L., Gardner, J. N., Vuataz, F. D., and Grigsby, C. O., 1988, The hydrothermal outflow plume of Valles Caldera, New Mexico, and a comparison with other outflow plumes: Journal of Geophysical Research, v. 93, no. B6, p. 6041-6058.

Goldberg, A., and Glaubig, R. A., 1988a, Anion sorption on a calcareous, montmorillonitic soil - Arsenic: Soil Science Society of America Journal, v. 52, p. 1297-1399.

Goldberg, S., 1986, Chemical modeling of arsenate adsorption on aluminum and iron oxide minerals: Soil Science Society of America Journal, v. 50, p. 1154-1157.

Goldberg, S., and Glaubig, R. A., 1988b, Anion sorption on a calcareous, montmorillonitic soil - selenium: Soil Science Society of America Journal, v. 52, p. 954-958.

Goldberg, S., and Johnston, C. T., 2001, Mechanisms of arsenic adsorption on amorphous oxides evaluated using macroscopic measurements, vibrational spectrosopy, and surface complexation modeling: Journal of Colloid and Interface Science, v. 234, p. 204-216.

Goldberg, S. A., and Steltenpohl, M. G., 1990, Timing and characteristics of Paleozoic deformation and metamorphism in the Alabama inner piedmont: American Journal of Science, v. 290, p. 1169-1200.

Goldfarb, R., Groves, D. I., and Gardoll, S., 2001, Orogenic gold and geologic time: A global synthesis: Ore Geology Reviews.

Goldhaber, M. B., 1983, Experimental study of metastable sulfur oxyanion formation during pyrite oxidation at pH 6-9 and 30 deg C: American Journal of Science, v. 238, p. 193-217.

Goldhaber, M. B., Bigelow, R. C., Hatch, J., and Pashin, J. C., 2000, Distribution of a suite of elements including arsenic and mercury in Alabama coal: USGS MF 2333.

Goldhaber, M. B., Church, S. E., Doe, B. R., Aleinikoff, J. N., Brannon, J. C., Podosek, F. A., Mosier, E. L., Taylor, C. D., and Gent, C. A., 1995, Lead and sulfur isotope investigation of Paleozoic sedimentary rocks from the southern Midcontinent of the United States: Implications for paleohydrology and ore genesis of the Southeast Missouri lead belts: Economic Geology, v. 90, no. 7, p. 1875-1910.

Goldhaber, M. B., Irwin, E., Atkins, B., Lee, L., Black, D. D., Zappia, H., Hatch, J. R., Pashin, J. C., Barwick, L. H., Cartwright, W. E., Sanzolone, R., Ruppert, L. F., Kolker, A., and Finkelman, R., 2001, Arsenic in stream sediments of northern Alabama: U.S. Geological Survey Miscellaneous Field Studies Map MF-2357.

Goldhaber, M. B., Rowan, E. L., Hatch, J. R., Zartman, R. E., Pitman, J. K., and Reynolds, R. L., 1994, The Illinois Basin as a flow path for low temperature hydrothermal fluids: *in* Proceedings, Illinois Basin energy and mineral resources workshop, United States, p. 10-12.

Golding, L. A., Timperley, M. H., and Evans, C. W., 1997, Non-lethal responses of the freshwater snail Potamopyrgus antipodarum to dissolved arsenic: Environmental Monitoring and Assessment, v. 47, p. 239-254.

Goleva, G. A., 1974, Metal content of hydrotherms in areas of active volcanism. In Hydrothermal Mineral-forming Solutions in the Areas of Active Volcanism, Nauka, Novosibiriskby: New Delhi, Amerind Publishing Co, 113-115 p.

Gooch, F. A., and Whitfield, J. E., 1888, Analysis of waters of the Yellowstone National Park with an account of the methods of analysis employed: U.S. Geological Survey Bulletin 47, 84 p.

Goodbred, S. L., and Kuehl, S. A., 2000, The significance of large sediment supply, active tectonism, and eustasy on margin sequence development: Late Quaternary stratigraphy and evolution of the Ganges-Brahmaputra delta: Sedimentary Geology, v. 133, p. 227-248.

Gotkowitz, M. B., Schreiber, M. E., and Simo, J. A., 2001, Contrasts in the geologic and hydrochemical occurrences of arsenic contamination of groundwater in eastern Wisconsin: *in* 47th Institute on Lake Superior Geology, p. 28.

Gotkowitz, M. B., Schreiber, M. E., and Simo, T., 2000, Delineating causes of arsenic contamination of groundwater, eastern Wisconsin: EOS Transactions, AGU Fall Meeting Supplement, v. 81, p. 552.

Grantham, D. A., and Jones, J. F., 1977, Arsenic contamination of water wells in Novia Scotia: American Water Works Association, v. 69, no. 12, p. 653-657.

Grauch, V. J. S., Gillespie, C. L., and Keller, G. R., 1999, Discussion of new gravity maps for the Albuquerque Basin area, in Albuquerque Geology: *in* New Mexico Geological Society Fiftieth Annual Field Conference, p. 119-124.

Grauch, V. J. S., Sawyer, D. A., Keller, G. R., and Gillespie, C. L., 2001, Contributions of gravity and aeromagnetic studies to improving the understanding of subsurface hydrogeology, Middle Rio Grande Basin, New Mexico: *in* Cole, J. C., ed., Fourth annual workshop of U.S. Geological Survey Middle Rio Grande Basin Study: Albuquerque, New Mexico, U.S. Geological Survey Open-File Report 00-488, p. 3-4.

Gregg, J. M., and Shelton, K. L., 1989, Geochemical and petrographic evidence for fluid sources and pathways during dolomitization and lead-zinc mineralization in Southeast Missouri: *in* Proceedings, Mississippi valley-type sulfide diagenesis and its relation to carbonate paragenesis and basin hydrologic systems: A special session of Annual midyear meeting, Austin, Texas, p. 153-175.

Gregg, J. M., Shelton, K. L., and Bauer, R. M., 1992, Geochemical and fluid inclusion evidence for regional alteration of Upper Cambrian carbonates by basinal fluids in southern Missouri source rocks in the southern Midcontinent: *in* 1990 symposium on source rocks in the southern Midcontinent, Norman, OK, p. 313-320.

Grenthe, I., Fuger, J., Konings, R. J. M., Lemire, R. J., Muller, A. B., Nguyen-Trung Cregu, C., and Wanner, H., 1992, Chemical Thermodynamics of Uranium, North-Holland: Amsterdam, 715 p.

Grossl, P. R., and Sparks, D. L., 1995, Evaluation of contaminant ion adsorption/desorpton on goethite using pressure-jump relaxation kinetics: Geoderma, v. 67, p. 87-101.

Grove, D. B., and Stollenwerk, K. G., 1987, Chemical reactions simulated by ground-water-quality models: Water Resources Bulletin, v. 23, p. 601-615.

Groves, D. I., Goldfarb, R. J., Gebre-Mariam, M., Hageman, S. G., and Robert, F., 1998, Orogenic gold deposits: A proposed classification in the context of their crustal distribution and relationship to other gold deposit types: Ore Geology Reviews, v. 302, p. 7-27.

Gruebel, K. A., Davis, J. A., and Leckie, J. O., 1988, The feasibilty of using sequential extraction techniques for arsenic and selenium in soils and sediments: Soil Science Society of America Journal, v. 52, p. 390-397.

Gudmundsson, B. T., and Arnorsson, S., in press, Geochemical monitoring of the Krafla and Namafjall geothermal areas, North Iceland: Geothermics.

Gulens, J., Champ, D. R., and Jackson, R. E., 1979, Influence of Redox Environments on the Mobility of Arsenic in Ground Water: *in* Jenne, E. A., ed., Chemical modeling in aqueous systems, p. 81-95.

Gulledge, J. H., and O'Conner, J. T., 1973, Removal of arsenic (V) from water by adsorption on aluminum and ferric hydroxides: Journal - American Water Works Association, v. 65, p. 548-552.

Guo, H. R., Chen, C. J., and Greene, H. L., 1994, Arsenic in drinking water and cancers: a brief descriptive view of Taiwan studies: *in* Chappell, W. R., Abernathy, C. O., and Cothern, C. R., eds., Arsenic Exposure and Health: Northwood, Science and Technology Letters, p. 129-138.

Gupta, S. K., and Chen, K. Y., 1978, Arsenic removal by adsorption: Journal of Water Pollution Control Federation, v. 50, p. 493-506.

Gurian, P. L., Small, M. J., Lockwood, J. R., III;, and Schervish, M. J., 2001a, Addressing uncertainty and conflicting cost estimates in revising the arsenic MCL: Environmental Science & Technology, v. 35, no. 22, p. 4414-4420.

Gurian, P. L., Small, M. J., Lockwood, J. R., and Schervish, M. J., 2001b, Benefit-cost estimation for alternative drinking water maximum contaminant levels: Water Resources Research, v. 37, no. 8, p. 2213-2226.

Guthrie, G. M., and Lesher, C. M., 1989, Geologic setting of lode gold deposits in the northern Piedmont and Brevard Zone, Alabama: U.S. Geological Survey Bulletin 136.

Haack, S. K., and Rachol, C. M., 2000a, Arsenic in ground water in Lapeer County, Michigan: U.S. Geological Survey Fact Sheet 129, 2 p.

-, 2000b, Arsenic in ground water in Livingston County, Michigan: U.S. Geological Survey Fact Sheet 130, 2 p.

-, 2000c, Arsenic in ground water in Genesse County, Michigan: U.S. Geological Survey Fact Sheet 127, 2 p.

-, 2000d, Arsenic in ground water in Huron County, Michigan: U.S. Geological Survey Fact Sheet 128, 2 p.

-, 2000e, Arsenic in ground water in Sanilac County, Michigan: U.S. Geological Survey Fact Sheet 132, 2 p.

-, 2000f, Arsenic in ground water in Shiawassee County, Michigan: U.S. Geological Survey Fact Sheet 131, 2 p.

-, 2000g, Arsenic in ground water in Tuscola County, Michigan: U.S. Geological Survey Fact Sheet FS 133, 2 p.

-, 2000h, Arsenic in ground water in Washtenaw County, Michigan: U.S. Geological Survey Fact Sheet FS 134, 2 p.

Haack, S. K., and Trecanni, S. L., 2000, Arsenic concentration and selected geochemical characteristics of ground water and aquifer material in southeastern Michigan: U.S. Geological Survey Water Resources Investigations Report 00-4171, 38 p.

Halcrow and DHV, 1995, National Minor Irrigation Development Project - Baseline Survey Water Quality Part 1: Sir William Halcrow and Partners Ltd, DHV Consultants BV.

Hall, G. E. M., Pelchat, J. C., and Gauthier, G., 1999, Stability of inorganic arsenic(III) and arsenic(V) in water samples: Journal of Analytical Spectrometry, v. 14, p. 205-213.

Hallberg, R. O., and Martinell, R., 1976, Vyredox - in situ purification of groundwater: Ground Water, v. 14, p. 88-93.

Halter, W. E., and Pfeifer, H. R., 2001, Arsenic(V) adsorption onto a-Al_2O_3 between 25 and 70°C: Applied Geochemistry, v. 16, p. 793-802.

Hamdy, A. A., and Gissel-Nielsen, G., 1977, Fixation of selenium by clay minerals and iron oxides: Z.eitschrift fur Pflanzenernachrung und Bodenkunde, v. 140, p. 63-70.

Hanor, J. S., 1997, Controls on the solubilization of lead and zinc in basinal brines, Carbonate-hosted lead-zinc deposits, Special Publication, Sangster, D.F., ed., Society of Economic Geologists, p. 483-500.

Harrison, J. B., and Berkheiser, V., 1982, Anion interactions with freshly prepared hydrous iron oxides: Clays and Clay Minerals, v. 30, p. 97-102.

Hatch, J. R., 1983, Geochemical processes that control minor and trace element composition of United States coals: *in* Shanks, W. C., III, ed., Cameron Volume on Unconventional Mineral Deposits, American Intstitute of Mining Engineers, p. 89-98.

Hatch, J. R., Gluskoter, H. J., and Lindahl, P. C., 1976, Sphalerite in coals from the Illinois Basin: Economic Geology, v. 71, no. 3, p. 613-624.

Hatcher, R. D., Jr., Thomas, W. A., Geiser, P. A., Snoke, A. W., Mosher, S., and Wiltschko, D. V., 1989, Alleghanian Orogen The Appalachian-Ouachita Orogen in the United: *in* The Geology of North America, Boulder, Colorado, p. 233-318.

Haury, V., Jann, S., Kofod, M., Scholz, C., and Isenbeck-Schroter, M., 2000, Redox-induced species distribution of arsenic in a suboxic groundwater environment - column experiments: *in* Proceedings of the International Conference on Groundwater Research, Copenhagen, Denmark, p. 197-198.

Hawley, J. W., and Haase, C. S., 1992, Hydrogeologic framework of the northern Albuquerque Basin: Socorro, New Mexico, Bureau of Mines and Mineral Resources Open-File Report 387.

Hay, R. L., Lee, M., Kolata, D. R., Matthews, J. C., and Morton, J. P., 1988, Episodic potassic diagenesis of Ordovician tuffs in the Mississippi Valley area: Geology, v. 16, p. 743-747.

Hayes, K. F., 1987, Equilibrium, spectroscopic, and kinetic studies of ion adsorption at the oxide/aqueous interface: Ph.D. Dissertation: Stanford University.

Hayes, K. F., Redden, G., Ela, W., and Leckie, J. O., 1990, Surface complexation models: An evaluation of model parameter estimation using FITEQL and oxide mineral titration data: Journal of Colloid and Interface Science, v. 142, p. 448-479.

Hayes, T. M., and Boyce, J. B., 1982, Extended x-ray absorption fine structure spectroscopy: Solid State Physics, v. 37, p. 173-348.

Hayes, T. S., Palmer, J. R., Pratt, W. P., Krizanich, G. W., Whitfield, J. W., and Seeger, C. M., 1997, Cross sections showing stratigraphic and depositional lithofacies of Upper Cambrian rocks and the relation of lithofacies to potential for mississippi valley-type mineralization in the Harrison 1 degrees X 2 degrees quadrangle, Missouri and Arkansas (folio of the Harrison 1 degrees X 2 degrees quadrangle, Missouri and Arkansas): *in* Miscellaneous Field Studies Map - U. S. Geological Survey.

He, Z., Gregg, J. M., Shelton, K. L., and Palmer, J. R., 1997, Sedimentary facies control of fluid flow and mineralization in Cambro-Ordovician strata, southern Missouri basin-wide diagenetic patterns: *in* SEPM Conference on Basin-wide diagenetic patterns: Integrated petrologic, geochemical, and hydrologic consideration, Lake Ozark, Missouri, United States, p. 81-99.

Hearn, P. P., Jr., Sutter, J. F., and Belkin, H. E., 1987, Evidence for late-Paleozoic brine migration in Cambrian carbonate rocks of the Central and Southern Appalachians; implications for mississippi valley-type sulfide mineralization: Geochimica et Cosmochimica Acta, v. 51, no. 5, p. 1323-1334.

Hearn, P. P., and Sutter, J. F., 1985, Authigenic potassium feldspar in Cambrian carbonates; evidence of Alleghanian brine migration: Science, v. 228, no. 4707, p. 1529-1531.

Heinrich, C. A., and Eadington, P. J., 1986a, Thermodynamic predictions of the hydrothermal chemistry of arsenic and their significance for the paragenetic sequence of some cassiterite-arsenopyrite-base metal sulphide deposits: Economic Geology, v. 81, p. 511-529.

-, 1986b, Thermodynamic predictions of the hydrothermal chemistry of arsenic, and their significance for the paragenetic sequence of some cassiterite-arsenopyrite-base metal sulfide deposits: Economic Geology, v. 81, p. 511-529.

Helgen, S. O., and Moore, J. N., 1996, Natural background determination and impact quantification in trace metal-contaminated river sediments: Environmental Science & Technology, v. 30, no. 1, p. 129-135.

Helz, G. R., Tossell, J. A., Charnock, J. M., Pattrick, R. A. D., Vaughn, D. J., and Garner, C. D., 1995, Oligomerization in As(III) sulfide solution: Theoretical constraints and spectroscopic evidence: Geochimica et Cosmochimica Acta, v. 59, p. 4591-4604.

Hem, J. D., 1985, Study and interpretation of the chemical characteristics of natural water: U.S. Geological Survey Water Supply Paper 2254, 264 p.

Henley, R. W., 1985a, Applied chemistry in the exploration and development of New Zealand geothermal systems: New Zealand Journal of Technology, v. 1, p. 207-221.

-, 1985b, The geothermal framework of epithermal deposits. In Geology and Geochemistry of Epithermal Systems: Reviews in Economic Geology, v. 2, p. 1-24.

Hering, J. G., Pen-Yuan, C., Wilke, J. A., and Elimelech, M., 1997, Arsenic removal from drinking water during coagulation: Journal of Environmental Engineering, p. 800-807.

Hess, R. E., and Blanchar, R. W., 1976, Arsenic stability in contaminated soils: Soil Science Society of America Journal, v. 40, p. 847-852.

Heyl, A. V., 1983, Geologic characteristics of three major Mississippi Valley districts: *in* International Conference on Mississippi Valley-Type lead-zinc deposits, Rolla, Missouri, United States, p. 27-60.

Hiemstra, T., and Van Riemsdijk, W. H., 1999, Surface structural ion adsorption modeling of competitive binding of oxyanions by metal (hydr)oxides: Journal of Colloid and Interface Science, v. 210, p. 182-193.

Hingston, F. J., Posner, A. M., and Quirk, J. P., 1971, Competitive adsorption of negatively charged ligands on oxide surfaces: Discussions, Faraday Society, v. 52, p. 334-342.

-, 1972, Anion adsorption by goethite and gibbsite: 1. The role of the proton in determining adsorption envelopes: Journal of Soil Science, v. 23, p. 177-192.

Hinkle, S. R., and Polette, D. J., 1999, Arsenic in ground water of the Willamette Basin, Oregon: U.S. Geological Survey, 98-4205, 28 p.

Hinman, N. W., and Linstrom, R. F., 1996, Seasonal changes in silica deposition in hot spring systems: Chemical Geology, v. 132, p. 237-246.

Hirner, A. V., Krupp, R. E., Gainsford, A. R., and Staerk, H., 1990, Metal-organic associations in sediments-II: Algal mats in contact with geothermal waters: Applied Geochemistry, v. 5, p. 507-513.

HLA, 1986, Milltown reservoir feasibility study: Denver, Colorado, Harding Lawson Associates.

-, 1987, Volume I, Milltown reservoir data report supplemental investigations conducted under the feasibility study: Denver, Colorado, Harding Lawson Associates.

Hochella, M. F., Jr., 1995, Mineral surfaces: Their characterization and their chemical, physical, and reactive nature: *in* Vaughan, D. J., and Pattrick, R. A. D., eds., Mineral Surfaces: London, Chapman and Hall, p. 17-60.

Hofstra, A. H., and Cline, J., S., 2000, Characteristics and Models for Carlin-Type Gold Deposits: SEG Reviews, v. 13, p. 163-220.

Hornberger, M. I., Lambing, J. H., Luoma, S. N., and Axtmann, E. V., 1997, Spatial and temporal trends of trace metals in surface water, bed sediment, and biota of the upper Clark Fork basin, Montana, 1985-1995: U.S. Geological Survey Open-File Report, 97-669, 127 p.

Hower, J. C., Wild, G. D., Pollock, J. D., Trinkle, E. J., Bland, A. E., and Fiene, F. L., 1990, Petrography, geochemistry, and mineralogy of the Springfield (western Kentucky No. 9) coal bed: The Journal of Coal Quality, v. 9, no. 3, p. 90-100.

Hsia, T. H., Lo, S. L., and Lin, C. F., 1992, As(V) adsorption on amorphous iron oxide: Triple layer modelling: Chemosphere, v. 25, p. 1825-1837.

Hsia, T. H., Lo, S. L., Lin, C. F., and Lee, D. Y., 1994, Characterization of arsenate adsorption on hydrous iron oxide using chemical and physical methods: Colloids and Surfaces A: Physiochemical and Engineering Aspects, v. 85, p. 1-7.

Hsu, K. H., Froines, J. R., and Chen, C. J., 1997, Studies of arsenic ingestion from drinking water in northeastern Taiwan: chemical speciation and urinary metabolites: *in* Abernathy, C. O., Calderon, R. L., and Chappell, W. R., eds., Arsenic Exposure and Health Effects: London, Chapman and Hall, p. 190-209.

Huber, G. R., Sacher, M., Vollman, A., Huber, H., and Rose, D., 2000, Respiration of arsenate and selenate by hyperthermophilic archea: Systematic and Applied Microbiology, v. 23, p. 305-314.

Hultgren, R., Desai, P. D., Hawkins, D. T., Gleiser, M., Kelley, K. K., and Wagman, D. D., 1973, Arsenic. Selected Values of the Thermodynamic Properties of the Elements: American Society for Metals, p. 39-46.

Hydrometrics, 1985, Water resources impacts resulting from proposed rehabilitation of the Milltown Hydroelectric project.: Helena, Montana, Hydrometrics.

Ilchik, R. P., and Barton, M. D., 1997, An amagmatic origin of carlin-type gold deposits: Economic Geology and the Bulletin of the Society of Economic Geologists, v. 92, no. 3, p. 269-288.

Imes, J. L., and Emmett, L. F., 1994, Geohydrology of the Ozark Plateaus aquifer system in parts of Missouri, Arkansas, Oklahoma, and Kansas, U.S. Geological Survey Professional Paper, p. D1-D127.

Iqbal, S. Z., 2001, Arsenic contamination in Pakistan.: *in* UN-ESCAP Meeting: Geology and Health., Bangkok, Thailand.

Isenbeck-Schröter, M., 1995, Transportconditions of heavy metals and oxoanions - column experiments and their modeling: Ber. Geowissenschaften Univ. Bremen (in German), v. 67, p. 182.

Isensee, A. R., Kearney, P. C., Woolson, E. A., Jones, G. E., and Williams, V. P., 1973, Distribution of alkyl arsenicals in model ecosystem: Environmental Science & Technology, v. 7, p. 841-845.

ISSI Consulting Group, The Cadmus Group, and ICF Consulting, 2000, Arsenic occurrence in public drinking water supplies: Prepared for the U.S. Environmental Protection Agency under contract nos. 68-C7-0005 and 68-C-99-206., EPA-815-R-00-023, 138 p.

Ivakin, A. A., Vorob'eva, S. V., and Gertman, E. M., 1979a, Determination of the second and third dissociation constants of arsenious acid: Russian Journal of Inorganic Chemistry, v. 24, p. 36-40.

Ivakin, A. A., Vorob'eva, S. V., Gertman, E. M., and Voronova, E. M., 1976, Acid-base equilibria and self-association in arsenious acid solutions: Zhurnal Neorganicheskoy Khimiy, v. 21, p. 442-448.

Ivakin, A. A., Vorobeva, S. V., Gorelov, A. M., and Gertmann, E. M., 1979b, Solubility of arsenic(III) sulphide in aqueous NaCl solutions: Russian Journal of Inorganic Chemistry, v. 24, p. 1089-1091.

Jacobs, L. W., Syers, J. K., and Keeney, D. R., 1970, Arsenic sorption by soils: Soil Science Society of America, Proceedings, v. 34, p. 750-754.

Jain, A., and Loeppert, R. H., 2000, Effect of competing anions on the adsorption of arsenate and arsenite by ferrihydrite: Journal of Environmental Quality, v. 29, p. 1422-1430.

Jain, A., Raven, K. P., and Loeppert, R. H., 1999, Arsenite and arsenate adsorption on ferrihydrite: Surface charge reduction and net OH- release stoichiometry: Environmental Science & Technology, v. 33, p. 1179-1184.

Jain , C. K., and Ali, I., 2000, Arsenic: Occurrence, toxicity and speciation techniques: Water Resources, v. 34, p. 4304-4312.

Jenne, E. A., 1976, Trace Element Sorption by Sediments and Soils -- Sites and Processes: *in* Chappel, W. R., and Peterson, K. K., eds., Molybdenum in the Environment, Dekker, p. 425-553.

Jewell, P. W., and Parry, W. T., 1988, Geochemistry of the Mercury gold deposit (Utah, U.S.A.): Chemical Geology, v. 69, p. 245-265.

Ji, G., and Silver, S., 1995, Bacterial resistance mechanisms for heavy metals of environmental concern: Journal of Industrial Microbiology, v. 14, p. 61-75.

Johnson, D. L., 1971, Simultaneous determination of arsenate and phosphate in natural waters: Environmental Science & Technology, v. 5, p. 411-414.

Johnson, G. K., Papatheodorou, G. N., and Johnson, C. E., 1980, The enthalpies of formation and high temperature thermodynamic functions of the arsenic sulfides (As_4S_4) and (As_2S_3): Journal of Chemical Thermodynamics, v. 12, p. 545-57.

Jones, B., Renaut, R. W., and Rosen, M. R., 1997, Biogenicity of silica precipitates around geysers and hot-spring vents, North Island, New Zealand: Journal of Sediment Research, v. 67, p. 88-104.

Joshi, A., and Chaudhuri, M., 1996, Removal of arsenic from ground water by iron coated sand: Journal of Environmental Engineering, v. 122, p. 769-771.

Junta, J. L., and Hochella, M. F. J., 1994, Manganese (II) oxidation at mineral surfaces: A microscopic and spectroscopic study: Geochimica et Cosmochimica Acta, v. 58, p. 4985-4999.

Kanivetsky, R., 2000, Arsenic in Minnesota Ground Water: Hydrogeochemical Modeling of the Quaternary Buried Artesian Aquifer and Cretaceous Aquifer Systems: University of Minnesota, 55, 23 p.

Kaplan, I. R., and Rittenberg, S. C., 1964, Microbiological fractionation of sulphur isotopes: Journal of General Microbiology, v. 34, p. 195-212.

Karagas, M. R., Le, C. X., Morris, S., Blum, J. D., Lu, X., Spate, V., Carey, M., Stannard, V., Klaue, B., and Tosteson, T. D., 2001, Markers of low level arsenic exposure for evaluating human cancer risks in a US population: International Journal of Occupational Medicine and Environmental Health, v. 14, no. 2, p. 171-175.

Karayigit, A. I., Spears, D. A., and Booth, C. A., 2000, Antimony and arsenic anomalies in the coal seams from the Gokler Coalfield, Gediz, Turkey: International Journal of Coal Geology, v. 44, no. 1, p. 1-17.

Karpov, G. A., and Naboko, S. I., 1990, Metal contents of recent thermal waters, mineral precipitates and hydrothermal alteration in active geothermal fluids, Kamchatka: Journal of Geochemical Exploration, v. 36, p. 57-71.

Keating, E. H., and Bahr, J. M., 1998, Reactive transport modeling of redox geochemistry: Approaches to chemical disequilibrium and reaction rate estimation at a site in northern Wisconsin: Water Resources Research, v. 34, p. 3573-3584.

Kelley, V. C., 1936, Occurrence of claudetite in Imperial County, California: American Mineralogist, v. 21, p. 137-138.

Kent, D. B., Abrams, R. H., Davis, J. A., Coston, J. A., and LeBlanc, D. R., 2000, Modeling the influence of variable pH on the transport of zinc in a contaminated aquifer using semiempirical surface complexation models: Water Resources Research, v. 36, p. 3411-3425.

Kent, D. B., Davis, J. A., Anderson, L. C. D., and Rea, B. A., 1995, Transport of chromium and selenium in a pristine sand and gravel aquifer: Role of adsorption processes: Water Resources Research, v. 31, p. 1041-1050.

Kent, D. B., Niedan, V. W., Isenbeck-Schroter, M., Stadler, S., Jann, S., Hohn, R., and Davis, J. A., 2001, The influence of oxidation-reduction and adsorption reactions on arsenic transport in the oxic, suboxic, and anoxic zones of a mildly acidic sand and gravel aquifer: *in* U. S. Geological Survey Workshop on Arsenic in the Environment, Denver, Colorado.

Keon, N. E., Swartz, C. H., Brabander, D. J., Myneni, S. C., and Hemond, H. F., 2000, Evaluation of arsenic mobility in sediments using a validated extraction method: EOS Transactions, AGU Fall Meeting Supplement, v. 81, p. 526.

Kernodle, J. M., McAda, D. P., and Thorn, C. R., 1995, Simulation of ground-water flow in the Albuquerque Basin, central New Mexico, 1901-1994, with projections to 2020: U.S. Geological Survey, 114 p.

Kerrich, R., and Cassidy, K. F., 1994, Temporal relationships of lode gold mineralization to accretion, magmatism, metamorphism, and deformation -- Archean to present: A review: Ore Geology Reviews, v. 9, p. 263-310.

Khadka, M. S., 1993, The groundwater quality situation in alluvial aquifers of the Kathmandu Valley, Nepal.: AGSO Journal of Australian Geology and Geophysics, v. 14, p. 207-211.

Khan, A. A., 1995, Arsenic contamination in groundwater, its causes and mitigation: a geological perspective: *in* International Conference on Arsenic in Ground Water: Cause, Effect and Remedy, Calcutta, India, p. 43.

Khan, A. A., Imam, B., Akhter, S. H., Hasan, M. A., and Ahmed, K. M. U., 1998, Subsurface investigation in the arsenic problem areas of Rajarampur, Chanlai and Baragharia; Nawabganj District, Bangladesh: Dhaka, Geohazards Research Group, Department of Geology, University of Dhaka.

Khramova, G. C., 1974, Effects of intensified activity of the Ebeko Volcano on the composition of the water of Goryachoe Lake. In Hydrothermal Mineral-forming Solutions in the Areas of Active Volcanism, Nauka, Novosibirisk.: *in* S.I., N., ed.: New Delhi,, Amerind Publishing Co., p. 88-94.

Kim, M.-J., 1999, Arsenic dissolution and speciation in groundwater of southeast MichiganUniversity of Michigan, 201 p.

Kim, M.-J., Nriagu, J. O., and Haack, S. K., 2000, Carbonate ions and arsenic dissolution by groundwater: Environmental Science & Technology, v. 34, no. 14, p. 3094-3100.

Kinniburgh, D. G., 2001, Sorption and transport: *in* Kinniburgh, D. G., and Smedley, P. L., eds., Arsenic contamination of groundwater in Bangladesh: Vol. 2: Final Report. British Geological Survey Technical Report WC/00/19: Keyworth, p. 211-228.

Kinniburgh, D. G., Gale, I. N., Smedley, P. L., Darling, W. G., West, J. M., Kimblin, R. T., Parker, A., Rae, J. E., Aldous, P. J., and O'Shea, M. J., 1994, The effects of historic abstraction of groundwater from the London Basin aquifers on groundwater quality: Applied Geochemistry, v. 9, no. 2, p. 175-196.

Kinniburgh, D. G., and Smedley, P. L., 2001, Arsenic contamination of groundwater in Bangladesh, Vol 3 Hydrochemical Atlas, . British Geological Survey Technical Report WC/00/19: Keyworth, p. 124.

Kipp, K. L., and Parkhurst, D. L., in press, Parallel processing for PHAST - A 3D reactive-transport simulator: *in* XIV International Conference on Computational Methods in Water Resources, Delft, The Netherlands.

Kirschning, H.-J., and Plieth, K., 1955, Electrochemical studies of the conversion point of cubic and monoclinic arsenic modifications: Zeitschrift fur Anorganische und Allgemeine Chemie, v. 280, p. 346-352.

Klemm, W., Spitzer, H., and Niermann, H., 1963, Investigations of some half-metals: Angewandte Chemie, v. 72, p. 985-994.

Kneebone, P. E., and Hering, J. G., 2000, Behaviour of arsenic and other redox-sensitive elements in Crowley Lake, CA: A reservoir in the Los Angeles aqueduct system: Environmental Science & Technology, v. 34, p. 4307-4312.

Koga, A., 1961, Distribution of gold in hot springs at Beppu (in Japanese): Journal of the Chemical Society of Japan, v. 82, p. 1476-1478.

Kolker, A., Cannon, W. F., Westjohn, D. B., and Woodruff, L. G., 1998a, Arsenic-rich pyrite in the Mississippian Marshall Sandstone: Source of anomalous arsenic in southeastern Michigan ground water: *in* Geological Society of America Annual Meeting, Toronto, Ontario, p. A-58.

-, 1998b, Arsenic-rich pyrite in the Mississippian Marshall Sandstone: Source of anomalous arsenic in southeatern Michigan ground water: Geological Society of America, 1998 Abstracts with Programs, v. 30, no. 7, p. A-59.

Kolker, A., Cannon, W. F., Woodruff, L. G., Westjohn, D. B., Haack, S. K., and Kim, M.-J., 1999, Arsenic in southeastern Michigan ground water: Results of USGS test drilling: EOS, American Geophysical Union Transactions, v. 80, no. 17, p. S146 - S147.

Kolker, A., and Finkelman, R. B., 1998, Potentially hazardous elements in coal: Modes of occurrence and summary of concentration data for coal components: Coal Preparation, v. 19, p. 133-157.

Konhauser, K. O., and Ferris, F. G., 1996, Diversity of iron and silica precipitation by microbial mats in hydrothermal waters, Iceland: Implications for Precambrian iron formations: Geology, v. 24, p. 323-326.

Korte, N., 1991, Naturally occurring arsenic in groundwaters of the midwestern United States: Environmental Geology and Water Sciences, v. 18, no. 2, p. 137-141.

Koterba, M. T., Wilde, F. D., and Lapham, W. W., 1995, Ground -water data-collection protocols and procedures for the National Water-Quality Assessment Program: Collection and documentation of water quality samples and related data: U.S. Geological Survey Open-File Report 95-399, 113 p.

Kotlyar, B. B., Sinder, D. A., Jachens, R. C., and Theodore, T. G., 1998, Regional Analysis of the Distribution of Gold Deposits in Northeast Nevada Using NURE Arsenic Data and Geophysical Data: *in* Tosdal, R. M., ed., Contributions to the Gold Metallogeny of Northern Nevada, US Geological Survey, p. 234-242.

Krohelski, J. T., 1986, Hydrogeology and ground-water use and quality, Brown County, Wisconsin: Wisconsin Geological and Natural History Survey Information Circular, 57 p.

Krupp, R. E., 1990, Comment on "As(III) and Sb(III) sulfide complexes: An evaluation of stoichiometry and stability from existing experimental data by N.F. Spycher and M.H. Reed: Geochimica et Cosmochimica Acta, v. 54, p. 3239-3240.

Krupp, R. E., and Seward, T. M., 1990, Transport and deposition of metals in the Rotokawa geothermal system, New Zealand: Mineralium Deposita, v. 25, p. 73-81.

Kucharski, R., 1996, Possible occurrence of an intrusive body in deep Paleozoic basement of the Olkusz district, Poland

Carbonate-hosted zinc-lead deposits in the Silesian-Cracow area, Poland, p. 163-165.

Kuo, T.-L., 1968, Arsenic content of artesian well water in endemic area of chronic arsenic poisoning: Reports of the Institute of Pathology of the National Taiwan University, v. 20, p. 7-13.

Kuwabara, J. S., Chang, C. Y., and Pasilis, S. P., 1990, Effects of benthic flora on arsenic transport.: Journal Environmental Engineering, v. 116, p. 394-409.

Kuzel, L. S., 1993, The role of Fe and Mn oxy-hydroxides in contaminated coarse-grained river sediment: M.S. Thesis: The University of Montana, 87 p.

La Force, M. J., Hansel, C. M., and Fendorf, S., 2000, Arsenic speciation, seasonal transformations, and co-distribution with iron in a mine waste-influenced palustrine emergent wetland: Environmental Science & Technology, v. 34, p. 3937-3943.

Ladeira, A. C. Q., Ciminelli, V. S. T., Duarte, H. A., Alves, M. C. M., and Ramos, A. Y., 2001, Mechanism of anion retention from EXAFS and density functional calculations: Arsenic (V) adsorbed on gibbsite: Geochimica et Cosmochimica Acta, v. 65, p. 1211-1217.

Lambing, J. H., 1991, Water-quality and transport characteristics of suspended sediment and trace elements in streamflow of the upper Clark Fork basin from Galen to Missoula, Montana, 1985-1990: U.S. Geological Survey Open-File Report, 91-4139, 73 p.

Langmuir, D., 1997, Aqueous Environmental Geochemistry: New Jersey, Prentice-Hall, 600 p.

Langner, H. W., and Inskeep, W. P., 2000, Microbial reduction of arsenate in the presence of ferrihydrite: Environmental Science & Technology, v. 34, p. 3131-3136.

Langner, H. W., Jackson, C. R., McDermot, T. R., and Inskeep, W. P., 2001, Rapid oxidation of arsenite in a hot spring microbial ecosystem: Environmental Science & Technology, v. 35, p. 3302-3309.

Lapham, D. M., Barnes, J. H., Downey, W. F., Jr., and Finkelman, R. B., 1980, Mineralogy associated with burning anthracite deposits of eastern Pennsylvania: Mineral Resources Report, v. 78, p. 1-82.

Laverman, A. M., Blum, J. S., Schaefer, J. K., Phillips, E. J. P., Lovley, D. R., and Oremland, R. S., 1995, Growth of strain SES-3 with arsenate and other diverse electron acceptors: Applied and Environmental Microbiology, v. 61, p. 3556-3561.

Leach, D., Viets, J. G., Foley-Ayuso, N., and Klein, D. P., 1995, Mississippi Valley-type Pb-Zn deposits: *in* du Bray, E. A., ed., Preliminary compilation of descriptive geoenvironmental mineral deposit models: USGS Open-File Report 95-831, p. 234-243.

Leach, D. L., Plumlee, G. S., Hofstra, A. H., Landis, G. P., Rowan, E. L., and Viets, J. G., 1991, Origin of late dolomite cement by CO2-saturated deep basin brines; evidence from the Ozark region, central United States: Geology, v. 19, no. 4, p. 348-351.

Leach, D. L., and Rowan, E. L., 1986, Genetic Link between Ouachita foldbelt tectonism and the Mississippi Valley-Type lead zinc deposits of the Ozarks: Geology, v. 14, p. 932-935.

Leach, D. L., and Sangster, D. F., 1993, Mississippi valley-type lead-zinc deposits: *in* Kirkham, R. V., Sinclair, W. D., Thorpe, R. I., and Duke, J. M., eds., Mineral deposit modeling, Geological Association of Canada, p. 289-314.

Leach, D. L., Viets, J. G., Kozlowski, A., and Kibitlewski, S., 1997, Geology, geochemistry, and genesis of the Silesia-Cracow zinc-lead district, southern Poland Carbonate-hosted lead-zinc deposits: *in* Sangster, D. F. e., ed., Special Publication, Society of Economic Geologists, p. 144-170.

Lee, L., and Goldhaber, M. B., 2001, The distribution of MVT-related metals in acid insoluble residues of paleozoic rocks in the Ozark plateaus region of the United States: U.S. Geological Survey Open-File Report, 01-0042, 35 p.

Lee, R. C. L., 2000, The effect of Mississippi Valley-type mineralization on the natural background chemistry of groundwater in the Ozark Plateaus region of the United States: M.S. Thesis: Colorado School of Mines, 210 p.

Lesher, C. M., Cook, R. B., and Dean, L. S., 1989, Gold deposits of Alabama: Geological Society of Alabama, 229 p.

Levine, I. N., 1975, Molecular Spectroscopy: New York, John Wiley and Sons, 491 p.

Lico, M. S., and Seiler, R. L., 1994, Ground-water quality and geochemistry, Carson Desert, western Nevada: U.S.Geological Survey Open-File Report 94-31, 91 p.

Liger, E., Charlet, L., and Van Cappellen, P., 1999, Surface catalysis of Uranium(VI) reduction by iron(II): Geochimica et Cosmochimica Acta, v. 63, p. 2939-2955.

Lin, Z., and Puls, R. W., 2000, Adsorption, desorption and oxidation of arsenic affected by clay minerals and aging process: Environmental Geology, v. 39, p. 753-759.

Lindgren, W., 1928, Mineral Deposits: New York, McGraw-Hill Book Co., 1049 p.

Lindorf, D. E., Feld, J., and Connelly, J., 1987, Groundwater sampling procedures guidelines: Wisconsin Department of Natural Resources Publication PUBL-WR-153-87, 44 p.

Lively, R. S., Morey, G. B., and Mossler, J. H., 1997, Element distribution patterns in the Ordovician Galena Group, southeastern Minnesota: Indicators of fluid flow and provenance of terrigenous material: Economic Geology, v. 92, no. 3, p. 351-359.

Livesey, N. T., and Huang, P. M., 1981, Adsorption of arsenate by soils and its relation to selected chemical properties and anions: Soil Science, v. 131, p. 88-94.

Lo, M.-C., Hsen, Y.-C., and Lin, K.-K., 1977, Second report on the investigation of arsenic content in underground water in Taiwan: Taiwan Provincial Institute of Environmental Sanitation, Taichung, Taiwan.

Lockwood, J. R., III, Schervish, M. J., Gurian, P., and Small, M. J., 2001, Characterization of arsenic occurrence in US drinking water treatment facility source waters: Department of Statistics, Carnegie Mellon University, 700, 32 p.

Loehr, T. M., and Plane, R. A., 1968, Raman spectra and structures of arsenious acid and arsenites in aqueous solution: Inorganic Chemistry, v. 7, p. 1708-1714.

Long, H. K., and Farrar, J. W., 1995, Report on the U.S. Geological Survey's evaluation program for standard reference samples distributed in October 1994; T-131 (trace constituents), T-133 (trace constituents), M-132 (major constituents), N-43 (nutrients), N-44 (nutrients), P-23 (low ionic strength) and Hg-19 (mercury): U.S. Geological Survey, 95-0117, 139 p.

Longtin, J. P., 1988, Occurrence of radon, radium, and uranuim in groundwater: Journal of the American Water Works Association, v. 80, no. 7, p. 84-93.

Lowson, R. T., 1982, Aqueous pyrite oxidation by molecular oxygen: Chemical Reviews, v. 82, p. 461-497.

Lozinsky, R. P., 1988, Stratigraphy, sedimentology, and sand petrology of the Santa Fe Group and pre-Santa Fe Tertiary deposits in the Albuquerque Basin, central New Mexico: Ph.D. dissertation: Socorro, New Mexico Institute of Mining and Technology, 298 p.

Lu, F. J., Shih, S. R., Lui, T. M., and Shown, S. H., 1990a, The effect of fluorescent humic substances existing in the well water of Blackfoot disease endemic areas in Taiwan on prothrombin time and activated partial thromboplastin time in vitro: Thrombosis Research, v. 57, p. 747-753.

Lu, G., Marshak, S., and Kent, D. V., 1990b, Characteristics of magnetic carriers responsible for Late Paleozoic remagnetization in carbonate strata of the mid-continent, U.S.A.: Earth and Planetary Sciences Letters, v. 99, p. 351-361.

Lumsdon, D. G., Fraser, A. R., Russell, J. D., and Livesey, N. T., 1984, New infrared band assignments for the arsenate ion adsorbed on synthetic goethite (α-FeOOH): Journal of Soil Science, v. 35, p. 381-386.

Luo, Z. D., Zhang, Y. M., Ma, L., Zhang, G. Y., He, X., Wilson, R., Byrd, D. M., Griffiths, J. G., Lai, S., He, L., Grumski, K., and Lamm, S. H., 1997, Chronic arsenicism and cancer in Inner Mongolia - consequences of well-water arsenic levels greater than 50 mg l^{-1}: *in* Abernathy, C. O., Calderon, R. L., and Chappell, W. R., eds., Arsenic Exposure and Health Effects: London, Chapman and Hall, p. 55-68.

Ma, H. Z., Xia, Y. J., Wu, K. G., Sun, T. Z., and Mumford, J. L., 1999, Human exposure to arsenic and health effects in Bayingnormen, Inner Mongolia: *in* Chappell, W. R., Abernathy, C. O., and Calderon, R. L., eds., Arsenic Exposure and Health Effects. Proceedings of the Third International Conference on Arsenic Exposure and Health Effects: Amsterdam, Elsevier, p. 127-131.

Macy, J. M., Nunan, K., Hagen, K. D., Dixon, D. R., Harbour, P. J., Cahill, M., and Sly, L. I., 1996, Chrysiogenes arsenatis gen. nov., sp. nov., a new arsenate-respiring bacterium isolated from gold mine wastewater: Systematic Biology, v. 46, p. 1153-1157.

Macy, J. M., Santini, J. M., Pauling, B. V., O'Neil, A. H., and Sly, L. I., 2000, Two new arsenate/sulfate-reducing bacteria: Mechanisms of arsenate reduction: Archives of Microbiology, v. 173, p. 49-57.

Maeda, S., Ohki, A., Kusadome, K., Kuroiwa, T., Yoshifuku, I., and Naka, K., 1992, Bioaccumulation of arsenic and its fate in a freshwater food chain: Applied Organometallic Chemistry, v. 6, p. 213-219.

Maeda, S., Wada, H., Kuneda, K., Onoue, M., Ohki, A., Higashi, S., and Takeshita, T., 1987, Methylation of inorganic arsenic by arsenic-tolerant freshwater algae: Journal - Organometallic, v. 1, p. 465-472.

Mahmood, S. N., Naeem, S., Siddiqui, I., and Khan, F. A., 1998, Studies on physico-chemical nature of ground water of Korangi/Landhi (Karachi). Journal of the Chemical Society of Pakistan, v. 19, p. 42-48.

Mai, H., and Dott, R. H. J., 1985, A subsurface study of the St. Peter Sandstone in southern and eastern Wisconsin: Wisconsin Geological and Natural History Survey Information Circular, 47 p.

Manceau, A., 1995, The mechanism of anion adsorption on iron oxides: Evidence for the binding of arsenate tetrahedra on free Fe(O,OH)6 edges: Geochimica et Cosmochimica Acta, v. 59, p. 3647-3653.

Manceau, A., and Combes, J., 1988, Structure of Mn and Fe oxides and hydroxides: A topological approach by EXAFS.: Physics and Chemistry of Minerals, v. 15, p. 283-295.

Manceau, A., Gorshkov, A. I., and Drits, V. A., 1992a, Structural chemistry of Mn, Fe, Co, and Ni in manganese hydrous oxides: Part I. Information from XANES spectroscopy: American Mineralogist, v. 77, p. 1133-1143.

-, 1992b, Structural chemistry of Mn, Fe, Co, and Ni in manganese hydrous oxides: Part II. Information from EXAFS spectroscopy and electron and X-ray diffraction: American Mineralogist, v. 77, p. 1144-1157.

Mandal, B. K., Chowdhury, T. R., Samanta, G., Basu, G. K., Chowdhury, P. P., Chanda, C. R., Lodh, D., Karan, N. K., Dhar, R. K., Tamili, D. K., Das, D., Saha, K. C., and Chakraborti, D., 1996, Arsenic in groundwater in seven districts of West Bengal, India - the biggest arsenic calamity in the world: Current Science, v. 70, no. 11, p. 976-986.

Manning, B. A., Fendorf, S. E., and Goldberg, S., 1998, Surface structures and stability of arsenic(III) on goethite: Spectroscopic evidence for inner-sphere complexes: Environmental Science & Technology, v. 32, p. 2383-2388.

Manning, B. A., Fendorf, S. E., and Suarez, D. L., 2002, Arsenic(III) oxidation and As(V) adsorption reactions on synthetic birnessite: Environmental Science & Technology, v. 36, p. 976-981.

Manning, B. A., and Goldberg, S., 1996a, Modeling arsenate competitive adsorption on kaolinite, montmorillonite and illite: Clays and Clay Minerals, v. 44, p. 609-623.

-, 1996b, Modeling arsenate competitive adsorption on kaolinite, montmorillonite and illite, Clays and Clay Minerals, 44:609-623.: Soil Science Society of America Journal, v. 60, p. 121-131.

-, 1997a, Adsorption and stability of arsenic(III) at the clay mineral-water interface: Environmental Science & Technology, v. 31, no. 7, p. 2005-2011.

-, 1997b, Arsenic(III) and arsenic(V) adsorption on three California soils: Soil Science, v. 162, no. 12, p. 886-895.

Manning, C. E., and Ingebritsen, S. E., 1999, Permeability of the continental crust: Implications of geothermal data and metamorphic systems: Reviews of Geophysics, v. 37, p. 127-150.

Marquez-Zavalia, F., Craig, J. R., and Solberg, T. N., 1999, Duranusite, product of realgar alteration, Mina Capillitas, Argentina: Canadian Mineralogist, v. 37, p. 1255-1259.

Masscheleyn, P. H., Delaune, R. D., and Patrick, W. H., Jr., 1991, Effect of redox potential and pH on arsenic speciation and solubility in a contaminated soil: Environmental Science & Technology, v. 25, no. 8, p. 1414-1418.

Mathur, S. S., 1995, Development of a database for ion sorption on goethite using surface complexation modeling: M.S. Thesis: Carnegie Mellon University, 178 p.

Matis, K. A., Zouboulis, A. I., and Valtadorou, A. V., 1999, Sorption of As(V) by goethite particles and study of their flocculation: Water, Air, and Soil Pollution, v. 111:, p. 297-316.

Matisoff, G., Khourey, C. J., Hall, J. F., Varnes, A. W., and Strain, W. H., 1982a, The nature and source of arsenic in northeastern Ohio groundwater: Ground Water, v. 20, no. 4, p. 446-456.

Matisoff, G. C., Khorey, J., Hall, J. F., Varnes, A. W., and Strain, W. H., 1982b, The nature and source of arsenic in northeastern Ohio ground water: Ground Water, v. 20, no. 4, p. 446-456.

Matthess, G., 1981, In situ treatment of arsenic contaminated groundwater: *in* Quality of groundwater; International symposium, Noordwijkerhout, Netherlands, p. 99-104.

Maurer, D. K., 2001, Hydrogeology and geochemistry of Fallon basalt and adjacent aquifers, and potential sources of basalt recharge, in Churchill County, Nevada: U.S. Geological Survey Water-Resources Investigations Report 01-4130, 72 p.

Maurer, D. K., and Welch, A. H., 1996, Hydrogeology and potential effects of changes in water use, Carson Desert Agricultural Area, Churchill County, Nevada: U.S. Geological Survey Water Supply Paper 2436, 106 p.

McArthur, J. M., Ravenscroft, P., Safiullah, S., and Thirlwall, M. F., 2001, Arsenic in groundwater: testing pollution mechanisms for sedimentary aquifers in Bangladesh: Water Resources Research, v. 37, p. 109-117.

McBride, B. C., and Wolfe, R. S., 1971, Biosynthesis of dimethylarsine by methanobacterium: Biochemistry, v. 10, p. 4312-4317.

McCabe, C., Jackson, M., and Saffer, B., 1989, Regional patterns of magnetite authigenesis in the Appalachian Basin: Implications for the mechanism of Late Paleozoic remagnetization: Journal of Geophysical Research, v. 94, no. B8, p. 10429-10443.

McKenzie, E. J., Brown, K. L., Cady, S. L., and Campbell, K. A., 2001, Trace metal chemistry and silicification of microorganisms in geothermal sinter, Taupo Volcanic Zone, New Zealand: Geothermics, v. 30, p. 483-502.

McKenzie, R., 1980, The adsorption of lead and other heavy metals on oxides of Mn and Fe.: Australian Journal of Soil Research, v. 18, p. 61-73.

McKenzie, R. M., 1989, Manganese oxides and hydoxides: *in* Dixon, J. B., and Weel, S. B., eds., Minerals in soil environments: Madison, Soil Science Society of America, p. 439-465.

McKibben, M. A., and Barnes, H. L., 1986, Oxidation of pyrite in low temperature acidic solutions; rate laws and surface textures: Geochimica et Cosmochimica Acta, v. 50, no. 7, p. 1509-1520.

McLaren, S. J., and Kim, N. D., 1995, Evidence for a seasonal fluctuation of As in NZ's longest river and the effect of treatment on concentrations in drinking water: Environmental Pollution, v. 90, p. 67-93.

McNeill, L. S., and Edwards, M., 1995, Soluble arsenic removal at water treatment plants: Journal - American Water Works Association, no. April, p. 105-113.

Melamed, R., Jurinak, J. J., and Dudley, L. M., 1995, Effect of adsorbed phosphate on transport of arsenate through an oxisol: Soil Science Society of America Journal, v. 59, p. 1289-1294.

Meng, X., Bang, S., and Korfiatis, G. P., 2000, Effects of silicate, sulfate, and carbonate on arsenic removal by ferric chloride: Water Research, v. 34, p. 1255-1261.

Menzies, A. W. C., and Potter, P. D., 1912, The two-component system: Water, arsenic pentoxide: Journal of the American Chemical Society, v. 34, p. 1452-1469.

Meyerhoff, R., 1996, Planning and application of in situ iron and manganese treatment of groundwater: Ph.D. Dissertation: University of Stuttgart, Germany (in German), 139 p.

Mielke, R. E., Southam, G., and Nordstrom, D. K., 2000, Arsenic resistant/oxidising bacteria in acidic geothermal environments, Yellowstone National Park, Wyoming. Abstracts with Programs: *in* Annual Meeting Geological Society of America, p. A-190.

Miller, J. D., and Kent, D. V., 1988, Regional trends in the timing of Alleghanian remagnetization in the Appalachians: Geology, v. 16, p. 588-591.

Miller, L. D., Goldfarb, R. J., Gehrels, G. E., and Snee, L. W., 1994, Genetic links among fluid cycling, vein formation, regional deformation, and plutonism in the Juneau gold belt, southeastern Alaska: Geology, v. 22, p. 203-206.

Minnesota Department of Natural Resources, 2000, Minnesota's water supply--Natural conditions and human impacts, 19 p.

Minnesota Pollution Control Agency, 1998, Baseline water quality of Minnesota's principal aquifers: Minnesota Pollution Control Agency, 299 p.

Mironova, G. D., and Zotov, A. V., 1980, Solubility studies of the stability of As(III) sulfide complexes at 90°C: Geochemistry International, v. 17, p. 46-54.

Mironova, G. D., Zotov, A. V., and Gulko, N. I., 1984, Determination of the solubility of orpiment in acid solutions at 25 - 150°C: Geochemistry International, v. 21, p. 53-59.

-, 1990, The solubility of orpiment in sulfide solutions at 25 - 150°C, and the stability of arsenic sulfide complexes: Geochemistry International, v. 27, p. 61-73.

Mok, W. M., Riley, J. A., and Wai, C. M., 1988, Arsenic speciation and quality of groundwater in a lead-zinc mine, Idaho: Water Research (Oxford), v. 22, no. 6, p. 769-774.

Montanez, I. P., 1994, Late diagenetic dolomitization of Lower Ordovician, upper Knox carbonates; a record of the hydrodynamic evolution of the southern Appalachian Basin: AAPG Bulletin, v. 78, no. 8, p. 1210-1239.

Moore, J. N., 1993, Contaminant mobilization resulting from redox pumping in a metal-contaminated river reservoir system, Montana, U.S.A, Environmental Chemistry Lakes and Reservoirs: Chemical Series, American Chemical Society, p. 451-471.

-, 1994, Contaminant mobilization resulting from redox pumping in a metal-contaminated river-reservoir system, Environmental Chemistry of Lakes and Reservoirs: Advances in Chemistry Series, p. 451-471.

Moore, J. N., Ficklin, W. H., and Johns, C., 1988, Partitioning of arsenic and metals in reducing sulfidic sediments: Environmental Science & Technology, v. 22, no. 4, p. 432-437.

Moore, J. N., and Luoma, S. N., 1990, Hazardous wastes from large-scale metal extraction: a case study: Environmental Science & Technology, v. 24, p. 1279-1285.

Moore, J. N., Walker, J. R., and Hayes, T. H., 1990, Reaction scheme for the oxidation of As(III) to As(V) by birnessite: Clays and Clay Minerals, v. 38, p. 549-555.

Morel, F. M. M., and Hering, J. G., 1993, Principles and Applications of Aquatic Chemistry: New York, Wiley-Interscience, 374 p.

Morel, F. M. M., Yeasted, J. G., and Westall, J. C., 1981, Adsorption Models: A Mathematical Analysis in the Framework of General Equilibrium Calculations: *in* Anderson, M. A., and Rubin, A. J., eds., Adsorption of Inorganics at Solid-Liquid Interfaces: Ann Arbor, Michigan, Ann Arbor Science, p. 263-294.

Morrison, G. M. P., 1989, Trace element speciation and its relationship to bioavailability and toxicity in aquatic samples: *in* Batley, G. E., ed., Trace Element Speciation: Analytical Methods and Problems: Boca Raton, Florida, CRC Press, p. 25-41.

Morse, J. W., 1994, Interactions of trace metals with authigenic sulfide minerals: Implications for their bioavailability: Marine Chemistry, v. 46, p. 1-6.

Moses, C. O., Nordstrom, D. K., Herman, J. S., and Mills, A. L., 1987, Aqueous pyrite oxidation by dissolved-oxygen and by ferric iron: Geochimica et Cosmochimica Acta, v. 51, no. 6, p. 1561-1571.

Murphy, J. W., and Riley, J. P., 1962, A modified single-solution method for the determination of phosphate in natural waters: Analytica Chimica Acta, v. 37, p. 31-36.

Murray, J. W., 1975, The interaction of metal ions at the manganese dioxide-solution interface: Geochimica et Cosmochimica Acta, v. 39, p. 505-510.

Myers, M. B., and Felty, E. J., 1970, Heats of fusion of the A_2VB3VI compounds As_2S_3, As_2Se_3, As_2Te_3, and Sb_2S_3: Journal of the Electrochemical Society, v. 117, p. 818-820.

Myneni, S. C. B., 1995, Oxyanion-mineral surface interaction in alkaline environments: AsO_4 and CrO_4 sorption and desorption in ettringite: Ph.D. Dissertation: Ohio State University, 250 p.

Myneni, S. C. B., Traina, S. J., Logan, T. J., and Waychunas, G. A., 1997, Oxyanion behavior in alkaline environments: sorption and desorption of arsenate in ettringite: Environmental Science & Technology, v. 31, p. 1761-1768.

Myneni, S. C. B., Traina, S. J., Waychunas, G. A., and Logan, T. J., 1998, Experimental and theoretical vibrational spectroscopic evaluation of arsenate coordination in aqueous solutions, solids, and at mineral-water interfaces: Geochimica et Cosmochimica Acta, v. 62, p. 3285-3300.

Nag, J. K., Balaram, V., Rubio, R., Alberti, J., and Das, A. K., 1996, Inorganic arsenic species in groundwater: A case study from Purbasthali (Burdwan), India: Journal of Trace Elements in Medicine and Biology, v. 10, no. 1, p. 20-24.

National Research Council, 1999, Arsenic in Drinking Water: Washington, DC, National Academy Press, 310 p.

Naumov, G. B., Ryzhenko, B. N., and Khodakovsky, I. L., 1974, Handbook of Thermodynamic Data, NTIS, U.S. Government Printing Office, 226-722 p.

Nesbitt, H. W., Canning, G. W., and Bancroft, G. M., 1998, XPS study of reactive dissolution of 7-A birnessite by H3AsO3, with constraints on reaction mechanism: Geochimica et Cosmochimica Acta, v. 62, p. 2097-2110.

Nesbitt, H. W., Muir, I. J., and Pratt, A. R., 1995, Oxidation of arsenopyrite by air and air-saturated, distilled water, and implications for mechanism of oxidation: Geochimica et Cosmochimica Acta, v. 59, p. 1773-1786.

Nesbitt, H. W., and Reinke, M., 1999, Properties of As and S at NiAs, NiS, and Fe1-xS surfaces, and reactivity of niccolite in air and water: American Mineralogist, v. 84, p. 639-649.

Newman, D. K., Ahmann, D., and Morel, F. M. M., 1998, A brief review of microbial arsenate respiration: Geomocrobiology Journal, v. 15, p. 255-268.

Newman, D. K., Beveridge, T. J., and Morel, F. M. M., 1997a, Precipitation of arsenic trisulfide by Desulfotomaculum auripigmentum: Applied and Environmental Microbiology, v. 63, p. 2022-2088.

Newman, D. K., Kennedy, E. K., Coates, J. D., Ahmann, D., Ellis, D. J., Lovley, D. R., and Morel, F. M. M., 1997b, Dissimilatory arsenate and sulfate reduction in Desulfotomaculum auripigmentum sp. nov.: Archives of Microbiology, v. 168, p. 380-388.

Nicholson, R. V., Gillham, R. W., and Reardon, E. J., 1988, Pyrite oxidation in carbonate-buffered solution. 1. Experimental kinetics: Geochimica et Cosmochimica Acta, v. 52, no. 5, p. 1077-1085.

Nickson, R., McArthur, J., Burgess, W., Ahmed, K. M., Ravenscroft, P., and Rahman, M., 1998, Arsenic poisoning of Bangladesh groundwater: Nature, v. 395, no. 6700, p. 338.

Nickson, R. T., McArthur, J. M., Ravenscroft, P., Burgess, W. G., and Ahmed, K. M., 2000, Mechanism of arsenic release to groundwater, Bangladesh and West Bengal: Applied Geochemistry, v. 15, no. 4, p. 403-413.

Nicolli, H. B., Suriano, J. M., Peral, M. A. G., Ferpozzi, L. H., and Baleani, O. A., 1989, Groundwater contamination with arsenic and other trace-elements in an area of the Pampa, Province of Cordoba, Argentina: Environmental Geology and Water Sciences, v. 14, no. 1, p. 3-16.

Nielsen, M. G., Stone, J. R., Hansen, B. P., and Nielsen, J. P., 1995, Geohydrology, water quality, and conceptual model of the hydrologic system, Saco landfill area, Saco Maine: U.S. Geological Survey Water-Resources Investigations Report, 95-4027, 94 p.

Nimick, D. A., 1994, Arsenic transport in surface and ground water in the Madison and Upper Missouri River valleys, Montana: *in* American Geophysical Union 1975 Fall Meeting, p. 247.

-, 1998, Arsenic hydrogeochemistry in an irrigated river valley -- a reevaluation: Ground Water, v. 36, p. 743-753.

Nimick, D. A., Moore, J. N., Dalby, C. E., and Svka, M. W., 1998, The fate of geothermal arsenic in the Madison and Missouri Rivers, Montana and Wyoming: Water Resources Research, v. 34, p. 3051-3067.

Nishimura, T., and Robins, R. G., 1998, A re-evaluation of the solubility and stability regions of calcium arsenites and calcium arsenates in aqueous solution at 25°C: Mineral Processing and Extractive Metallurgy Reviews, v. 18, p. 283-308.

Noguchi, K., and Nakagawa, R., 1969, Arsenic and arsenic-lead sulfide sediments from Tamagawa Hot Springs, Akita Prefecture: Proceedings of the Japan Academy, v. 45, p. 45-50.

Nordstrom, D. K., 1982, Aqueous pyrite oxidation and the consequent formation of secondary iron minerals, Acid Sulfate Weathering: Madison, Wisconsin, Soil Science Society of America, p. 37-56.

-, 2000a, Advances in the hydrogeochemistry and microbiology of acid mine waters: International Geology Review, v. 42, p. 490-515.

-, 2000b, Thermodynamic properties of environmental arsenic species: Limitations and needs: *in* Proceedings, Society Mining Metallurgy & Exploration Meeting, Salt Lake City, USA.

Nordstrom, D. K., Alpers, C. N., Ptacek , C. J., and Blowes, D. W., 2000, Negative pH and extremely acidic mine waters from Iron Mountain, California: Environmental Science & Technology, v. 34, p. 254-258.

Nordstrom, D. K., Ball, J. W., and McCleskey, R. B., 2001, Processes governing arsenic concentrations in thermal waters of Yellowstone National Park: *in* U.S. Geological Survey Workshop on Arsenic in the Environment.

Nordstrom, D. K., and Munoz, J. L., 1994, Geochemical Thermodynamics, Blackwell Science, 493 p.

Nordstrom, D. K., and Southam, G., 1997, Geomicrobiology of sulfide mineral oxidation, Geomicrobiology: Interactions Between Microbes and Minerals: Reviews in Mineralogy, p. 361-390.

NRECA, 1997, Report of study of the impact of the Bangladesh rural electrification program on groundwater quality: NRECA International Ltd, 124 p.

Nriagu, J. O., 1994, Arsenic in the environment, Part I: Cycling and characterization: New York, John Wiley & Sons, 430 p.

Ohio Environmental Protection Agency, 2000, Ohio's ground water quality: 2000 305(b) report: Division of Drinking and Ground Waters, 55 p.

-, 2001a, Arsenic in Ohio's ground water: Columbus, Division of Drinking and Ground Waters (on line).

-, 2001b, Ground water in Ohio: Division of Drinking and Ground Waters, 4 p.

Ohle, E. L., 1990, A comparison of the Old Lead Belt and the New Lead Belt in southeast Missouri: Economic Geology, v. 85, no. 8, p. 1894-1895.

-, 1997, Significant events in the geological understanding of the Southeast Missouri lead district: *in* Sangster, D. F., ed., Carbonate-hosted lead-zinc deposits, Society of Economic Geologists Special Publication, p. 1-7.

Oliver, N. H. S., 1996, Review and classification of structural controls on fluid flow during regional metamorphism: Journal of Metamorphic Geology, no. 14, p. 477-492.

Oman, C. L., Hassenmueller, W. H., and Bragg, L. J., 1992, Indiana Coal and Associated Rock Samples Collected from 1975-1977: U. S. Geological Survey Open File Report, 93-0111, 133 p.

O'Neil, P. E., Harris, S. C., Mettee, M. F., Shepard, T. E., and McGregor, S. W., 1993, Surface discharge of wastewaters from the production of methane from coal seams in Alabama; the Cedar Cove model: Geological Survey of Alabama, 155, 259 p.

O'Reilly, S. E., Strawn, D. G., and Sparks, D. L., 2001, Residence time effects on arsenate adsorption/desorption mechanisms on goethite: Soil Science Society of America Journal, v. 65, p. 67-77.

Ortiz, D., Lange, K., and Beal, L., 1999, Water resources data, New Mexico, water year 1998- Volume 1. The Rio Grande Basin, the Mimbres River Basin, and the Tularosa Valley Basin: U.S. Geological Survey Water-Data Report NM-98-1, 404 p.

Osborne, F. H., and Ehrlich, H. L., 1976, Oxidation of arsenite by a soil isolate of Alcaligenes: Journal of Applied Bacteriology, v. 41, p. 295-305.

Oscarson, D. W., Huang, P. M., Defosse, C., and Herbillon, A., 1981a, Oxidative power of Mn(IV) and Fe(III) oxides with respect to As(III) in terrestrial and aquatic environments: Nature, v. 291, p. 50-51.

Oscarson, D. W., Huang, P. M., and Liaw, W. K., 1980, The oxidation of arsenite by aquatic sediments: Journal of Environmental Quality, v. 9, p. 700-703.

-, 1981b, Role of manganese in the oxidation of arsenite by freshwater lake sediments: Clays and Clay Minerals, v. 29, p. 219-225.

Oscarson, D. W., Huang, P. M., Liaw, W. K., and Hammer, U. T., 1983, Kinetics of oxidation of arsenite by various manganese dioxides: Soil Science Society of America Journal, v. 47, p. 644-648.

Ostergren, J. D., Jr, Parks, G. A., and Tingle, T. N., 1999, Quantitative speciation of lead in selected mine tailings form Leadville, CO: Environmental Science & Technology, v. 33, p. 1627-1635.

Paktunc, A. D., Laflamme, J. H. G., Riveros, P. A., and Deschenes, G., 2000, Arsenic mineralogy and bench-scale cyanidation testwork on composite materials from the Ketza River mine site, Yukon.: CANMET, MMSL 2000-017 (CR).

Palache, C., Berman, H., and Frondel, C., 1944, Dana's System of Mineralogy: New York, John Wiley and Sons, 834 p.

Panno, S. V., Hackley, K. C., Cartwright, K., and Liu, C. L., 1994, Hydrochemistry of the Mahomet Bedrock valley aquifer, east-central Illinois: Indicators of recharge and ground-water flow: Ground Water, v. 32, p. 591-604.

Park, C. F., Jr., and MacDiarmid, R. A., 1975, Ore Deposits: San Francisco, California, W.H. Freeman, 530 p.

Parker, D. R., Chaney, R. L., and Norvell, W. A., 1995, Chemical equilibrium models: Applications to plant nutrition research: *in* Loeppart, R. H., Schwab, A. P., and Goldberg, S., eds., Chemical Equilibrium and Reaction Models, Soil Science Society of America, Special Publication, No. 42, p. 163-200.

Parkhurst, D. L., and Appelo, C. A. J., 1999, User's guide to PHREEQC (version 2): U.S. Geological Survey Water-Resources Investigations Report 99-4259, 312 p.

Parkhurst, D. L., Engesgaard, P., and Kipp, K. L., 1995, Coupling the geochemical model PHREEQC with a 3D multi-component solute-transport model: *in* Fifth Annual V.M. Goldschmidt Conference, University Park, Pennsylvania, p. 77.

Parkhurst, D. L., and Plummer, L. N., 1993, Geochemical models: *in* Alley, W. M., ed., Regional Ground-Water Quality: New York, Van Nostrand Reinhold, p. 199-225.

Parks, G. A., 1965, The isoelectric points of solid oxides, solid hydroxides, and aqueous hydroxo complex systems: Chemical Reviews, v. 65, p. 177-198.

-, 1990, Surface energy and adsorption at mineral-water interfaces: An introduction: *in* Hochella, M. F., Jr., and White, A. F., eds., Mineral-Water Interface Geochemistry: Washington, D.C., Mineralogical Society of America, p. 133-175.

Parks, G. A., and de Bruyn, P. L., 1962, The zero point of charge of oxides: Journal of Physical Chemistry, v. 66, p. 967-973.

Paronikyan, V. O., and Matevosyan, A. S., 1965, Native arsenic from Amas deposits of Armenian S.S.R.: Doklady Akademii Nauk Armyanskoy, v. 41, p. 240-243.

Paschke, S. S., Harrison, W. J., and Walton-Day, K., 2001, Effects of acidic recharge on groundwater at the St. Kevin Gulch site, Leadville, Colorado: Geochemistry: Exploration, Environment, Analysis, v. 1, no. 1, p. 3-14.

Pashin, J. C., 1994, Coal-body geometry and synsedimentary detachment folding in Oak Grove coalbed methane field, Black Warrior Basin, Alabama: AAPG Bulletin, v. 78, no. 6, p. 960-980.

-, 1998, Stratigraphy and structure of coalbed methane reservoirs in the United States: An overview: International Journal of Coal Geology, v. 35, p. 1-4.

Pashin, J. C., Carrol, R. E., Hatch, J. R., and Goldhaber, M. B., 1999, Mechanical and thermal control of cleating and shearing in coal: Examples from the Alabama coalbed methane fields, USA: *in* Mastalerz, M., Glikson, M., and Golding, S. D., eds., Coalbed Methane: Scientific, Environmental, and Economic Evaluation: Dordrecht, Kluwer Academic Publishers, p. 305-328.

Pashin, J. C., Carroll, R. E., Barnett, R. L., and Beg, M. A., 1995, Geology and coal resources of the Cahaba coal field: Geological Survey of Alabama, 163, 49 p.

Paukov, I. E., Nogteva, V. V., and Strelkov, P. G., 1969, The true heat capacity of crystalline black phosphorous and metallic arsenic in the temperature range 13-300°K: Russian Journal of Physical Chemistry, v. 43, p. 773-777.

Pauling, L., 1970, General Chemistry: San Francisco, 959 p.

Pearce, J., Kinniburgh, D. G., Smedley, P. L., Ahmed, K. M., and Rahman, M., 2001, Mineralogy and sediment chemistry: *in* Kinniburgh, D. G., and Smedley, P. L., eds., Arsenic contamination of groundwater in Bangladesh: Keyworth, British Geological Survey, Keyworth, p. 187-210.

Pelczar, J. S., 1996, Groundwater chemistry of wells exhibiting natural arsenic contamination in east-central Wisconsin: M.S. Thesis: University of Wisconsin-Green Bay, 206 p.

Peralta, G. L., Kirk, D. W., Graydon, J. W., and Seyfried, P. L., 1996, Assessing the environmental impact of geothermal residues: Canadian Waste Management Research, v. 16, p. 225-232.

Perry, M. J., and Mackun, P. J., 2001, Population change and distribution, 1990 to 2000: U.S. Census Bureau, C2KBR/01-2, 7 p.

Peryea, F. J., 1991, Phosphate-induced release of arsenic from soils contaminated with lead arsenate: Soil Science Society of Americal Journal, v. 55, p. 1301-1306.

-, 1998, Historical use of lead arsenate insecticides, resulting soil contamination and implications for soil remediation: *in* 16th World Congress of Soil Science (CD Rom), Montpellier, France.

Peryea, F. J., and Kammereck, R., 1997, Phosphate-enhanced movement of arsenic out of lead arsenate-contaminated topsoil and through uncontaminated subsoil: Water, Air and Soil Pollution, v. 93, no. 1-4, p. 243-254.

Peters, S. C., Blum, J. D., Klaue, B., and Karagas, M. R., 1999, Arsenic occurrence in New Hampshire drinking water: Environmental Science & Technology, v. 33, no. 9, p. 1328-1333.

PHED, 1991, Arsenic pollution in groundwater in West Bengal. Final Report: Public Health Engineering Department, Public Health Department, Government of West Bengal.

PHED/UNICEF, 1999, Joint Plan of Action to Address Arsenic Contamination of Drinking Water.

Phillips, S. E., and Taylor, M. L., 1976, Oxidation of arsenite to arsenate by Alcaligenes faecalis: Applied and Environmental Microbiology, v. 32, p. 392-399.

Pierce, M. L., and Moore, C. B., 1982, Adsorption of arsenite and arsenate on amorphous iron hydroxide: Water Research, v. 16, p. 1247-1253.

Pitman, J. K., and Spoeetl, C., 1996, Origin and timing of carbonate cements in the St. Peter Sandstone, Illinois Basin: Evidence for a genetic link to Mississippi Valley-type mineralization: *in* Crossey, L. J., Loucks, R., and Totten, M. W., eds., Siliciclastic diagenesis and fluid flow: Concepts and applications: Special Publication, Society for Sedimentary Geology, p. 187-203.

Plumlee, G. S., Leach, D. L., Hofstra, A. H., Landis, G. P., Rowan, E. L., and Viets, J. G., 1994, Chemical reaction path modeling of ore deposition in Mississippi Valley-type Pb-Zn deposits of the Ozark region, U.S. Midcontinent: Economic Geology, v. 89, no. 6, p. 1361-1383.

-, 1995, Chemical reaction path modeling of ore deposition in Mississippi Valley-type Pb-Zn deposits of the Ozark region, U.S. Midcontinent; reply: Economic Geology, v. 90, no. 5, p. 1346-1349.

Plummer, L. N., Bexfield, L. M., Anderholm, S. K., Sanford, W. E., and Busenberg, E., 2001, Geochemical characterization of ground-water flow in parts of the Santa Fe Group aquifer system, Middle Rio Grande Basin, New Mexico, U.S. Geological Survey Middle Rio Grande Basin Study-Proceedings of the Fourth Annual Workshop: *in* Cole, J. C., ed.: Albuquerque, New Mexico, U.S. Geological Survey Open-File Report 00-488, p. 7-10.

-, in press, Geochemical characterization of ground-water flow in the Santa Fe Group aquifer system, Middle Rio Grande Basin, New Mexico, U.S. Geological Survey Water-Resources Investigations Report.

Plummer, L. N., Busenberg, E., Sanford, W. E., Bexfield, L. M., Anderholm, S. K., and Schlosser, P., 1997, Tracing and dating young ground water in the Middle Rio Grande Basin, Albuquerque, New Mexico: Geological Society of America Abstracts with Programs, v. 29, no. 6, p. 135-136.

Pokrovski, G., Gout, R., Schott, J., Zotov, A., and Harrichoury, J., 1996, Thermodynamic properties and stoichiometry of As(III) hydroxide complexes at hydrothermal conditions: Geochimica et Cosmochimica Acta, v. 60, p. 737-749.

Popova, V. I., and Polyakov, V. O., 1985, Uzonite As4S5 - a new arsenic sulfide from Kamchatka: Zapiski Vsesoyuznogo Mineralogicheskogo Obshchestva, p. 369-373.

Popova, V. I., Popov, V. A., Clark, A. H., Polyakov, V. O., and Borisovskii, S. E., 1986, Alacranite As8S9 - a new mineral: Zapiski Vsesoyuznogo Mineralogicheskogo Obshchestva, v. 115, p. 360-368.

Post, J. E., and Bish, D. L., 1988, Ritveld refinement of the todorokite structure: American Mineralogist, v. 73, p. 861-869.

Post, J. E., and Veblen, D. R., 1990, Crystal structure determination of synthetic sodium, magnesium, and potassium birnessite using TEM and the Rietveld method: American Mineralogist, v. 75, p. 477-489.

Postma, D., and Appelo, C. A. J., 2000, Reduction of Mn-oxides by ferrous iron in a flow system: Column experiment and reactive transport modeling: Geochimica et Cosmochimica Acta, v. 64, p. 1237-1247.

Postma, D., Boesen, C., Kristiansen, H., and Larsen, F., 1991, Nitrate reduction in an unconfined sandy aquifer - water chemistry, reduction processes, and geochemical modeling: Water Resources Research, v. 27, no. 8, p. 2027-2045.

Pratt, W. P., 1992, Maps showing areal extent of selected Paleozoic shales in the northern Midcontinent, U.S.A.: U. S. Geological Survey Miscellaneous Field Studies Map, scale 1:1,000,000.

Pratt, W. P., and Goldhaber, M. B., 1990, Mineral Resource Potential of the Midcontinent: *in* U. S. Geological Survey-Missouri Geological Survey symposium on Mineral-resource potential of the Midcontinent, St. Louis, Missouri, p. 42.

Pratt, W. P., Hayes, T. S., Erickson, R. L., Berendsen, P., and Kisvarsanyi, E. B., 1993, Assessment of the Joplin 1 degree by 2 degrees quadrangle, Kansas and Missouri, for Mississippi Valley-type deposits and other minerals: U. S. Geological Survey Miscellaneous Field Studies Map MF-2125-E, scale 1:250,000.

Prior, H. L., and Williams, G. A., 1996, Some mines and minerals of Morgan County, Missouri: Rocks and Minerals, v. 71, p. 102.

Puls, R. W., and Powell, R. M., 1992, Transport of inorganic colloids through natural aquifer material: Implications for contaminant transport: Environmental Science & Technology, v. 26, p. 614-621.

Qing, H., and Mountjoy, E. W., 1994, Origin of dissolution vugs, caverns, and breccias in the Middle Devonian Presqu'ile barrier, host of Pine Point mississippi valley-type deposits: Economic Geology and the Bulletin of the Society of Economic Geologists, v. 89, no. 4, p. 858-876.

Rahim, S., 1998, Chemical quality of groundwater of Rawalpindi and Islamabad region: M.S. Thesis: National University of Sciences and Technology, Pakistan.

Rahman, M., and Hossain, K. S., 1999, Study Report on Source Detection of Arsenic Contamination in Bangladesh Groundwater & Mitigation Measures: Bangladesh Water Development Board, 54 p.

Rankin, A., 1995, Hydrothermal orefields and ore fluids. Milestones in geology: *in* Le Bas, M., ed., Memoir - Geological Society of London, p. 237-247.

Rankin, A. H., 1990, Fluid inclusions associated with oil and ore in sediments.: *in* Ala, M., Hatamian, H., Hobson, G. D., King, M. S., and Williamson, I., eds., Seventy-five years of progress in oil field science and technology: London, United Kingdom, p. 113-124.

Rard, J. A., and Archer, D. G., 1995, Isopiestic investigation of the osmotic and activity coefficients of aqueous NaBr and the solubility of $NaBr.2H_2O$(cr) at 298.15 K. Thermodynamic properties of the $NaBr+H_2O$ system over wide ranges of temperature and pressure: Journal of Chemical and Engineering Data, v. 40, p. 170-185.

Raven, K. P., Jain, A., and Loeppert, R. H., 1998, Arsenite and arsenate adsorption on ferrihydrite: Kinetics, equilibrium, and adsorption envelopes: Environmental Science & Technology, v. 32, p. 344-349.

Ray, S. P. S., 1997, Arsenic in groundwater in West Bengal.: *in* Consultation on arsenic in drinking water and resulting arsenic toxicity in India and Bangladesh, New Delhi, p. 13.

Reay, P. F., 1972, The accumulation of arsenic from arsenic-rich natural waters by aquatic plants: Journal of Applied Ecology, v. 9, p. 557-565.

Reeder, R. J., Lamble, G. M., Lee, J. F., and Staudt, W. J., 1994, Mechanism of SeO42- substitution in calcite: An XAFS study: Geochimica et Cosmochimica Acta, v. 58, p. 5639-5646.

Renaut, R. W., Jones, B., and Rosen, M. R., 1996, Primary silica oncoids from Orakeikorako Hot Springs, North Island, New Zealand: Palaios, v. 11, p. 446-458.

Ressler, T., Wong, J., Roos, J., and Smith, I. L., 2000, Quantitative speciation of Mn-bearing particulates emitted from autos burning (methylcyclopentadienyl)manganese tricarbonyl-added gasolines using XANES spectroscopy: Environmental Science & Technology, v. 34, p. 950-958.

Reuther, R., 1992, Arsenic introduced into a littoral freshwater model ecosystem: Sci. Total Env., v. 115, p. 219-237.

Reynolds, J. C., 2001, Distributed *in situ* gas measurements for the analysis and modeling of biogeochmical changes in the Clark Fork River: M.S.: The University of Montana, 118 p.

Reynolds, R. L., and Goldhaber, M. B., 1978, Recognition of oxidized sulfide minerals as an exploration guide for uranium: Journal of Research of the U. S. Geological Survey, v. 6, no. 4, p. 483-488.

RGAG & RGACG, 2000, Arsenic contamination in groundwater and hydrogeological background in Samta village, western Bangladesh: Sub-group for Arsenic contamination, Research Group for Applied Geology & Miyazaki University Research Group for Arsenic Contamination in Groundwater, 32 p.

Richardson, S., and Vaughan, D. J., 1989, Arsenopyrite: a spectroscopic investigation of altered surfaces: Mineralogical Magazine, v. 53, p. 223-229.

Rickman, D. L., 1981, A thermochemical study of the ore deposits of the Milliken Mine, New Lead Belt, Missouri: Ph.D. Dissertation: University of Missouri, Rolla, Rolla, MO, United States, 336 p.

Riewe, T., Weissbach, A., Heinen, L., and Stoll, R. C., 2000, Naturally occurring arsenic in well water in Wisconsin: Water Well Journal, p. 24-29.

Ritchie, J. A., 1961, Arsenic and antimony in some New Zealand thermal waters: New Zealand Journal of Science, v. 4, p. 218-229.

Roberts, A. C., Ansell, H. G., and Bonardi, M., 1980, Pararealgar a new polymorph of arsenic sulfide, from British Columbia: Canadian Mineralogist, v. 18, p. 525-527.

Roberts, W. L., and Rapp, G., Jr., 1965, Mineralogy of the Black Hills: South Dakota School Mines Technical Bulletin 18, 268 p.

Robertson, F. N., 1989, Arsenic in groundwater under oxidizing conditions, south-west United States: Environmental Geochemistry and Health, v. 11, no. 3-4, p. 171-185.

Robie, R. A., and Hemingway, B. S., 1995, Thermodynamic properties of minerals and related substances at 298.15 K and 1 bar (105 Pascals) pressure and at higher temperatures: U.S. Geological Survey Bulletin 2131, 461 p.

Robie, R. A., Hemingway, B. S., and Fisher, J. R., 1978, Thermodynamic properties of minerals and related substances at 298.15 K and 1 bar (105 pascals) pressure and at higher temperatures: U.S. Geological Survey Bulletin 1452, 456 p.

Robie, R. A., and Waldbaum, D. R., 1968, Thermodynamic properties of minerals and related substances at 298.15 K (25.0oC) and one atmosphere (1.013 bars) pressure and at higher temperatures: U.S. Geological Survey Bulletin 1259, 256 p.

Robinson, B. H., Brooks, R. R., Outred, H. A., and Kirkman, J. H., 1995, Mercury and arsenic in trout from the Taupo Volcanic Zone and Waikato River, North Island: New Zealand Chem. Speciation & Bioavail, v. 7, p. 27-32.

Robinson, G. R., Ayotte, J. D., and Ayuso, R. A., 2000, Arsenic in bedrock, ground water, and soils in New England: EOS Transactions, AGU Fall Meeting Supplement, v. 81, p. 524.

Rochette, E. A., Bostick, B. C., Li, G., and Fendorff, S., 2000, Kinetics of arsenate reduction by dissolved sulfide: Environmental Science & Technology, v. 34, p. 4714-4720.

Rochette, E. A., Li, G. C., and Fendorf, S. E., 1998, Stability of arsenate minerals in soil under biotically generated reducing conditions: Soil Science Society of America Journal, v. 62, p. 1530-1537.

Rogers, R. J., 1989, Geochemical comparison of ground water in areas of New England, New York, and Pennsylvania: Ground Water, v. 27, no. 5, p. 690-712.

Rosen, B. P., Bhattacharjee, H., and Shi, W., 1995, Mechanisms of metalloregulation of an anion-translocating ATPase: Bioenergetics and Biomembranes, v. 27, p. 85-91.

Roth, W. A., and Schwartz, O., 1926, Physical-chemical properties of solutions of germanium dioxide (and arsenic trioxide): Bericht, v. 59, p. 338-348.

Rothbaum, H. P., and Anderton, B. H., 1976, Removal of silica and arsenic from geothermal discharge waters by precipitation of useful calcium silicates: *in* 2nd United Nations Symposium on the Development and Uses of Geothermal Resources, p. 1417-1425.

Rott, U., and Friedle, M., 1999, Subterranean removal of arsenic from groundwater: *in* W. R. Chappell, C. O. A., and Calderon, R. L., eds., Arsenic Exposure and Health Effects: Oxford, United Kingdom, Elsevier Science Ltd., p. 389-396.

Rott, U., and Lamberth, B., 1993, Groundwater clean up by in situ treatment of nitrate, iron and manganese: Water Supply 11, p. 143-156.

Rott, U., and Meyerhoff, R., 1996, In-situ-Treatment of Arsenic in Groundwater: *in* IWSA International Workshop: Natural origin inorganic micropollutants, Vienna.

Rott, U., Meyerhoff, R., and Bauer, T., 1996, In situ treatment of groundwater with increased concentrations of iron, manganese and arsenic: Wasser-Abwasser (in German), v. 137, p. 358-363.

Rowan, E. L., 1986, Cathodoluminescent zonation in hydrothermal dolomite cements: Relationship to Mississippi Valley-type Pb-Zn mineralization in southern Missouri and northern Arkansas.: *in* Hagi, R. D., ed., Process Mineralogy VI, Metallurgical Society, p. 69-87.

Rowan, E. L., and Goldhaber, M. B., 1996, Fluid inclusions and biomarkers in the Upper Mississippi Valley zinc-lead district: Implications for the fluid-flow and thermal history of the Illinois Basin: U.S. Geological Survey Bulletin, 20994-F.

Rowan, E. L., Goldhaber, M. B., and Hatch, J. R., 2001, The role of regional fluid flow in the Illinois basin's thermal history: Constraints from fluid inclusions and maturity of Pennsylvanian coals: AAPG Bulletin.

Rowan, E. L., Goldhaber, M. B., and Hatch, R. R., 1996, Constraints on the thermal and burial history of the Illinois Basin from fluid inclusions and thermal maturity of organic matter: *in* Geological Society of America, 28th annual meeting, p. 387.

Rowan, E. L., and Leach, D. L., 1989, Constraints from fluid inclusions on sulfide precipitation mechanisms and ore fluid migration in the Viburnum Trend lead district, Missouri: Economic Geology, v. 84, no. 7, p. 1948-1965.

Roy, W. R., Hassett, J. J., and Griffin, R. A., 1986a, Competitive coefficients for the adsorption of arsenate, molybdate, and phosphate mixtures by soils: Soil Science Society of America Journal, v. 50, p. 1176-1182.

-, 1986b, Competitive interactions of phosphate and molybdate on arsenate adsorption: Soil Science, v. 142, p. 203-210.

Rumble III, D., 1994, Water circulation in metamorphism: Journal of Geophysical Research, v. 99, no. B8, p. 15499-15502.

Runkel, R. L., Kimball, B. A., McKnight, D. M., and Bencala, K. E., 1999, Reactive solute transport in streams: a surface complexation approach for trace metal sorption: Water Resources Research, v. 35, p. 3829-3840.

Ryan, P. B., Huet, N., and MacIntosh, D. L., 2000, Longitudinal investigation of exposure to arsenic, cadmium, and lead in drinking water: Environmental Health Perspectives, v. 108, no. 8, p. 731-735.

Ryden, J. C., Syers, J. K., and Tillman, R. W., 1987, Inorganic anion sorption and interactions with phosphate sorption by hydrous ferric oxide gel: Journal of Soil Science, v. 38, p. 211-217.

Ryker, S. J., 2001, Mapping arsenic in ground water: Geotimes Newsmagazine of the Earth Sciences, v. 46, no. 11, p. 34-36.

Ryling, R. W., 1961, A preliminary study of the distribution of saline water in the bedrock aquifers of eastern Wisconsin: Wisconsin Geological and Natural History Survey Information Circular 5, 23 p.

Rytuba, J. J., 1995, Cenozoic metallogeny of California: *in* Geology and ore deposits of the American Cordillera; Symposium Proceedings, Reno, Nevada, p. 803-822.

Saad, D. L., 1996, Ground-water quality in the western part of the Cambrian -Ordovician Aquifer in the western Lake Michigan drainages, Wisconsin and Michigan: U.S. Geological Survey Water-Resources Investigation Report 96-4231, 40 p.

Sadiq, M., 1997, Arsenic chemistry in soils: An overview of thermodynamic predictions and field observations: Water, Air and Soil Pollution, v. 93, no. 1-4, p. 117-136.

Sadiq, M., and Lindsay, W. L., 1981, Selection of standard free energies of formation for use in soil chemistry,: Colorado State University Experiment Station Techical Bulletin 134, Arsenic Supplement, 39 p.

Sakamoto, H., Kamada, M., and Yonehara, 1988, The contents and distribution of arsenic, antimony and mercury in geothermal waters: Bulletin of the Chemical Society of Japan, v. 61, p. 3471-3477.

Sakata, M., 1987, Relationship between adsorption of arsenic(III) and boron by soil and soil properties: Environmental Science & Technology, v. 21, p. 1126-1130.

Salomaa, P., Hakala, R., Vesala, S., and Aalto, T., 1969, Solvent deuterium isotope effects on acid-base reactions. Part III. Relative acidity constants of inorganic oxyacids in light and heavy water. Kinetic applications: Acta Chemica Scandinavica, v. 23, p. 2116-2126.

Sato, M., 1960, Oxidation of sulfide ore bodies: I. Geochemical environments in terms of Eh and pH: Economic Geology, v. 55, p. 928-961.

Savage, K. S., Tingle, T. N., O'Day, P. A., Waychunas, G. A., and Bird, D. K., 2000, Arsenic speciation in pyrite and secondary weathering phases, Mother Lode Gold District, Tuolumne County, California: Applied Geochemistry, v. 15, p. 1219-1244.

Sayers, D. E., and Bunker, B. A., 1988, Data analysis: *in* Koningsberger, D. C., and Prins, R., eds., X-ray Absorption: Principles, Applications, Techniques of EXAFS, SEXAFS, and XANES: New York, John Wiley and Sons.

Schedl, A., McCabe, C., Montanez, I. P., Fullagar, P. D., and Valley, J. W., 1992, Alleghanian regional diagenesis: A response to the migration of modified metamorphic fluids derived from beneath the Blue Ridge-Piedmont thrust sheet: Journal of Geology, v. 100, p. 339-352.

Schindler, P. W., 1981, Surface Complexes at Oxide-Water Interfaces: *in* Anderson, M. A., and Rubin, A. J., eds., Adsorption of Inorganics at Solid-Liquid Interfaces, p. 1-49.

Schindler, P. W., and Stumm, W., 1987, The Surface Chemistry of Oxides, Hydroxides, and Oxide Minerals: *in* Stumm, W., ed., Aquatic Surface Chemistry, Wiley Interscience, p. 83-110.

Schlottmann, J. L., and Breit, G. N., 1992, Mobilization of As and U in the Central Oklahoma Aquifer, USA: *in* Kharaka, Y. K., and Maest, A. S., eds., Water-Rock Interaction, Balkema, Rotterdam, p. 835-838.

Schombel, L. F., 1994, Bedrock and alluvial aquifer saturated thickness isopach maps: *in* Wossner, W. W., ed.: Missoula, Montana, Montana Department of Justice, Natural Resource Damage Litigation Program, p. 130.

Schreiber, M. E., Simo, J. A., and Freiberg, P. G., 2000, Stratigraphic and geochemical controls on naturally occurring arsenic in groundwater, eastern Wisconsin, USA: Hydrogeology Journal, v. 8, p. 161-176.

Schrenk, M. O., Edwards, K. J., Goodman, R. M., Hamers, R. J., and Banfield, J. F., 1998, Distribution of Thiobacillus ferrooxidans and Leptospirillum ferrooxidans: Implications for generation of acid mine drainage: Science, v. 279, no. 5356, p. 1519-1522.

Schwertman, U., and Taylor, R. M., 1989, Iron Oxides: *in* Dixon, J. B., and Weed, S. B., eds., Minerals in Soil Environments: Madison, Wisconsin, Soil Science Society of America, p. 379-438.

Scotese, C. R., Bambach, R. K., Barton, C., Van der Voo, R., and Ziegler, A. M., 1979, Paleozoic base maps: Journal of Geology, v. 87, no. 3, p. 217-277.

Scott, M. J., and Morgan, J. J., 1995, Reactions at Oxide Surfaces. 1. Oxidation of As(III) by Synthetic Birnessite: Environmental Science & Technology, v. 29, no. 8, p. 1898-1905.

Serfes, M. E., Spayd, S. E., Herman, G. C., and Monteverde, D. E., 2000, Arsenic occurrence, source and possible mobilization mechanisms in ground water of the Piedmont physiographic province in New Jersey: EOS Transactions, AGU Fall Meeting Supplement, v. 81, p. 525.

Sergeyeva, E. I., Khodakovskiy, I. L., and Vernadskiy, V. I., 1969, Physiochemical conditions of formation of native arsenic in hydrothermal deposits: Geochemistry International, v. 6, p. 681-694.

Seward, T. M., 1991, The hydrothermal geochemistry of gold: *in* Foster, R. P., ed., Gold metallogeny and exploration, Blackie and Son, Glasgow, United Kingdon, p. 37-62.

Shannon, W. T., Owers, W. R., and Rothbaum, H. P., 1982, Pilot scale solids/liquid separation in hot geothermal discharge waters using dissolved air flotation: Geothermics, v. 11, p. 43-58.

Shelton, K. L., Bauer, R. M., and Gregg, J. M., 1993, Fluid-inclusion studies of regionally extensive epigenetic dolomites, Bonneterre Dolomite (Cambrian), southeast Missouri: Evidence of multiple fluids during dolomitization and lead-zinc mineralization: Geological Society of America Bulletin, v. 105, no. 7, p. 972-978.

Shevenell, L., Goff, F., Vuataz, F., Trujillo, P. E., Jr., Counce, D., Janik, C. J., and Evans, W., 1987, Hydrogeochemical data for thermal and nonthermal waters and gases of the Valles Caldera-southern Jemez Mountains region, New Mexico: Los Alamos National Laboratory Report LA-10923-OBES, 100 p.

Shifflett, J., in preparation, Sediment release from Milltown Reservoir during the 1997 runoff: M.S. Thesis: University of Montana.

Shippers, A., and Jorgensen, B. B., 2001, Oxidation of pyrite and iron sulfide by manganese dioxide in marine sediments: Geochimica et Cosmochimica Acta, v. 65, p. 915-922.

Shoen, R., White, D. E., and Hemley, J. J., 1974, Argillization by descending acid at Steamboat Springs, Nevada: Clays and Clay Minerals, v. 22, p. 1-22.

Siegel, D. I., 1989, Geochemistry of the Cambrian-Ordovician aquifer system in the northern midwest: U.S. Geological Survey Professional Paper 1405-D, 76 p.

Simmons, S. F., and Browne, P. R. L., 2000, Hydrothermal minerals and precious metals in the Broadlands-Ohaaki Geothermal systems: Implications for understanding low-sulfidation epithermal environments: Economic Geology, v. 95, p. 971-999.

Simo, J. A., Freiberg, P. G., and Freiberg, K. S., 1996, Geological constraints on arsenic in groundwater with applications to groundwater modeling: Water Resources Center Groundwater Research Report GRR 96-01, 60 p.

Singer, P. C., and Stumm, W., 1970, Acidic mine drainage: the rate-determining step: Science, v. 167, p. 1121-1123.

Singh, D. B., Prasad, G., Rupainwar, D. C., and Singh, V. N., 1988, As(III) removal from aqueous solution by adsorption: Water, Air & Soil Pollution, v. 42, p. 373-386.

Slattery, M., Kenah, C., Slattery, L., and Musser, K., 2000, Occurrence and release of ground water arsenic in public water supply wells in Ohio: *in* 2000 Midwest Ground Water Conference, Columbus, Ohio.

Sloss, L. L., 1963, Sequences in the cratonic interior of North America: Geological Society of America Bulletin, v. 74, no. 2, p. 93-114.

Smedley, P. L., Edmunds, W. M., and Pelig-Ba, K. B., 1996, Mobility of arsenic in groundwater in the Obuasi gold-mining area of Ghana: some implications for human health: *in* Appleton, J. D., Fuge, R., and McCall, G. J. H., eds., Environmental Geochemistry and Health, Geological Society Special Publication No. 113, p. 163-181.

Smedley, P. L., and Kinniburgh, D. G., 2001, Source and behaviour of arsenic in natural waters, U.N. Synthesis Report on Arsenic in Drinking Water: Geneva, World Health Organization, p. 1-61.

-, 2002, A review of the source, behaviour and distribution of arsenic in natural waters: Applied Geochemistry, v. 17, no. 5, p. 517-568.

Smedley, P. L., Kinniburgh, D. G., Milne, C., Huq, S. I., and Ahmed, K. M., 2001a, Changes with time: piezometer monitoring: *in* Kinniburgh, D. G., and Smedley, P. L., eds., Arsenic contamination of groundwater in Bangladesh, British Geological Survey Technical Report WC/00/19, Keyworth, p. 175-185.

Smedley, P. L., Kinniburgh, D. G., Milne, C., Trafford, J. M., Huq, S. I., and Ahmed, K. M., 2001b, Hydrogeochemistry of three Special Study Areas: *in* Kinniburgh, D. G., and Smedley, P. L., eds., Arsenic contamination of groundwater in Bangladesh, British Geological Survey Technical Report WC/00/19, Keyworth, p. 105-149.

Smedley, P. L., Nicolli, H. B., Macdonald, D. M. J., Barros, A. J., and Tullio, J. O., 2002, Hydrogeochemistry of arsenic and other inorganic constituents in groundwaters from La Pampa, Argentina: Applied Geochemistry, v. 17, no. 3, p. 259-284.

Smedley, P. L., Zhang, M., Zhang, G. Y., and Luo, Z. D., 2001c, Arsenic and other redox-sensitive elements in groundwater from the Huhhot Basin, Inner Mongolia: *in* Cidu, R., ed., Water Rock Interaction: Lisse, A.A. Balkema, p. 581-584.

-, in press, Mobilisation of arsenic and other trace elements in fluviolacustrine aquifers of the Huhhot Basin, Inner Mongolia: Applied Geochemistry.

Smith, A. L., Lingas, E. O., and Rahman, M., 2000, Contamination of drinking-water by arsenic in Bangladesh: a public health emergency: Bulletin of the World Health Organization, v. 78, p. 1093-1103.

Smith, B. C., 1996, Fundamentals of Fourier Transform Infrared Spectroscopy: New York, CRC Press, 202 p.

SOES/DCH, 2000, Groundwater arsenic contamination in Bangladesh: Calcutta and Dhaka, School of Environmental Studies, Jadavpur University, Calcutta, India and Dhaka Community Hospital, Dhaka, Bangladesh.

Sohrin, Y., Matsui, M., Kawashima, M., Hojo, M., and Hasegawa, H., 1997, Arsenic biogeochemistry affected by eutrophication in Lake Biwa, Japan: Environmental Science & Technology, v. 31, p. 2712-2720.

Solley, W. B., Pierce, R. R., and Perlman, H. P., 1988, Estimated Use of Water in the United States in 1985: U.S. Geological Survey Circular 1004, 82 p.

-, 1998, Estimated use of water in the United States in 1995: U.S. Geological Survey, 1200, 71 p.

Sonderegger, J. L., and Ohguchi, T., 1988, Irrigation related arsenic contamination of a thin, alluvial aquifer, Madison River Valley, Montana, U.S.A.: Environmental Geology and Water Sciences, v. 11, p. 153-161.

Soussan, T., 2001, Bill would void arsenic standard, Albuquerque Journal: Thursday, February 1, 2001, p. D3.

Sposito, G., 1984, The Surface Chemistry of Soils, Oxford University Press, 234 p.

Spycher, N. F., and Reed, H. M., 1989a, As(III) and Sb(III) sulfide complexes: An evaluation of stoichimetry and stability from existing experimental data: Geochimica et Cosmochimica Acta, v. 53, p. 2185-2194.

-, 1989b, Evolution of a Broadlands-type epithermal ore fluid along alternative P-T paths: Implications for the transport and deposition of base, precious and volatile metals: Economic Geology, v. 84, p. 328-359.

Stanton, M. R., Grimes, D. J., Sanzolone, R. F., and Sutley, S. J., 1998, Preliminary geochemical data from Santa Fe Group sediments in the 98th St. drill core, Middle Rio Grande Basin, near Albuquerque, New Mexico: U.S. Geological Survey Open-File Report 98-230, 59 p.

Stanton, M. R., Sanzolone, R. F., Sutley, S. J., Grimes, D. J., and Meier, A. M., 2001a, Mineralogical and geochemical constraints on Fe and As residence and mobility in the Albuquerque Basin-Examples from basin sediments and volcanic rocks: *in* Cole, J. C., ed., U.S. Geological Survey Middle Rio Grande Basin Study: Albuquerque, New Mexico, U.S. Geological Survey Open-File Report 00-488, p. 45-46.

Stanton, M. R., Sanzolone, R. F., Sutley, S. J., Grimes, D. J., Meier, A. M., and Lamothe, P. J., 2001b, Abundance, residence, and mobility of arsenic in Santa Fe Group sediments, Albuquerque Basin, New Mexico: Rocky Mountain and South-Central Sections: Geological Society of America, 2001 Abstracts with Programs, v. 33, no. 5, p. A-2.

Stanton, W. I., 1991, The habitat and origin of lead ore in Grebe Swallet Mine, Charterhouse-on-Mendip, Somerset: Proceedings - University of Bristol Spelaeological Society, v. 19, no. 1, p. 43-65.

Stauffer, R. E., and Thomson, J. M., 1984, Arsenic and antimony in geothermal waters of Yellowstone National Park, Wyoming, USA: Geochimica et Cosmochimica Acta, v. 48, p. 2547-2561.

Stein, H. J., Markey, R. J., Morgan, J. W., Hannah, J. L., and Zak, K., 1997, Re-Os dating of shear-hosted Au deposits using molybdenite: *in* Papunen, H., ed., Mineral Deposits: Research and Exploration--Where Do They Meet?: Rotterdam, Balkema, p. 313-317.

Steltenpohl, M. G., Heatherington, A. L., and Guthrie, G. M., 1995, Geologic setting and Alleghanian development of the Alabama and southwest Georgia inner and southern piedmonts: *in* Guthrie, G. M., ed., The timing and tectonic mechanisms of the Alleghanian orogeny, Alabama piedmont: A guidebook for the thirty second annual field trip of the Alabama Geological Society.

Stenger, D. P., Kesler, S. E., Peltonen, D. R., and Tapper, C. J., 1998, Deposition of gold in carlin-type deposits; the role of sulfidation and decarbonation at Twin Creeks, Nevada: Economic Geology and the Bulletin of the Society of Economic Geologists, v. 93, no. 2, p. 201-215.

Steuer, J. J., and Hunt, R. J., 2001, Use of a watershed-modeling approach to assess hydrologic effects of urbanization, North Fork Pheasant Branch basin near Middleton, Wisconsin: U.S. Geological Survey Water-Resources Investigations Report, 01–4113, 49 p.

Stojanovic, M., 1982, Crystallographic study of realgar and orpiment: Zapisnici Srpsko Geolosko Drustvo, p. 51-53.

Stollenwerk, K. G., 1995, Modeling the effects of variable groundwater chemistry on adsorption of molybdate: Water Resources Research, v. 31, p. 347-357.

-, 1996, Simulation of phosphate transport in sewage-contaminated groundwater, Cape Cod, Massachusetts: Applied Geochemistry, v. 11, p. 317-324.

-, 1998, Molybdate transport in a chemically complex aquifer: Field measurements compared with solute transport model predictions: Water Resources Research, v. 34, p. 2727-2740.

Stolz, J. F., and Oremland, R. S., 1999, Bacterial respiration of arsenic and selenium: FEMS Micobiology Letters, v. 23, p. 615-627.

Stone, A. T., and Ulrich, H., 1989, Kinetics and reaction stoichiometry in the reductive dissolution of manganese(IV) dioxide and Co(III) oxide by hydroquinone: Journal of Colloid Interface Science, v. 102, p. 509-522.

Stowell, H. H., Guthrie, G. M., and Lesher, C. M., 1989, Metamorphism and gold mineralization in the Wedowee Group, Goldville District, northern Piedmont, Alabama: Geological Survey of Alabama, Report 136, 133-158 p.

Stowell, H. H., Lesher, C. M., Green, N., Peng, S., Guthrie, G., and Sinha, A. K., 1996, Metamorphism and gold mineralization in the Blue Ridge, southernmost Appalachians: Economic Geology, v. 91, p. 1115-1144.

Strickland, J. D. H., and Parson, T. R., 1968, A practical handbook of seawater analysis: Canada, Fisheries Research Board, Bulletin, v. 167, p. 50.

Stumm, W., 1992, Chemistry of the solid-water interface: New York, Wiley Interscience, 428 p.

Stumm, W., Hohl, H. H., and Dalang, F., 1976, Interaction of metal ions with hydrous oxide surfaces: Croat. Chem. Acta, v. 48, p. 491-504.

Stumm, W., Huang, C. P., and Jenkins, S. R., 1970, Specific chemical interactions affecting the stability of dispersed system: Croat. Chem. Acta, v. 42, p. 223-244.

Stumm, W., Kummert, R., and Sigg, L., 1980, A ligand exchange model for the adsorption of inorganic and organic ligands at hydrous oxide interfaces: Croat. Chem. Acta, v. 53, p. 291-312.

Stumm, W., and Morgan, J. J., 1981, Aquatic Chemistry, John Wiley & Sons, 780 p.

-, 1996, Aquatic Chemistry: Chemical Equilibria and Rates in Natural Waters, John Wiley and Sons, 1022 p.

Su, C., and Suarez, D. L., 1997, In situ infrared speciation of adsorbed carbonate on aluminum and iron oxides: Clays and Clay Minerals, v. 45, p. 814-825.

Suarez, D. L., Goldberg, S., and Su, C., 1999, Evaluation of oxyanion adsorption mechanisms on oxides using FTIR spectroscopy and electrophoretic mobility: *in* Sparks, D. L., and Grundl, T. J., eds., Mineral-Water Interfacial Reactions, American Chemical Society, p. 137-178.

Suk, D., Peacor, D. R., and Van der Voo, R., 1990, Replacement of pyrite framboids by magnetite in limestone and implications for palaeomagnetism: Nature, v. 345, no. 6276, p. 611-613.

Sullivan, K. A., and Aller, R. C., 1996, Diagenetic cycling of arsenic in Amazon shelf sediments: Geochimica et Cosmochimica Acta, v. 60, no. 9, p. 1465-1477.

Sun, G., Pi, J., Li, B., Guo, X., Yamauchi, H., and Yoshida, T., 2001, Arsenic exposure and health effects: *in* Abernathy, C. O., Calderon, R. L., and Chappell, W. R., eds., Arsenic Exposure and Health Effects IV, Elsevier, p. 79-85.

Sun, X., and Doner, H., 1998, Adsorption and oxidation of arsenite on goethite: Soil Science, v. 163, p. 278-287.

Sun, X., and Doner, H. E., 1996, An investigation of arsenate and arsenite bonding structures on goethite by FTIR: Soil Science, v. 161, p. 865-872.

Sun, X., Doner, H. E., and Zavarin, M., 1999, Spectroscopy study of arsenite [As(III)] oxidation on Mn-substituted goethite: Clays and Clay Minerals, v. 47, p. 474-480.

Sverjensky, D. A., and Garven, G., 1992, Geochemistry; Tracing great fluid migrations: Nature, v. 356, no. 6369, p. 481-482.

Swedlund, P. J., and Webster, J. G., 1999a, Adsorption and polymerisation of silicic acid on ferrihydrite, and its effect on arsenic adsorption: Water Research, v. 33, p. 3413-3422.

-, 1999b, Hydrous ferric oxide adsorption of arsenic from geothermal bore effluents: the effects of moni-silic acid: Water Research, v. 33, p. 3413-3422.

Switzer, B. J., Bindi, A. B., Buzzelli, J., Stolz, J. F., and Oremland, R. S., 1998, *Bacillus arsenicoselenatis,* sp. nov., *and Bacillus selentireducens,* sp nov.: two haloalkaliphiles from Mono Lake California that respire oxyanions of selenium and arsenic.: Archives of Microbiology, v. 171, p. 19-30.

Symons, D. T. A., Sangster, D. F., and Leach, D. L., 1997, Paleomagnetic dating of Mississippi Valley-type Pb-Zn-Ba deposits: *in* Sangster, D. F., ed., Carbonate-hosted lead-zinc deposits, Special Publication - Society of Economic Geologists, p. 515-526.

Tandukar, E. N., 2001, Scenario of arsenic contamination in groundwater in Nepal.: Department of Water Supply and Sewerage (DWSS), Nepal, 6 p.

Tasneem, M. A., 1999, Impact of agricultural and industrial activities on groundwater quality in Kasur area.: The Nucleus, Quarterly Journal of the Pakistan Atomic Energy Commission, v. 36.

Taylor, J. B., Bennett, S. L., and Heyding, R. D., 1965, Physical properties of α and β arsenic: Journal of Physics and Chemistry of Solids, v. 26, p. 69-74.

Taylor, R. M., 1987, Non-silicate Oxides and Hydroxides: *in* Newman, A. C. D., ed., Chemistry of Clays and Clay Minerals, Wiley Interscience, p. 129-201.

Tazaki, K., 1995, Electron microscopic observations of biomineralization in biomats from hot springs: Journal of the Geological Society of Japan, v. 101, p. 304-314.

Tempel, R. N., Shevenell, L. A., Lechler, P., and Price, J., 1999, Geochemical modeling approach to predicting arsenic concentrations in a mine pit lake: Applied Geochemistry, v. 15, p. 475-492.

Tessier, A., Campbell, P. G. C., and Bisson, M., 1979, Sequential extraction procedure for the speciation of particulate trace metals: Analytical Chemistry, v. 51, p. 844-851.

Thanabalasingam, P., and Pickering, W. F., 1986a, Arsenic sorption by humic acids: Environmental Pollution Series B, v. 12, p. 233-246.

-, 1986b, Effect of pH on interaction between As(III) or As(V) and Manganese Dioxide: Water, Air, and Soil Pollution, p. 205-216.

Thomas, W. A., 1989, The Appalachian-Ouachita Orogen beneath the Gulf Coastal Plain between the outcrops in the Appalachian and Ouachita Mountains: *in* Hatcher, R. D. J., Thomas, W. A., and Viele, G. W., eds., The Appalachian-Ouachita orogen in the United States: Boulder Colorada, Geological Society of America, p. 537-554.

Thomson, J. M., 1979, Arsenic and fluoride in the upper Madison River system: Firehole and Gibbon Rivers and their tributaries, Yellowstone National Park, Wyoming and Southeast Montana: Environmental Geology, v. 3, p. 13-21.

-, 1985, Chemistry of thermal and non-thermal springs in the vicinity of Lassen Volcanic National Park: Journal of Volcanology and Geothermal Research, v. 25, p. 81-104.

Thorn, C. R., McAda, D. P., and Kernodle, J. M., 1993, Geohydrologic framework and hydrologic conditions in the Albuquerque Basin, central New Mexico: U.S. Geological Survey, 93-4149, 106 p.

Thurman, E. M., 1985, Organic Geochemistry of Natural Waters, Kluwer, 497 p.

Timperley, M. H., and Huser, B. A., 1996, Inflows of geothermal fluid chemicals to the Waikato River catchment, New Zealand: New Zealand Journal of Marine and Freshwater Research, v. 30, p. 525-535.

Titan, 1994, Milltown reservoir sediments operable unit, working draft, final remedial investigation report Volume 1, Prepared for Atlantic Richfield Company, Prepared by: Bozeman, Montana, Titan Environmental Corporation.

-, 1995, Milltown Reservoir Sediments Operable Unit Final Draft Remedial Investigation Report: Anaconda, ARCO.

-, 1996, Milltown Reservoir Sediments Site Draft Feasibility Study Report, ARCO.

Titus, F. B., 1961, Ground-water geology of the Rio Grande trough in north-central New Mexico, with sections on the Jemez Caldera and Lucero Uplift, Guidebook of the Albuquerque country: *in* Northrop, S. A., ed., 12th Field Conference, New Mexico Geological Society, p. 186-192.

Toran, L., 1987, Sulfate contamination in groundwater from a carbonate-hosted mine: Journal of Contaminant Hydrology, v. 2, no. 1, p. 1-29.

Toran, L., and Harris, R. F., 1989, Interpretation of sulphur and oxygen isotopes in biological and abiological sulfide oxidation: Geochimica et Cosmochimica Acta, v. 53, no. 2341-2348.
Tossell, J. A., 1997a, Theoretical studies on arsenic oxide and hydroxide species in minerals and in aqueous solution: Geochimica et Cosmochimica Acta, v. 61, p. 1613-1623.
-, 1997b, Theoretical studies on arsenic oxide and hydroxide species in minerals and in aqueous solutions: Geochimica et Cosmochimica Acta, v. 61, p. 1613-1623.
Trafford, J. M., Lawrence, A. R., Macdonald, D. M. J., Van Dan, N., and Tran, D. N. H. N. T., 1996, The effect of urbanisation on the groundwater quality beneath the city of Hanoi, Vietnam.: British Geological Survey Technical Report WC/96/22, Keyworth.
Trainer, F. W., 1974, Ground water in the southwestern part of the Jemez Mountains volcanic region, New Mexico, 25th Field Conference, New Mexico Geological Society, p. 337-345.
-, 1975, Mixing of thermal and nonthermal waters in the margin of the Rio Grande Rift, Jemez Mountains, New Mexico, New Mexico Geological Society 26th Field Conference, p. 213-218.
-, 1984, Thermal mineral springs in Cañon de San Diego as a window into Valles Caldera, New Mexico, 35th Field Conference, New Mexico Geological Society, p. 249-255.
Trainer, F. W., Rogers, R. J., and Sorey, M. L., 2000, Geothermal hydrology of Valles Caldera and the southwestern Jemez Mountains, New Mexico: U.S. Geological Survey Water-Resources Investigations Report, 00-4067, 115 p.
Trentelman, K., Stodulski, L., and Pavlosky, M., 1996, Characterization of pararealgar and other light-induced transformation products from realgar by Raman microspectroscopy: Analytical Chemistry, p. 1755-1761.
Tseng, W. P., Chu, H. M., How, S. W., Fong, J. M., Lin, C. S., and Yeh, S., 1968, Prevalence of skin cancer in an endemic area of chronic arsenicism in Taiwan: Journal of the National Cancer Institute, v. 40, p. 453-463.
Tsuda, T., Ogawa, T., Babazono, A., Hamada, H., Kanazawa, S., Mino, Y., Aoyama, H., Yamamoto, E., and Kurumatani, N., 1992, Historical cohort studies in three arsenic poisoning areas in Japan: Applied Organometallic Chemistry, v. 6, no. 4, p. 333-341.
Turekian, K. K., and Wedepohl, K. H., 1961, Distribution of the elements in some major units of the earth's crust: Geological Society of America Bulletin, v. 72, no. 2, p. 175-191.
U.S. Bureau of the Census, 1999, Housing: Then and now, 50 years of decennial censuses, Historical census of housing: Washington, D.C., U.S. Bureau of the Census.
U.S. Environmental Protection Agency, 1976, Quality Criteria for Water, 286 p.
-, 2001a, Drinking water standard for arsenic: USEPA Fact Sheet 815-F- 00-105, 3 p.
-, 2001b, Factoids: Drinking water and ground water statistics for 2000: U.S. Environmental Protection Agency, EPA 816-K-01-004, 10 p.
-, 2002, Drinking water from household wells: U.S. Environmental Protection Agency, EPA 816-K-02-003, 19 p.
Udaloy, A. G., 1988, Arsenic mobilization in response to the draining and filling of the reservoir at Milltown, Montana: M.S. Thesis: University of Montana, 140 p.
Umitsu, M., 1993, Late Quaternary sedimentary environments and landforms in the Ganges Delta: Sedimentary Geology, v. 83, p. 177-186.
Upadhyay, S. K., 1993, Use of groundwater resources to alleviate poverty in Nepal: policy issues.: *in* Kahnert, F., and Levine, G., eds., Groundwater irrigation and the rural poor: options for development in the Gangetic Basin: Washington D.C., World Bank.
USEPA, 1972, Water Quality Criteria 1972, U.S. Environmental Protection Agency, p. unpaginaged.
-, 1986, Quality Criteria for Water 1986: U.S. Environmental Protection Agency, 440/5-86-001.

-, 1994, Methods for the determination of metals in environmental samples-Supplement 1: EPA-600/R-94-111.

UWEIL, 1994, Pretreatment of sulphates and sulfides, Technical Procedure 30.0: University of Waterloo, 30 p.

Vairavamurthy, M. A. e., Schoonen, M. A. A. e., Eglinton, T. I. e., Luther, G. W., III (editor), and Manowitz, B. e., 1994, Geochemical transformations of sedimentary sulfur: 208th national meeting of the American Chemical Society, v. 612, p. 467.

Van Beek, C. G. E. M., 1980, A model for the induced removal of iron and manganese from groundwater in the aquifer: *in* 3rd Water-Rock Interaction Symposium, Edmonton, Canada, p. 29-31.

Van der Hoek, E., Bonouvrie, P. A., and Comans, R. N. J., 1994, Sorption of As and Se on mineral components of fly ash: relevance for leaching processes: Applied Geochemistry, v. 9, p. 403-412.

Van Geen, A., Robertson, A. P., and Leckie, J. O., 1994, Complexation of carbonate species at the goethite surface: implications for adsorption of metal ions in natural waters: Geochimica et Cosmochimica Acta, v. 58, p. 2073-2086.

Vergasova, L. P., 1983, Fumarole incrustations of lava flows of the effusive-explosive period of the Great Tolbachik fissure eruption: Vulkanologiya i Seysmologiya, v. 6, p. 75-87.

Vesala, A., and Saloma, E., 1977, Determination of basicity constants by potentiostatic titration: Finnish Chemical Letters, p. 160-163.

Viets, J. G., and Leach, D. L., 1990, Genetic implications of regional and temporal trends in ore fluid geochemistry of Mississippi Valley-type deposits in the Ozark region: Economic Geology, v. 85, no. 4, p. 842-861.

Voigt , D. E., and Brantley, S. L., 1997, Chemical fixation of arsenic in contaminated soils: Applied Geochemistry, v. 11, p. 633-643.

Vorobeva, S. V., Ivakin, A. A., Gorelov, A. M., and Gertman, E. M., 1977, Thio-complexes of Arsenic(III) in solution: Russian Journal of Inorganic Chemistry, v. 22, p. 1479-1481.

Voss, R. L., and Hagni, R. D., 1985, The application of cathodoluminescence microscopy to the study of sparry dolomite from the Viburnum Trend, southeast Missouri: *in* Hausen, D. M., and Kopp, O. C., eds., Mineralogy: Applications to the minerals industry; proceedings of the Paul F. Kerr memorial symposium: New York, NY, United States, p. 51-68.

Wagman, D. D., Evans, W. H., Parker, V. B., Halow, I., Bailey, S. M., and Schumm, R. H., 1968, Selected Values of Chemical Thermodynamic Properties. Tables for the First Thirty-Four Elements in the Standard Order of Arrangement: NBS Technical Note 270-3, 264 p.

Wagman, D. D., Evans, W. H., Parker, V. B., Schumm, R. H., Halow, I., Bailey, S. M., Churney, K. L., and Nuttall, R. L., 1982, The NBS Tables of Chemical Thermodynamic Properties. Selected Values for Inorganic and C1 and C2 Organic Substances in SI Units: Journal of Physical and Chemical Reference Data, v. 11, p. 392.

Wang, G., 1984, Arsenic poisoning from drinking water in Xinjiang: Chinese Journal of Preventative Medicine, v. 18, p. 105-107.

Wang, G. Q., Huang, Y. Z., Xiao, B. Y., Qian, X. C., Yao, H., Hu, Y., Gu, Y. L., Zhang, C., and Liu, K. T., 1997, Toxicity from water containing arsenic and fluoride in Xinjiang: Fluoride, v. 30, no. 2, p. 81-84.

Wang, L., and Huang, J., 1994, Chronic arsenism from drinking water in some areas of Xinjiang, China: *in* Nriagu, J. O., ed., Arsenic in the Environment, Part II: Human Health and Ecosystem Effects, John Wiley, p. 159-172.

WAPDA/EUAD, 1989, Booklet on hydrogeological map of Pakistan 1:2,000,000.scale: Lahore, Water and Power Development Authority and Environment & Urban Affairs Division, Government of Pakistan.

Ward, D. M., Ferris, M. J., Nold, S. C., and Bateson, M. M., 1998, Natural view of microbial biodiversity within hot spring cyanobacterial mat communities: Microbio. & Molec. Bio. Rev, v. 62, p. 1353-1370.

Warner, K. L., 1998, Water-quality assessment of the lower Illinois River Basin--Environmental setting: U.S. Geological Survey, 97-4165, 50 p.

-, 2001, Arsenic in glacial drift aquifers and the implication for drinking water-- Lower Illinois River Basin: Ground Water, v. 39, no. 3, p. 433-442.

Wauchope, R. D., 1975, Fixation of arsenical herbicides, phosphates and arsenate in soils: Journal of Environmental Quality, v. 14, p. 355-358.

Wauchope, R. D., and McDowell, L. L., 1984, Adsorption of phosphate, arsenate, methanearsonate, and cacodylate by lake and stream sediments: Comparisons with soils: Journal of Environmental Quality, v. 13, p. 499-504.

Waychunas, G. A., 1991, Crystal Chemistry of Oxides and Hydroxides: *in* Lindsley, D. H., ed., Oxide Minerals: Petrologic and Magnetic Significance, Mineralogical Society of America, p. 11-68.

Waychunas, G. A., Davis, J. A., and Fuller , C. C., 1995, Geometry of sorbed arsenate on ferrihydrite and crystalline FeOOH: Re-evaluation of EXAFS results and topological factors in predicting sorbate geometry, and evidence for monodentate complexes: Geochimica et Cosmochimica Acta, v. 59, p. 3655-3661.

Waychunas, G. A., Fuller, C. C., Rea, B. A., and Davis, J. A., 1996, Wide angle X-ray scattering (WAXS) study of "two-line" ferrihydrite structure; effect of arsenate sorption and counterion variation and comparison with EXAFS results: Geochimica et Cosmochimica Acta, v. 60, p. 1765-1781.

Waychunas, G. A., Rea, B. A., Fuller, C. C., and Davis, J. A., 1993, Surface chemistry of ferrihydrite: Part 1. EXAFS studies of the geometry of coprecipitated and adsorbed arsenate.: Geochimica et Cosmochimica Acta, v. 57, p. 2251-2269.

Waypa, J. J., Elimelech, M., and Hering, J. G., 1997, Arsenic removal by RO and NF membranes: Journal - American Water Works Association, v. 89, p. 102-114.

Weaver, T. R., and Bahr, J. M., 1991, Geochemical evolution in the Cambrian-Ordovician sandstone aquifer, eastern Wisconsin .1. Major ion and radionuclide distribution: Ground Water, v. 29, no. 3, p. 350-356.

Webster, J. G., 1990, The solubility of As_2S_3 and speciation of As in dilute and sulfide-bearing fluids at 25°C and 90°C: Geochimica et Cosmochimica Acta, v. 54, p. 1009-1017.

-, 1999, The source of arsenic (& other elements) in the Marbel-Matingao River catchment, Mindanao, Philippines: Geothermics, v. 28, p. 95-111.

Webster-Brown, J. G., 2000, Chemical contaminants and their effects. In Environmental Safety and Health Issues in Geothermal Development.: *in* World Geothermal Congress, Kazuno, Japan.

Weissberg, B. G., 1969, Gold-silver ore-grade precipitates from New Zealand thermal waters: Economic Geology, v. 64, p. 95-108.

Weissberg, B. G., Dickson, F. W., and Tunell, G., 1966, Solubility of orpiment (As_2S_3) in Na_2S-H_2O at 50-200°C and 100-1500 bars, with geological applications: Geochimica et Cosmochimica Acta, v. 30, p. 815-827.

Welch, A. H., 1994, Ground-water quality and geochemistry in Carson and Eagle Valleys, western Nevada and eastern California: U.S. Geological Survey, 93-33, 99 p.

-, 1998, Arsenic in ground water of the United States; processes leading to widespread high concentrations: *in* International Conference on arsenic pollutioin of ground water in Bangladesh: Causes, effects and remedies, Dhaka, Bangladesh, p. 146.

-, 1999, Arsenic in ground water supplies of the United States: *in* Chappell, W. R., Abernathy, C. O., and Calderon, R. L., eds., Arsenic Exposure and Health Effects, Elsevier, p. 416.

Welch, A. H., Lawrence, S. J., Lico, M. S., Thomas, J. M., and Schaefer, D. H., 1997, Ground-water quality assessment of the Carson River basin, Nevada and California; results of investigations, 1987-91: U. S. Geological Survey Water-Supply Paper, 0886-9308, 93 p.

Welch, A. H., Lico, M. S., and Hughes, J. L., 1988, Arsenic in ground water of the western United States: Ground Water, v. 26, no. 3, p. 333-347.

Welch, A. H., Westjohn, D. B., Helsel, D. R., and Wanty, R. B., 2000, Arsenic in ground water of the United States: Occurrence and geochemistry: Ground Water, v. 38, no. 4, p. 589-604.

Weller, W. W., and Kelley, K. K., 1964, Low-temperature heat capacities and entropies at 298.15 °K of sulfides of arsenic, germanium, and nickel: U.S. Bureau of Mines Report Investigations 6511, 7 p.

Wentz, D. A., Bonn, B. A., Carpenter, K. D., Hinkle, S. R., Janet, M. L., Rinella, F. A., Uhrich, M. A., Waite, I. R., Laenen, A., and Bencala, K. E., 1998, Water quality in the Willamette Basin, Oregon, 1991-95: U.S. Geological Survey Circular 1161, 34 p.

Westall, J. C., 1982, FITEQL: A program for the determination of chemical equilibrium constants from experimental data: Oregon State University Technical Report 82-02, 61 p.

Westall, J. C., and Hohl, H., 1980, A comparison of electrostatic models for the oxide/solution interface: Advances in Colloid Interface Science, v. 12, p. 265-294.

Westjohn, D. B., and Weaver, T. L., 1995, Configuration of freshwater/saline-water interface and geologic controls on distribution of freshwater in a regional aquifer system, central Lower Peninsula of Michigan: U.S. Geological Survey, Water Resources Investigations Report 94-4242, 44 p.

-, 1998, Hydrogeologic framework of the Michigan Basin Regional Aquifer System: U.S. Geological Survey Professional Paper 1418, 47 p.

Whelan, J. F., Cobb, J. C., and Rye, R. O., 1988, Stable isotope geochemistry of sphalerite and other mineral matter in coal beds of the Illinois and Forest City basins: Economic Geology, v. 83, no. 5, p. 990-1007.

White, D. E., 1967, Mercury and base-metal deposits with associated thermal and mineral waters: *in* Barnes, H., ed., Geochemistry of Hydrothermal Ore Deposits: New York, Holt, Rinehart, and Winston, p. 575-631.

-, 1968, Environments of generation of some base-metal ore deposits: Economic Geology, v. 63, p. 301-335.

-, 1974, Diverse origins of hydrothermal fluids: Economic Geology, v. 69, p. 954-973.

-, 1981, Active geothermal systems and hydrothermal ore deposits: Economic Geology, v. 75th Anniversary Volume, p. 392-423.

White, D. E., Fournier, R. O., and Heropoulos, C., 1992, Gold and other minor elements associated with the hot springs and geysers of Yellowstone National Park, Wyoming, supplemented with data from Steamboat Springs, Nevada: United States Geological Survey Bulletin 2001, 19 p.

White, D. E., Muffler, L. J. P., and Truesdell, A. H., 1971, Vapour-dominated hydrothermal systems compared with hot-water systems: Economic Geology, v. 66, p. 75-97.

White, D. E., and Waring, G. A., 1963, Volcanic emanations: *in* Fleischer, M., ed., Data of Geochemistry, U.S. Geological Survey Professional Paper 440-K, p. 29.

WHO, 1993, Guidelines for Drinking Water Quality, 2nd Ed, Volume 1 Recommendations: World Health Organization, 188 p.

Wiersma, C. L., and Rimstidt, J. D., 1984, Rates of reaction of pyrite and marcasite with ferric iron at pH 2: Geochimica et Cosmochimica Acta, v. 48, no. 1, p. 85-92.

Wiese, R. G., Jr., Muir, I. J., and Fyfe, W. S., 1990, Trace element siting in iron sulfides from coal determined by secondary ion mass spectrometry: Energy Sources, v. 12, no. 3, p. 251-264.

Wijnja, H., and Schulthess, C. P., 2000, Interaction of carbonate and organic anions with sulfate and selenate adsorption on an aluminum oxide: Soil Science Society of America Journal, v. 64, p. 898-908.

Wilde, F. D., and Radtke, D. B., 1998, National Field Manual for the collection of water-quality data, U.S. Geological Survey Techniques of Water-Resources Investigations, Book 9, Chapter A6, p. 128.

Wilke, J. A., and Hering, J. G., 1998, Rapid oxidation of geothermal arsenic(III) in streamwaters of the eastern Sierra Nevada: Environmental Science & Technology, v. 32, p. 657-662.

Wilkie, J. A., and Hering, J. G., 1996, Adsorption of arsenic onto hydrous ferric oxide: effects of adsorbate/adsorbent ratios and co-occurring solutes: Colloids and Surfaces A, v. 107, p. 97-110.

Willett, I. R., Chartres, C. J., and Nguyen, T. T., 1988, Migration of phosphate into aggregated particles of ferrihydrite: Journal of Soil Science, v. 39, p. 275-282.

Williams, M., 1997, Mining-related arsenic hazards: Thailand case-study: British Geological Survey, WC/97/49, 36 p.

-, 2001, Arsenic in mine waters: an international study: Environmental Geology, v. 40, p. 267-278.

Williams, M., Fordyce, F., Paijitprapapon, A., and Charoenchaisri, P., 1996, Arsenic contamination in surface drainage and groundwater in part of the southeast Asian tin belt, Nakhon Si Thammarat Province, southern Thailand: Environmental Geology, v. 27, no. 1, p. 16-33.

Winston, R. B., 1990, Vitrinite reflectance of Alabama's bituminous coal: Geological Survey of Alabama Circular, 139, 54 p.

Wisconsin Department of Natural Resources, 1997, Status of groundwater in Wisconsin, WDNR: Bureau of Drinking Water and Ground Water, PUBL-DG-043-97, 51 p.

-, 1999, Answers to your questions about groundwater, Bureau of Drinking Water and Ground Water.

Woessner, W. W., 1995, Milltown Groundwater Injury Assessment report: Department of Justice, Natural Resource Damage Litgation Program, 36 p.

Woessner, W. W., Moore, J. N., Johns, C., Popoff, M. A., Sartor, L. C., and Sullivan, M., 1984, Arsenic source and water supply remedial action study, Milltown, Montana: Final report: Department of Geology, University of Montana, 448 p.

Woessner, W. W., and Popoff, M. A., 1982, Hydrogeologic survey of Milltown, Montana and vicinity: Department of Geology, University of Montana, 130 p.

Woo, N. C., and Choi, M. J., 2001, Arsenic and metal contamination of water resources from mining wastes in Korea: Environmental Geology, v. 40, p. 305-311.

Wood, S. A., Tait, C. D., and Janecky, D. R., 1998, A Raman spectroscopic study of thioarsenite and arsenite species in low-temperature aqueous solutions: *in* Proceedings, 9th Water-Rock Interaction Symposium, Taupo, New Zealand, p. 863-866.

Woodard and Curran, I., 1998, Final Remedial Investigation Report of Saco Municipal Landfill Superfund Site, Saco Maine: Portland, Maine, Woodard and Curran, Inc., p. variously paginated.

WRUD, 2001, Preliminary study on arsenic contamination in selected areas of Myanmar: Water Resources Utilization Department, Ministry of Agriculture and Irrigation.

Wu, Y., Hagni, R. D., and Paarlberg, N., 1996, Silver distribution in iron sulfides at the Buick and Brushy Creek mines, Viburnum Trend, southeast Missouri: *in* Sangster, D. F., ed., Carbonate-hosted lead-zinc deposits: Special Publication #4, Society of Economic Geologists.

Xu, H., Allard, B., and Grimvall, A., 1988, Influence of pH and organic substance on the adsorption of As(V) on geologic materials: Water, Air, and Soil Pollution, v. 40, p. 293-305.

Yardley, B., W.D., 1997, The evolution of fluids through the metamorphic cycle: *in* Yardley, B. J. a. B., ed., Fluid Flow and Transport in Rocks: Mechanisms and Effects: London, Chapman and Hall, p. 101-121.

Yates, D. E., Levine, S., and Healy, T. W., 1974, Site-binding model of the electrical double layer at the oxide/water interface: Chemical Society, Faraday Transactions, v. 1, no. 70, p. 1807-1818.

Yeh, G. T., and Tripathi, V. S., 1991, A model for simulating transport of reactive multispecies components: Model development and demonstration: Water Resources Research, v. 27, p. 3075-3094.

Yokoyama, T., Takahashi, Y., and Tarutani, T., 1993, Simultaneous determination of arsenic and arsenious acids in geothermal water: Chemical Geology, v. 103, p. 103-111.

Young, C. A., and Robins, R. G., 2000, The solubility of As2S3 in relation to the precipitation of arsenic from process solutions: *in* Young, C. A., ed., Minor Elements 2000: Littleton, Colorado, Society of Mining, Metallurgy, and Exploration, Inc., p. 381-391.

Yu, S.-C., and Zoltai, T., 1972, Crystallography of a high-temperature phase of realgar: American Mineralogical Journal, v. 57, p. 1873-1876.

Zacek, V., and Ondrus, P., 1997, Mineralogy of recently formed sublimates from Katerina colliery in Radvanice, eastern Bohemia, Czech Republic: Bulletin of the Czech Geological Survey, v. 72, p. 289-302.

Zachara, J. M., Girvin, D. C., Schmidt, R. L., and Resch, C. T., 1987, Chromate adsorption on amorphous iron oxyhydroxide in the presence of major groundwater ions: Environmental Science & Technology, v. 21, p. 589-594.

Zhang, E., and Davis, A., 1993, Coalification patterns of the Pennsylvanian coal measures in the Appalachian foreland basin, western and south-central Pennsylvania: Geological Society of America Bulletin, v. 105, p. 162-174.

Zhang, Y., Lockwood, J. R., III, Schervish, M. J., Gurian, P., and Small, M. J., 2001, Hierarchical modeling of arsenic concentration at entry points in US public drinking water supplies: Carnegie Mellon University, 739, 12 p.

Zheng, C., and Bennett, G. D., 1995, Applied Contaminant Transport Modeling, Theory and Practice: New York, Van Nostrand Reinhold, 440 p.

Zobrist, J., Dowdle, P. R., Davis, J. A., and Oremland, R. S., 1998, Microbial arsenate reduction vs arsenate sorption: Experiments with ferrihydrite suspensions.: Mineralogical Magazine, v. 62A, p. 1707-1708.

-, 2000, Mobilization of arsenite by dissimilatory reduction of adsorbed arsenate: Environmental Science & Technology, v. 34, no. 22, p. 4747-4753.

Zuena, A. J., and Keane, N. W., 1985, Arsenic contamination in private potable well: *in* U.S. EPA National Conference on Environmental Engineering, Northeastern University of Boston, Massachusetts, p. 717-725.

INDEX